AF556890

LLC

Dedicated

To

Professor S. S. Raghuvanshi

September 4, 1932 – December 9, 2000

A Distinguished Teacher
and
Outstanding Plant Geneticist

Molecular Plant Breeding: Principle, Method and Application

Editors

R.K. Singh
Division of Crop Improvement
Indian Institute of Sugarcane Research
Lucknow - 226 002 (UP), India

Rajesh Singh
Department of Genetics & Plant Breeding
Institute of Agricultural Sciences
Banaras Hindu University
Varanasi - 221 005 (UP), India

Guoyou Ye
Bundoora Centre, Biosciences Division
Department of Primary Industries and
Molecular Plant Breeding Research Centre
1 Park Drive, Bundoora, Vic 3086, Australia

A. Selvi
Division of Crop Improvement
Sugarcane Breeding Institute
Coimbatore - 641 007 (Tamil Nadu), India

G.P. Rao
Sugarcane Research Station
Kunraghat, Gorakhpur - 273 008 (UP), India

2010

Studium Press LLC, U.S.A.

Published by

STUDIUM PRESS, LLC
P.O. Box-722200, Houston, Texas-77072, USA
Tel. : 713-541-9400; Fax: 713-541-9401
E-mail: studiumpress@studiumpress.com

ISBN: 1-933699-49-3

Composed & Designed at :
Panacea Computers
Lucknow, India
E-mail : prasgupt@rediffmail.com

Printed at: Thomson Press (India) Ltd.

Secretary, Department of Agricultural Research & Education and Director General, Indian Council of Agricultural Research
Ministry of Agriculture, Krishi Bhavan,
New Delhi - 110 001

Dr. Mangala Rai

FOREWORD

Conventional plant breeding is primarily based on selection of superior individuals among segregating progenies. Although significant strides have been made in crop improvement through conventional plant breeding, difficulties are often encountered during this process, primarily due to genotype-environment interactions. Besides, testing procedures may be some times difficult, unreliable or expensive due to the nature of the target traits or the environment. With the application of DNA marker technology, several types of molecular breeding strategies are now available to the plant breeders and geneticists, to over come the problems faced during conventional breeding. Impressive advancesin molecular genetics over the last two decades have provided us with a range of tools and techniques for analyzing genomes. Molecular markers, an important out growth of modern genetics, are now available in abundance in both plants and animals. Innovative techniques making use of these markers have facilitated analysis of genetic variation reliably and efficiently at the DNA level. Besides, their significant contribution to the basic understanding of the genomes, DNA based markers have shown their value in mapping genes conferring economically important traits in various crop plants. Recent successes in gene tagging, QTL mapping, gene pyramiding, and marker-assisted selection, shall form the foundation of 'precision breeding'. At last it can be concluded that the era of DNA markers and molecular breeding strategies are stadily advancing towards a phase where crops would have increased productivity and all its traits, beneficial to both the farmers and the consumer, could be manipulated.

The present comprehensive compendium attempts to provide an overall up-to-date accounts of recent advances in molecular breeding, molecular markers in plants, high throughput detection of PCR products and SNPs, mapping populations in plants, tagging and mapping of qualitative as well as quantitative traits in plants, association mapping in crops, statistical analysis of molecular data, genomics and gene based markers for crop improvement, *in situ* hybridization - an important tool in cyto-molecular mapping, molecular markers in biosystematics, marker-assisted gene pyramiding in cultivar development, marker-assisted recurrent backcrossing in cultivar development, formulating marker-assisted selection strategies using computer simulation, improvement of quality traits and molecular and physiological approaches for drought management in crops have been included.

I hope that this informative book entitled "**Molecular Plant Breeding: Principle, Method and Application**" will be of interest to the researchers, academicians, students and other stakeholders in molecular marker technologies and molecular breeding strategies in crop improvement. The editors must be complemented for their outstanding efforts in bringing out this timely reference book.

14 January, 2009 (MANGALA RAI)

New Delhi

Preface

Conventional plant breeding is primarily based on phenotypic selection of superior individuals among segregating progenies resulting from hybridization. Although significant strides have been made in crop improvement through phenotypic selections for agronomically important traits, considerable difficulties are often encountered during this process, primarily due to genotype-environment interactions. Besides, testing procedures it may be difficult, unreliable or expensive due to the nature of the target traits (e.g. abiotic stresses) or the target environment. With the advent of DNA based technology, several types of DNA markers and molecular breeding strategies are now available to plant breeders and geneticists, helping them to overcome the problems faced during conventional breeding. The potential benefits of using markers linked to gene of interest in breeding programs have been obvious for many decades. The use of markers in the identification of Mendelian components underlying both simple and complex agronomic traits is currently a key component of almost all crop breeding programs. Recent advances in plant molecular genetics have lead to the development of enormous molecular markers in all major crop species. Even for under studied minor crops a good number of valuable markers can be easily developed, thanks for the rapid development in gene sequencing, functional genomics and comparative genetics. The developments of highly automated genotyping a system not only improve the process and quality, but also dramatically reduce the cost of genotyping. The availability of abundant molecular markers greatly facilitates the construction of highly saturated linkage maps, which in turn speeds up the establishment of marker-trait associations and the quality of established association. Significant progress has also been made in developing efficient statistical methods for the identification of QTL using typical mapping populations, mixed model-based approaches for mapping using multiple populations and approaches accounting for epistatic or genotype by environment interactions. More recently, association mapping using populations consisted of cultivars, breeding lines and germplasm collections used in breeding program has started to be used in the establishment of marker-trait association.

With all these technical and methodological developments genetic information about the key agronomic traits has accumulated in an unprecedented speed. The challenge for plant breeders now, is to determine how best this multitude of information can be utilized in the improvement of crop performance and production. At last it can be concluded that genomic especially functional genomics era is steadily advancing towards a phase where crops would have increased productivity with all its characteristics, beneficial to both the farmers and the consumer, would have been manipulated, in the time of world globalization.

The purpose of bringing out this publication is to provide latest available advancement on molecular markers, development of markers from non - coding and coding regions of genome and their application in crop improvements with classical examples. The book comprises of 16 chapters contributed by the experienced and recognized experts all around the world, whom have deep knowledge and research experience in plant molecular breeding. The different chapter covers current status of our knowledge about an overall up-to-date progress of recent advances in plant molecular breeding, molecular markers in plants, high throughput detection of PCR products and SNPs , mapping populations in plants, tagging and mapping of qualitative as well as quantitative traits in plants, association mapping, statistical analysis of molecular data, genomics and gene based markers for crop improvement, *in situ* hybridization - an important tool in cyto-molecular mapping, markers in biosystematics, marker-assisted gene pyramiding in cultivar development, marker-assisted recurrent backcrossing, formulating marker-assisted selection strategies using computer simulation, maker assisted selection for quality traits and molecular and physiological approaches for drought management in crops have been included. The book is very useful for the researchers and students in the field of genetics and breeding who have been working or planned to work/ involved may aware of the impact of different aspects of molecular breeding in crop improvement.

We also expect that this issue will be of valuable reference for students, scientists and researchers in genetics and molecular breeding. We express our gratitude to all the esteemed authors for their valuable contributions and inputs to this volume and for their compliance with the reviewing process. We also acknowledge Mr. Prashant Gupta, Panacea Computers, Lucknow, India for nicely composing the text and Studium Press LLC, Houston, USA, who have done an excellent job in compiling the chapters and publishing this book on schedule. We hope that this book will also be useful to everyone interested in plant genomics, molecular breeding, bioinformatics, applied plant sciences and biotechnology.

Editors

About the Editors

R.K. Singh

Dr. R.K. Singh, M.Sc., Ph. D. was born on 31st March 1962. He did his B.Sc., M.Sc., and Ph.D. from Lucknow University, Lucknow. He joined U. P. Council of sugarcane Research as Scientific Officer (Genetics) and served for eleven years and he was also in-charge of Genetics and Cytogenetics section. After that he joined Indian Council of Agricultural Research (ICAR) as Senior Scientist (Genetics) at National Research Centre for Soybean, Indore. Presently, he is working at Division of Crop Improvement, Indian Institute of Sugarcane Research, Lucknow in the field of molecular breeding and transgenic. He has been awarded SERC visiting fellowship (DST) and National Biotechnology Associateship (DBT) in the year 1998 and 2000 in the field of Basic Molecular Biology, respectively and worked in the laboratory of National reputes. He worked on molecular markers for genetic diversity, tagging and mapping of gene/QTLs in the crops like rice, soybean and now on sugarcane. He was one of the pioneer molecular breeders in soybean crop in India. He was Principal Investigator in the Networking projects launched by ICAR on the molecular breeding and development of transgenic and functional genomics for soybean crop. He has more than sixty research papers to his credit published in the journals of national and international repute. He guided many M.Sc. students in the field of molecular breeding. He also visited Germany for learning advance technique in the field of Cytogenetics in the year 1994. He is member of several academic societies and editor of journal. Recently, he has co-edited a book on Plant Genomics and Bioinformatics.

Rajesh Singh

Dr Rajesh Singh is currently Course Coordinator of Plant Biotechnology at Rajiv Gandhi South Campus and Professor of Genetics & Plant Breeding in the Department of Genetics & Plant Breeding, Institute of Agricultural Sciences, Banaras Hindu University, Varanasi, Uttar Pradesh. Professor Singh also served as Senior Scientist at Indian Institute of Sugarcane Research, Lucknow and at Vivekanand Parvatiya Krishi Anusandhan Sansanthan, Indian Council of Agriculture Research, Almora for almost 8 years. Prof Singh also served as Assistant Professor and lecturer for over 8 years at Banaras Hindu University and G B Pant university of Agriculture and

Technology, Pant Nagar. Uttarnchal. Prof Singh having over 20 years research experience in field of maize breeding and biotechnology (molecular breeding), rice/hybrid rice and sugarcane breeding and crop genetic resources. Prof Singh successfully completed about 15 projects with total outlay of about 5 crores of national and international importance. Prof Singh has developed and released 16 crop varieties of rice and maize of national importance. Prof. Singh successfully initiated and executed biotechnological programmes at VPKAS, Almora, Indian Institute of Sugarcane Research, and BHU, Varanasi. Initiated programmes on mapping of quantitative trait loci for northern corn leaf blight (NCLB) in maize and marker aided selection for Quality Protein Maize (QPM) , red rot and top borer in sugarcane etc. Prof Singh has over 120 publications to his credit including 3 books/bulletin. Prof Singh is having about 15 years teaching experience at under graduate and postgraduate levels. He is recipient of several awards including SEE Fellow Award. Prof Singh did his graduation and post graduation form Haryana Agriculture University, Hisar and Ph.D form ND University of Agriculture and Technology, Faizabad, U.P., India.

Guoyou Ye

Guoyou Ye, Ph.D., is a senior scientist with the Biosciences Research Division, Department of Primary Industries Victoria, Australia. His major research activities are in the area of quantitative plant breeding including classical and modern methods for genetic analysis of qualitative and quantitative traits, analysing genotype-by-environment interaction, designing conventional and molecular breeding strategies, and developing simulation tools for molecular plant breeding. He has considerable experiences in genetics and breeding studies of both self-pollinated and open-pollinated species. He is currently a key member of several molecular crop breeding programs in wheat, barley, canola, ryegrass and white clover. He has published in many areas of plant genetics and breeding in reputed research journals.

A. Selvi

Dr A. Selvi is a Senior scientist in Plant biotechnology with the Sugarcane Breeding Institute, Coimbatore, Tamil Nadu, India. She did her B.Sc in Agriculture and her M Sc and PhD in Plant Biotechnology at the Tamil Nadu Agricutural University, Coimbatore. She was awarded the fellowship from the Department of Biotechnology, to pursue her post graduation in Biotechnology. She was selected in the first batch of Plant biotechnology scientists by the Agricutural Service Recruitment Board and joined the Agricultural Research Service in 1994. Since then she is serving at the Sugarcane Breeding Institute in the DBT lab. She is a pioneer in sugarcane biotechnology in the country and has worked extensively on sugarcane genomics. Her pioneering work includes diversity studies in *Saccharum* complex and commercial sugarcane cultivars, establishing methodologies for fingerprinting of sugarcane varieties, assessing the genomic constitution of commercial varieties, development of genus and species specific markers and establishing methodologies for identifying intergeneric and interspecific hybrids of sugarcane, mapping and tagging genes for several disease and pest resistance, map based cloning of rust resistance, smut resistance and sugarcane yellow leaf virus resistance genes in sugarcane. She has also developed microsatellite markers and cytoplasmic markers for sugarcane that would be useful for various molecular studies. She has worked in leading labs in India and abroad. She was awarded the fellowship by Agropolis Platform of genomics, of the French government in 2002 to pursue research at CIRAD on sugarcane genomics. She was also awarded the DBT overseas associateship from the Department of Biotechnology, India for the year 2006-2007 for which she went back to CIRAD to work on map based cloning of smut resistance genes. She has been involved in several projects that were funded by the DBT. She is serving on the editorial board of several national and international journals and she has more than sixty research papers to her credit that are published in highly rated journals.

G.P. Rao

Dr. G.P. Rao is working as Officer-in-charge, Sugarcane Research Station, Gorakhpur,India. He did his Ph.D. in Plant Virology in 1987 from Gorakhpur

University on pea viruses. Dr. Rao is a Fellow of Indian Phytopathological Society and Sugarcane Technologists Association of India. He has published 10 edited books, 5 text books and 100 research publications to his credit. He is editor-in-chief of "Sugar Tech", an international journal of Sugarcane crops and related industries. He is also regular member of American Phytopathological Society, British Society for Plant Pathology, Indian Virologial Society and American Microbiological Society. He has been awarded several national and international academic fellowships and honours for doing valuable research in plant pathology, molecular biology, postdoctoral work and attending international workshops and conferences. He has worked and visited many scientific laboratories of USA, France, Scotland, UK, Sri Lanka, Yugoslavia, Singapore, Malaysia, Australia, Columbia, China and Thailand. He has been involved in the research on characterization and management of sugarcane pathogens for the last 20 years. Currently he is working on characterization and management of sugarcane diseases of South East Asia.

Abbreviations

AFLP	Amplified Fragment Length Polymorphism
AMOVA	Analysis of Molecular Variance
AP-PCR	Arbitrarily Primed –Polymerase Chain Reaction
BSA	Bulk Segregant Analysis
CAPS	Cleaved Amplified Polymorphic Sequence
CIM	Composite Interval Mapping
COS	Conserved Orthologous Set
CRB	Conventional Recurrent Backcrossing
DALP	Direct Amplification of Length Polymorphism
DArT	Diversity Array Technology
DGGE	Denaturing Gradient Gel Electrophoresis
DOP-PCR	Degenerate Oligonucleotide Primed Polymerase Chain Reaction
EB	Ethedium Bromide
EST	Expressed Sequence Tag
FISH	Fluorescent *In Situ* Hybridization
FM	Functional Marker
GFP	Green Fluorescent Protein
GISH	Genomic *In Situ* Hybridization
GMM	Genic Molecular Marker
GTM	Gene targeted Marker
IRAP	Inter Retrotransposon Amplified Polymorphism
ISH	*In Situ* Hybridization
ISSR	Inter Simple Sequence Repeat
LOD	Log of Odds
MAAP	Multiple Arbitrary Amplicon Profiling
MABC	Marker Assisted Backcrossing
MAFS	Marker Assisted Foreground Selection
MALDI	TOF-MS-Matrix Assisted Laser Desorption/Ionization Time-of-Flight-Mass Spectrometry
MARB	Marker Assisted Recurrent Backcrossing
MAS	Marker Assisted Selection/Marker Aided Selection
MIM	Multiple Interval Mapping
NBS	Nucleotide Binding Site

PCR	Polymorphic Chain Reaction
QTL	Quantitative Trait Locus
RAMPO	Randomly Amplified Microsatellite Polymorphism
RAPD	Random Amplified Polymorphic DNA
RBIP	Retrotransposon-Based Insertional Polymorphism
REMAP	Retrotransposon-Microsatellite Amplified Polymorphism
RFLP	Restriction Fragment Length Polymorphism
RGA	Resistant Gene Analog
SAMPL	Selectively Amplified Microsatellite Polymorphic Locus
SCAR	Sequence Characterized Amplified Region
SDS	Sodium Dodecyl Sulfate
SNP	Single Nucleotide Polymorphism
SPAR	Single Primer Amplification Reaction
SRAP	Sequence Related Amplified Polymorphism
SSCP	Single Strand Conformation Polymorphism
SSR	Simple Sequence Repeat
STMS	Sequence Tagged Microsatellite Site
STS	Sequence Tagged Site
TGGE	Thermal Gradient Gel Electrophoresis
TILLING	Targeting Induced Local Lesions in Genome
TRAP	Target Region Amplification Polymorphism

Contents

Molecular Plant Breeding: Principle, Method and Application
Eds : R.K. Singh, Rajesh Singh, Guoyou Ye, A. Selvi and G.P. Rao
Studium Press LLC, Texas, USA, 2009, pp. 1-36

CHAPTER 1

Advances in Molecular Plant Breeding: An Introduction

GYAN P. MISHRA[1], R.K. SINGH[2], PRADEEP K. NAIK[3], RAGHWENDRA SINGH[1] and SHASHI BALA SINGH[1]*

ABSTRACT

Many interesting developments in the field of molecular plant breeding are currently ongoing which are accompanied with new and exciting challenges, demanding involvement of expertise from other disciplines, including statistics, bioinformatics and economics. The main focus of research conducted these days is on using molecular technologies to more efficiently develop improved breeds of commercial crop varieties. Molecular markers assay variation in the nuclotide sequences of DNA and they are different types *viz.* Restriction fragment length polymorphism (RFLP), random amplified polymorphic DNA (RAPD), sequence tagged microsatellite site (STMS), Amplified length polymorphism (AFLP) and single nuclotide polymorphism (SNP). These markers have been used to construct linkage maps, gene tagging, gene search, as well as for addressing issues of gene resource management. Considerable research is being carried out towards characterizing a number of quantitative trait loci (QTL), determining their numbers and locations on the genomes and the size of their effect. Current research is concentrating on validating QTL and testing their usefulness for the marker-aided selection of plants. An expanding research area is determining genes controlling different traits using genomics approach. Gene transformation systems are being developed in many plants. Rice, maize, *Arabidopsis*, etc. are being developed as a model species for molecular breeding.

Key Words: Molecular markers, Gene tagging, QTL, Marker aided selection

[1]*Defence Institute of High Altitude Research, DRDO, C/o 56 APO, Leh, India*
[2]*Indian Institute of Sugarcane Research, Lucknow, UP, India*
[3]*Jaypee University of Information Technology, Waknaghat, Solan, HP, India*
**Corresponding author e-mail:gyan.gene@gmail.com*

1.0 INTRODUCTION

Plant breeding has been in rapid transition as more molecular tools are applied to commonly accepted field techniques. Predictably, the scientific development of this field is headed toward sequence-based knowledge which should improve both reliability and adoption in agriculture. This chapter describes the application of molecular technologies to plant breeding with more emphasis on marker-assisted breeding. One weakness in the application of molecular breeding is the general lack of better phenotypic measurement tools for components of yield in agronomic crops. This weakness occurs, in large part, because detailed, precise phenotypic measurements are not available in agronomic crop species for many traits (*e.g.*, drought tolerance, standability), and it represents a severe limitation to the application of molecular breeding at this time. With the development of molecular markers, genetic maps have been constructed in plant species which allow for the localization of major loci and QTLs controlling agronomical trait variation. Molecular markers have been widely used for the introgression of major loci, but marker-assisted selection for quantitative trait breeding is used less.

2.0 MOLECULAR MARKERS SYSTEMS

It reveals variation in genomic DNA sequence. The first generation of molecular marker, RFLP, was based on DNA-DNA hybridization and was slow and expensive. The invention of the polymerase chain reaction (PCR) to amplify short segments of DNA gave rise to a second generation of faster and less expensive PCR-based markers (Table 1). Among the techniques that are particularly promising are Amplified Fragment Length Polymorphism (AFLPs), Random Amplified Polymorphic DNA (RAPDs), Microsatellites, Sequence Characterized Amplified Regions (SCARs) or Sequence Tagged Sites (STS), etc.

Table 1. Type of different markers and their behavior.

Marker	PCR Based	Polymorphism	Dominance
RFLP	No	Low-medium	Co-dominant
RAPD	Yes	Medium-high	Dominant
SSR	Yes	High	Co-dominant
AFLP	Yes	High	Dominant / Co-dominant
SNP	Yes	Extremely-high	Co-dominant / Dominant

Molecular markers are being used extensively to investigate the genetic basis of agronomic traits and to facilitate the transfer and accumulation of desirable traits between breeding lines. A number of techniques have been particularly useful for genetic analysis. These markers are often used in conjunction with bulk segregant analysis and detailed genetic maps. It provides a very efficient method of characterizing and locating natural and induced mutated alleles at genes controlling interesting agricultural traits. Markers have also been used to identify the genes underlying quantitative variation for height, maturity, disease resistance and yield in virtually all major crops. In particular, the PCR-based techniques have been useful in the assessment of biodiversity, the study of plant and pathogen populations and their interactions; and identification of plant varieties and cultivars. Amplified DNA techniques have produced sequence-tagged sites that serve as landmarks for genetic and physical mapping. It is envisioned that emerging oligonucleotide-based technologies derived from the use of hybridization arrays, the so-called DNA chips and oligonucleotide arrays, will become important in future genomic studies. However, many of these are still under development, are proprietary, or require the use of expensive equipment, and are therefore neither suitable nor cost-effective for adequate transfer to developing countries. However, techniques are continuously changing and evolving, so technology transfer needs to keep pace with current developments in genomics.

In the last decade, use of DNA markers for the study and improvement of traits has become routine, and has revolutionized the crop breeding strategies. Increasingly, techniques are being developed to more precisely, quickly and cheaply improve the genetic structure of different crops. These techniques have changed the standard equipment of many labs, forcing most plant breeders to be trained in DNA data generation and interpretation. As we enter the post-genomics era, the need for genetic markers does not diminish, even in the species with fully sequenced genomes.

3.0 OVERVIEW OF MOLECULAR TECHNOLOGIES

Molecular markers should not be considered as normal genes, as these do not have any biological effect, but should be treated as landmarks in the genome. These rely on DNA assay, in contrast to morphological markers, based on visible traits, and biochemical markers based on proteins produced by genes. Molecular markers are generally employed to detect and understand the naturally occurring polymorphism in DNA, which forms the basis for designing the strategies to exploit it for applied genetic purposes. Due to the rapid developments in the field of molecular genetics, a variety of different techniques have emerged to analyze genetic variation during the last few decades (Whitkus *et al.*, 1994; Karp, *et al.*, 1996, 1997a,b; Parker *et al.*, 1998; Schlötterer, 2004). These genetic markers may differ with respect to important features, such as genomic abundance, level of polymorphism detected, locus specificity, reproducibility, technical requirements and financial investment. The main marker technologies that have been widely applied during the last decades are summarized in Table 2. The information provided by the markers for the breeder will vary

Table 2. Main characteristics of major types of molecular markers.

Characteristic	RFLP	SSR	AFLP	RAPD	SNP
Type of visualization	Single locus	Single locus	Multi-loci	Multi-loci	Single locus
Allelism	Co-dominant	Co-dominant	Dominant	Dominant	Co-dominant
Type of polymorphism	Sequence	No. of repeats	Sequence	Sequence	Sequence
Level of polymorphism	Good	Excellent	Excellent	Good	Excellent
Polymorphism at the locus	2 to 5 alleles	Multiple alleles	Presence/ absence	Presence/ absence	2 alleles
Quantity of DNA needed	Large	Small	Small	Small	Small
Quality of DNA needed	Good	No restrictions	Good	Good	Good
Reproducibility	Good	Good	Good	Low	Good
Time	Long	Fast, once markers are developed	Fast	Fast	Fast, once markers are developed
Cost	Expensive	Average	Average	Average	Expensive
Technical difficulty	High	Low	Medium	Medium	High

depending on the type of marker system used. Each one has its advantages and disadvantages and in near future other systems will also likely to be developed. Various types of molecular markers are utilized to evaluate DNA polymorphism. Table 3 gives the details of four major classes of DNA polymorphism. No marker is superior to all others for a wide range of applications. The most appropriate genetic marker will depend on the specific application, the presumed level of polymorphism, the presence of sufficient technical facilities and know-how, time constraints and financial limitations. The markers can be used for both qualitative and quantitative traits for the improvement of trait of interest (Fig. 1).

3.1 General Applications of Different Marker Systems

Molecular markers have been looked upon as tool for a large number of applications ranging from localization of a gene to improvement of plant varieties by marker-assisted selection (MAS). These have also become extremely popular for phylogenetic analysis adding new dimensions to the evolutionary theories. Genome analysis based on molecular markers has generated a vast amount of information and a number of databases are being generated to preserve and popularize it. In general it is used for different and diverse purposes like formation of comparative maps,

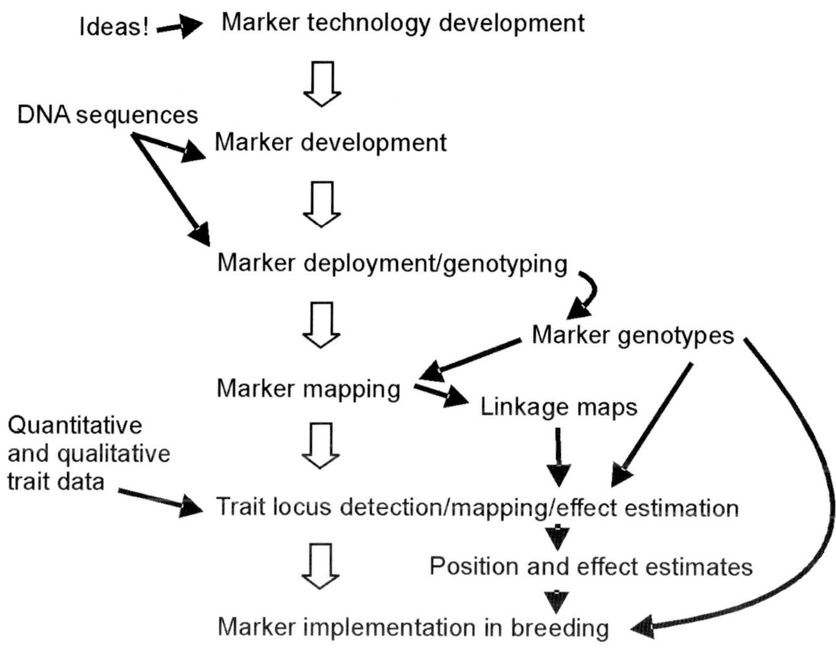

Fig 1. Schematic presentation of markers in plant breeding (source : Mather, 2006)

Table 3. Four classes of DNA polymorphisms

Class	Size of locus	Number of Alleles	Number of loci In population	Rate of Mutations	Use	Method of detection
SNP	Single base pair	2	100 million	10^{-9}	Linkage mapping	PCR followed by ASO hybridization of primer extention
Microsatellite	30-300 bp	2-10	200,000	10^{-3}	Linkage mapping	PCR and gel electro-phoresis
Multilocus minisatellite	1-20 kb	2-10	30,000	10^{-3}	DNA finger-printing	Southern blot and hybridization
Small changes in DNA content (Deletions and Duplications	1-100 bp	2	N/A	$<10^{-9}$	Linkage mapping	PCR and gel electro-phoresis

framework maps, genetic maps, breeding, varietal/line identification (multiplexing of probes necessary), marker-assisted selection, F_1 identification, diversity studies, novel allele detections, gene tagging, bulk segregant analysis, map-based gene cloning, seed testing, fingerprinting, framework/region specific mapping, novel allele detections, high-resolution mapping, very fast mapping, region-specific marker saturation, alien gene introduction, etc.

These techniques help in direct selection of many desired characters simultaneously using F_2 and back-cross populations, near isogenic lines, doubled haploids and recombinant inbred lines. Molecular markers have already played a major role in the genetic characterization and improvement of many crop species. They have also contributed to and greatly expanded our abilities to assess biodiversity, reconstruct accurate phylogenetic relationships, and understand the structure, evolution and interaction of plant populations. Molecular markers are required in a broad spectrum of gene screening approaches, ranging from gene-mapping within traditional 'forward'-genetics approaches through trait identification studies to genotyping and haplotyping studies. The availability of genetic markers is fundamental within plant biology and plant breeding. There are a wide range of uses and applications for molecular markers, but most are associated with the map-based cloning of individual genes, the characterization of quantitative multi-gene traits etc. With the subsequent association of genes and their related markers they become additionally valuable within the context of both genotyping and haplotyping.

4.0 MARKER ASSISTED SELECTION

Marker assisted selection is perhaps the first direct application of molecular markers for the improvement of any crop. High density linkage map and markers tightly linked to the gene/loci affecting the trait of agronomic importance are the two major pre-requisites for this to happen. Molecular marker assisted selection, often simply referred to as marker assisted selection (MAS) involves selection of plant carrying genomic regions that are involved in the expression of traits of interest through molecular markers.

Advantages: DNA markers are not affected by the environmental factors, heritability of traits, number of genes involved and gene interaction. Hence these are more reliable. These markers can be scored at seedling stage. This is especially advantageous when selecting for traits which are expressed only at later stage of development. By selecting at seedling stage considerable time, labour and space can be saved.

Applications: Linkage is sought between DNA marker and agronomical important traits such as resistance to pathogens, insects, nematodes, tolerance to abiotic stress and quality parameters. Instead of selecting for a trait, the breeder can select for a marker that can be easily detected and early in selection scheme. The use of DNA markers for the indirect selection offer the greatest benefits for quantitative traits with low heritability as these are the most difficult characters to asses in the field experiments. Obviously the development of marker assisted assay for such traits is extra difficult and most costly due to the expensive phenotypic assays. With the development and availability of an array of molecular markers and dense molecular maps in crop plants, MAS has become possible for traits governed by major genes as well as quantitative trait loci (QTLs). Thus marker assisted selection has become a promising and potent approach for integrating biotechnology with conventional and traditional breeding.

4.1 Molecular Markers for Diversity Studies

Molecular markers can be used to:

- Estimate genetic diversity between varieties or between populations.
- Compare populations with reference points (*e.g.* groups of lines developed by classical breeding programmes, sets of local commercial varieties, or reference core collections).

This information can be used in programmes for population improvement through recurrent selection to maximize diversity within populations by crossing lines that are genetically the most distant, to manage pedigrees, or adjust the long-term strategies of improvement programmes.

4.1.1 Number of Markers and Its Density for Diversity Study

No definitive rule exists for choosing the number of markers to use in a diversity study, but for rice, between 12 and 24 are normally used. This is a safe range according to Barry (2000), who studied the diversity of rice varieties in Guinea and showed that similar structural patterns were obtained from as many as 12 to as few as six microsatellite markers. One marker per chromosome or chromosome arm is a reasonable marker density for ensuring the best possible coverage of the genome. Such density will also prevent bias due to the presence of genes involved in gametic or zygotic selection, or controlling traits under selection in a specific chromosome. Among the possible markers for each arm, those that reveal the largest number of alleles in the studied background should be selected as priority. Those with a very high mutation rate should be avoided as they may blur results.

Among the molecular markers microsatellites or simple sequence repeats (SSRs) has merits because it reveals high degree of allelic variation due to variation in number of repeat motifs at a locus caused by replication slipage and/or unequal crossing over during meosis (Goldstein and Schlotterer, 1999). Microsatellite markers have gained considerable importance in plant genetics and breeding owing to their many desirable genetic attributes including hypervariability, wide genomic distribution, co-dominant inheritance resproducibility, mutli-allelic nature, and chromosome specific location. These markers are amenable to high thoroughput genotyping and are thus suitable for paternity determination, construction of high density genome maps, mapping of useful genes, markers-assisted selection, and for establishing genetic and evolutionary relationships (Parida *et al.*, 2008).

4.2 Possible Use in Population Improvement Programmes

4.2.1 Choosing parents and developing the base population

The choice of parents to constitute a gene pool depends on the objectives and strategy of each breeding programme. Both the number and nature of the parents are important. A large number of parents is not, in itself, an indicator of a broad genetic base because some may share part or most of their genomes. The genetic relatedness between parents is, as a result, an important, although complex, aspect to take into account. So far, genetic relatedness has been evaluated on the basis of line pedigree (Graterol, 2000). Rice is strongly structured, and varieties belonging to the same varietal group are genetically closer than those varieties belonging to different groups (Glaszmann, 1987). Equal contribution of both parents is seldom true in a context of selection where plant breeders favour an ideotype.

Molecular markers are more informative. By determining the alleles present at different markers for the set of varieties to compare, we can establish a matrix of similarities that offers a truer picture of the relationships between varieties. The classic indices of similarity for comparisons between varieties are those of Dice and Jaccard for co-dominant markers, and Sokal and Michener for dominant markers (Perrier *et al.*, 2003). After transformation of the similarity matrix to obtain distances, we can produce a geometric representation of the distance between varieties through various clustering techniques such as the unweighted pair-group method, using the arithmetic average (UPGMA), or the neighbour-joining tree method (NJ tree). Results are usually represented as dendrograms. Lorenzen *et al.* (1995) provide recent examples of evaluating relatedness based on molecular markers between genotypes used in classic breeding programmes for soybean; Soleimani *et al.* (2002) for durum wheat; and Lima *et al.* (2002) for sugar cane.

4.2.2 Optimizing the Number of Recombination Cycles

An important question that arises when constituting a gene pool or, later in the recurrent selection process, when crossing selected progenies, is how many intercrossing cycles are needed to better recombine the parents' alleles. The objective in this case is panmixia. On a theoretical basis, Hanson (1959) concluded that three crossing cycles were needed. Marín-Garavito (1994), working on a recurrent selection population, did not find differences among the phenotypes of the first and third cycles of recombination. Such a question can also be answered by using co-dominant molecular markers.

For each marker, the allelic and genotypic frequencies in the populations and Nei's diversity index (Nei, 1972) can be computed. These data can be used to analyse deviations from panmixia, using the Wright fixation index. To further check the efficiency of recombination in breaking unfavourable linkages, one strategy would be to select additional markers close to the markers under study. The segregation of this specific group of markers can then be compared in successive recombination cycles. The evolution of linkage disequilibrium between markers of increasing physical distances can be evaluated over several cycles. The recombinations, if they occur normally, should progressively break linkages, and linkage disequilibrium should decrease from cycle to cycle.

4.2.3 Analyzing Effect of Population Improvement on Genetic Diversity

Several projects of recurrent selection, based on evaluations of $S_{0:2}$ families, are being carried out to improve various traits of agronomic importance. Examples include resistance to blast (Guimarães, 2000), tolerance of acid soils (Ospina *et al.*, 2000) and yield. Progress made for the trait or traits under selection is usually evaluated after a few cycles of recurrent selection. This is done by comparing phenotypes of the initial population with those of the progeny from the last cycle. Deliberate selection reduces genetic diversity to a level that depends on the initial variability and intensity of selection. Such reduction of diversity may not be harmful if the alleles eliminated are those that are unfavorable.

Molecular markers can be used to evaluate the degree of reduction of total diversity by comparing the diversity between successive recurrent cycles of the population. The markers can be chosen according to the allelic diversity in the sub-species used to construct the population.

4.2.4 Locating the Male-Sterility Gene and Evaluating Linkage Drag

The male-sterility gene from 'IR36' has been used in most rice populations to facilitate crossing. Molecular markers can be used to locate such gene in the rice genome. Identifying a co-dominant marker closely linked to the male-sterility gene would allow distinguishing heterozygotes from homozygotes for the male-sterility gene among male-fertile plants extracted from populations. Such selection would accelerate line fixation during line extraction. The easiest method for finding such markers is through bulked segregant analysis (BSA) (Michelmore *et al.*, 1991). In this method, two separate DNA bulks are constituted: one from the pooled DNA of 10 male-sterile plants and the other from 10 male-fertile plants. The two pools are tested for allelic differences with a set of markers. Any method that enables rapid testing of a large number of markers is advisable. Using BSA approach Gyan *et al.* (2003) has identified one STMS markers (RM258) on chromosome 10 linked to fertility restore gene in basmati rice at 9.5 cM.

4.2.5 Marker-Assisted Selection for Recurrent Selection Populations

One major use of molecular markers is to locate genes that control agronomic traits of interest on dense genetic maps and identify those markers closely linked with these genes. To facilitate mapping, specific bi-allelic populations, resulting from crosses between pure lines that correspond to the simpler genetic situation, are developed. De Koeyer *et al.* (2001) used RFLP markers to identify QTLs linked with yield, plant height, and flowering in a recurrent selection population of oat. The locations of the QTLs were consistent with those detected in a classic population of recombinant inbred lines.

Once markers tightly linked with traits of interest are identified, the next step will be the manipulation of the traits. Identification of QTL is based on linkage disequilibrium, which is artificially increased through hybridization. Intercrossing induces recombinations and reduces linkage disequilibrium in a recurrent selection population. Thus, the positions of the QTLs need to be regularly re-evaluated, not for every cycle but at least every 2 years (Hospital *et al.*, 1997). This problem can be overcome using polymorphic functional markers/gene based markers. But this implies extensive work to identify such genes through positional cloning or the candidate gene approach.

5.0 GENEBANK MANAGEMENT

The realization that the world was rapidly losing much of its agro-biodiversity led to a global effort to collect and conserve germplasm. An increasing awareness of the narrow genetic base of crops in advanced agriculture and potential susceptibility to crop failures (National Research Council, 1972) further stimulated the efforts to collect, and to develop a system of national and international Genebanks. A major question facing genebank managers concerns which material needs to be included in a collection to conserve a representative sample of the total genetic diversity range of a crop. Schoen and Brown (1993) document how molecular marker techniques provide the optimal strategy to sample materials from populations of wild relatives to maximize the number of alleles. Hamilton (1994) argues that simple marker diversity is an insufficient measure of maintenance of diversity, and suggests that more detailed studies of genetic correlations of quantitative genetic variation and gene and environment interaction are needed.

5.1 Relationship of Molecular Variation To Geography

The literature sometimes shows association of DNA-based relationships to geography and sometimes fails to show such associations. Whitkus *et al.* (1998) examined these alternative hypotheses using RAPD markers from a wide variety of geographically diverse wild and cultivated populations. Their results supported a single origin of modern cultivars in South America. However, they also discovered relationship of distinct populations in ancient Mayan groves in southern Mexico to wild Mesoamerican populations. Other studies showing a relationship of molecular variation to geography or ecology are Espejo-Ibañez *et al.* (1994), Garvin and Weeden (1994), Yang *et al.* (1996), Lanner *et al.* (1997), Nevo *et al.* (1997, 1998) Fahima *et al.* (1998), and Zerega (2004). However, there is often not a congruence of genetic distance and geographic distance as shown in Maass and Ocampo (1995), Lanner *et al.* (1996), Ursla *et al.* (1997), Varghese *et al.* (1997), Comes and Abbott (1998), Freville *et al.* (2001) and del Rio and Bamberg (2002).

5.2 Marker Data for Germplasm Acquisition

Marker studies addressing the distribution of genetic variation within and between populations have been used to guide the acquisition of new material for germplasm conservation. Steiner *et al.* (1998) used RAPDs to measure genetic diversity of germplasm collections of *Trifolium incarnatum* L. In combination with data on pedigree, they documented low diversity in the existing cultivars and identified populations that need additional collecting. Other studies addressing the application of marker data for germplasm acquisition include Brown and Munday (1982), Murphy and Philips (1993), Lamboy *et al.* (1994), Tsegaye *et al.* (1996), Nebauer *et al.* (1999), Maquet *et al.* (1996, 1997) and Zoro Bi *et al.* (1998). In summary, molecular data are useful to guide genebank curators in decisions concerning acquisition in cases where molecular variation is associated with geographic variation.

5.3 Characterization of Germplasm

5.3.1 Systematics

Taxonomic questions are addressed by almost every molecular marker class from microsatellites useful at the species level to DNA sequences useful at the generic and family levels. For example, Roa *et al.* (1997) used phenetic analyses of AFLPs to investigate the origin of cassava (*Manihot esculenta* Crantz) relative to four other wild species and two non-cultivated subspecies of *M. esculenta*. These subspecies were thought to be progenitors of the cultivars or escapes from cultivation. Molecular markers provide tools to test the taxonomic validity of species and can provide species-specific diagnostic markers. For example, Martin *et al.* (1997) used RAPDs to re-identify a collection of oat accessions. Lee *et al.* (1996) used RAPDs to discriminate *Brassica* L. varieties. Species-specific markers also have been found with minisatellites (Baurens *et al.*, 1997, banana), and Zhou *et al.* (1997) used minisatellite sequences to discover genome-specific fragments in oat, but not cultivar-specific fragments. McGregor *et al.* (2002) used AFLPs to characterize 314 accessions of the wild potato subgroup *Solanum* L. series *Acaulia* Juz.

5.3.2 Fingerprinting studies

"Fingerprinting" is an attempt to discover cultivar-specific molecular markers that aid their identification. Certain cultivars are economically very important and form the backbone of regionally important industries. Potato is a clonal crop with many morphologically similar cultivars. Schneider and Douches (1997) were able to distinguish 24 of 40 potato cultivars with 7 SSR primer pairs. When the SSR data were combined with the tuber morphology, only five pairs of cultivars could not be distinguished. These many studies clearly document the use of SSRs for fingerprinting genebank accessions.

5.3.3 Distinctness, uniformty and stability (DUS) testing

A variety is traditionally described by a set of morphological characteristics, which are used to establish its distinctness, uniformity and stability (DUS). The DUS testing involves comparison of new and existing varieties by the recording of a number of phenotypic characters called descriptors by growing the varieties side by side. In view of the fact that the number of existing varieties in crop is quite large, the traditional method of DUS testing becomes time-consuming and expensive, requiring large areas of land and skilled personnel often making subjective decisions. Many of the characters used are multigenic or quantitative and their expression is altered by enviromental factors, which essentially requires replication of observations. Further, the number of characters may not be sufficient for discrimination of all the extant and new varieties. There are thus compelling reasons to find more rapid and cost-effective procedures to augement this approach. Molecular markers assay variation in the nucleotide sequences of DNA. Such differnces remain unaffected across different growth stages, seasons, locations and agronomic practices. Varietal identification and differentiation therefore become more reproducible and objective. The characteristics of a new variety based on DNA patterns can therefore be compared with those in common knowledge in any part of the world for establishing its uniqueness. Among the molecular markers, sequence tagged microsatellite sites (STMS) are highly polymorphic, reproducible, co-dominant and distributed throughout the genome. Singh *et al.* (2004) has proved the suitability of STMS markers in estabilishing distinctness, uniformty and stability in aromatic rice. The working group of Biochemical and Molecular Techniques (BMT) of the International Union for the Protection of New Varieties of Plants (UPOV) has in fact identified STMS as the most widely used marker system for plant variety characterization (UPOV, 2002). It was however, emphasized by this group that prior to the use of molecular markers in DUS testing, it would be essential to evaluate two of the important aspects : (i) correspondence between molecular patterns and phenotypic characteristics, and (ii) the uniformity and stability of the same molecular characteristics as used for distinctness (UPOV, 2002).

6.0 HYBRID AND HETEROSIS

6.1 Putative natural hybrids

Hybridization is thought to be a major evolutionary force at both the diploid and polyploid levels (Rieseberg, 1995). Rieseberg and Ellstrand (1993) showed that hybrids are no more likely to display intermediate character states than parental ones, and can additionally express an array of transgressive or novel traits. The utility of molecular markers to investigate hybrids decreases with time of divergence, as species-specific markers from both parents can be disrupted through recombination, and both can mutate to new markers. Molecular markers present powerful new tools to reinvestigate hypotheses of hybridization. One method is to search for markers in the hybrid that are found in both putative parents.

6.2 Parental Contributions of Artificial Hybrids

DNA markers are very useful for confirming hybridity of artificial sexual hybrids or somatic fusion hybrids. Molecular markers are especially useful when hybridity is questioned by morphological reasons or for early screening of large putative hybrid populations. Thieme *et al.* (1997) used isozyme and RAPD data to screen hundreds of first generation somatic fusion hybrids and later sexual backcross generations between cultivated and wild species of potato, to distinguish hybrid from non-hybrid progeny. Lee *et al.* (1998) used RAPDs to disprove hybridity in putative *Saccharum* L. and *Erianthus* Michx. hybrids. Yamada *et al.* (1997) used cpDNA and mtDNA to investigate symmetric somatic fusion hybrids of cultivated potato and one of its wild relatives. The results suggested that chloroplast genomes of the respective parents segregated randomly while the mitochondrial genomes favoured the cultivated species parent.

6.3 Molecular Diversity and Heterosis

Heterosis, or hybrid vigour, originally referred to the selective superiority of heterozygotes regarding continuously variable characters of size, yield and vigour. Information on the genetic diversity in a crop species is important for selection of parental strains and in the prediction of hybrid performance especially in crops in which hybrids are commercially important. The various steps involved in hybrid breeding programs such as making several crosses and screening the combinations for superior performance and heterosis, are very costly, laborious, and time consuming. Hence if heterosis can be predicted before making the crosses, then the number of crosses to be performed and the progeny to be screened can be considerably reduced. The genetic distance estimated through molecular markers distributed throughout the genome could provide a mean to predicting hybrid performance prior to making and evaluating a actual cross. Various investigators are trying to correlate genetic diversity, as quantified by DNA markers, to predict hybrid performance, in various hybrid breeding programs because the level of genetic diversity between the parents has been proposed as a possible predictor of heterosis. Smith *et al.* (1990) showed a strong correlation of genetic distance and heterosis in maize. Melchinger *et al.* (1994) reported that genetic distances of inbred lines of maize, as assayed *via* RFLP data, were not correlated with heterosis for yield. Paz and Veilleux (1997) reported that in crosses between a diploid cultivated potato species with diploid clones of other complex interspecific hybrids of potato, the greatest total tuber yield was associated with diverse parents as measured by RAPDs. Manjarrez-Sandoval *et al.* (1997) compared RFLP similarity estimates of the parents, to yield in soyabean, and also found a positive correlation.

7.0 GENE SEARCH BASED ON MOLECULAR MARKERS

Complex traits such as yield are controlled by polygenes that are prone to the influence of the environment. Since each gene makes only a small contribution to the total variation of the trait, identifying individual polygenes has proved difficult in the past. Consequently, breeding for polygenic trait has been difficult. The past two decades have witnessed explosive developments in molecular markers technology. These markers are DNA sequences of specific size and/ or specific polynucleotides, and correspond to genomic sequences. They have fixed locations on the genome and follow patterns of Mendelian inheritance. Thus these markers can be regarded as specific phenotypes. Unlike the traditional traits, these markers are not influenced by the environment and hence can be easily identified at all stages. The major advantage of these markers is that they are not limited in number and can be used to mark every locus along the entire genome.

7.1 Tagging Traits with Molecular Markers

By studying the co-segregation of a trait and a molecular marker(s), one can identify the molecular marker(s) linked with the trait of interest. Thus the gene(s) controlling a trait can be tagged with specific molecular marker(s). Molecular tagging has come in handy to identify novel genes as well as to pyramid desirable genes. For example breeding for disease resistance, identification and tracking of different resistance genes for a given disease is difficult because presence of one resistance gene masks the effect of others. In earlier years special cytogenic stocks (*e.g.* monosomic lines in wheat) were used to identify and pyramid multiple disease resistance genes. However, such stocks are difficult to build and maintain, and were available only in few species. Using molecular markers different resistance genes can be tagged and followed during breeding. The recent plant breeding literature is replete with the reports of molecular markers linked to various traits such as disease resistance, male fertility restoration (of CMS lines), pest tolerance, quality traits etc. Thus molecular markers are aiding new discoveries and use of new genes. Another major advantage is that desirable genes can be assembled without evaluating for the trait. This is particularly advantageous where phenotyping of the trait (*e.g.* disease or pest resistance, drought, salinity, yield) is difficult (Chopra, 2006).

7.2 Gene Search Based on Crop Circles

Detailed linkage maps have been constructed using molecular markers in different crops in recent years. Comparative analysis of the maps based on common markers has revealed highly conserved nature of marker positions and order among different species. This is termed as synteny. For example, all members of the *Graminae* family share significant synteny, which is depicted in what is now termed "Crop Circles". This synteny has emerged as an important tool for gene discovery. Finished rice genome sequence now serves as reference for discovery of corresponding new genes in other related species. For example, Chen *et al.* (2003) found four QTLs for rice blast, which were also found in barley. Two of the barley genes had conserved isolate specificity while the other two had partial specificity. Similarly QTL for shattering resistance and plant height were found conserved among grasses. Synteny is not only seen at the macro level but also at the micro level *i.e.* in short stretches of a few hundred kb. Such microsynteny is particularly exploited for cloning genes through chromosome walking. For example verbalization genes *VRN1* and *VRN2* of wheat were cloned using this strategy. The map-based cloning is highly relevant in crops with large sized genomes such as barley, wheat and corn. For example stem rust resistance gene of barley *Rpg* 1 was cloned based on microsynteny of the region with rice although rice dose not have this gene (Kilian *et al.*, 1997). The *Lr*21 gene of bread wheat was cloned by shuttle mapping between diploid and hexaploid wheat (Huang *et al.*, 2003). Recently, Gualtieri *et al.* (2006) reported microsynteny between rice chromosome 11 and apomictic species region of *Cenchrus ciliaris* and *Pennisetum squalatum*. Similarly, Calderini *et al.* (2006) found microsynteny between *Paspalum simplex* apomixis locus and rice chromosome 12. These findingings will speed up identification of gene(s) controlling apomixis. However, microsynteny is not strictly conserved and may show variation apparently at random. Therefore, a case-by-case examination is essential.

7.3 Gene Search+ Based on Sequence Information

A very large number of genes involved in plant growth, morphogenesis and development have been cloned. Similarly, genes mediating various metabolic pathways have been characterized using different approaches and in different organisms. The database of sequences of genes, cDNA etc. is growing rapidly and is available in public domain. Information about these sequences has become an important tool for gene discovery.

7.4 Gene Orthologs, Analogs and Aallele Mining

Analysis of a number of genes sequences coding for proteins has revealed high degree of nucleotide/amino acid homology among genes of related functions. In particular, conserved sequence domains are frequent among different classes of genes. This is expected in the light of evolution of different life forms from common ancestors. Hence, using sequence information from one species, orthologous genes from related species have been cloned through PCR and other techniques.

Plant disease resistance genes fall under 5 categories, namely the NBS-LRR, protein kinase, eLRR-TM, LRR and toxin reductase. The largest group of plant defense genes belong to NBS-LRR class. These proteins have nucleotide binding site (NBS) and C-terminal leucin rich repeats (LRR). Using sequence information of these conserved domains, similar genes (analogs) can be easily isolated by PCR cloning and tested for their novelty. Such genes are termed as resistant gene analogs (RGAs). A number of such RGAs have been cloned in different crop species.

Similarly, using sequence information of specific genes, various allelic forms can be amplified and characterized. Since most of the inter-varietal differences originate due to allelic differences, allele mining has become a rapid method of search for new genes.

7.5 Component Search

So far we have considered gene search with reference to traits they confer on the individuals that carry them. However using knowledge of metabolic pathways, r-DNA and plant transformation technologies, and engineering novel traits has become feasible by putting together components from different, and even unrelated, sources. Therefore modern gene search is not only about full length genes (including regulatory elements) but also about components. The significance of this approach is best illustrated in Golden Rice story.

Rice seeds do not accumulate Vitamin A (or more precisely its precursor, beta carotene) even though a part of the synthetic pathway of Vit. A is present in rice. In order to enrich rice grains with Vit. A, additional genes necessary for beta carotene gene were sourced from an unrelated plant (*psy* gene encoding for phytoene synthase from Daffodil) and a bacterium (*Crt1* gene encoding carotene desaturase from *Erwinia uredovora*). Also a gene lycopene beta cyclase from *Narcissus pseudonarcissus* was introduced to convert lycopene to beta carotene. To ensure that beta carotene accumulates in the endosperm, these genes were linked to seed specific glutelin promoter. When this gene cassette was introduced into rice, the resulting seeds accumulated beta carotene in the seeds. Thus was engineered a new trait in rice (Ye *et al.*, 2000). Recently, further improvements have been made to elevate the expression levels of beta carotene from 1.6 µg/g to 30 µg/g of seed (Al-Babili and Beyer, 2005). This later development was made possible by replacing *psy* gene of daffodil with its counterpart from maize. This again illustrates the point of knowledge based gene search in modern plant breeding. Similarly, using bacterial RNAse gene bar, driven by an anther (tapetum) specific promoter, male sterility was engineered in *Brassica* for hybrid seed production (Mariani *et al.*, 1990). There are several similar cases where metabolic engineering has provided novel compounds of pharmaceutical and industrial value.

This component search is highly knowledge intensive and requires convergence of interdisciplinary expertise and utilizes a range of modern tools and techniques related to gene expression. It is a telling example not only of the unification of entire biological system but also of how knowledge and interdisciplinary science can mould into societal gods of high value.

7.6 High Throughput Gene Discovery Base on Gene Expression Analysis

In recent years, complete genome sequences have become available for several organisms. Using bioinformatic tools and other data, probable coding regions have been specified. For example, in *Arabidopsis* a total of 25,498 genes have been predicted. Similarly, rice genome is estimated to code for a total of 37,544 genes. These coding region sequences can be arranged individually on a slide in arrays. Such microarray slides or gene chips are now commercially available. By hybridizing cDNA derived from contrasting treatments one can detect specific changes in gene expression. For example, in a study with rice, Rabbani *et al.* (2003) examined changes in gene expression in rice exposed to four different abiotic stresses, namely, drought, salinity, cold and abscisic acid treatments. In total 73 genes were found to be differentially expressed in response to these stresses. Of these, 58 cases were novel, not reported previously in rice. 15 genes were common to all the four stresses. Further, 51 of these genes were also known to be induced by similar stresses in Arabidopsis. This study demonstrates the power of new technology for gene identification. Also, the common gene expression responses of Arabidopsis and rice to abiotic stresses underscores the exportability of information gained from model species to the other species (Chopra, 2006).

8.0 MARKER-ASSISTED BACKCROSS AND QTL MAPPING

8.1 Marker-Assisted Backcross (MABC)

The efficiency and complexity of MAS depend on the genetic nature of the trait (monogenic or polygenic). For monogenic traits, marker-assisted backcross (MABC) is the most straightforward strategy, whereas for polygenic traits various strategies are available. The principle of MABC for a single gene is quite simple. First, molecular markers tightly linked to the target gene must be identified, allowing assessment of the presence of the introgressed gene ("foreground selection"). Other markers are also used in order to accelerate the return to the recipient parent genotype at other loci ("background selection"). Background selection is based not only on markers located on the chromosomes carrying the gene to introgress (carrier chromosome), but also on non- carrier chromosomes. Markers devoted to background selection on a carrier chromosome allow the identification of individuals for which recombination events took place on one or both sides of the gene, in order to reduce the length of the donor type segment of genome dragged along with the gene (Young and Tanksley, 1989). Background selection on non-carrier chromosomes was investigated by Hospital *et al.* (1992). Visscher *et al.* (1996) investigated both foreground and background selection. In three generations of MABC, isogenicity is higher than that obtained by classical methods. By comparison, traditional approaches would require approximately two more generations to obtain such an isogenicity (Hospital *et al.*, 1992).

8.2 QTL Mapping

Molecular markers can be used to locate important genes or QTLs in the genome. Once tagged by tightly linked markers, valuable alleles of the genes can be manipulated indirectly by selecting specific alleles at the flanking markers. For this purpose, markers with known positions spaced at regular intervals throughout the genome are needed. For their technical simplicity, microsatellite markers are currently the ones most chosen. AFLPs should be selected for quick mapping or saturating a map where microsatellites or RFLPs have left gaps.

8.2.1 Theoretical Investigations on MAS for QTLs

Traits showing quantitative variation are usually controlled by a number of genes (quantitative trait loci, QTL), each with variable effect. Due to the genetic complexity of such traits, several QTLs with small effects must be simultaneously manipulated. As for monogenic traits, MABC is the most effective strategy when a small number of QTLs (less than five), coming all from the same parent, are transferred. In practice, the position of the QTL is not precisely estimated and the true position of the QTL is unknown, but is supposed to be within a confidence interval. From this confidence interval length, Hospital and Charcosset (1997) deduced the number of markers and their position relative to the estimated position of the QTL, in order to insure an optimal control of the QTL. However, as the minimum number of individuals that should be genotyped at each generation depends on (i) the confidence interval length, (ii) the number of markers and (iii) the number of QTLs. It seems illusive to transfer more than four or five QTLs with this simultaneous design unless a very large population can be considered, or the precision of the QTL location is very high.

8.2.2 Advanced Backcross QTL Analysis

This analysis is another strategy tailored for the discovery and transfer of valuable QTL alleles from unadapted donor lines into established elite inbred lines (Tanksley and Nelson, 1996). The QTL analysis is delayed until an advanced generation (BC_3 or BC_4), and during the development of this generation, negative selection is pursued to reduce the frequency of deleterious donor alleles. The advantages of using BC_3 /BC_4 populations include reducing linkage drag and epistatic effects and decreasing the amount of time later needed to develop QTL-NILs (Fulton *et al.*, 1997).

8.2.3 Pyramidal Design

When the number of QTLs to introgress becomes important, Hospital and Charcosset (1997) proposed to use a pyramidal design. QTLs are first monitored one by one by MABC, to benefit from a higher background selection intensity, and then the selected individuals are crossed, to accumulate favorable alleles at the QTLs in the same genotype. According to predictions, the pyramidal design should provide approximately the same efficiency as the simultaneous design with almost one third of the individuals.

8.2.4 Index Method

When favorable alleles come from different sources, van Berloo and Stam (1998) proposed an index method to select among recombinant inbred lines for crossing, to obtain a single genotype containing as many favorable quantitative trait alleles as possible. The index is constructed for each pair of lines based on the genotype of the markers flanking the putative QTLs. Plants showing the optimal index are crossed together. The use of genetic markers to improve populations was proposed using a statistical approach based on an index combining phenotypic and marker information (Lande and Thompson, 1990).

9.0 MAP-BASED CLONING

A traditional strategy for cloning a gene requires prior knowledge about the biochemistry of a gene or its products including the gene's mRNA, a heterologous cloned copy of the gene from another species or differential nucleic acids preparations. Map-based cloning or positional cloning also known as reverse genetics is a strategy which does not require prior knowledge about the gene or its products; instead, this cloning strategy requires knowledge about the chromosomal location of the gene. The first step of map-based position cloning is to identify molecular marker that lies close to gene of interest. The identification of tightly linked markers and their relative order, in the vicinity of target gene, is feasible due to the availability of high density molecular maps and several regional targeting procedures. The next step is to screen a large insert genomic library (BAC or YAC) to isolate clones that hybridize with tightly linked marker probes. The markers most tightly linked to the gene are used to screen a large fragment library (YAC or BAC). The positive clones are supposed to contain the target gene. The step that follows termed as chromosome walking. This procedure involves creating new markers (usually sequences at the end of the clone) and screening segregating population with these new markers. Often this population is large (1000-3000 individuals).The process is repeated with markers designed from different clones in order to achieve co-segregation of markers with the gene of interest. Co-segregation means that whenever one allele of the gene is expressed, the markers associated with that allele are also present and there is no recombination between gene and the markers. As target gene flanked on a single clone between two markers, the DNA fragments between the flanking markers are cloned. Bioinformatics analysis of the clone may reveal relevant candidates. To decide between the candidates, the most direct method is transformation, whenever possible.

10.0 DEVELOPMENTS IN FUNCTIONAL GENOMICS

Simply defined, functional genomics is a scientific approach that seeks to identify and define the function of genes, and uncover when and how genes work together to produce traits. Current structural genomic approaches (*i.e.*, mapping) generally focus on traits controlled by one or only a few genes, and often they only provide information regarding the location of a gene or genes. Although obtaining location information is a critical first step, functional genomics goes further to examine the interrelationships and interactions between thousands of genes to determine when and why certain traits are expressed, which sets of genes are specifically responsible for that expression, and under what conditions. This information equips scientists to create varieties with exact combinations of traits. Whether the ultimate goal is improved drought tolerance, enhanced nutritional composition, or higher yield potential, functional genomics will play an increasingly importantly role in helping scientists achieve their aims.

10.1 Allele Mining

Identification and access to allelic variation that affects the plant phenotype is of the utmost importance for the utilization of genetic resources. Allele mining is a research field aimed at identifying allelic variation of relevant traits within genetic resources collections. For identified genes of known function and basic DNA sequence, genetic resources collections may be screened for allelic variation by *e.g.* the "tiling strategy" using DNA chip technology (Lemieux *et al.,* 1998). In that approach the basic DNA sequence of a gene is spotted on a chip in the form of large series of sequence overlapping probes consisting of 15–20 bases. Each base position in a fluorescently labeled sample is then interrogated for the presence of point mutations by monitoring hybridization signals with the spotted probes. Because the sequence of samples is determined in comparison with the primary composition of a gene, this method is also known as "re-sequencing". With this method new point mutations, in relatively large DNA fragments, can be detected. As an example, the tiling strategy has been used by the International Rice Research Institute (IRRI) to identify favorable alleles related to tolerance to biotic and abiotic stress factors in rice.

10.2 Association Genetics

In allele mining studies, allelic variation is analyzed for identified genes whose function and basic DNA sequence are known and whose map position in the genome will generally have been determined. On the contrary, in association genetic studies no prior information about the genes of interest is available, but associations between genetic markers and the considered traits are simply derived from observational research. Association genetics focuses on the identification of correlations between phenotypic traits and genetic markers with the aim to identify and locate the underlying genes in the genome (association mapping). Association genetics originated in human genetic studies focusing on the application of significant associations between marker data and diseases for mapping and diagnostic purposes (Peroutka, 1997). Recently, association genetics has gained increasing interest from plant geneticists (Buckler and Thornsberry, 2002; Rafalski, 2002). The rationale behind association genetics is that, in general, alleles at different loci are expected to be randomly associated into genotypes, or in other words, to occur in linkage equilibrium. The more adjacent two loci are, the lower the probability of chromosomal recombination occurring between the loci during meiosis. However, given sufficient numbers of regeneration cycles, once established linkage disequilibria will eventually be disrupted by recombination even in the case of tightly linked loci. Linkage disequilibria are therefore expected to be readily lost from populations, the rate of decay being determined by the recombination fraction and the number of generations elapsed. However, selection tends to accumulate favourably interacting alleles and hence opposes the decay of linkage disequilibria.

10.3 Comparative Genomics

Comparative genomics focuses on the integration of genome information derived from different species with the aim to obtain more insight in the genetic organization of traits through the identification of conserved mechanisms (Laurie and Devos, 2002). This research field has emerged from previous findings of comparative mapping, by which conservation of tracts of collinear markers have been investigated not only among members of the same botanical family, but also among different species, and often led to better understanding of genome evolution. Comparative mapping has shown that a large proportion of the markers and genes are indeed located at comparable positions in the genome (synteny). Together with the availability of an increasing number of genome sequences, including those of known genes, the conservation of gene sequences and their functions among species have also been investigated and are used to further develop the knowledge obtained from previous genetic linkage maps for different species.

10.3.1 Conserved Ortholog Set (COS)

These markers are conserved markers that may serve as anchors for map development (Fulton *et al.*, 2002). To study synteny between different living organisms it is very important to select the set of conserved ortholog markers (genes). Because of multiple gene duplications it is hard to distinguish orthologs from paralogs. Recently fully sequenced *Arabidopsis* genome revealed about 3,000 single copy genes distributed evenly over all five chromosomes. These unique genes distributed over genome regardless of segmental duplications regions. Using lettuce/sunflower as well as tomato and corn ESTs a set of more than 2,000 conserved ortholog markers were selected and characterized using *Arabidopsis* genes as candidates for COS (Conserved Ortholog Set) markers. By computational approach, only those sequences were selected which have just a single BLAST hit to *Arabidopsis* genome.

10.3.2 Degenerate Oligonucleotide Primed-PCR (DOP-PCR)

DOP-PCR uses partially degenerated primers for polymorphism detection (Telenius *et al.*, 1992). Because no prior sequence information is required, DOPPCR is considered useful when crops are involved for which no, or limited sequence information is available. Comparison of the genetic maps of different species will reveal information on chromosome evolution and the common genetic control of traits in different organisms. Comparative maps are being developed for various important crops and are expected to facilitate the understanding of the genetic organization of traits in less studied organisms. This may also reveal novel alleles for relevant genes that subsequently may be exploited in crop improvement.

11.0 FOUNDATIONS FOR MOLECULAR BREEDING

11.1 Integration of Genomics and Genetics

It is expected that the foundation for studying natural genetic variation at the DNA sequence level for traits in elite breeding germplasm pools and the design of successful molecular breeding strategies will involve integration of structural and functional genomics technologies within the framework of classical trait mapping methods. The availability of physical maps for a number of the important crop plants and the complete genome sequence for the model plant *Arabidopsis thaliana* (L.) Heynh. (The Arabidopsis Genome Initiative, 2000) and the crop plant rice (*Oryza sativa*; Goff *et al.*, 2002; Yu *et al.*, 2002) has enabled alignment of genetic maps with physical sequence and thus opportunity to target sequence data from genomic regions for gene discovery and gene function analysis. Further advances in technologies for the study of gene expression and protein interactions have opened up glimpses of some of the gene networks that are involved in the relationships between inter individual DNA sequence variation and trait phenotypic variation in elite breeding populations.

11.1.1 Some Current Considerations

Much of the outcomes from plant genomics efforts to date have involved large-scale data generation of a narrow sample of genotypic variation and its systematic organization, combined with descriptive efforts to annotate the features observed in the organized data. Comparative approaches have been used to initiate candidate gene searches from model to target species. Moving from this broad comparative genomics view of gene discovery to a breeding strategy view that is focused on understanding the detailed organization of extant allelic variation for multiple traits within a selected elite germplasm pool, presents significant challenges.

11.2 Power of Selection

An important consideration in the design of any selection process, be it molecular or phenotypic, is that you get what you select for. Direct selection on the phenotype of the end-product traits of commercial significance is a relatively robust, albeit in some cases slow, approach when genetic variation exists for the target traits. Replacing or augmenting this system with a knowledge-based approach that targets selection at the level of DNA sequence variation will also rapidly bring about genetic changes. The rigor of the associations we develop between population level sequence variation and phenotypic variation will determine the robustness of this molecular breeding approach. We will be continually forced to refine our knowledge of trait genetics and gene-to-phenotype associations.

12.0 OUTCOME OF GENOMICS RESEARCH

Without appropriate investment into crop genomics research, we will always lack detailed knowledge in two areas critical for successful breeding: (i) the structural organization and functional properties of genetic variation for traits and (ii) the influences that plant breeding strategies have on the genetic variation that is widely used in agriculture. Without detailed knowledge in both of these domains, it is difficult to answer many of the important questions asked in breeding programs: (i) how sustainable is a breeding effort in the long term, (ii) how robust are the products of a given breeding program, and (iii) how important is it to and what are the appropriate procedures to maintain and utilize genetic resources?

Genomics gives us some new perspectives on genetic variation. With access to data and information at the sequence level, our views of what contributes to the natural genetic variation that resides within the germplasm pools developed by breeding programs are changing. The traditional view of genetic variation as a function of loci with fixed effects acting in a predominantly additive manner is challenged by many of the properties of genes that are observed using genomics technologies. An important component of experimental evidence indicates that gene regulation is an important source of genetic variation. What appears to be linear when examined at the phenotypic level is not necessarily linear at the level of the gene network (Peccoud *et al.*, 2004). This creates a complex situation where many of the effects of genes can be highly context dependent. Therefore, the genetic background and environments within which the genes are studied will influence the estimates of the effects of the genes. Plant breeders have always been exposed to this phenomenon but have never had the tools to investigate its genetic basis. However, strategies based on manipulation of the genotype at the molecular level will only be able to utilize the currently available experimental information from statistically determined genotype-phenotype associations. Reminiscent of much of the debate around the effects of epistasis on response to selection, this forces the plant breeding community to ask specific questions of the implications of molecular breeding strategies for both short-term and long-term genetic improvement of complex traits.

12.1 Designing Molecular Breeding Strategies

Today, we can consider coordinated development of three themes and associated research paths arising from research over the last 10 to 15 years.

1. There is a growing knowledge base of the genetic architecture for some traits and how genetic variation is organized within unimproved and elite germplasm pools. Further work in this area will require the integration of genomics technologies with the study of genetic variation to conduct focused gene-to-phenotype studies. This will require fundamental questioning and in some cases refinement of our models of genetic variation. Development of our bioinformatics and computational modeling tools will be necessary.
2. High throughput genetic profiling of individuals for key regions of the genome is now feasible for elite and unimproved germplasm pools. High performance management, manipulation, analysis and interpretation of molecular and phenotypic data will continue to be areas of research priority.
3. Determining the power of molecular and conventional breeding strategies to achieve directed phenotypic changes for simple to complex traits.

High performance computing and simulation is being used to complement theoretical and experimental investigations. There is a clear need for further research into the appropriate statistical and biological modeling procedures for determining and testing gene-to-phenotype associations for complex traits.

13.0 OTHER ASPECTS OF MOLECULAR BREDING STRATEGY

13.1 Gene Machines

More direct clues about the function of individual genes are derived from the large-scale insertional mutagenesis of genomes using well-defined transposons, the 'gene machines' (also known as reverse genetics). The 'machine' consists of an active and well characterized transposon (known DNA sequence), a large population of plants (approximately 40,000 individuals) segregating for the transposon, DNA samples and progeny of each individual in the population and a partial DNA sequence of the genes of interest (the targets of the new insertions by the transposon). The population is created in such a way that the transposon inserts into many new sites in the genome. Some insertions into or near genes will modify an aspect of gene expression, possibly affecting gene function and resulting in an observable change in phenotype (a mutant). Since the transposon's DNA sequence is known, the new site of insertion is effectively identified or 'tagged'. Using the DNA sequences of the transposon and the partial sequences of the gene(s) of interest, the entire population is surveyed by PCR and/or hybridization for new insertions into DNA sequences related to the gene(s) of interest and for evidence of cosegregation of the new insertions and a mutant phenotype. DNA fragments that cosegregate with the new mutants may be readily cloned, sequenced, analyzed and identified regarding their information content and relation to the change in phenotype. Many of the DNA fragments will be pieces of genes and will provide the first clues as to the role of that gene in the context of that organism. The details of various strategies for transposon tagging have been summarized (Sundaresan, 1996; Maes *et al.,* 1999) and the plans and background of a maize-specific transposon-tagging gene machine have been made available (http://www.zmdb.iastate.edu/zmdb/nsf_grant_online.html).

13.2 Amplification of Ancient DNA

The DNA persisting in specimens of paleontological, archaeological or forensic interest is inevitably damaged. D'Abbadie *et al.* (2007) described a strategy for the recovery of genetic information from damaged DNA. By molecular breeding of polymerase genes from the genus *Thermus* [*Taq* (*Thermus aquaticus*), *Tth* (*Thermus thermophilus*) and *Tfl* (*Thermus flavus*)] and compartmentalized self-replication selection, they evolved polymerases that can extend single, double and even quadruple mismatches, process non-canonical primer-template duplexes and bypass lesions found in ancient DNA, such as hydantoins and abasic sites. Applied to the PCR amplification of 47,000–60,000-year-old cave bear DNA, these outperformed *Taq* DNA polymerase by up to 150% and yielded amplification products at sample dilutions at which *Taq* did not. These results demonstrate that engineered polymerases can expand the recovery of genetic information from Pleistocene specimens and may benefit genetic analysis in paleontology, archeology and forensic medicine.

13.3 Cotton Breeding for Drought and Salinity

Plant stress caused by drought and salinity are among the major constraints on crop production and food security worldwide. Breeding programs to improve crop yield in dry and saline environments have progressed slowly due to our limited understanding of the underlying physiological, biochemical, developmental, and genetic mechanisms that determine plant responses to these forms of stress, as well as to technical difficulties in combining favorable alleles to create the improved high yielding genotypes needed for these environments. Cotton is an especially appropriate system for research into the molecular basis of plant response to water deficit and salinity, as it originates from wild perennial plants adapted to semi-arid, sub-tropical environments which experienced periodic drought and temperature extremes that are associated with soils with high salt content. The current primary molecular breeding approaches include transgenic modification and quantitative trait mapping with marker-assisted selection. The preliminary work in QTL mapping for drought response and the relationships of the QTLs with the drought-associated measurements is developing a foundation for understanding and using the molecular basis of drought tolerance. QTL mapping for salt tolerance is not moving apace. Using and/or regulating transgene effects on the plant responses to drought and salinity has shown success and will continue to increase our understanding of the complexity of plant's physiological pathways. Improvements in all areas of molecular breeding are almost certain, but the most effective improvements will come from exploiting our improved understanding of the genetic architecture (Jenks *et al.*, 2007).

13.4 Molecular Breeding for Rainfed Lowland Rice

The conventional breeding program takes at least 15 years for releasing new cultivars. New breeding strategy can be established to shorten period for cultivar improvement by using marker-assisted selection (MAS), rapid generations advance (RGA), early generation testing in multi-locations for grain yield and qualities, etc. Four generations of MAS backcross breeding were conducted to transfer gene and QTL for conferring resistance/tolerance to bacterial blight (BLB), submergence (SUB), brown plant-hopper (BPH) and blast (BL) into Kao Dawk Mali 105 (KDML105) a line from Thailand. Selected backcross lines, introgressed with target gene/QTL, were tolerant to submergence, and resistant to BLB, BPH and BL. The agronomic performance and grain quality of these lines were as good as or better than

KDML105. Approximately 15 populations based on crosses between donor lines selected for drought tolerance and recipient lines selected for yield and quality traits of acceptance to farmer/markets are being developed in Thailand, Laos and Cambodia (Toojinda *et al.*, 2004).

14.0 PLANT MARKERS – A DATABASE OF PREDICTED MOLECULAR MARKERS

PlantMarkers is a genetic marker database that contains a comprehensive pool of predicted molecular markers. Researchers have adopted contemporary techniques to identify putative single nucleotide polymorphism (SNP), simple sequence repeat (SSR) and conserved orthologue set markers. A systematic approach to identify as broad a range of putative markers has been undertaken by screening the available open Sputnik unigene consensus sequences from over 50 plant species. A web presence at http://markers.btk.fi/ provides functionality so that a user may search for species-specific markers on the basis of many specific criteria not limited to non-synonymous SNPs segregating between different varieties or measured polymorphic SSRs. Feedback forms are provided with all sequence entries to enable inclusion of, for example, map location for markers validated by the research community (Rudd *et al.*, 2005)

15.0 CONCLUSIONS

Marker-based selection could be more efficient than purely phenotypic selection in quite large populations and for traits showing relatively low heritability. Regardless of the ultimate reason for the application of molecular markers for molecular breeding, there is a general need for both high-density and uniform maps that represent whole genomes. While techniques for the classification and typing of alleles have become amenable to high-throughput screening, *e.g.* microarray-based genotyping, the more important marker discovery steps require significant investments in both time and money. Single nucleotide polymorphism (SNP) markers, for example, have enjoyed massive popularity through their high density within the genome and their ease of characterization. The identification of these markers, however, requires access to reliable DNA sequence from the complete range of plants strains/varieties or ecotypes that will subsequently be used. With the availability of PCR-based markers, which are simple to use, the cost of different marker techniques is progressively decreasing and may become affordable in the framework of breeding programmes. Their use in applied breeding programmes can range from facilitating the appropriate choice of parents for crosses, to mapping/tagging of gene blocks associated with economically important traits (often termed "quantitative trait loci" (QTLs). Molecular breeding applications must not only respond to the challenge of improving food security and fostering socio-economic development, but in doing so, promote the conservation, diversification and sustainable use of plant genetic resources for food and agriculture. Through comparative genomics, molecular markers can be used in ways that allow us to more effectively discover and efficiently exploit biodiversity and the evolutionary relationships between organisms. Substantial progress has been made in recent years in mapping, tagging and isolating many agriculturally important genes using molecular markers due in large part to improvements in the techniques that have been developed to help find markers of interest.

New developments in genomic research have given access to an enormous amount of sequence information as well as new insights on the function and interaction of genes and the evolution of functional domains, chromosomes and genomes. Such information can help in comparative genetic mapping and linkage analysis of useful agricultural traits. Similarly, comparative analyses of sequence information in the growing databases now publicly available on the World Wide Web will be an invaluable resource for formulation of different strategies for molecular breeding. This will have an increasingly important role in the evolution of newer ways for plant breeding.

16.0 REFERENCES

Al-Babili, S. Beyer, P. 2005. Golden Rice — Five years on the road — five years to go? Trends in Plant Science, 10:565-573.

Badan, A.C. de Carvalho, Geraldi, I.O., Guimarães, E.P. and Ospina, Y. 1998. Estimativa do tamanho efetivo de amostra ideal para caracterizar uma população de arroz para as características peso de panículas e peso de 100 grãos. Genetics and Molecular Biology, 21:246.

Barry, M.B. 2000. Diversité génétique de variétés locales de riz en Guinée. Analyse de la variabilité intra-variétale au moyen de marqueurs microsatellites. Rennes, France, Université de Rennes. 30 pp. (DEA thesis).

Baurens, F., Noyer, J.L., Lanaud, C. and Lagoda, P.J.L. 1997. Assessment of a species specific element (Brep 1) in banana. Theoretical and Applied Genetics, 95:922–931.

Brown, A.H.D. and Munday, J. 1982. Population-genetic structure and optimal sampling of land races of barley from Iran. Genetica, 58:85–96.

Buckler, E.S. and Thornsberry, J.M. 2002. Plant molecular diversity and applications to genomics. Current Opinions in Plant Biolology, 5:107–111.

Calderini, O., Chang, S.B., de Jong, H., Busti, A., Paolocci, F., Arcioni, S., de Vries, S.C., Abma-Henkens, M.H., Klein Lankhorst, R.M., Donnison, I.S. and Pupilli, F. 2006. Molecular cytogenetics and DNA sequence analysis of an apomixis-linked BAC in *Paspalum simplex* reveal a non pericentromere location and partial microcolinearity with rice. Theoretical andApplied Genetics 112:1179-1191.

Chen, H.L., Wang, S.P., Xing, Y.Z., Xu, C.G., Patrick, M.H. and Zhang, Q.F. 2003. Comparative analyses of genomic locations and race specificities of loci for quantitative resistance to *Pyricularia grisea* in rice and barley. Proceedings of the National Academy of Sciences, USA, 100:2544-2549.

Chopra, V.L. 2006. The search of new genes then and now. 13th B.P. Pal Memorial Lecture, May 26, 2006. IARI. 6-11.

Comes, H.P. and Abbott, R.J. 1998. The relative importance of historical events and gene flow on the population structure of a Mediterranean ragwort, *Senecio gallicus* (Asteraceae). Evolution, 52:355–367.

D'Abbadie, M., Hofreiter, M., Vaisman, A., Loakes, D., Gasparutto, D., Cadet, J., Woodgate, R., Pääbo, S. and Holliger, P. 2007. Molecular breeding of polymerases for amplification of ancient DNA. Nature Biotechnology, 25:939 – 943.

De Koeyer, D.L., Philips, R.L. and Stuthman, D.D. 2001. Allelic shifts and quantitative trait loci in a recurrent selection population of oat. Crop Science, 41:1228-1234.

Del Rio, A.H. and Bamberg, J.B. 2002. Lack of association between genetic and geographical origin characteristics for the wild potato *Solanum sucrense*. American Journal of Potato Research, 79:335–338.

Espejo-Ibañez, M.C., Sanchez, M.P., Sanchez, M.D. and Yelamo, M.D. 1994. Isoenzymatic variability in seeds of some Spanish common beans (*Phaseolus vulgaris* L. Leguminosae): relation to their domestication centers. Biochemical Systematics and Ecology, 22:827–833.

Fahima, T., Roeder, M.S., Grama, A. and Nevo, E. 1998. Microsatellite DNA polymorphism divergence in *Triticum dicoccoides* accessions highly resistant to yellow rust. Theoretical and Applied Genetics, 96:187–195.

Freville, H., Justy, F. and Olivieri, I. 2001. Comparative allozyme and microsatellite population structure in a narrow endemic plant species, *Centaurea corymbosa* Pourret (Asteraceae). Molecular Ecology, 10:879–889.

Fulton, T.M., van der Hoeven, R., Eannetta, N.T. and Tanksley, S.D. 2002. Identification, analysis and utilization of conserved ortholog set markers for comparative genomics in higher plants. Plant Cell, 14:1457–1467.

Fulton, T.M., Beck-Bunn, T., Emmatty, D., Eshed, Y., Lopez, J., Petiard, V., Uhlig, J., Zamir, D. and Tanksley. S.D. 1997. QTL analysis of an advanced backcross of *Lycopersicon peruvianum* to the cultivated tomato and comparisons with QTLs found in other wild species. Theoretical andApplied Genetics, 95:881-894.

Garvin, D.F. and Weeden, N.F. 1994. Isozyme evidence supporting a single geographic origin for domesticated tepary bean. Crop Science, 34:1390–1395.

Geraldi, I.O. and Souza, Jr, C.L. de. 2000. Muestreo genético para programas de mejoramiento poblacional. *In* E.P. Guimarães, ed. Avances en el mejoramiento poblacional en arroz, pp. 9-20. Santo Antônio de Goiás, Brazil, Embrapa Arroz e Feijão.

Glaszmann, J.C. 1987. Isozymes and classification of Asian rice varieties. Theoretical andApplied Genetics, 74:21-30.

Goff, S.A. *et al.* 2002. A draft sequence of the rice genome (*Oryza sativa* L. spp. *japonica*). Science, 296:92–100.

Goldstein, D.B. and Schlotterer, C. 1999. Microsatellites : evolution and applications oxford University Press, Oxford.

Graterol, M.E.J. 2000. Caracterización de poblaciones e introducción de variabilidad genética para iniciar un programa de mejoramiento poblacional del arroz en Venezuela. *In* E.P. Guimarães, ed. *Avances en el mejoramiento poblacional en arroz*, pp. 87-103. Santo Antônio de Goiás, Brazil, Embrapa Arroz e Feijão.

Gualtieri, G., Conner, J.A., Morishige, D.T., Moore, L.D., Mullet, J.E. and Ozias-Akins, P.O. 2006. A segment of the Apospory Specific Genomic Region (ASGR) is highly microsyntenic not only between the apomicts *Pennisetum squamulatum* and *Cenchrus ciliaris*, but also with a rice chromosome 11 centromeric-proximal genomic region. Plant Physiology, 140:936-971.

Guimarães, E.P. 2000. Mejoramiento poblacional del arroz en America Latina: Dónde estamos y para dónde vamos. *In* E.P. Guimarães, ed. *Avances en el mejoramiento poblacional en arroz*, pp. 299-311. Santo Antônio de Goiás, Brazil, Embrapa Arroz e Feijão.

Hamilton, M.B. 1994. *Ex situ* conservation of wild plant species: time to assess the genetic assumptions and implications of seed banks. Conservation Biology, 8:39–49.

Hanson, W.D. 1959. The breakup of initial linkage blocks under selection mating systems. Genetics, 44:857-868.

Hospital, F. and Charcosset, A. 1997. Marker-assisted introgression of Quantitative Trait Loci. Genetics, 147:1469-1485.

Hospital, F., Chevalet, C. and Mulsant, P. 1992. Using markers in gene introgression breeding programs. Genetics, 132:1199-1210.

Hospital, F., Moreau, L., Lacoudre, F., Charcosset, A. and Gallais, A. 1997. More on the efficiency of marker-assisted selection. Theoretical andApplied Genetics, 95:1181-1189.

Huang, L., Brooks, S.A., Li, W., Fellers, J.P., Trick, H.N. and Gill, B.S. 2003. Map-based cloning of leaf rust resistance gene *Lr21* from the large and polyploid genome of bread wheat. Genetics, 164:655-664.

Jenks, M.A., Hasegawa, P.M. and Jain, S.M. 2007. Advances in Molecular Breeding Toward Drought and Salt Tolerant Crops. Dordrecht (Netherlands), Springer, Accession No: 415202, ISBN 978-1-4020-5577-5 , Call No: 631.5 J42 (RR) 775-796.

Karp, A., Edwards, K.J., Bruford, M., Funk, S., Vosman, B., Morgante, M., Seberg, O., Kremer, A., Boursot, P., Arctander, P., Tautz, D. and Hewitt, G.M. 1997a. Molecular technologies for biodiversity evaluation: opportunities and challenges. Nature Biotechnology, 15:625–628.

Karp, A., Kresovich, S., Bhat, K.V., Ayad, W.G. and Hodgkin, T. 1997b. Molecular tools in plant genetic resources conservation: a guide to the technologies IPGRI Technical Bulletin No. 2. International Plant Genetic Resources Institute, Rome, Italy.

Karp, A., Seberg, O. and Buiatti, M. 1996. Molecular techniques in the assessment of botanical diversity. Annals of Botany, 78:143–149.

Kilian, A., Chen, J., Han, F., Steffenson, B. and Kleinhofs, A. 1997. Towards map-based cloning of the barley stem rust resistance genes *Rpg1* and *rpg4* using rice as an intergenomic cloning vehicle. Plant Molecular Biology, 35:187-195.

Lamboy, W.F., McFerson, J.R., Westman, A.L. and Kresovich, S. 1994. Application of isozyme data to the management of the United States national *Brassica oleracea* L. genetic resources collection. Genetic Resources and Crop Evolution, 41:99–108.

Lande, R. and Thompson, R. 1990. Efficiency of marker-assisted selection in the improvement of quantitative traits. Genetics, 124:743-756.

Lanner, H.C., Bryngelsson, T. and Gustafsson, M. 1997. Relationships of wild *Brassica* species with chromosome number 2n = 18, based on RFLP studies. Genome, 40:302–308.

Lanner, H.C., Gustafsson, M., Falt, A.S. and Bryngelsson, T. 1996. Diversity in natural populations of wild *Brassica oleracea* as estimated by isozyme and RAPD analysis. Genetic Resources and Crop Evolution, 43:13–23.

Laurie, D.A. and Devos, K.M. 2002. Trends in comparative genetics and their potential impacts on wheat and barley research. Plant Molecular Biology, 48:729–740.

Lecomte, L., Hospital, F., Buret, M. and Causse, M. 2004. Recent Advances in Molecular Breeding: The Example of Tomato Breeding for Flavor Traits. Proc. XXVI IHC – Advances in Vegetable Breeding, Eds. J.D. McCreight and E.J. Ryder, Acta Hort. 637.

Lee, D.J., Berding, N., Jackes, B.R. and Bielig, L.M. 1998. Isozyme markers in *Saccharum* spp. hybrids and *Erianthus arundinaceus* (Retz.) Jeswiet. Australian Journal of Agricultural Research. 49:915–921.

Lee, D.J., Reeves, J.C. and Cooke, R.J. 1996. DNA profiling and plant variety registration: 1. The use of random amplified DNA polymorphisms to discriminate between varieties of oilseed rape. Electrophoresis, 17:261–265.

Lemieux, B., Aharoni, A. and Schena, M. 1998. Overview of DNA chip technology. Molecular Breeding, 4:277–289.

Lima, M.L.A., García, A.A.F., Oliveira, K.M., Matsuoka, S., Arizono, H., Souza, Jr, C.L. de and Souza, A.P. de. 2002. Analysis of genetic similarity detected by AFLP and coefficient of parentage among genotypes of sugar cane. Theor. Appl. Genet., 104:30-38.

Lorenzen, L.L., Boutin, S., Young, N., Specht, J.E. and Shoemaker, R.C. 1995. Soybean pedigree analysis using map-based molecular markers, I: Tracking RFLP markers in cultivars. Crop Science, 35:1326-1336.

Maass, B.L. and Ocampo, C.H. 1995. Isozyme polymorphism provides fingerprints for germplasm of *Arachis glabrata* Bentham. Genetic Resources and Crop Evolution, 42:77–82.

Maes, T., P. de Keukeleire and T. Gerats. 1999. Plant tagnology. Trends Plant Sci., 4:90-96.

Manjarrez-Sandoval, P., Carter, Jr. T.E., Webb, D.M. and Burton, J.W. 1997. Heterosis in soybean and its prediction by genetic similarity measures. Crop Science, 23:1443–1452.

Maquet, A., Zoro Bi, I.Z., Delvaux, M., Wathelet, B. and Baudoin, J.P. 1997. Genetic structure of a Lima bean base collection using allozyme markers. Theoretical and Applied Genetics, 95:980–991.

Maquet, A., Zoro Bi, I.Z., Rocha, O.J. and Baudoin, J.P. 1996. Case studies on breeding systems and its consequences for germplasm conservation. Genetic Resources and Crop Evolution, 43:309–318.

Mariani, C., De Beuckeleer, M., Truettner, J., Leemans, J. and Goldberg, R.B. 1990. Induction of male sterility in plants by a chimaeric ribonuclease gene. Nature, 347: 737-741.

Marín-Garavito, J.M. 1994. Efecto de número de ciclos de recombinación en la variabilidad de poblaciones de arroz *(Oryza sativa L.)*. Palmira, Colombia, Facultad de Ciencias Agropecuarias, Universidad Nacional de Colombia. 50 pp. (BSc thesis).

Martin, C., Juliano, A., Newbury, H.J., Lu, B.R., Jackson, M.T. and Ford Lloyd, B.V. 1997. The use of RAPD markers to facilitate the identification of *Oryza* species within a germplasm collection. Genetic Resources and Crop Evolution, 44:175–183.

Mather, D. 2006. Molecular Plant Breeding CRC, Australian Centre for Plant functional Genomics. PTY.LTD.

McGregor, C.E., van Treuren, R., Hoekstra, R. and van Hintum, T.J.L. 2002. Analysis of the wild potato germplasm of the series *Acaulia* with AFLPs: implications for *ex situ* conservation. Theoretical and Applied Genetics, 104:146–156.

Melchinger, A.E., Graner, A., Singh, M. and Messmer, M.M. 1994. Relationships among European barley germplasm. I. Genetic diversity among winter and spring cultivars revealed by RFLPs. Crop Science, 34:1191–1199.

Michelmore, R.W., Paran, I. and Kesseli, R.V. 1991. Identification of markers linked to disease resistance genes by bulked segregant analysis: a rapid method to detect markers in specific genomic regions using segregating populations. Proceedings of the National Academy of Sciences, USA, 88:9282-9832.

Mishra, G.P., Singh, R.K., Mohapatra, T., Singh, A.K., Prabhu, K.V., Zaman, F.U. and Sharma, R.K. 2003. Molecular mapping of a gene for fertility restoration of wild abortive (WA) cytoplasmic male sterility using a Basmati rice restorer line. Journal of Plant Biochemistry and Biotechnology, 12:37-42.

Murphy, J.P. and Philips, T.D. 1993. Isozyme variation in cultivated oat and its progenitor species, *Avena sterilis* L. Crop Science, 33:1366–1372.

National Research Council. 1972. Genetic vulnerability of major crops. National Academy of Sciences, Washington, D.C., USA.

Nebauer, S.G., del Castillo-Agudo, L. and Segura, J. 1999. RAPD variation within and among natural populations of outcrossing willow-leaved foxglove (*Digitalis obscura* L.). Theoretical and Applied Genetics, 98:985–994.

Nei, M. 1972. Genetic distance between populations. Am. Nat., 106:283-292.

Nevo, E., Apelbaum Elkaher, I., Garty, J. and Beiles, A. 1997. Natural selection causes microscale allozyme diversity in wild barley and a lichen at 'Evolution Canyon', Mt. Carmel, Israel. Heredity, 78:373–382.

Nevo, E., Baum, B., Beiles, A. and Johnson, D.A. 1998. Ecological correlates of RAPD DNA diversity of wild barley, *Hordeum spontaneum*, in the Fertile Crescent. Genetic Resources and Crop Evolution, 45:151–159.

Ospina, Y., Châtel, M. and Guimarães, E.P. 2000. Mejoramiento poblacional del arroz de sabanas. *In* E.P. Guimarães, ed. *Avances en el mejoramiento poblacional en arroz*, pp. 241-254. Santo Antônio de Goiás, Brazil, Embrapa Arroz e Feijão.

Parida, S.K., Kalia, S.K., Sunit, K., Dalal, V., Hemprabha, G., Selvi, A., Pandit, A., Singh, A., Gaikewad, K., Sharma, T.R., Srivastava, P.S., Singh, N.K., and Mohapatra, T. 2008. Informative genomic microsatellite markers for efficient genotyping applications in sugarcane. Theoretical and Applied Genetics (on line).

Parker, P.G., Snow, A.A., Schug, M.D., Booton, G.C. and Fuerst, P.A. 1998. What molecules can tell us about populations: choosing and using a molecular marker. Ecology, 79:361–382.

Paz, M.M. and Veilleux, R.E. 1997. Genetic diversity based on randomly amplified polymorphic DNA (RAPD) and its relationship with the performance of diploid potato hybrids. Journal of the American Society of Horticultural Science, 122:740–747.

Peccoud, J.K., Velden, V., Podlich, D.W., Winkler, C.R., Arthur, W.L. and Cooper, M. 2004. The selective values of alleles in a molecular network model are context dependent. Genetics, 166:1715–1725.

Peroutka, S.J. 1997. The medical utility of genomics data in neuropsychiatry: mutational genetics versus association genetics. Current Opinions in Biotechnology, 8:688–691.

Perrier, X., Flori, A. and Bonnot, F. 2003. Les méthodes d'analyse des données. In P. Hamon, M. Seguin, X. Perrier, J.C. Glaszmann, eds. *Genetic diversity of cultivated tropical plants*, pp. 31-64. Enfield, NH, USA, Science Publishers and CIRAD.

Rabbani, M.A., Maruyama, K., Abe, H., Khan, M.A., Katsura, K., Ito, Y., Yoshiwara, K. and Seki, M. 2003. Monitoring Expression Profiles of Rice Genes under Cold, Drought, and High-Salinity Stresses and Abscisic Acid Application Using cDNA Microarray and RNA Gel-Blot Analyses. Plant Physiology, 133:1755-1767.

Rafalski, J.A. 2002. Novel genetic mapping tools in plants: SNPs and LD-based approaches. Plant Science, 162:329–333.

Rieseberg, L.H. and Ellstrand, N.C. 1993. What can molecular and morphological markers tell us about plant hybridization? Critical Reviews in Plant Science, 12:213–241.

Rieseberg, L.H. 1995. The role of hybridization in evolution: old wine in new skins. American Journal of Botany, 82:944–953.

Roa, A.C., Maya, M.M., Duque, M.C., Tohme, J., Allem, A.C. and Bonierbale, M.W. 1997. AFLP analysis of relationships among cassava and other *Manihot* species. Theoretical and Applied Genetics, 95:741–750.

Rudd, S., Schoof, H. and Mayer, K. 2005. PlantMarkers—a database of predicted molecular markers from plants. Nucleic Acids Research, 33:D628-D632.

Schlötterer, C. 2004. The evolution of molecular markers—just a matter of fashion? Nature Reviews Genetics, 5:63–69.

Schneider, K. and Douches, D.S. 1997. Assessment of PCR-based simple sequence repeats to fingerprint North American potato cultivars. American Potato Journal, 74:149–160.

Schoen, D.J. and Brown, A.H.D. 1993. Conservation of allelic richness in wild crop relatives is aided by assessment of genetic markers. Proceedings of the National Academy of Sciences USA, 90:10623–10627.

Smith, O.S., Smith, J.S.C., Bowen, S.L., Tenborg, R.A. and Wall, S.J. 1990. Similarities among a group of elite maize inbreds as measured by pedigree, F1 grain yield, grain yield, heterosis and RFLPs. Theoretical and Applied Genetics, 80:833–840.

Singh, R.K., Sharma, R.K., Singh, A.K., Singh, V.P., Singh, N.K., Tiwari, S.P. and Mohapatra, T. 2004. Suitability of mapped sequence tagged microsatellite site markers for establishing distinctness, uniformity and stability in aromatic rice. Euphytica, 135(2):135-144.

Soleimani, V.D., Baum, B.R. and Johnson, D.A. 2002. AFLP and pedigree-based genetic diversity estimates in modern cultivars of durum wheat. Theoretical and Applied Genetics, 104: 350-357.

Steiner, J.J., Piccioni, E., Falcinelli, M. and Liston, A. 1998. Germplasm diversity among cultivars and the NPGS crimson clover collection. Crop Science, 38:263–271.

Sundaresan, V. 1996. Horizontal spread of transposon mutagenesis: new uses for old elements. Trends Plant Sci., 6:184-190.

Tanksley, S.D. and Nelson, J.C. 1996. Advanced backcross QTL analysis: a method for the simultaneous discovery and transfer of valuable QTLs from unadapted germplasm into elite breeding lines. Theoretical and Applied Genetics, 92:191–203.

Telenius, H., Carter, N.P., Bebb, C.E., Nordenskjold, M., Ponder, B.A.J. and Tunnacliffe, A. 1992. Degenerate oligonucleotide-primed PCR: general amplification of target DNA by a single degenerate primer. Genomics, 13:718–725.

Thieme, R., Darsow, U., Gavrilenko, T., Dorokhov, D. and Tiemann, H. 1997. Production of somatic hybrids between *S. tuberosum* L. and late blight resistant Mexican wild potato species. Euphytica, 97:189–200.

Toojinda, T., Tragoonrung, S., Vanavichit, A., Siangliw, J.L., Pa-In, N., Jantaboon, J., Siangliw, M. and Fukai, S. 2004. Molecular breeding for rainfed lowland rice in the Mekong Region. New directions for a diverse planet: Proceedings of the 4th International Crop Science Congress, Brisbane, Australia, 26 Sep – 1 Oct 2004.

Tsegaye, S., Tesemma, T. and Belay, G. 1996. Relationships among tetraploid wheat (*Triticum turgidum* L) landrace populations revealed by isozyme markers and agronomic traits. Theoretical and Applied Genetics, 93:600–605.

UPOV-BMT, 2002. BMT/36/10 Progress report of the 36th session of the technical committee, the technical working parties and working group on biochemical and molecular techniques and DNA-profiling in particular. Geneva.

Ursla, F.W.M., Hayward, M.D. and Kearsey, M.J. 1997. Isozyme and quantitative traits polymorphisms in European provenances of perennial ryegrass (*Lolium perenne* L.). Euphytica, 93:263–269.

Van Berloo, R. and Stam, P. 1998. Marker-assisted selection in autogamous RIL populations: a simulation study. Theoretical and Applied Genetics, 96:147-154.

Varghese, Y.A., Knaak, C., Sethuraj, M.R. and Ecke, W. 1997. Evaluation of random amplified polymorphic DNA (RAPD) markers in *Hevea brasiliensis*. Plant Breeding, 116:47–52.

Visscher, P.M., Haley, C.S. and Thompson, R. 1996. Marker-assisted introgression in backcross breeding programs. Genetics, 144:1923-1932.

Whitkus, R., de la Cruz, M., Mota Bravo, L., Gomez Pompa, A. and De la Cruz, M. 1998. Genetic diversity and relationships of cacao (*Theobroma cacao* L.) in southern Mexico. Theoretical and Applied Genetics, 96:621–627.

Whitkus, R., Doebley, J. and Wendel, J.F. 1994. Nuclear DNA markers in systematics and evolution. In DNA-based markers in plants (Advances in cellular and molecular biology of plants, vol. 1.) (RL Phillips and IK Vasil, eds.). Kluwer Academic Publishers, Dordrecht, The Netherlands,116–141.

Yamada, T., Misoo, S., Ishii, T., Ito, Y., Takaoka, K. and Kamijima, O. 1997. Characterization of somatic hybrids between tetraploid *Solanum tuberosum* L. and dihaploid *S. acaule*. Breeding and Science, 47:229–236.

Yang, W.P., De Oliveira, A.C., Godwin, I., Schertz, K. and Bennetzen, J.L. 1996. Comparison of DNA marker technologies in characterizing plant genome diversity: variability in Chinese sorghums. Crop Science, 36:1669–1676.

Ye, X., Al-Babili, S., Kloti, A., Zhang, J., Lucca, P., Beyer, P. and Potrykus, I. 2000. Engineering provitamin A (b-carotene) biosynthetic pathway into (carotenoid-free) rice endosperm. Science, 287:303-305.

Young, N.D. and Tanksley, S.D. 1989. RFLP analysis of the size of chromosomal segments retained around the *Tm-2* locus of tomato during backcross breeding. Theoretical and Applied Genetics, 77:353-359.

Yu, J. *et al.* 2002. A draft sequence of the rice genome (*Oryza sativa* L. spp. *indica*). Science 296:79–92.

Zerega, N.J.C., Ragone, D. and Motley, T.J. 2004. Complex origins of breadfruit (*Artocarpus altilis*, Moraceae): implications for human migrations in Oceania. American Journal of Botany, 91:760–766.

Zhou, Z., Bebeli, P.J., Somers, D.J. and Gustafson, J.P. 1997. Direct amplification of minisatellite-region DNA with VNTR core sequences in the genus *Oryza*. Theoretical and Applied Genetics, 95:942–949.

Zoro Bi, I., Maquet, A., Degreef, J., Wathelet, B. and Baudoin, J.P. 1998. Sample size for collecting seeds in germplasm conservation: the case of the Lima bean (*Phaseolus lunatus* L.). Theoretical and Applied Genetics, 97:187–194.

Molecular Plant Breeding: Principle, Method and Application
Eds : R.K. Singh, Rajesh Singh, Guoyou Ye, A. Selvi and G.P. Rao
Studium Press LLC, Texas, USA, 2009, pp. 37-80

CHAPTER

2

Molecular Markers in Plants

R.K. SINGH[1*], *GYAN P. MISHRA*[2], *ANIL KANT*[3] *and SHASHI BALA SINGH*[2]

ABSTRACT

Advances in biotechnology have resulted in a large variety of molecular marker systems and enhanced opportunities for automation of the majority of the techniques, resulting in a wealth of information. PCR was a major breakthrough for molecular markers in that for the first time, any genomic region could be amplified and analyzed in many individuals without the requirement for cloning and isolating large amounts of ultra-pure genomic DNA. Molecular markers are now common tools that are applied in many aspects of breeding strategies. The markers can be classified in different ways as hybridization based markers and polymerase chain reaction based markers with respect to basic principle involved. Molecular markers differ in many qualities and must therefore be carefully chosen and analyzed differently with their differences in mind. However, the use of molecular markers requires resources of time and funds that should not be taken for granted. Due to the developments in the detection techniques it is particularly useful in diagnostic applications, such as the screening of samples for the presence or incorporation of favourable traits, the detection of pathogens and diseases in plants and the screening of plant material for the presence of transgenic elements.

Key Words: Molecular markers, RAPD, AFLP, SSR, MAS

[1]*Indian Institute of Sugarcane Research, Lucknow - 226002, UP, India*
[2]*Defence Institute of High Altitude Research, DRDO, C/o 56 APO, Leh, India*
[3]*Jaypee University of Information Technology, Waknaghat, Solan, HP, India*
* *Corresponding author e-mail: ikrps@yahoo.com*

1.0 INTRODUCTION

All living things are made-up of cell(s) that are under control of naturally programmed genetic material called nucleic acids. These nucleic acids (DNA/RNA) are is turn made-up of long chains of nitrogenous bases linked through phosphate ions. Only a small fraction of DNA sequences typically make up genes, *i.e.* code for proteins, while remaining and major share of DNA represent non-coding sequences whose role is not clearly understood. The genetic material is organized into sets of chromosomes, and entire set is called genome. In a diploid individual, there are two alleles of every gene one from each parent. A molecular marker is a DNA sequence which is readily detected and whose inheritance can be easily monitored. These identifiable DNA sequences are found at specific locations of genome, and transmitted by law of inheritance from one generation to the next (Fig. 1).

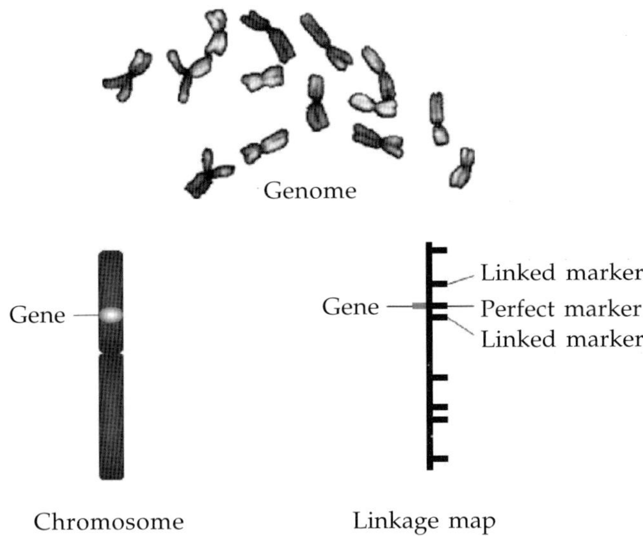

Fig. 1. Schematic presentation of genome, chromosome, gene, and linked markers (Source: Mather, 2006)

Markers are the traits which can be used to distinguish or differentiate between the populations under study. Markers are broadly divided into two major category *i.e.* morphological and molecular. Morphological markers are those traits which are visible to the naked eyes while molecular markers are those which expresses at the molecular level (*i.e.* DNA or proteins). Problem with the morphological markers are that they are very few in numbers and are not always desired for the breeding purpose. In contrast molecular markers are more robust, independent of environmental conditions and are more in numbers. Molecular markers are genetic tools that allow us to examine differences between individuals at many randomly spaced positions across an entire genome. These markers are used to tag the likely position of genes controlling traits on the chromosomes. Identifying the position of genes then allows them to be physically isolated and characterized. Markers are also ideal for examining the relationships among individuals, populations or phylogenetic taxa.

2.0 DETECTING DNA POLYMORPHISMS

Because any DNA molecule greater than 10 base pairs contains essentially the same mass-to-charge ratio, any procedure that separates the molecules based on mass alone will be useful to uncover DNA polymorphisms. Currently, gel electrophoresis is the most often used procedure to detect these polymorphisms. But as we move into the post genomic era, techniques that rapidly screen large numbers of samples are emerging. The most widely applied procedure of recent time is capillary array electrophoresis. In the near future, matrix-assisted laser desorption/ ionization time-of-flight mass spectrometry (MALDI-TOF MS) may replace many of the current procedures.

2.1 Gel Electrophoresis

Gel electrophoresis is most widely adapted technique for detecting polymorphism. Samples are loaded into a gel and allowed to migrate in an electric field. Since DNA is negatively charged, the samples are loaded near the negative pole, and they migrate toward the positive pole. Separation of the molecules is strictly based on size. The smallest fragments move farther in the gel because they can navigate through the small pores in the gel better than large molecules.

The two gel matrices used to separate the molecules are agarose and polyacrylamide. The separation capabilites of each molecule is function of the concentration of the polymer in the gel. The resolution that can be obtained with different polymer concentrations have been shown is Table 1.

We can choose the type of polymer to use based on the range of fragment that need to be distinguished. For RFLP and RAPD procedures, agarose is the polymer of choose. Because microsatellites and AFLPs procedures generate smaller fragments for comparison, polyacrylamide is typically used.

Following gel electrophoresis, the polymorphism is revealed using a detection agent. The dye Ethidium Bromide (EB) is typically used to reveal RAPD polymorphisms in agarose gels. Silver nitrate is procedure used to

Table 1. Relationship between per cent (%) of gel and resolution obtained (bp) of amplified fragments (Source : Mc Clean, 1998)

Agarose		Polyacrylamide	
%	Resolution (kb)	%	Resolution (bp)
0.9	0.5 - 0.7	3.5	1000 – 2000
1.2	0.4 - 6.0	5.0	80 – 500
1.5	0.2 - 3.0	8.0	60 – 400
2.0	0.1 - 2.0	12.0	440 – 200

detect polymorphisms in polyacrylamide gels. This has been used for both microsatellites and AFLPs. With these marker systems, it is also an option to include a radio-labeled nucleotide during the Polymerase Chain Reaction (PCR) step. If this option is chosen, the gel is used to expose autoradiographic film. The film is then analyzed to uncover the polymorphism. Laser technology has also been applied to the microsatellite marker system. When using this approach, the primer is labeled with a fluorescent dye. Samples are then separated in a polyacrylamide gel. As the samples flow through the bottom of the gel, fragments are detected by a laser that detects the presence of the fluor. Computer programs output data in a form that can be analyzed. Detecting RFLPs requires the use of a southern hybridization procedure.

2.2 Capillary Array Electrophoresis

Small DNA fragments can be separated rapidly in narrow capillaries. Because heat is lost rapidly from these thin capillaries, molecules can be separated rapidly using high voltage. This greatly speeds the processing of samples. Recently, instruments that contain capillaries arrays have reached the market. These arrays consist of eight (Beckman-Coulter) or 96 (Applied Biosystems) cappillaries. These machines simultaneously introduce samples into each of the arrays. Laser technology is then used to detect the samples as they exit the capillary.

2.3 Matrix-Assisted Laser Desorption/Ionization Time-of-Flight Mass Spectrometry (MALDI-TOF MS)

MALDI-TOF MS is the newest procedure in use to detect microsatellite polymorphisms. In general, mass spectrometry involves the ionization of a molecule, accelerating the ions in an electric field, and passing the ions through a magnetic field. MALDI-TOF is a recent improvement of mass spectrometry. The molecule to be analyzed is mixed with a matrix that does not ionize, but aids in the ionization of large molecules (up to 100,000 daltons) such as DNA. Time-of-flight allows for high throughput (and thus high sensitivity) of ions. Samples can be analyzed in a couple of seconds using this procedure.

A basic procedure is as follows: PCR amplification is performed with microsatellite specific primers. It is best if the primers are located as close to the repeat as possible. This ensures that smaller molecules are amplified. The primer contains an attached biotin molecule. The product is recovered from the PCR reaction with streptavidin magnetic beads. The product is mixed with the matrix and analyzed. In this method, both alleles of heterozygous individuals are easily scored. In some cases, the detected alleles, differ even by a single copy of a tetranucleotide repeat (McClean, 1998).

3.0 MARKER TYPES, PROPRETIES AND THEIR APPLICATIONS

The molecular markers can be classified in different ways. It may be classified as hybridization based markers and PCR based markers with respect to basic principle involved.

3.1 Properties of Ideal Genetic Markers

Ideal genetic marker possesses certain unique properties, *viz.* no detrimental effect on phenotype, co-dominant in expression, single copy, economic to use, highly polymorphic, easily assayed, multi-functional, highly available (un-restricted use), genome-specific in nature (especially when working with polyploids), can be multiplexed and its ability to be automated.

3.2 Genomic Abundance

The number of markers that can be generated is determined mainly by the frequency at which the sites of interest occur within the genome. RFLPs and AFLPs generate abundant markers due to the large number of restriction enzymes available and the frequent occurrence of their recognition sites within genomes. Within eukaryotic genomes, microsatellites have also been found to occur frequently. RAPD markers are even more abundant because numerous random sequences can be used for primer construction. In contrast, the number of allozyme markers is restricted due to the limited number (about 30) of enzyme detection systems available for analysis. If, in addition to genomic abundance, genome coverage is also sought, caution should be taken in marker selection. While some markers are known to be scattered quite evenly across the genomes, others, sometimes cluster in certain genomic regions. For example, clustering of AFLP markers has been reported in centromeric regions of *Arabidopsis thaliana* (Alonso-Blanco *et al.,* 1998), soybean (Young *et al.,* 1999) and rye (Saal and Wricke, 2002).

3.3 Level of Polymorphism

The resolving power of genetic markers is determined by the level of polymorphism detected, which is determined by the mutation rate at the genomic sites involved. Variation at allozyme loci is caused by point mutations, which occur at low frequency ($<10^{-6}$ per meiosis). Moreover, only mutations modifying the net electric charge and conformation of proteins can be detected, reducing the resolving power of allozymes. In contrast, mutations at minisatellite and microsatellite loci, mainly due to changes in the number of repeat units of the core sequence, have been estimated to occur at the relatively high frequency of 10–3–10–2 and 10–5–10–2 per meiosis, respectively (Jarne and Lagoda, 1996).

Higher resolving power is required when samples are more closely related. For example, analyses within species or among closely related species may call for fast evolving markers such as microsatellites. However if the objective is to study genetic relatedness at higher taxonomic levels (such as congeneric species), AFLPs or RFLPs may be a better choice because co-migrating, fast-evolving markers will have less chance of being homologous. A primary guiding principle in marker selection is that more conservative markers (those having slower evolutionary rates) are needed with increasing evolutionary distance and vice-versa.

3.4 Co-dominance of Alleles

Codominant markers are markers for which both alleles are expressed when co-occurring in an individual. Therefore, with codominant markers, heterozygotes can be distinguished from homozygotes, allowing the determination of genotypes and allele frequencies at loci. Codominant markers are preferred for most applications. The majority of codominant markers are single locus markers and hence the degree of information per assay is usually lower compared to the multilocus techniques.

3.5 Advantages of Molecular Markers

The molcular markers freedom from environmental and pleiotropic effects. It does not exhibit phenotypic plasticity, while morphological and biochemical markers can vary in different environments. DNA characters have a much better chance of providing homologous traits. Most morphological or biochemical markers, in contrast, are under polygenic control, and subject to epistatic control and environmental modification (plasticity). Unlimited numbers of independent markers are available, unlike morphological or biochemical data. DNA characters can be more easily scored as discrete states of alleles or DNA base pairs, while some morphological, biochemical and field evaluation data must be scored as continuously variable characters that are less amenable to robust analytical methods. Many molecular markers are selectively neutral. However, molecular markers differ in many qualities and must therefore be carefully chosen and analyzed differently with their differences in mind.

4.0 BIOCHEMICAL MARKERS

4.1 Allozymes

They are allelic variants of enzymes encoded by structural genes. Enzymes have a net electric charge, depending on the stretch of amino acids comprising the protein. When a mutation in the DNA results in an amino acid being replaced, the net electric charge of the protein may be modified, and the overall shape (conformation) of the molecule can change. Because changes in electric charge and conformation can affect the migration rate of proteins in an electric field, allelic variation can be detected by gel electrophoresis. Usually two, or sometimes even more loci can be distinguished for an enzyme and these are termed isoloci. Therefore, allozyme variation is often also referred to as isozyme variation (Kephart, 1990; May 1992).

Advantages: Simple, does not require DNA extraction or the sequence information, primers or probes, quick and easy to use. Simple analytical procedures, allow some allozymes to be applied at relatively low costs, depending on the enzyme staining reagents used. They are co-dominant markers with high reproducibility. Zymograms (the banding pattern of isozymes) can be readily interpreted.

Disadvantages: Relatively low abundance and low level of polymorphism. In addition, their selective neutrality may be in question (Berry and Kreitman, 1993; Hudson *et al.*, 1994; Kreiger and Ross, 2002). They may be affected by environmental conditions. For example, the banding profile obtained for a particular allozyme marker may change depending on the type of tissue used for the analysis (*e.g.* root vs. leaf). This is because a gene that is being expressed in one tissue might not be expressed in other tissues.

Applications: Allozymes have been applied in many population genetics studies, including measurements of out-crossing rates (Erskine and Muehlenbauer, 1991), (sub) population structure and population divergence (Freville *et al.*, 2001). They have been used, often in concert with other markers, for fingerprinting purposes (Tao and Sugiura, 1987; Maass and Ocampo, 1995), and diversity studies (Lamboy *et al.*, 1994; Ronning and Schnell, 1994), to study interspecific relationships (Garvin and Weeden 1994), the mode of genetic inheritance (Warnke *et al.*, 1998), and allelic frequencies in germplasm collections over serial increase cycles in germplasm banks (Reedy *et al.*, 1995), and to identify parents in hybrids (Parani *et al.*, 1997).

5.0 DNA MARKERS

5.1 Hybridization Based Markers

In hybridization based markers, DNA profiles are visualized by hybridizing the restriction enzyme digested DNA to a labeled probe, which can be DNA fragment of known origin or sequence. Restriction fragment length polymorphism (RFLP) was the first hybridization based marker system. Later on many marker probes such as microsatellites, minisatellites and STS were developed initially as hybridization based markers. With the development of easier cloning and sequencing techniques and availability of sequence databases these markers have been converted in to PCR based markers which are easier to assay.

5.1.1 Restriction Fragment Length Polymorphism (RFLP)

From the historical viewpoint, RFLPs were the first marker system developed to reveal the polymorphism at DNA level (Botstein *et al.*, 1980). This technology was first developed in the 1980s for use in human genetic applications and was later applied to plants. The technique involves digesting DNA with a restriction enzyme. DNA fragments are separated by agarose gel electrophoresis and are detected by subsequent Southern blot hybridization to a labeled DNA probe (Fig. 2). Labeling of the probe may be performed with a radioactive isotope or with alternative non-radioactive stains, such as digoxigenin or fluorescein. RFLPs are bands that correspond to DNA fragments, usually within the range of 2–10 kb, that have resulted from the digestion of genomic DNA with restriction enzymes. Probes are generated through the construction of genomic or complementary DNA (cDNA) libraries and therefore may be composed of a specific sequence of unknown identity (genomic DNA) or part of the sequence of a functional gene (exons only, cDNA). The hybridization results can be visualized by autoradiography (if the probes are radioactively labelled), or using chemiluminesence (if non-radioactive, enzyme-linked methods are used for probe labeling and detection). Any of the visualization techniques will give the same results. The visualization techniques used will depend on the laboratory conditions. Given that single-copy probes are selected, RFLP markers are locus specific and co-dominant that are capable of scoring both alleles at a locus, thus heterozygotes are always recognizable as such. Their polymorphism is due to variations in the restriction sites. Sequence variation due to substitution, addition, deletion or sequence rearrangements affects the occurrence (absence or presence) of endonuclease recognition sites, is considered to be main cause of length polymorphism. RFLP markers converted to PCR based markers *e.g.* STS, ASAP, EST and SSCP. All these marker systems are discussed in detail in the following sections.

Advantages: Unlimited number of loci, moderately polymorphic, co-dominant, many detection systems, can be converted to SCARs, robust in usage, good use of probes from other species, detects in related genomes, no sequence information required and is highly reproducible.

Disadvantages: Labour intensive and technically demanding, fairly expensive, large quantity of DNA needed, often very low levels of polymorphism, can be slow (often long exposure times), needs considerable degree of skill. In general, if research is conducted with poorly studied groups of wild species or crops (orphan crops), suitable probes may not yet be available, so considerable investments are needed for development. Moreover, large quantities (1–10 µg) of purified, high molecular weight DNA are required for each DNA digestion. RFLPs are not amenable to automation and collaboration among research teams requires distribution of probes.

Applications: RFLPs can be applied in diversity and phylogenetic studies ranging from individuals within populations or species, to closely related species. RFLPs have been widely used in gene mapping studies because of their high genomic abundance due to the ample availability of different restriction enzymes and random distribution throughout the genome (Neale and Williams, 1991). Botstein *et al.* (1980) used RFLP markers for the first time in the construction of genetic map. These are very reliable marker in linkages analysis and can determine if linked gene is present in homozygous or heterozygous stage due to their codominant behaviour. They also have been used to investigate relationships of closely related taxa (Miller and Tanksley, 1990; Lanner *et al.*, 1997), as fingerprinting tools (Fang *et al.*, 1997), for diversity studies, and for studies of hybridization and introgression, including studies of gene flow between crops and weeds (Brubaker and Wendel, 1994; Clausen and Spooner, 1998; Desplanque *et al.*, 1999).

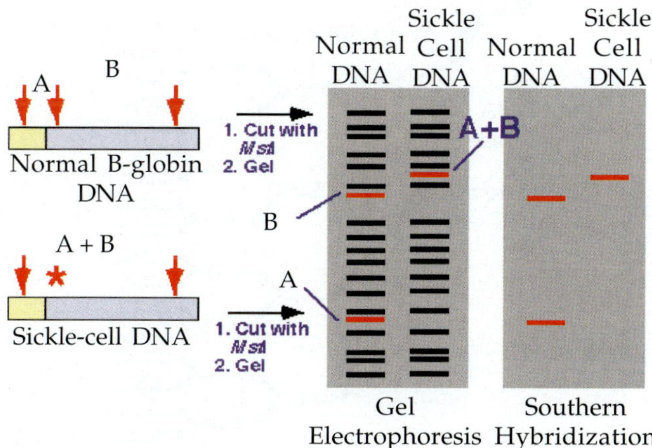

Fig. 2. RFLP analysis of normal and sickle-cell DNA (Source : Sangitha, 2006)

5.1.2 VNTR / Minisatellites

Minisatellite or variable number tandem repeats (VNTRs) analysis is quite similar to RFLPs, which involves digestion of genomic DNA with restriction endonucleases, but minisatellites are conceptually very different class of marker. This method has been discussed in detail in the section 'sequence dependent markers'.

5.2 Polymerase Chain Reaction (PCR) Based Markers

5.2.1 Polymerase Chain Reaction

The polymerase chain reaction (PCR) is basically a technique for *in vitro* amplification of specific DNA sequences by the simultaneous primer extension of complementary DNA strands. It is a cyclic process, which is repeated 25 to 45 times. One cycle consists of three basic steps with characteristic reaction temperatures: 1. Denaturation of the double stranded DNA, 2. Annealing of primers to complementary sequence on template and, 3. Extension of primers by DNA-polymerase. The amplified products we can see in the gel in the form of bands. The amplification of any region is in exponential manner (Fig. 3).

PCR-based markers involve *in vitro* amplification of particular DNA sequences or loci with the help of specially or arbitrarily chosen oligonucleotide primers and a thermostable DNA polymerase enzyme. The amplified products are separated electrophoretically and banding patterns are detected by different methods such as staining and autoradiography.

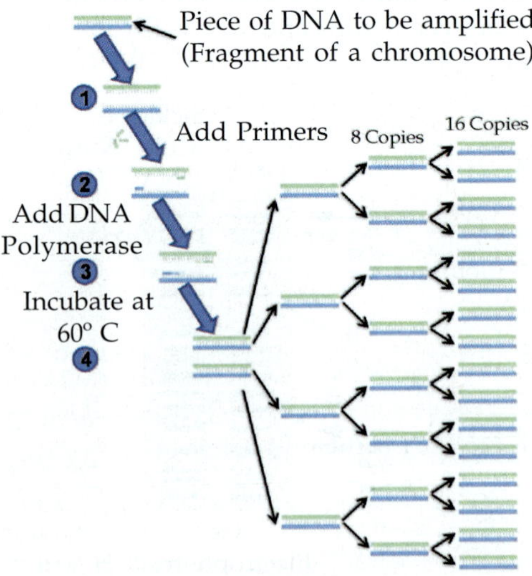

Fig. 3. Steps of PCR amplification of DNA (Source : Rediscovering Biology, 1997)

5.2.2 Types of PCR based molecular markers

The PCR based markers can be divided in two categories *viz.*, sequence arbitrary markers and sequence dependent markers. The commonest PCR-based markers are amplified fragment length polymorphisms (AFLPs), random amplified polymorphic DNA (RAPD) and simple sequence repeats (SSRs), also known as microsatellites. Single-nucleotide polymorphisms (SNPs) are the newest type. No absolute hierarchy exists between the marker types-all having advantages and disadvantages. The requirements of different marker systems vary depending on the type and nature of marker(s) (Table 2). The choice of a type of marker therefore will depend on the user's objectives and thus be determined case by case. Features and implementation of different marker techniques are also very specific (Tables 3 and 4) and it depends on the marker type under study.

Table 2. Requirements for different marker(s) systems

Marker/technique	Amount/quality of DNA required	DNA sequence required	Radioactive detection	Gel system
RFLP	High/High	No	Yes/No	Agarose
RAPD	Low/Low	No	No	Agarose
SSR	Low/Medium	Yes	No	Acrylamide/Agarose
ISSR	Low/Medium	Yes/No	No	Acrylamide/Agarose
AFLP	Low/High	No	Yes/No	Acrylamide
IRAP/REMAP	Low/Medium	Yes	No	Acrylamide/Agarose
Additional marker systems				
Morphological	No	No	No	None
Protein/isozyme	No	No	No	Agarose/Acrylamide
STS/EST	Low/High	Yes	Yes/No	Acrylamide/Agarose
SNP	Low/High	Yes	No	Sequencing required
Microarray	Low/High	Yes	No	None
SCARS/CAPS	Low/High	Yes	Yes/No	Agarose

Table 3. Features of different marker techniques

Marker/technique	PCR-based	Polymorphism (abundance)	Dominance
RFLP	No	Low-Medium	Co-dominant
RAPD	Yes	Medium-High	Dominant
SSR	Yes	High	Co-dominant
ISSR	Yes	High	Dominant
AFLP	Yes	High	Dominant
IRAP/REMAP	Yes	High	Co-dominant
Additional marker systems			
Morphological	No	Low	Dominant/Recessive/Co-dominant
Protein/isozyme	No	Low	Co-dominant
STS/EST	Yes	High	Co-dominant/Dominant
SNP	Yes	Extremely High	Co-dominant
SCARS/CAPS	Yes	High	Co-dominant
Microarray		High	

5.2.3 Sequence Arbitrary Markers

The sequence arbitrary markers do not require prior sequence information and DNA is amplified by oligonucleotide primers designed at random. The substantial sequence information is limited in plant species and obtaining sequence information for most of the species may be cost prohibitive. Therefore development and use of sequence arbitrary markers has occurred primarily with plant species to reveal polymorphism. The popularity of sequence arbitrary markers is driven by their near absence of development cost and ability to readily detect polymorphism. However, by cloning and sequencing of polymorphic fragments, sequence arbitrary markers can be converted into sequence dependent markers, which are more reproducible with many other advantages. Thus sequence arbitrary PCR-based markers are cost-effective and it makes sense to rely on them for initial survey for polymorphism or linkage with specific traits. Once a polymorphic PCR product is identified to be polymorphic or linked with gene of interest, it can be isolated, cloned and sequenced. Such a product can be used for development of SCAR marker or as a standard RFLP marker by using it as probe. Some major sequence arbitrary markers are as follows.

Table 4. Implementation of different marker(s) techniques

Marker/ techniques	Development costs per data point	Running costs (Lab/Crops)	Portability
RFLP	Medium	High	High/High
RAPD	Low	Low	Low/Low
SSR	High	Medium	High/Low
ISSR	Low	Low	High/Low
AFLP	Medium-High	Low	High/Low
IRAP/REMAP	High	Medium	High/Low
Additional marker systems			
Morphological	Depends	Depends	Limited to breeding aims
Protein and isozyme	High	Medium	High/High
SCARS/CAPS	High	Medium	High/Low
STS/EST	High	Medium	Medium/High
SNP	High	Medium-Low	Unknown
Microarray	Medium	Low	Unknown

5.2.3.1 *Random Amplified Polymorphic DNA (RAPD)*

RAPDs are produced by PCR using genomic DNA and arbitrary primers (Welsh and McClelland, 1990; Devos and Gale, 1992) that are generally 10 base pairs (bp) long and defined at random. These oligonucleotides serve as both forward and reverse primer, and are usually able to amplify fragments from 1–10 genomic sites simultaneously. The principle involved in generating RAPDs is that, a single, short oligonucleotide primer, which binds to many different loci, is used to amplify random sequences from a sequence (complementary to that of primer) will occur in the genome, on opposite DNA strands in opposite orientation within a distance that is readily amplifiable by PCR. Amplified fragments, usually within the 0.5–5 kb size range, are separated by agarose gel electrophoresis, and polymorphisms are detected, after ethidium bromide staining, as the presence or absence of bands of particular sizes. Several bands are visualized simultaneously (Fig. 5). These markers are dominant, and their polymorphism is due to sequence variations in the primer hybridization site, but they can also be generated by length differences in the amplified sequence between primer annealing sites. Although attractive for their simplicity, problems of reproducibility limit interest in their use, compared with other types of markers (Jones *et al.*, 1997). PCR products can be separated by gel electrophoresis.

Advantages: They are quick and easy to assay. Low quantities of template DNA are required, usually 5–50 ng per reaction. No sequence data for primer construction are needed. RAPDs have a very high genomic abundance and are randomly distributed throughout the genome. Fairly cheap, good polymorphism and it can be automated.

Disadvantages: Low reproducibility within and between laboratories (Schierwater and Ender, 1993) and precautions are needed to avoid contamination of DNA samples because short random primers are used. They are not locus-specific, band profiles cannot be interpreted in terms of loci and alleles (dominance of markers), and similar sized fragments may not be homologous (fragment allelism). Cannot be used across populations nor across species, often see multiple loci and are dominant marker.

Applications: RAPDs have been used for many purposes, ranging from studies at the individual level (*e.g.* genetic identity) to studies involving closely related species. RAPD can also be applied for gene mapping studies to fill gaps not covered by other markers (Williams *et al.*, 1990; Hadrys *et al.*, 1992).

5.2.3.2 Variants of the RAPD marker

1. ***Arbitrarily Primed Polymerase Chain Reaction (AP-PCR)***: It uses longer arbitrary primers than RAPDs. It is based upon use of single, sequence arbitrary primer for DNA amplification reaction. This is a special case of RAPD wherein discrete amplification patterns are generated by employing single primer of 10-15 bp in amplification of genomic DNA (Welsh and McCelland, 1990). The thermal cycling profile begins with one or two cycles of incorporating a low annealing temperature followed by high stringency annealing temperature and radiolabeled dCTP to label the newly synthesized fragments. The DNA fragments are size separated on polyacrylamide gel and

visualized *via* radiography. Because AP-PCR is also based upon DNA amplification polymorphism, concerns over allelism exist. It is unknown whether AP-PCR reactions are infact more robust. A variation of AP-PCR is known as direct amplification of length polymorphisms (DALP). It uses an arbitarily primed PCR (AP-PCR) to produce genomic fingerprints and to sequence DNA polymorphisms.

2. ***DNA Amplification Fingerprinting (DAF)***: It uses a single short oligonucleotide of 5-8 bp at high concentration, in combination with either low or high stringency annealing temperature, high resolution fragment separation on polyacrylamide gel and high resolution fragment detection (Silver Staining) (Caetano-Anolles *et al.*, 1991). The resulting DNA fragment pattern is highly complex. In spectrum of product obtained, simple patterns are useful as genetic markers for mapping, while more complex patterns are useful for DNA fingerprinting. DAF share common drawbacks of RAPD and AP-PCR namely, preponderance of dominant marker loci and unknown allelism between fragments of equal molecular weight. DAF has not been used in a large number of labs and therefore issue of reproducibility has not been well addressed. As with all sequence arbitrary DNA amplification methods consistent results can only be obtained by careful attention of optimization and standardization of protocols.

3. ***Randomly amplified microsatellite polymorphisms (RAMPO)***: In this PCR-based strategy, genomic DNA is first amplified using arbitrary (RAPD) primers. The amplified products are then electrophoretically separated and the dried gel is hybridized with microsatellite oligonucleotide probes. Several advantages of oligonucleotide fingerprinting, RAPD and microsatellite-primed PCR are thus combined, these being the speed of the assay, the high sensitivity, the high level of variability detected and the non-requirement of prior DNA sequence information. This technique has been successfully employed in the genetic fingerprinting of tomato, kiwi fruit and closely-related genotypes of *Datura bulbifera* (Richardson *et al.* 1995).

4. ***Multiple Arbitrary Amplicon Profiling (MAAP)***: is the collective term for techniques that uses single arbitrary primers.

5.2.3.3 Amplified Fragment Length Polymorphism (AFLP)

This marker system (Vos *et al.*, 1995) is a sequence arbitrary amplification-based method. From the technological viewpoint, AFLPs as markers are intermediate between RFLPs and microsatellites. The technique combines the use of restriction enzymes and PCR amplification. AFLPs are

DNA fragments (80–500 bp) obtained from digestion with restriction enzymes, followed by ligation of oligonucleotide adapters to the digestion products and selective amplification by the PCR. The DNA is cut with two restriction enzymes, one being a frequent cutter and the other an infrequent cutter. This is followed by ligation of adapters, including restriction motifs, and a two-step PCR amplification of selected fragments (Fig. 6). The latter procedure uses primers composed of the adapters and 1 to 3 selected nucleotides (Vos *et al.*, 1995). It limits the number of fragments to a resolvable range. The PCR-amplified fragments can then be separated by gel electrophoresis and banding patterns visualized (Fig. 7). A range of enzymes and primers are available to manipulate the complexity of AFLP fingerprints to suit application. Care is needed in selection of primers with selective bases.

On an AFLP gel, many bands are simultaneously visualized. Each band is assimilated to an allele at a specific locus that has only two allelic

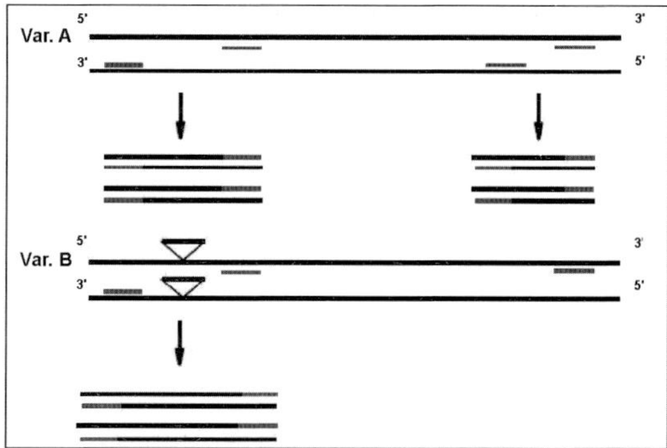

Fig. 4. The scheme above shows RAPD variation between varieties A and B. In variety A there are 4 primer binding sites resulting in two RAPD products; variety B lacks one of the binding sites resulting in only one RAPD marker being produced (Source : FAO/IAEA, 2002)

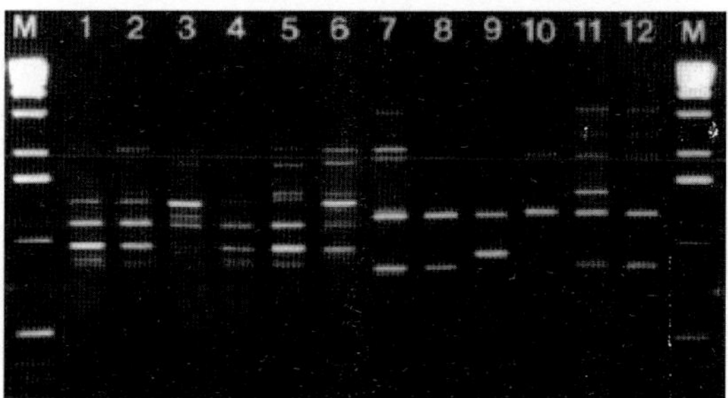

Fig. 5. DNA fingerprinting of 12 plants of *Rhododendron* with a random primer set (Source : Iqbal, 2006)

forms: presence or absence of the band. The AFLP banding profiles are the result of variations in the restriction sites or in the intervening region. The AFLP technique simultaneously generates fragments from many genomic sites (usually 50–100 fragments per reaction) that are separated by polyacrylamide gel electrophoresis and then scored.

AFLP markers are mainly scored as dominant. However, by use of automatic gel scanner heterozygote may be distinguished from homozygote based on band intensity differences, which facilitates the scoring of many AFLPs as codominant markers. Polymorphism is due to point mutations in the restriction sites. They are probably not distributed at random throughout the genome, as they sometimes cluster in some areas. However, even with such disadvantages, they are very useful for simultaneously observing a large number of polymorphic markers. The AFLP pattern is moderately complex. However, pattern complexity and concomitantly number of observed polymorphisms is dependent upon both primer and restriction endonuclease selection. The number of fragments detected for a given set of restriction endonuclease used can be tuned by varying the number of selective nucleotides in the primer combination. This flexibility is an attractive feature of AFLP.

Selective Fragment Length Amplification (SFLA) and Selective Restriction Fragment Amplification (SRFA) are synonyms sometimes used to refer to AFLPs.

5.2.3.3.1 Variant of AFLP

Selectively Amplified Microsatellite Polymorphic Locus (SAMPL): It is a variation of the AFLP technique which amplifies microsatellite loci by using a single AFLP primer in combination with a primer complementary to compound microsatellite sequences, which do not require prior cloning and characterization. It is considered more applicable to intraspecific than to interspecific studies due to frequent null alleles. Comparative qualities of marker techniques DNA provides many advantages that make it especially attractive in studies of diversity and relationships.

Advantages: The advantages by AFLP includes high genomic abundance, considerable reproducibility, many informative bands per reaction, wide range of applications, small DNA quantities required, no sequence data for primer construction are required. AFLPs may not be totally randomly distributed around the genome as clustering in certain genomic regions, such as centromere, has been reported for some crops (Alonso-Blanco *et al.*, 1998; Young *et al.*, 1999; Saal and Wricke, 2002). It can be adapted for different uses, *e.g.* cDNA-AFLP. AFLPs can be analyzed on automatic sequencers, but software problems concerning the scoring of AFLPs are encountered on some systems.

Disadvantages: Need for purified, high molecular weight DNA, the dominance of alleles, and the possible non-homology of comigrating fragments belonging to different loci, etc. are some of the demerits of AFLP. In addition, due to the high number and different intensity of bands per primer combination, there is the need to adopt certain strict but subjectively determined criteria for acceptance of bands in the analysis. AFLP bands are not always independent *e.g.* in case of an insertion between two restriction sites the amplified DNA fragment results in increased band size. Moreover, AFLP is technically more demanding and relatively complex.

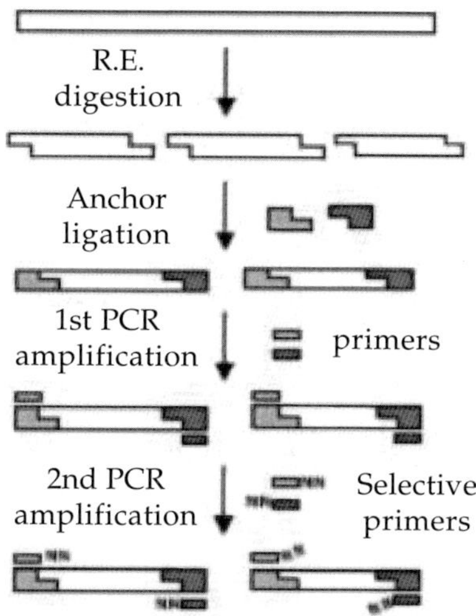

Fig. 6. Mechanism of band amplification in AFLP (Source : Tsai, 2007)

Applications: AFLPs can be applied in studies involving genetic identity, parentage and identification of clones and cultivars, and phylogenetic studies of closely related species. Their high genomic abundance and generally random distribution throughout the genome make AFLPs a widely valued technology for gene mapping studies (Vos *et al.*, 1995).

5.2.3.4 *Single-Strand Conformation Polymorphism (SSCP)*

SSCP relies on intra strand (single strand) differences (conformation) in DNA of different sequence. SSCPs are DNA fragments of about 200–800 bp amplified by PCR using specific primers of 20–25 bp. Gel electrophoresis of single-strand DNA is used to detect nucleotide sequence variation among the amplified fragments. The electrophoretic mobility of single-strand DNA depends on the secondary structure (conformation) of the molecule, which is changed significantly with mutation. Thus, it provides a method to detect nucleotide variation among DNA samples without having to perform

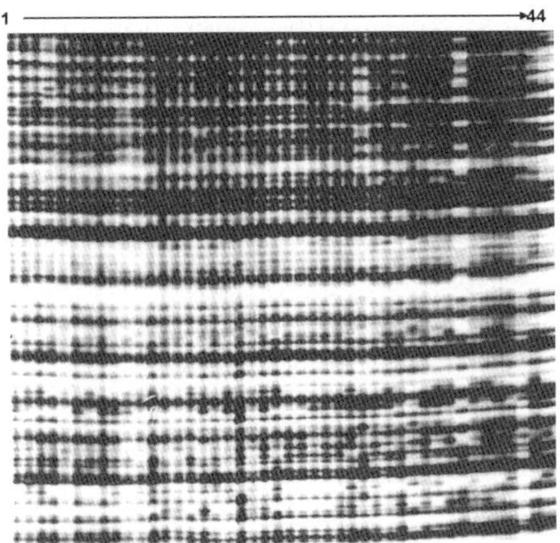

Fig. 7. AFLP banding pattern of 44 soybean genotypes (Source: Singh *et al.*, 2006)

sequence reactions. In SSCP the amplified DNA is first denatured, and then subjected to non-denaturing gel electrophoresis.

Related techniques to SSCP are Denaturing Gradient Gel Electrophoresis (DGGE), Thermal Gradient Gel Electrophoresis (TGGE) and Heteroduplex Analysis (HA).

1. ***Denaturing Gradient Gel Electrophoresis (DGGE):*** It uses double stranded DNA which is converted to single stranded DNA in an increasingly denaturing physical environment during gel electrophoresis. It detects mutations (small deletions and insertions, point mutations) by separating PCR amplified DNA fragments (different from wild type in melting behaviour) on a denaturing gradient gel. It is having high probability of detection of DNA sequence differences but involves high equipment costs too.
2. ***Thermal Gradient Gel Electrophoresis (TGGE):*** It uses temperature gradients to denature double stranded DNA during electrophoresis. It uses heat as a source of energy to make the hydrogen bonds thermodynamically unstable. DNA or RNA fragments with point mutations will show different melting behaviour, thus different from wild type DNA fragments of the same length but different sequence can be separated *e.g.* HA (heteroduplex analysis).
3. ***Heteroduplex Analysis (HA):*** DNA of a putative heteroallelic sample that contains wild type *AA* and mutant aa is multiplied by PCR. Additional denaturation and rehybridization gives 4 different DNA duplices:
   - Homoduplices *AA* and *aa*
   - Heteroduplices *Aa* and *aA*

Each heteroduplex contains a mismatch base pair which lowers the melting temperature. Any mismatches in the heteroduplex DNA will ultimately lead to the different 3D structure of homoduplex DNA which lowers the mobility on gel. This reduced mobility is proportional to the degree of divergence of the DNA sequences.

Advantages: In SSCP analysis there is codominance of alleles and the low quantities of template DNA required (10–100 ng per reaction), because technique is PCR-based.

Disadvantages: In SSCP analysis we need sequence data to design PCR primers and highly standardized electrophoretic conditions in order to obtain reproducible results. Furthermore, some mutations may remain undetected, and hence absence of mutation cannot be proven.

Applications: SSCPs have been used to detect mutations in genes using gene sequence information for primer construction (Hayashi, 1992).

5.2.3.5 Allele-specific associated primers (ASAPs)

To obtain an allele-specific marker, specific allele (either in homozygous or heterozygous state) is sequenced and specific primers are designed for amplification of DNA template to generate a single fragment at stringent annealing temperatures. These markers tag specific alleles in the genome and are more or less similar to SCARs (Gu *et al.*, 1995).

5.2.3.6 Inter Simple Sequence Repeats (ISSR)

ISSRs are DNA fragments of about 100–3000 bp located between adjacent, oppositely oriented microsatellite regions. ISSR fingerprinting was developed such that no sequence knowledge was required. These markers involve polymerase chain reaction (PCR) amplification of inter sequence repeat DNA using a single primer composed of microsatellite sequences. Primers based on repeat sequences, such as (CA)n, can be made with a degenerate 3′-anchor, such as $(CA)_8$ RG or $(AGC)_6$TY. An unlimited number of primers can be designed for various combinations of di-, tri-, tetra and pentanucleotides with an anchor made up of few bases and can be exploited for a broad range of applications in plant species. The resultant PCR amplifies the sequence between two SSR, yielding multilocus marker system useful for fingerprinting, diversity analysis and genome mapping (Zietkiewicz *et al.*, 1994) (Fig. 8). PCR products may be radiolabeled with ^{32}P or ^{33}P *via* end labeling or PCR incorporation, and separated on polyacrycamaide gel prior to autoradiographic visualization. A typical reaction yield 20-100 bands per lane depending on species and primer (Godwin *et al.*, 1997). PCR products may be separated on agarose gel with ethidium bromide visualization, or polyacrylamide gel with silver staining techniques. The ISSR marker technique is nearly identical to RAPD expect that ISSR primer sequences are designed from microsatellite regions and the annealing temperatures used are higher which makes them more robust. The advantage of greater band amplification is augmented.

Techniques related to ISSR analysis are Single Primer Amplification Reaction (SPAR) and Directed Amplification of Minisatelliteregion DNA (DAMD).

1. ***Single Primer Amplification Reaction (SPAR):*** A PCR-based genotyping technique in which genomic template is amplified with a single primer. It uses a single primer containing only the core motif of a microsatellite.
2. ***Directed Amplification of Minisatelliteregion DNA (DAMD):*** This technique, introduced by Heath *et al.*(1993), has been explored as a means of generating DNA probes useful for detecting polymorphism. It uses a single primer containing only the core motif of a minisatellite. DAMD-PCR clones can yield individual-specific DNA fingerprinting pattern and thus have the potential as markers for species differentiation and cultivar identification (Somers *et al.,* 1996).

Advantages: No sequence data for primer construction, low quantities of template DNA are required (5–50 ng per reaction), randomly distributed throughout the genome, highly polymorphic, robust in usage and it can be automated.

Disadvantages: Non-homology of similar sized fragments, can have reproducibility problems, usually dominant and are species-specific.

Applications: Because of the multilocus fingerprinting profiles obtained, ISSR analysis can be applied in studies involving genetic identity, parentage, clone and strain identification, and taxonomic studies of closely related species. In addition, ISSRs are considered useful in gene mapping studies (Godwin *et al.,* 1997; Zietkiewicz *et al.,* 1994; Gupta *et al.,* 1994). ISSR markers have been compared favourably with RFLP and RAPD markers with application to DNA fingerprinting and diversity analysis in sorghum (Yang *et al.,* 1996), maize (Kantely *et al.,* 1995) and sweet potato (Huang and Sun, 2000). It is now clear that ISSR markers have great potential for studies of natural populations (Wolfe *et al.,* 1998). Rabeena *et al.* (2003) used ISSR markers for the construction of linkage map in lentil. Zehdi *et al.* (2004) used ISSR markers to assess genetic diversity among set of Tunisian date palm varieties. They used 12 ISSR primers of which 9 generated polymorphic bands with an average of 9.11 per primer and seven primers were sufficient to identify all 12 varieties. These results confirm hyper variability of ISSR markers and their usefulness to find out genetic polymorphism.

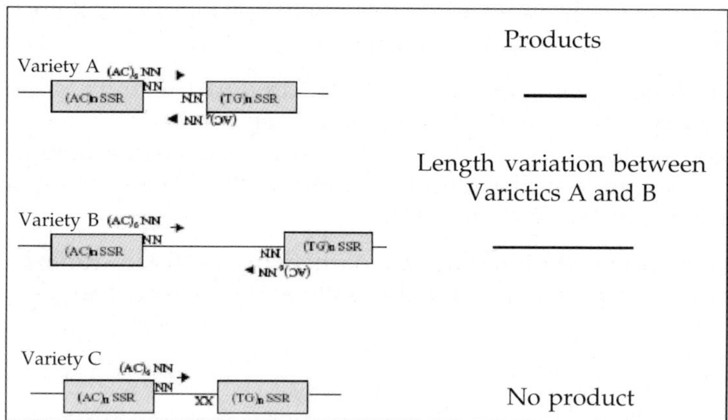

Fig. 8. The above scheme shows how sequence variation between two SSRs results in variation in PCR products in varieties A, B and C. The figure shows variation at only one ISSR locus, amplification of all compatible ISSR loci among the genomes of a range of varieties will result in complex, fingerprinting, banding patterns (Source : FAO/IAEA, 2002)

5.2.4 Sequence Dependent Markers

The sequence dependent markers require prior sequence information and amplify specific DNA fragments or sequences by primers specially designed for their amplification. Sequence dependent PCR-based marker systems are now being exploited in many plant species despite of their high initial development cost. Sequence dependent markers are generally co-dominant, highly reproducible, easily exchanged between labs and shows allelism. These features make them most popular marker system for plant genome analysis and marker assisted selection (MAS).

5.2.4.1 Minisatellites / VNTR

A repeat of 10-60 nucleotides is called a minisatellite or variable number of tandem repeats (VNTR). Minisatellite analysis, like RFLPs, also involves digestion of genomic DNA with restriction endonucleases, but minisatellites are conceptually very different class of marker. Therefore this is also a part of hybridization based marker. They consist of chromosomal regions containing tandem repeat units of a 10–60 base motif, flanked by conserved DNA restriction sites (Fig. 9). A minisatellite profile consisting of many bands, usually within a 4–20 kb size range, is generated by using common multilocus probes that are able to hybridize to minisatellite sequences in different species (Fig. 10). Variation in the number of repeat units, due to unequal crossing over or gene conversion, is considered to be the main cause of length polymorphisms. Minisatellite loci are also often referred to as Variable Number of Tandem Repeats (VNTR) loci.

Advantages: The main advantages of minisatellites are their high level of polymorphism and high reproducibility.

Disadvantages: Similar to RFLPs due to the high similarity in methodological procedures. Band profiles can not be interpreted in terms of loci and alleles and similar sized fragments may be non-homologous. In addition, the random distribution of minisatellites across the genome has been questioned (Schlötterer, 2004).

Applications: Minisatellites are particularly useful in studies involving genetic identity, parentage, clonal growth and structure, and identification

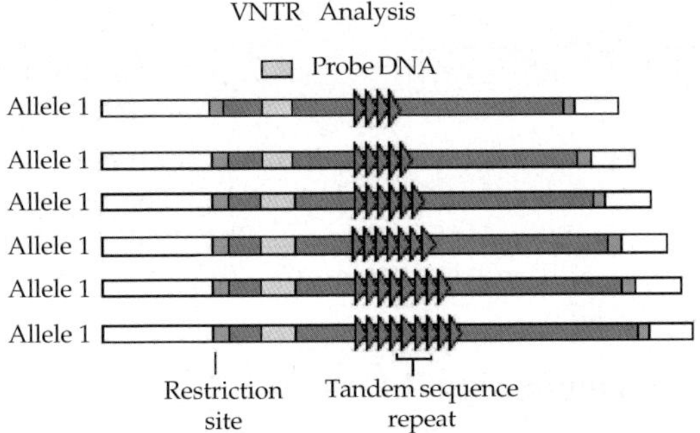

Fig. 9. Principle of VNTR analysis (Source : Sangitha, 2006)

of varieties and cultivars (Jeffreys *et al.*, 1985a,b; Zhou *et al.*, 1997), and for population-level studies (Wolff *et al.*, 1994). Minisatellites are of reduced value for taxonomic studies because of hypervariability.

5.2.4.2 *Simple Sequence Repeat Markers (SSR) or Microsatellites*

Microsatellites, like minisatellites, represent tandem repeats, but their repeat motifs are shorter (1–6 base pairs). It is now well known that 30-90%

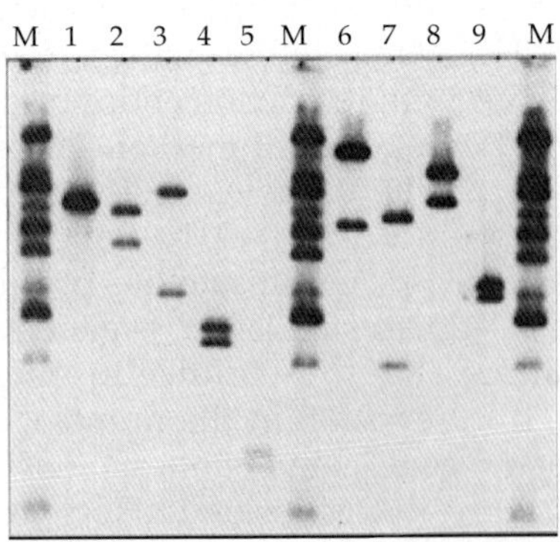

Fig. 10. Banding pattern in VNTR analysis (Source : Sangitha, 2006)

of the genome of virtually all eukaryotic species is constituted by regions of repetitive DNA, which are highly polymorphic in nature SSR (Simple Sequence Repeats) or microsatellites consists of tandomly repeated units of short nucleotide motifs. Di-, tri- or tetranucleotide repeats like $(CA)_n$, $(AAT)_n$ or $(GATA)_n$ are widely distributed throughout the genome of the plants and animals (Fig. 11). Compound repeats composed of two or more repeats motifs are also frequently found (Morgante and Vogel, 1996). Their polymorphism lies in the variation of the number of repeats, probably because of errors during replication. If nucleotide sequences in the flanking regions of the microsatellite are known, specific primers (generally 20–25 bp) can be designed to amplify the microsatellite by PCR.

SSR Identification: SSRs can be identified by searching among DNA databases (*e.g.* EMBL and Genebank), or alternatively small insert (200-600 bp) genomic DNA libraries can be produced and enriched for particular repeats (Powell *et al.*, 1996). From the sequence data, primer pairs (of about 20 bp each) can be designed (software programmes are available for this). In addition, primers may be used that have already been designed for closely related species.

SSR Detection: Polymerase slippage during DNA replication, or slipped strand mispairing, is considered to be the main cause of variation in the number of repeat units of a microsatellite, resulting in length polymorphisms that can be detected by either high percentage agarose or polyacrylamide gel electrophoresis (Fig. 12). However, accurate size measurement is difficult with agarose and polyacrilamide gels as these gels matrices do not allow single base pair resolution. Single nucleotide resolution of DNA fragments requires the use of denaturing Poly Acrylamide Gel Electrophoresis (PAGE) or capillary elecrophoresis. SSR amplification products are detected in PAGE by silver staining, radio labeling (primer or product) or fluorescent labeling (primer or product). The use of fluorescent labeled primers combined with an automated electorphoresis system, greatly simplifies the analysis of SSR allele size. The use of fluorescent primers enables PCR products to be detected without any post elecrophoresis treatments. Sample throughput can also be increased through multiplexing by use of different fluorescent tags on different PCR products.

Advantages: Co-dominance of alleles, high genomic abundance in eukaryotes and random distribution throughout the genome, with preferential association in low-copy regions (Morgante *et al.,* 2002). Low quantities of template DNA (10–100 ng per reaction) are required. Due to the use of long PCR primers, the reproducibility of microsatellites is high. SSR marker can be exchanged between researchers because each locus is defined by specific primer sequences. SSR assays are more robust than random amplified polymorphic DNA (RAPD) and more transferable than

AFLPs. Multiple microsatellites may be multiplexed during PCR or gel electrophoresis if the size ranges of the alleles of different loci do not overlap (Ghislain *et al.*, 2004). This decreases the analytical costs. Furthermore, the screening of microsatellite variation can be automated, if the use of automatic sequencers is an option.

Disadvantages: High development costs if adequate primer sequences for the species of interest are unavailable, making them difficult to apply to unstudied groups crops. Although they are in co-dominant markers, mutations in the primer annealing sites may result in the occurrence of null alleles (no amplification of the intended PCR product), which may lead to errors in genotype scoring. Homoplasy may occur at microsatellite loci due to different forward and backward mutations, which may cause underestimation of genetic divergence. A very common observation in microsatellite analysis is the appearance of stutter bands that are artifacts in the technique that occur by DNA slippage during PCR amplification.

Applications: In general, microsatellites show a high level of polymorphism. As a consequence, they are very informative markers that can be used for many population genetics studies, ranging from the individual level (*e.g.* clone and strain identification) to that of closely related species. Conversely, their high mutation rate makes them unsuitable for studies involving higher taxonomic levels. Microsatellites are also considered

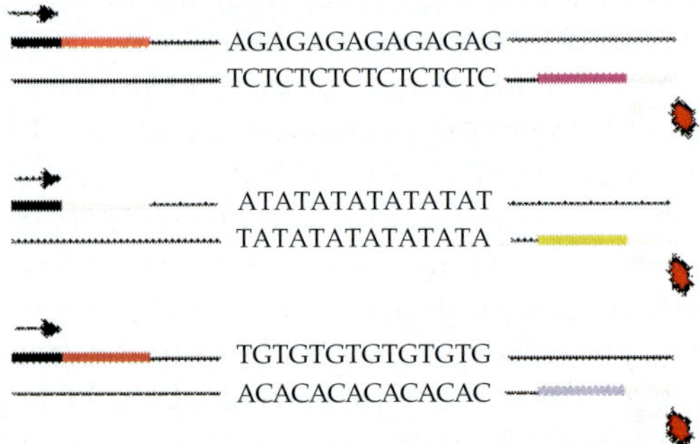

Fig. 11. Sequence information for SSR (Source : Sangitha, 2006)

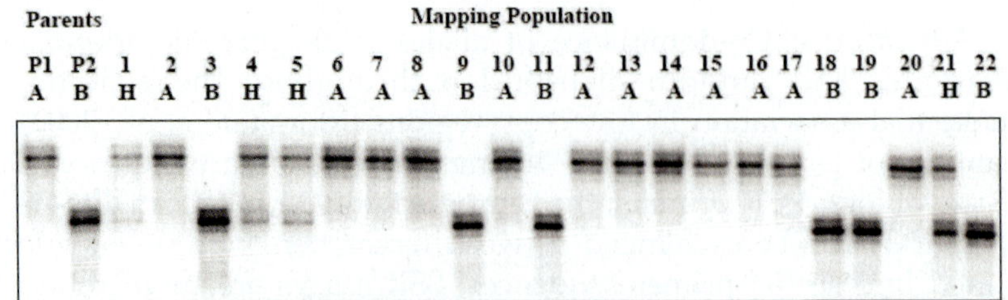

Fig. 12. Microsatellite markers or Simple Sequence Repeat (Source : Iqbal, 2006)

ideal markers in gene mapping studies (Hearne *et al.*, 1992; Morgante and Olivieri, 1993; Queller *et al.*, 1993; Jarne and Lagoda, 1996). The discovery of microsatellites or simple sequence repeat markers has significantly increased the marker density of linkage maps for some mammals and plants (Cregan *et al.*, 1999; Love *et al.*, 1990; Murray *et al.*, 1994).

5.2.4.3 STR Markers (Sequence tagged repeats)

This marker is related to VNTR. It is a PCR based marker and requires sequence knowledge of the VNTR regions. Regions flanking the VNTRs are often highly conversed which are used for primer synthesis. Therefore,

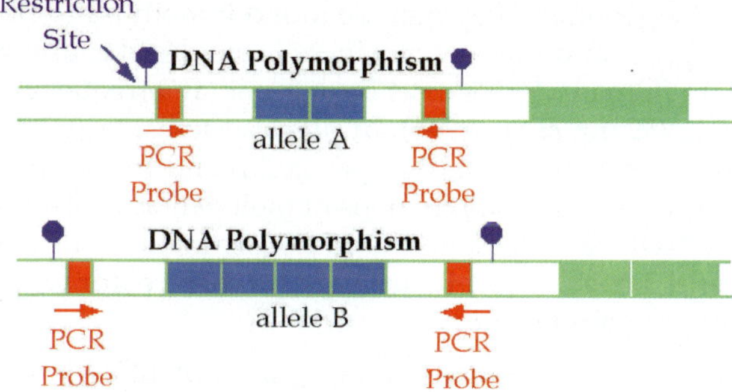

Fig. 13. STR analysis (Source : Sangitha, 2006)

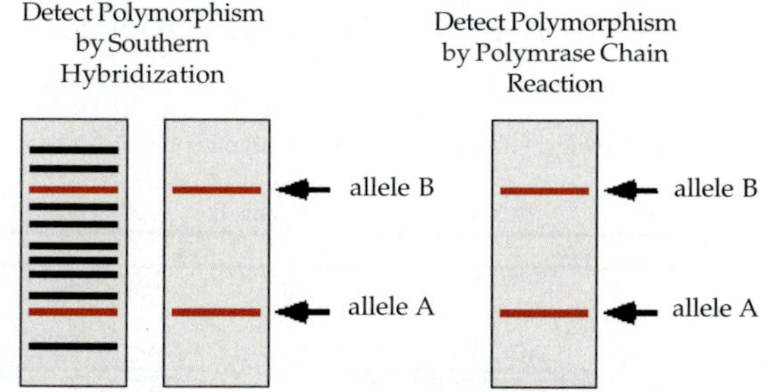

Fig. 14. STR polymorphism detection by southern and PCR (Source : Sangitha, 2006)

specific primers for PCR are designed to the conserved flanking regions and PCR reactions amplify the VNTR reAgion (Figs. 13 & 14). Reaction products are run on gel and different number of repeats = different length PCR products. It is faster than Southern blottinAg and can use much less sample. It can use partially degraded samples and not all VNTRs can be converted to STR. Small samples can skew results and more susceptible to contamination.

5.2.4.4 *Single Nucleotide Polymorphisms (SNPs)*

Single nucleotide polymorphism (SNP) markers are based on variation of single nucleotide between two or more genotypes or individuals. It corresponds to single-base substitutions in sequences. SNPs are the most abundant polymorphic marker with 2 – 3 polymorphic sites every kilobase (Cooper *et al.*, 1985). Originally discovered in humans, SNPs have now been developed for genotyping in plants. SNP technology is heavily dependent upon sequence data.

SNP hold tremendous potential for increasing genotype throughput and lowering cost per data point generated. SNPs are extremely abundant through out the genome. They may be found in both transcribed and non-transcribed regions and in some instances is the direct cause of phenotypic variation(s). In humans, SNPs have been found at a frequency greater than 1 per 1000 bps (Wang *et al.*, 1994). In plant systems, the SNPs seem to be more abundant. In wheat one SNP per 20 bp and in maize 1 SNP per 70 bp has been recorded in certain regions of their genomes. Thus one can only hope that SNP will be developed expeditiously in all major crops in a large scale and will be extensively utilized in future for variety of crop improvement programmes.

SNP Detection: Several methods are available for SNP detection including automated fluorescent sequencing denaturing high-performance liquid chromatography (DHPLC, Underhill *et al.*, 1996), DNA microarrays (Hacia and Collins, 1999), single-strand conformationl polymorphis-capillary electrophoresis (SSCP-CE, Ren, 2001; Fig. 15), microplate-array diagonal-gel

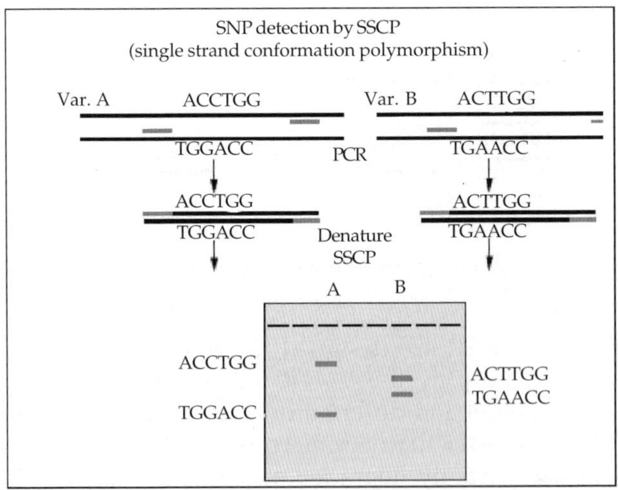

Fig. 15. The scheme above shows how SNP variation can be detected between varieties A and B (Source : FAO/IAEA, 2002)

electrophoresis (MADGE, Day *et al.*, 1998), EST datamining, overlapping sequences from BACs/PACs genomes of different individuals (Taillon – Miller *et al.*, 1998) and matrix assisted laser desorption/ionisation time-of-flight (MALDI-TOF, Griffin and Smith, 2000). Their detection and validation, demanding significant sequencing effort, are costly. Different techniques of genotyping are used (Syvänen, 2001), which can also be costly (*e.g.* mass spectrometry). Once these limitations are resolved, these markers will probably become widely used, because they provide detailed genomic information and have potential for automation.

Advantages: Robust in usage, polymorphisms are identifiable, different detection methods available, suitable for high throughput and it can be automated.

Disadvantages: Very high development costs, requires sequence information and it can be technically challenging.

Applications: These are interesting for their extremely high density (one every 300 to 1500 bp in humans), which permits monitoring linkage over very short physical distances (Brookes, 1999). However, very few have been developed for crops like rice. They can be used to create higher resolution maps and there is a greater likely hood of finding a SNP marker close to a gene or QTL of interest.

5.2.4.5 Sequence Tagged Sites (STS)

In this technique, tract of DNA from the clone that has been primarily mapped is sequenced (300-500 bp) and the site is now defined by this sequence and called as STS. A PCR assay for an STS would be implemented simply by synthesizing two short oligonucliotides complimentary to opposite strands at opposite ends of STS sequence and using them as primer in PCR. RFLP markers or probes can also be converted into STS markers by sequencing RFLP fragments and designing primers for their amplification.

Advantage: The tedious hybridization procedure involved in RFLP analysis can be avoided.

Disadvantage: Sequence data are needed for primer synthesis.

Application: This is especially important when linkage of RFLP markers and some specific trait has been already established. They can be easily integrated into plant breeding programmes by converting them into a PCR based assay.

5.2.4.6 Cleaved Amplified Polymorphic Sequence (CAPS)

CAPS are DNA fragments amplified by PCR using specific 20–25 bp primers, followed by digestion of the PCR products with a restriction enzyme.

Subsequently, length polymorphisms resulting from variation in the occurrence of restriction sites are identified by gel electrophoresis of the digested products. The technique is particularly important when PCR product does not reveal polymorphism but may differ at internal sites. CAPS have also been referred to as PCR-Restriction Fragment Length Polymorphism (PCR-RFLP).

Advantages: Need low quantities of template DNA (50–100 ng per reaction), co-dominance of alleles and are high reproducible. Does not include the laborious and technically demanding steps of Southern blot hybridization and radioactive detection procedures as needed in RFLP. Highly reliable, usually single locus and species-specific.

Disadvantages: In comparison with RFLP analysis, CAPS polymorphisms are more difficult to find because of the limited size of the amplified fragments (300–1800 bp). Furthermore, sequence data are needed to design the PCR primers.

Applications: CAPS markers have been applied predominantly in gene mapping studies (Akopyanz *et al.* 1992; Konieczny and Ausubel, 1993). Recently Sabatini *et al.* (2004) developed a CAPS marker linked to 'Ovate' gene in tomato (confer pear shape to tomatoes), which could be of great use for marker assisted selection in favour of this trait and speeding up introgression of genetic resistance in pear shaped ecotypes. Pear shaped tomatoes are liked in Italy for their typical taste and shape.

5.2.4.7 Sequence Characterized Amplified Region (SCAR)

Paran and Michelmore (1993) converted RAPD markers into SCARs. SCARs are DNA fragments amplified by the PCR using specific 15–30 bp primers, designed from nucleotide sequences established from cloned RAPD fragments linked to a trait of interest. These are similar to STS markers in construction and application. By using longer PCR primers, SCARs do not face the problem of low reproducibility generally encountered with RAPDs. Obtaining a co-dominant marker may be an additional advantage of converting RAPDs into SCARs, although SCARs may exhibit dominance when one or both primers partially overlap the site of sequence variation. SCARs are usually dominant markers but can also behave as co-dominant markers in some cases. Polymorphism can be deduced by either denaturing gel electrophoresis or single strand conformation polymorphism (Rafalski and Tingey, 1993).

Advantages: They are quick and easy to use. Have a high reproducibility and are locus-specific. Low quantities of template DNA are required (10–100 ng per reaction).

Disadvantages: Need for sequence data to design the PCR primers.

Applications: SCARs are locus specific and have been applied in gene mapping studies and marker assisted selection (Paran and Michelmore, 1993). Gygax *et al.* (2004) developed three SCAR markers from RAPD markers linked to apple scab resistant gene *vbj* which proved to be codominant in inheritance. Wen-Jei *et al.* (2004) developed a SCAR marker from RAPD marker linked to sex determination in *Eucommia ulmoides*.

5.2.4.8 Polymerase Chain Reaction (PCR) Sequencing

PCR was a major breakthrough for molecular markers in that for the first time, any genomic region could be amplified and analyzed in many individuals without the requirement for cloning and isolating large amounts of ultra-pure genomic DNA (Schlötterer, 2004). PCR sequencing involves determination of the nucleotide sequence within a DNA fragment amplified by the PCR, using primers specific for a particular genomic site. The method that has been most commonly used to determine nucleotide sequences is based on the termination of *in vitro* DNA replication. The procedure is initiated by annealing a primer to the amplified DNA fragment, followed by dividing the mixture into four subsamples. When a dideoxynucleotide is incorporated, DNA replication is terminated. Because each reaction contains many DNA molecules and incorporation of dideoxynucleotides occurs at random, each of the four subsamples contains fragments of varying length terminated at any occurrence of the particular dideoxy base used in the subsample. Finally, the fragments in each of the four subsamples are separated by gel electrophoresis.

Advantages: PCR sequencing provides the ultimate measurement of genetic variation. Universal primer pairs to target specific sequences in a wide range of species are available for the chloroplast, mitochondrial and ribosomal genomes. Highly reproducible and increased chances of detecting truly homologous differences. Low quantities of template DNA are required, *e.g.* 10–100 ng per reaction. Moreover, most of the technical procedures are amenable to automation.

Disadvantages: Low genome coverage and low levels of variation below the species level. In the event that primers for a genomic region of interest are unavailable, high development costs are involved. Analytical procedures are laborious and technically demanding.

Applications: PCR sequencing is most useful to address questions of interspecific and intergeneric relationships (Sanger *et al.*, 1977; Clegg, 1993). Until recently, chloroplast DNA and nuclear ribosomal DNA have provided the major datasets for phylogenetic inference because of the ease of obtaining data due to high copy number. Recently, single- to low-copy nuclear DNA markers have been developed as powerful new tools for phylogenetic analyses (Mort and Crawford, 2004; Small *et al.*, 2004).

6.0 RETROTRANSPOSON-BASED MARKERS

Retrotransposons consist of long terminal repeats (LTR) with a highly conserved terminus, which is exploited for primer design in the development of retrotransposon-based markers. Several of these elements have been sequenced and were found to display a high degree of heterogeneity and insertional polymorphism, both within and between species.

Applications: Because retrotransposon insertions are irreversible (Minghetti and Dugaiczyk, 1993; Shimamura *et al.*, 1997), they are considered particularly useful in phylogenetic studies. In addition, they are widespread throughout the genome of both plants and animals. Therefore it can be exploited in gene mapping studies, and they are frequently observed in regions adjacent to known plant genes. Several variations of retrotransposon-based markers exist *e.g.* S-SAP, IRAP, REMAP etc.

6.1 Sequence-Specific Amplified Polymorphism (S-SAP)

It is a dominant, multiplex marker system for the detection of variation in DNA flanking the retrotransposon insertion site. Retrotransposon containing fragments are amplified by PCR, using one primer designed from the conserved terminus of the LTR and one based on the presence of a nearby restriction endonuclease site. Experimental procedures resemble those used for AFLP analysis and they are usually dominant markers. Compared to AFLP, S-SAP generally yields fewer fragments but higher levels of polymorphism (Waugh *et al.*, 1997).

6.2 Retrotransposon-Microsatellite Amplified Polymorphism (REMAP)

The DNA sequences between the LTRs and adjacent microsatellites (SSRs) are amplified using appropriate primers. Any difference in DNA sequence between two genomes, detected by polymerase chain reaction-mediated amplification of the region between a long terminal repeat (LTR) of a retrotransposon and a nearby microsatellite (Kahl, 2001). They are dominant, multiplex marker systems that examine variation in retrotransposon insertion sites.

6.3 Inter-Retrotransposon Amplified Polymorphism (IRAP)

IRAP markers are generated by the proximity of two retrotransposons using outward facing primers annealing to their long terminal repeats (LTRs). They are dominant, multiplex marker systems that examine variation in retrotransposon insertion sites. IRAP as well as REMAP fragments can be separated by high-resolution agarose gel electrophoresis (Kalendar *et al.*, 1999).

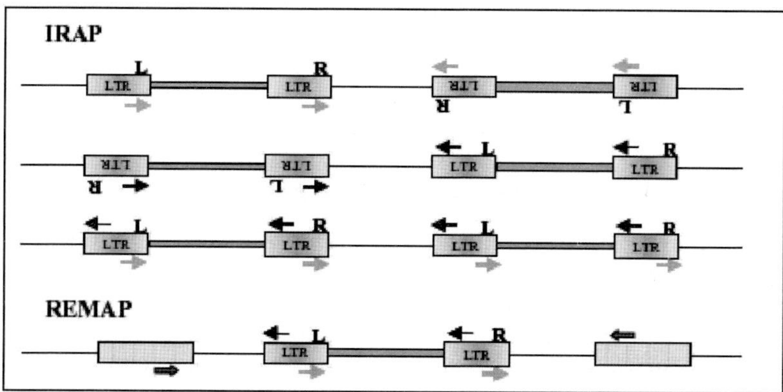

Fig. 16. Principle of the IRAP und REMAP strategy. **IRAP:** PCR primers facing outward from the 5′ (black arrows) and 3′ (gray arrows) ends of LTRs will amplify intervening DNA from the retrotransposon in any of the three possible orientations (tail-to-tail, head-to-head, head-to-tail). **REMAP:** LTR primers are used together with a primer consisting of simple sequence repeats (blank boxes) (Source: Kalendar *et al.*, 1999)

Advantages: Both REMAP and IRAP are highly polymorphic, species-specific, depends on the transposon, robust in usage and can be automated.

Disadvantages: In REMAP and IRAP the alleles cannot be detected and it can be technically challenging.

The principle of IRAP und REMAP is shown in Figure 16 below:

6.4 Retrotransposon-Based Insertional Polymorphism (RBIP)

It is a co-dominant marker system that uses PCR primers designed from the retrotransposon and its flanking DNA to examine insertional polymorphisms for individual retrotransposons. Presence or absence of insertion is investigated by two PCRs, the first using one primer from the retrotransposon and one from the flanking DNA, the second using primers designed from both flanking regions. Polymorphisms are detected by simple agarose gel-electrophoresis or by dot hybridization assays.

Advantage: It does not necessarily require a gel-based detection system but can easily be adapted to automated and gel-free procedures such as TaqMan or DNA chip technology in order to increase sample throughput (Flavell *et al.*, 1998).

Disadvantage: Drawback of the method is that sequence data of the flanking regions are required for primer design.

7.0 FUNCTIONAL MARKERS

The availability of sequence data of expressed DNA has enabled the development of markers that are physically associated with coding regions of the genome.

7.1 Expressed Sequence Tag (ETS) markers

These markers are similar to STS markers in principle but the EST markers represent only expressed or transcribed sequences. An EST is simply a segment of sequence from a cDNA clone that corresponds to a mRNA and the information generated is generally stored in databases. EST markers are generated by partial sequencing of random cDNA clones and designing primers for their polymerase chain reaction amplification or as a source of bands for CAPS markers. Alternatively, EST markers can also be developed by utilizing protein sequence database and genetic code.

Advantages: Fast, cDNA sequences, non-radioactive, small DNA quantities required, highly reliable, usually single-specific and it can be automated.

Disadvantages: Sequence information required and substantially decreased levels of polymorphism.

Applications: Once generated, they are useful in cloning specific genes of interest, synteny mapping of functional genes in various related organisms and tissue specific gene expression. ESTs allow for more information to be gained from QTL mapping.

7.2 EST-SSR and EST-SNP

The raw sequence information will also aid in screening for the occurrence of microsatellite sequences (EST-SSR) or single nucleotide polymorphisms (EST-SNP), after which markers can be developed that are targeted to transcribed regions of the genome. EST-SSRs have been applied successfully in the characterization of accessions of wheat (Eujayl *et al.*, 2002) and barley (Thiel *et al.*, 2003). EST-SNPs have been used to study functional diversity in maize (Rafalski, 2002).

7.3 cDNA-AFLP

cDNA-AFLP is commonly used in the identification of genetic polymorphisms between contrasting phenotypes under controlled conditions in order to facilitate the construction of linkage maps (*e.g.* Brugmans *et al.*, 2002), or to identify candidate genes. In diversity studies, the application of cDNA-AFLP should be limited to the identification and comparison of specific gene-related patterns, as in general cDNA differences may be caused by differences in the developmental stage of plants and the environmental conditions rather than by existing DNA polymorphisms.

7.4 Sequence-Related Amplified Polymorphism (SRAP)

A simple PCR-based marker technique that targets coding sequences in the genome is Sequence-Related Amplified Polymorphism (SRAP). SRAP uses forward primers consisting of an unspecific filler sequence of ten bases, the sequence CCGG and three selective nucleotides. Reverse primers also contain a filler sequence, but are followed by the sequence AATT and three selective nucleotides. The CCGG sequence is used to target GC-rich regions, such as exons in open reading frames, while the AATT sequence on the reverse primers is aimed at AT-rich regions, such as promoters and introns. The generally conserved nature of exon sequences, combined with the generally variable nature of introns, promoters and spacers, enables SRAP analysis to generate polymorphic bands. The use of selective nucleotides on the PCR primers results in the amplification of subsets of open reading frames that display multilocus band profiles following appropriate labeling of a primer, polyacrylamide gel electrophoresis and autoradiography. In *Brassica oleracea* L. sequence analysis revealed that 45% of the SRAP fragments could be matched with known genes. Furthermore, SRAP analysis could easily be applied successfully to other crops, such as potato, rice, lettuce and apple (Li and Quiros, 2001).

7.5 Target Region Amplification Polymorphism (TRAP)

A related technique that uses EST sequence information is Target Region Amplification Polymorphism (TRAP). For TRAP analysis, a fixed primer designed from a targeted EST sequence is combined with an arbitrary primer having an AT- or CG-rich core sequence. For different plant species TRAP revealed multiple scorable fragments, and the technique may be well suited for determining the genotypes of germplasm and tagging genes for traits of interest (Hu and Vick, 2003).

EST-based, cDNA-based and SRAP markers are common in that they target diversity in coding regions.

8.0 RESISTANCE BASED MARKERS

The finding that pathogen resistance genes share common sequence motifs, such as the nucleotide binding site (NBS) and leucine-rich repeat regions (LRR), have resulted in the development of molecular markers derived from these conserved regions. Appealing feature of such marker systems is that these are not restricted to known genes, but in principle are able to detect variation in novel resistance genes and analogs as well. Marker development may be limited to the gene sequences (Chen *et al.*, 1998;

Sicard *et al.*, 1999) or may be extended to the flanking DNA, *e.g.* by transformation into an SSAP like assay. This latter approach is provisionally termed NBS-directed profiling.

8.1 Resistance Gene Homologue Polymorphism (RGHP)

In this methodology, groups of resistance genes are targeted by PCR using primers aimed at conserved domains of resistance genes, such as the Leucine Rich Repeat (LRR) or the Nucleotide Binding Site (NBS), both involved in resistance mechanisms (Chen *et al.*, 1998; Sicard *et al.*, 1999). In NBS-directed profiling (NBS-DP), one primer is targeted to a conserved sequence of the NBS, while the other primer is based on the presence of a nearby endonuclease restriction site. Polyacrylamide gel electrophoresis and autoradiography are part of the NBS-DP technique, resulting in AFLP-like banding profiles that are scored as dominant markers.

Applications: The highly conserved nature of these targets allows the NBS-DP primers to be used beyond the species level. Because resistance genes have often originated from gene duplications, variation may also be traced in analogs of the gene.

9.1 DEVELOPMENTS IN DETECTION TECHNIQUES

TaqMan™: TaqMan™ is a probe used to detect specific sequences in PCR products by employing the 5′-3′ exonuclease activity of the Taq DNA polymerase. The TaqMan™ probe (20–30 bp), disabled from extension at the 3′ end, consists of a site-specific sequence labeled with a fluorescent reporter dye and a fluorescent quencher dye. During PCR the TaqMan™ probe hybridizes to its complementary single strand DNA sequence within the PCR target. When amplification occurs, the TaqMan™ probe is degraded due to the 5′-3′ exonuclease activity of Taq DNA polymerase, thereby separating the quencher from the reporter during extension. Due to the release of the quenching effect on the reporter, the fluorescence intensity of the light emission increases exponentially, the final level being measured by spectrophotometry after termination of the PCR. The TaqMan™ assay allows high sample throughput because no gel electrophoresis is required. It is also referred to as fluorogenic 5′ nuclease assay (Holland *et al.*, 1991; Lee *et al.*, 1993).

Applications: It offers a sensitive method to determine the presence or absence of specific sequences. Therefore, this technique is particularly useful in diagnostic applications, such as the screening of samples for the presence or incorporation of favourable traits, the detection of pathogens and diseases in plants and the screening of plant material for the presence of transgenic elements. When different probes are used which are able to discriminate between allelic variants, TaqMan™ behaves as a co-dominant marker.

9.2 Automated Sequencers

DNA sequencing was initially carried out through radiolabeling, polyacrylamide gel electrophoresis and visual reading of gels. Throughput of DNA samples has increased substantially by the use of automated sequencers and automated data processing. The high throughput capacity of these machines is achieved by the improved mechanical operation and detection systems, but also by allowing multiplexing of PCR reactions. Collected electronic data may then be processed with dedicated software packages.

Applications: Apart from sequence analyses, automated sequencers are also being used for the analysis of microsatellites, AFLPs and SNPs.

9.3 Microarray or DNA Chip Technology

Microarray or DNA chip technology is a high throughput screening technique based on the hybridization between oligonucleotide probes (genomic DNA or cDNA) and either DNA or mRNA. Chips may consist of arrays of amplified DNA immobilized on miniature glass or nylon substrates that are then tested by hybridization to a series of fluorescently labeled probes. More commonly however, arrays of oligonucleotide probes are synthesized (*e.g.* Affymetrix GeneChips), followed by the exposure to fluorescently labelled PCR samples. Hybridization signals are determined by laser technology from which data on sequence variation is obtained.

Advantages: Single base changes, highly abundant, highly polymorphic, co-dominant, small DNA quantities required, highly reliable, usually single locus, species-specific, suitable for high throughput, no gel system and it can be automated.

Disadvantages: Very high development and start-up costs, portability unknown.

Applications: Diagnostics, mutation and polymorphism detection, gene discovery, gene expression and gene mapping (Lemieux *et al.*, 1998; Ramsay 1998; Gibson, 2002). Wide application of DNA chips, such as the detection of SNPs between genotypes, has not yet found its way to genebanks, particularly because of the laborious sequencing methods and high costs involved.

9.4 Diversity Array Technology (DArT)

Recent development in the application of DNA chips to the analysis of genetic polymorphisms is the Diversity Array Technology (DArT) introduced by Jaccoud *et al.* (2001). First, a so-called 'diversity panel' is created consisting of arrays of DNA fragments originating from a group of diverse genotypes.

To construct a diversity panel, total genomic DNA is isolated from a group of genotypes, after which the DNA is pooled and digested by restriction enzymes. Enzyme-specific adapters are then ligated to the fragments, followed by reduction of the genomic complexity through PCR using primers with selective nucleotides. Subsequently, the DNA fragments are molecularly cloned, the inserts amplified by PCR and the amplified fragments (amplicons) arrayed onto DNA chips.

Applications: DArT does not require any sequence data and is considered an economical, high-throughput, robust marker detection technology with a high genome coverage that may have potential relevance to gene banks in germplasm characterization (Jaccoud *et al.*, 2001). The data are similar to AFLPs.

9.5 Pyrosequencing™

Pyrosequencing™ refers to sequencing by synthesis, a simple tool for accurate analysis of DNA sequences. It is a novel sequencing method of relatively short DNA templates based on real-time (quantitative) pyrophosphate release. A DNA fragment consisting of a sequencing primer hybridized to a single stranded DNA template is incubated with the enzymes DNA polymerase, ATP sulfurylase, firefly luciferase and a nucleotide degrading enzyme. The deoxynucleotides dATP, dCTP, dGTP and dTTP are added sequentially in an iterative manner. In case of complementarity to the base of the DNA template, each time the DNA polymerase incorporates a nucleotide to the new DNA strand, pyrophosphate is released in equal molarity to that of the incorporated deoxynucleotide. Pyrophosphate is then used as a substrate for the enzyme ATP sulfurylase converting pyrophosphate into ATP.

Subsequently, the concentration of ATP is detected by the enzyme luciferase as a visible and measurable real-time light signal. Between each addition of deoxynucleotides, unincorporated nucleotides and ATP are degraded by the nucleotide-degrading enzyme.

Applications: Sequences of up to 20 bases, such as those in which SNPs are found, can be accurately determined by pyrosequencing™. It can be successfully used for SNP and insertion/deletions detection, genotype identification and characterization, and in general for applications focusing on the variation of short DNA fragments. It can be used for accurate quantification of ratios of variant bases in a DNA sample.

10.0 CONCLUSION

High through-put, low cost, highly reproducible marker technology provides a powerful tool that can be applied to biological research, including that related to biodiversity and crop improvement. The most important issues in such studies are the biological questions to be addressed using whatever tools are most appropriate to provide timely and cost-effective answers. The marker system which offers comparative advantages relative to other possible approaches for addressing specific biological questions of interest can be chosen. One major limitation of DNA markers such as RFLPs is their high cost. However, PCR-based markers make DNA marker technology more efficient and cost effective by striking a balance between cost and information provided. Although biotechnology is becoming increasingly important in agriculture, the fact that over 50% of the agricultural productivity in the world has been achieved through traditional plant breeding hence, it should not be ignored. While DNA marker technology cannot replace conventional methods, it will certainly augment the efforts of plant breeders by providing new tools to ease many problems faced by breeders.

11.0 REFERENCES

Akopyanz, N., Bukanov, N., Westblom, T.U. and Berg, D.E. 1992. PCR-based RFLP analysis of DNA sequence diversity in the gastric pathogen *Helicobacter pylori*. Nucleic Acids Research, 20:6221–6225.

Alonso-Blanco, C., Peeters, A.J., Koornneef, M., Lister, C., Dean, C., van den Bosch, N., Pot, J. and Kuiper, M.T. 1998. Development of an AFLP based linkage map of Ler, Col and Cvi *Arabidopsis thaliana* ecotypes and construction of a Ler/Cvi recombinant inbred line population. Plant Journal, 14:259–271.

Berry, A. and Kreitman, M. 1993. Molecular analysis of an allozyme cline: alcohol l dehydrogenase in *Drosophila meloganaster* on the east coast of North America. Genetics, 134:869–893.

Botstein, D., White, R.L., Skolnick, M. and Davis, R.W. 1980. Construction of a genetic map in man using restriction fragment length polymorphisms. American Journal of Human Genetics, 32:314–331.

Brookes, A.J. 1999. The essence of SNPs. Gene, 234: 177-186.

Brubaker, C.L. and Wendel, J.F. 1994. Reevaluating the origin of domesticated cotton (*Gossypium hirsutum*; Malvaceae) using nuclear restriction fragment length polymorphisms (RFLPs). American Journal of Botany, 81:1309–1326.

Brugmans, B., Fernandez del Carmen, A., Bachem, C.W.B., van Os, H., van Eck, H.J. and Visser, R.G.F. 2002. A novel method for the construction of genome wide transcriptome maps. Plant Journal, 31:211–222.

Caetano-Anolles, G., Bassam, B.J. and Gresshoff, P.M. 1991. DNA amplification fingerprinting using very short arbitrary oligonucleotide primers. Bio-Technology, 9:292-305.

Chen, X.M., Line, R.F. and Leung, H. 1998. Genome scanning for resistance-gene analogs in rice, barley, and wheat by high-resolution electrophoresis. Theoretical and Applied Genetics, 97:345–355.

Clausen, A.M. and Spooner, D.M. 1998. Molecular support for the hybrid origin of the wild potato species *Solanum* × *rechei*. Crop Science, 38:858–865.

Clegg, M.T. 1993a. Chloroplast gene sequences and the study of plant evolution. Proceedings of the National Academy Sciences USA, 90:363–367.

Cooper, D.N., Smith, B.A., Cooke, H.J., Niemann, S. and Schmidtke, J. 1985. An estimate of unique DNA sequence heterozygosity in the human genome. Human Genetics, 69(3):201-205.

Cregan, P.B., Jarvik, T., Bush, A.L., Shoemaker, R.C., Lark, K.G., Kahler, A.L., Kaya, N., VanToai, T.T., Lohnes, D.G., Chung, J. and Specht, J.E. 1999 An integrated linkage map of soybean genome. Crop Science, 39, 1464-1490.

Day, I.N., Spanakis, E., Palamand, D., Weavind, G.P. and O'Dell, S.D. 1998. Microplatearrays diagonal-gel electrophoresis (DADGE) and melt-MADGE: Tool for molecular genetic epidemiology. Trends in Biotechnology, 16:187-290.

Desplanque, B., Boudry, P., Broomberg, K., Saumitou-Laprade, P., Cuguen, J. and van Dijk, H. 1999. Genetic diversity and gene flow between wild, cultivated and weedy forms of *Beta vulgaris* L. (Chenopodiaceae), assessed by RFLP and microsatellite markers. Theoretical and Applied Genetics, 98:1194–1201.

Devos, K.M. and M.D. Gale, 1992. The use of random amplified polymorphic DNA markers in wheat. Theoretical and Applied Genetics, 84:567-572.

Erskine, W. and Muehlbauer, F.J. 1991. Allozyme and morphological variability, outcrossing rate and core collection formation in lentil germplasm. Theoretical and Applied Genetics, 83:119–125.

Eujayl, I., Sorrells, M.E., Baum, M., Wolters, P. and Powell, W. 2002. Isolation of EST-derived microsatellite markers for genotyping the A and B genomes of wheat. Theoretical and Applied Genetics, 104:399–407.

Fang, D.Q., Roose, M.L., Krueger, R.R. and Federici, C.T. 1997. Fingerprinting trifoliate orange germplasm accessions with isozymes, RFLPs, and inter-simple sequence repeat markers. Theoretical and Applied Genetics, 95:211–219.

FAO/IAEA. 2002. Mutant Germplasm Characterization using Molecular Markers: A Manual. Prepared by the Joint FAO/IAEA Division of Nuclear Techniques in Food and Agriculture. Vienna, Training Course Series. 19, ISSN 1018–5518.

Flavell, A.J., Knox, M.R., Pearce, S.R. and Ellis, T.H.N. 1998. Retrotransposon-based insertion polymorphisms (RBIP) for high throughput marker analysis. Plant Journal, 16:643–650.

Freville, H., Justy, F. and Olivieri, I. 2001. Comparative allozyme and microsatellite population structure in a narrow endemic plant species, *Centaurea corymbosa* Pourret (Asteraceae). Molecular Ecology, 10:879–889.

Garvin, D.F. and Weeden, N.F. 1994. Isozyme evidence supporting a single geographic origin for domesticated tepary bean. Crop Science, 34:1390–1395.

Ghislain, M., Spooner, D.M., Rodríguez, F., Villamon, F., Núñez, C., Vásquez, C. and Bonierbale, M. 2004. Selection of highly informative and user-friendly microsatellites (SSRs) for genotyping of cultivated potato. Theoretical and Applied Genetics, 108:881–890.

Gibson, G. 2002. Microarrays in ecology and evolution: a preview. Molecular Ecology, 11:17–24.

Godwin, I.D., Aitken, E.A.B. and Smith, L.W. 1997. Application of inter simple sequence repeat (ISSR) markers to plant genetics. Electrophoresis, 18:1524–1528.

Griffin, T.J. and Smith, L.M. 2000. Single-nucleotide polymorphism analysis by MALDITOF mass spectrometry. Trends in Biotechnology, 18:77.

Gu, W.K., Waden, N.F., Yu, J. and Wallace, D.H. 1995. Large-scale,cost-effective screening of PCR products in marker-assisted selection applications. Theoretical and Applied Genetics, 91:465–470.

Gupta, M., Chyi, Y.S., Romero-Severson, J. and Owen, J.L. 1994. Amplification of DNA markers from evolutionarily diverse genomes using single primers of simple sequence repeats. Theoretical and Applied Genetics, 89:998–1006.

Gygax, M., Gianfranceschi, L., Liebhard, R., Kellerhals, M., Gessler, C. and Patocchi, A. 2004. Molecular markers linked to the apple scabe resistance gene Vbj derived from *Malus Baccata* jackii. Theoretical and Applied Genetics, 109:1702-1709.

Hacia, J.G. and Collins, F.S. 1999. Mutational analysis using oligonucleotide microarrays. Journal of Medical Genetics, 36: 730-736.

Hadrys, H., Balick, M. and Schierwater, B. 1992. Applications of random amplified polymorphic DNA (RAPD) in molecular ecology. Molecular Ecology, 1:55–63.

Hayashi, K. 1992. PCR-SSCP: A method for detection of mutations. Genetic Analysis Techniques and Applications, 9:73–79.

Heath, D.D., Iwama, G.K. and Devlin, R.H. 1993. Nucleic Acids Research, 21:5782–5785.

Hearne, C.M., Ghosh, S. and Todd, J.A. 1992. Microsatellites for linkage analysis of genetic traits. Trends in Genetics, 8:288–294.

Holland, P.M., Abramson, R.D., Watson, R. and Gelfland, D.H. 1991. Detection of specific polymerase chain reaction product by utilizing the 5'?3' exonuclease activity of *Thermus aquaticus* DNA polymerase. Proceedings of the National Academy of Sciences USA, 88:7276–7280.

Hu, J. and Vick, B.A. 2003. Target region amplification polymorphism: a novel marker technique for plant genotyping. Plant Molecular Biology Reporter, 21:289–294.

Huang, J.C. and Sun, M. 2000. Genetic diversity and relationships of sweet potato and its wild relatives in *Ipomoea sereies* Batatas, Convolvulaceae) as revealed by inter-simple sequence repeat (ISSR) and restriction analysis of chloroplast DNA. Theoretical and Applied Genetics, 100:1050-1060.

Hudson, R.R., Bailey, K., Skarecky, D., Kwaitowski, J. and Ayala, F.J. 1994. Evidence for positive selection in the superoxide dismutase (Sod) region of *Drosophila melanogaster*. Genetics, 136:1329–1340.

Iqbal, J.M. 2006. Molecular markers: Applications in forest breeding. The Institute for advanced learning and research. Sustainable Engineered Materials Institute, 230 Cheatham Hall (0323), Blacksburg, VA 24061. (http://www.semi.vt.edu).

Jaccoud, D., Peng, K., Feinstein, D. and Kilian, A. 2001. Diversity arrays: A solid state technology for sequence information independent genotyping. Nucleic Acids Research, 29:25.

Jarne, P. and Lagoda, P.J.L. 1996. Microsatellites, from molecules to populations and back. Trends Ecology and Evolution, 11:424–429.

Jeffreys, A.J., Wilson, V. and Thein, S.L. 1985a. Hypervariable "minisatellite" regions in human DNA. Nature, 314:67–73.

Jeffreys, A.J., Wilson, V. and Thein, S.L. 1985b. Individual-specific "fingerprints" of human DNA. Nature, 316:76–79.

Jones, C.J., Edwards, K.J., Castaglione, S., Winfield, M.O., Sala, F., van de Wiel, C., Bredemeijer, G., Vosman, B., Matthes, M., Daly, A., Brettschneider, R., Bettini, P., Buiatti, M., Maestri, E., Malcevschi, A., Marmiroli, N., Aert, R., Volckaert, G., Rueda, J., Linacero, R., Vazquez, A. and Karp, A. 1997. Reproducibility testing of RAPD, AFLP and SSR markers in plants by a network of European laboratories. Molecular Breeding, 3:381–390.

Kahl, G. 2001. The Dictionary of Gene Technology. Wiley-VCH.

Kalendar, R., Grob, T., Regina, M., Suoniemi, A. and Schulman, A. 1999. IRAP and REMAP: Two new retrotransposon-based DNA fingerprinting techniques. Theoretical and Applied Genetics, 98:704–711.

Kantely, R.V., Zeng, X.P., Bennetzen, J.L. and Zehr, B.E. 1995. Assesment of genetic diversity in dent and popcorn (*Zea Mays* L.) inbred lines using inter-simple sequences. Molecular Breeding, 1:365-373.

Kephart, S.R. 1990. Starch gel electrophoresis of plant isozymes: A comparative analysis of techniques. American Journal of Botany, 77:693–712.

Konieczny, A. and Ausubel, F.M. 1993. A procedure for mapping *Arabidopsis* mutations using co-dominant ecotype-specific PCR-based markers. Plant Journal, 4:403–410.

Kreiger, M. and Ross, K.G. 2002. Identification of a major gene regulating complex social behavior. Science, 295:328–332.

Lamboy, W.F., McFerson, J.R., Westman, A.L. and Kresovich, S. 1994. Application of isozyme data to the management of the United States national *Brassica oleracea* L. genetic resources collection. Genetic Resources and Crop Evolution, 41:99–108.

Lanner, H.C., Bryngelsson, T. and Gustafsson, M. 1997. Relationships of wild *Brassica* species with chromosome number 2n = 18, based on RFLP studies. Genome, 40:302–308.

Lee, L.G., Connell, C.R. and Bloch, W. 1993. Allelic discrimination by nick-translation PCR with fluorogenic probes. Nucleic Acids Research, 21:3761–3766.

Lemieux, B., Aharoni, A. and Schena, M. 1998. Overview of DNA chip technology. Molecular Breeding, 4:277–289.

Li, G. and Quiros, C.F. 2001. Sequence-related amplified polymorphism (SRAP), a new marker system based on a simple PCR reaction: its application to mapping and gene tagging in *Brassica*. Theoretical and Applied Genetics, 103:455–461.

Love, J.M., Knight, A.M., McAleer, M.A. and Todd, J.A. 1990. Towards construction of a high resolution map of the mouse genome using PCR-analyzed microsatellites. Nucleic Acid Research, 18:4123-4130.

Maass, B.L. and Ocampo, C.H. 1995. Isozyme polymorphism provides fingerprints for germplasm of *Arachis glabrata* Bentham. Genetic Resources and Crop Evolution, 42:77–82.

Mather, D. 2006. Molecular Plant Breeding CRC, Australian Centre for Plant Functional Genomics PTY LTD. (http://www.molecularplantbreeding.com)

May B. 1992. Starch gel electrophoresis of allozymes. In Molecular genetic analysis of populations: a practical approach (AR Hoelzel, ed.). Oxford University Press, Oxford, UK. pp. 1–27.

McClean, P.E. 1998. Plant Molecular Genetics. PLSC 731. (http://www.ndsu.nodak.edu).

Miller, J.C. and Tanksley, S.D. 1990. RFLP analysis of phylogenetic relationships and genetic variation in the genus *Lycopersicon*. Theoretical and Applied Genetics, 80:437–448.

Minghetti, P.P. and Dugaiczyk, A. 1993. The emergence of new DNA repeats and the divergence of primates. Proceedings of the National Academy of Sciences USA, 90:1872–1876.

Morgante, M. and Olivieri, A.M. 1993. PCR-amplified microsatellites as markers in plant genetics. Plant Journal, 3:175–182.

Morgante, M., Hanafey, H. and Powell, W. 2002. Microsatellites are preferentially associated with nonrepetitive DNA in plant genome. Nature Genetics, 30:194–200.

Morgante, M. and Vogel, J.M. 1996. Compound microsatellite primers for the detection of genetic polymorphism. World Patent organization, WO, 96/17082.

Mort, M.E. and Crawford, D.J. 2004. The continuing search: low-copy nuclear sequences for lower level plant molecular phylogenetic studies. Taxon, 53:257–261.

Murray, J.C., Buetow, K.H., Weber, J.L., Ludwingsen, S., Scherpbier-Heddema, T., Manion, F., Quillen, J., Sheffield, V.C., Sunden, S., Duyk, G.M., Weissenback, J., Gyapay, G., Dib, C., Morrissette, J., Lathrop, G.M., Vignal, A., White, R., Matsunami, N., Gerken, S., Melis, R., Albertseb, H., Plaetke, R., Odelberg, S., Ward, D., Dausset, J., Cohen, D. and Cann, H. 1994. A comprehensive human linkage map with cetimorgan density. Science, 265:2049-2054.

Neale, D.B. and Williams, C.G. 1991. Restriction fragment length polymorphism mapping in conifers and applications to forest genetics and tree improvement. Canadian Journal of Forest Research, 21:545–554.

Paran, I. and Michelmore, R.W. 1993. Development of reliable PCR based markers linked to downy mildew resistance genes in lettuce. Theoretical and Applied Genetics 85:985–993.

Parani, M., Singh, K.N., Rangasamy, S. and Ramalingam, R.S. 1997. Identification of *Sesamum alatum* × *Sesamum indicum* hybrid using protein, isozyme and RAPD markers. Indian Journal of Genetics and Plant Breeding, 57:381–388.

Powell, W., Machray, G. C. and Provan, J. 1996. Polymorphism revealed by simple sequence repeats. Trends in Plant Sciences, 1(7): 215-222.

Queller, D.C., Strassmann, J.E. and Hughes, C.R. 1993. Microsatellites and kinship. Trends in Ecology and Evolution, 8:285–288.

Rabeena, R., Ford, P. and Taylor, P.W.J. 2003. Construction of an intra-specific linkage map of lentil (*Lens culinaris* spp.culinaris). Theoretical and Applied Genetics, 107:910-916.

Rafalski, J.A. and Tingey, S.V. 1993 Genetic diagnostics in plant breeding: RAPDs, microsatellites and machines. Trends in Genetics, 9:275-280.

Rafalski, J.A. 2002. Novel genetic mapping tools in plants: SNPs and LD-based approaches. Plant Science, 162:329–333.

Ramsay G. 1998. DNA chips: State-of-the art. Nature Biotechnology, 16:40–44.

Rediscovering Biology (1997). Online text book, Unit 4, Microbial Diversity. (http://www.learner.org)

Reedy, M.E., Knapp, A.D. and Lamkey, K.R. 1995. Isozyme allelic frequency changes following maize (*Zea mays* L.) germplasm regeneration. Maydica, 40:269– 273.

Ren, J. 2001. High-throughput single-strand conformation polymorphism analysis by capilary electrophoresis. Journal of Chromatography B. Biomedical Sciences and Applications, 74:115-128.

Richardson, T., Cato, S., Ramser, J., Kahl, G. and Weising, K. 1995. Hybridization of microsatellites to RAPD: a new source of polymorphic markers. Nucleic Acids Research, 23:3798–3799.

Ronning, C.M. and Schnell, R.J.S. 1994. Allozyme diversity in a germplasm collection of *Theobroma cacao* L. Journal of Hereditity, 85:291–295.

Saal, B. and Wricke, G. 2002. Clustering of amplified fragment length polymorphism markers in a linkage map of rye. Plant Breeding, 121:117–123.

Sabatini, E., Rotino, G.L. and Acciarri, N. 2004. Development of A CAPS marker linked to the *ovate* gene in tomato (*L. esculentum* M.) and its possible use for the 'pear' shape marker assisted selection. Proceeding of the XLVIII Italian Society of Agricultural Genetics – SIFV – SIGA Joint Meeting Leoce Iitaly, 15-18 September 2004.

Sanger, F., Nicklen, S. and Coulson, A.R. 1977. DNA sequencing with chainterminating inhibitors. Proceedings of the National Academy of Sciences USA, 74:5463–5467.

Sangitha (2006) (http://www.slideshare.net).

Schierwater, B. and Ender, A. 1993. Different thermostable DNA polymerases may apply to different RAPD products. Nucleic Acids Research, 21:4647–4648.

Schlötterer, C. 2004. The evolution of molecular markers—just a matter of fashion? Nature Review Genetics, 5:63–69.

Shimamura, M., Yasue, H., Oshima, K., Abe, H., Kato, H., Kishiro, T., Goto, M., Munechika, I. and Okada, N. 1997. Molecular evidence from retrotransposons that whales form a clade within even-toed ungulates. Nature, 388:666–670.

Sicard, D., Woo, S.S., Arroyo-Garcia, R., Ochoa, O., Nguyen, D., Korol, A., Nevo, E. and Michelmore, R. 1999. Molecular diversity at the major cluster of disease resistance genes in cultivated and wild *Lactuca* spp. Theoretical and Applied Genetics, 99:405–418.

Singh, R.K., Bhatia,V.S., Singh, N.K. and Chauhan, G.S. 2006. Genetic diversity of photoperiod insensitive and sensitive genotypes of soybean as determined by AFLP. Crop Production in Stress Management: Genetic and Management Options,pp. 276-283.

Small, R.L., Cronn, R.C. and Wendel, J.F. 2004. Use of nuclear genes for phylogeny reconstruction in plants. Australian Systematic Botany, 17:145–170.

Somers, D. J., Zhou, Z., Bebeli, P. and Gustafson, J.P. 1996. Theoretical and Applied Genetics, 93:983–989.

Syvänen, A.C. 2001. Assessing genetic variation: genotyping SNPs. Nature Review Genetics, 12:930-942.

Taillon-Miller, P., Gu, Z., Li, Q., Hillier, L. and Kwok, P.Y. 1998. Overlapping genomic sequences: A treasure trove of single-nucleotide polymorphisms. Genome Research, 8 (7):748-754.

Tao, R. and Sugiura, A. 1987. Cultivar identification of Japanese persimmon by leaf isozymes. Hort Science, 22:932–935.

Thiel, T., Michalek, W., Varshney, R.K. and Graner, A. 2003. Exploiting EST databases for the development and characterization of gene-derived SSR-markers in barley (*Hordeum vulgare* L.). Theoretical and Applied Genetics, 106:411–422.

Tsai, C.J. 2007. DNA Markers & Genome Mapping, FW5085, Slide Show. (http://forest.mtu.edu/faculty/former/tsai/FW5085.html).

Underhill, P.A., Jin, L., Zemans, R., Oefner, P. J. and Cavalli-Sforza, L.L. 1996. A pre-Columbian Y chromosome-specific transition and its implications for human evolutionary history. Proceedings of the National Academy of Sciences USA, 93:196-200.

Vos, P., Hogers, R., Bleeker, M., Reijans, M., van de Lee, T., Hornes, M., Frijters, A., Pot, J., Peleman, J., Kuiper, M. and Zabeau, M. 1995. AFLP: a new technique for DNA fingerprinting. Nucleic Acids Research, 23:4407–4414.

Wang, Z., Weber, J.L., Zhong, G. and Tanksley, S.D. 1994. Survey of plant short tandem DNA repeats. Theoretical and Applied Genetics, 88:1–6.

Warnke, S.E., Douches, D.S. and Branham, B.E. 1998. Isozyme analysis supports allotetrapoloid inheritance in tetraploid creeping bluegrass (*Agrostis palustris* Huds.). Crop Science, 38:801–805.

Waugh, R., McLean, K., Flavell, A.J., Pearce, S.R., Kumar, A., Thomas, B.B.T. and Powell, W. 1997. Genetic distribution of Bare-1-like retrotransposable elements in the barley genome revealed by sequence-specific amplification polymorphisms (S-SAP). Molecular Genetics and Genomics, 253:687–694.

Welsh and McClelland, M. 1990 Fingerprinting genomes using PCR with arbitrary primers. Nucleic Acid Research, 18:7213-7218.

Williams, J.G.K., Kubelik, A.R., Rafalski, J.A. and Tingey, S.V. 1990. DNA polymorphism amplified by arbitrary primers are useful as genetic markers. Nucleic Acid Research, 18:6531-6535.

Wolfe, A.D. and Liston, A. 1998. Contributions of PCR-based methods to plant systematic and evolutionary biology. *In* Plant molecular systematics II (DE Soltis, PS Soltis and JJ Doyle, *eds.*). Kluwer Academic Publishers, Dordrecht, The Netherlands. pp. 43–86.

Wolff, K., Rogstad, S.H. and Schaal, B.A. 1994. Population and species variation of minisatellite DNA in *Plantago*. Theoretical and Applied Genetics, 87:733–740.

Wen-Jie, X., Bing-Wu, W. and Ke-Ming, C. 2004. RAPD and SCAR markers linked to sex determination in *Eucommia ulmoides* Olive. Euphytica, 136:233-238.

Yang, W.P., De Oliveira, A.C., Godwin, I., Schertz, K. and Bennetzen, J.L. 1996. Comparison of DNA marker technologies in characterizing plant genome diversity: variability in Chinese sorghums. Crop Science, 36:1669–1676.

Young, W.P., Schupp, J.M. and Keim, P. 1999. DNA methylation and AFLP marker distribution in the soybean genome. Theoretical and Applied Genetics, 99:785–792.

Zehdi, S., Sakka, H., Rhouma, A., Ould, Mohamed Salem, A., Marrakchi, M. and Trifi, M. 2004. Analysis of Tunisian date palm germplasm using simple sequence repeat primers. African Journal of Biotechnology, 3(4):215-219.

Zhou, Z., Bebeli, P.J., Somers, D.J. and Gustafson, J.P. 1997. Direct amplification of minisatellite-region DNA with VNTR core sequences in the genus *Oryza*. Theoretical and Applied Genetics, 95:942–949.

Zietkiewicz, E., Rafalski, A. and Labuda, D. 1994. Genome fingerprinting by simple sequence repeat (SSR)-anchored polymerase chain reaction amplification. Genomics, 20:176–183.

Molecular Plant Breeding: Principle, Method and Application
Eds : R.K. Singh, Rajesh Singh, Guoyou Ye, A. Selvi and G.P. Rao
Studium Press LLC, Texas, USA, 2009, pp. 81-99

CHAPTER

3

High Throughput Detection of PCR Products and SNPs for Molecular Breeding

GYAN P. MISHRA[1], ANIL KANT[2], SHASHI BALA SINGH[1] and R.K. SINGH[3]*

ABSTRACT

The efficiency of application of molecular markers could be improved if laboratory throughput is high and costs are low. An ideal system for high throughput testing would involve a 'homogenous PCR assay' *i.e.* one in which the process of amplification and detection are performed simultaneously. Single nucleotide polymorphism (SNP) have gained wide acceptance as genetic markers for use in linkage and association studies. Most of the methods are based on one or more of the following mechanisms like Allele specific hybridization (ASH), Flap endonuclease discrimination (FEN), Primer extension and Oligonucleotide ligation. In ASH, an allele specific probe is hybridized with target and then their hybridization is detected by various methods like DASH, dHPLC, RT-PCR, TaqMan assay, Molecular beacons, and Scorpion primers. The FEN discrimination assay is based on flap endonuclease which recognizes and cleaves 3D structures formed by hybridization of overlapping oligonucleotide probes, near the SNP site. Primer extension also known as mini-sequencing forms the basis of a number of methods for allelic discrimination. Oligonucleotide ligation chemistry is based on DNA ligase enzyme which closes down the nick in the double stranded DNA molecule by formation of phosphodiester bond between two adjacent nucleotides. It can also be suitably modified to develop multiplex systems. In this chapter we have discussed in detail the chemistry most suitable for high throughput genotyping and methods based on them.

Key Words: TaqMan assay, Molecular beacons, Flap endonucleas, Scorpion primers, Pyrosequencing, SNPs.

[1]*Defence Institute of High Altitude Research, DRDO, C/o 56 APO, Leh, 901205, India*
[2]*Lecturer, Jaypee University of Information Technology, Waknaghat, Solan, HP, 173215, India*
[3]*Indian Institute of Sugarcane Research, Lucknow, UP, India*
* *Corresponding author e-mail: anilkantv@gmail.com*

1.0 INTRODUCTION

The polymerase chain reaction (PCR) technique has proved to be invaluable tool for researches in diverse areas. Almost every marker system include polymerase chain reaction of DNA as one of its steps. Currently, the most common way to evaluate the outcome of PCR amplification and to detect specific DNA fragment, is to analyze the amplified product using gel electrophoresis once thermal cycling is complete. This process takes extra time, effort and also increases the chance of laboratory equipment or reagents being contaminated. The efficiency of application of molecular markers could be improved if laboratory throughput is high and costs are low. An ideal system for high throughput testing would involve a 'homogenous PCR assay' *i.e.* one in which the process of amplification and detection are performed simultaneously or at least the gel electrophoresis step could be replaced by any simpler step. A homogenous assay would allow specific PCR products to be monitored within a closed reaction tube which would allow post PCR automation.

Single nucleotide polymorphism (SNP) have gained wide acceptance as genetic markers for use in linkage and association studies. The whole genome analysis approaches, which involve large scale SNP genotyping, are expected to transform complex disease genetics and realize the potential of pharmacogenetics. To achieve this very high throughput SNP genotyping are required. The pharmaceutical industry, biotechnology companies and academic group alike aspire to carry out an increasing volume of SNP genotyping without increasing the cost of genetic research. To meet this demand, the search is on to find a genotyping technology able to deliver accurate genotyping at very high throughput and at low cost per genotype. In addition the desired platform should be simple, reliable and flexible with respect to design and development.

Higuchi *et al.* (1992, 1993) developed the first homogenous real time detection of PCR products. They exploited the fact that florescence of the intercalater dye Ethidium Bromide increases in the presence of double stranded DNA (dsDNA). By monitoring the florescence during amplification, they could continuously follow the progress and kinetics of a PCR reaction. Despite the obvious advantage of this intercalator dye system its major disadvantage is its non specificity i.e. it can not discriminate between the PCR fragment of interest and common PCR background. In recent years many non-gel based assays have been developed. Current technologies can deliver >99% accuracy at a rate of >100K genotype per day. Whilst this is a vast improvement on what was possible few years ago, we now seek a platform able to deliver >99% accuracy, at rate of >1 million genotypes per day at relatively lower cost per genotype. At the same time, DNA usage should be around 1ng per genotype or less. In this article we discuss the chemistries most suitable for high throughput genotyping and methods based on them.

2.0 HIGH THROUGHPUT ASSAY CHEMISTRIES

There are lots of chemistries available for allelic discrimination, some more suitable for high throughput SNP genotyping than others. The assay chemistry needs to be inexpensive, readily amenable to automation and simple with respect to assay design and development. Most of the methods are based on one or more of the following mechanisms:

1. Allele specific hybridization
2. Flap endonuclease discrimination
3. Primer extension
4. Oligonucleotide ligation

2.1. Allele Specific Hybridization (ASH)

In allele specific hybridization chemistry, an allele specific probe is hybridized with target and then their hybridization is detected by various methods. Dynamic allele specific hybridization (DASH), Denaturing high performance liquid chromatography (dHPLC), Real time PCR (TaqMan assay and Molecular beacons) and Scorpion primers are all example of ASH genotyping chemistry with high throughput potential.

2.1.1 Dynamic allele specific hybridization (DASH)

DASH involves the scoring of a variant position in a DNA sequence (an SNP or an insertion\deletion polymorphism) by annealing an allele specific oligonucleotide probe to a single stranded DNA target sequence containing the variation. The duplex status of probe target hybridization is tracked as temperature is increased dynamically. The basic strategy is exemplified in Fig.1. The basic process starts with creation of a single strand DNA target. This is done by performing a PCR reaction using two primers located on either side of an SNP, creating a product which is 45-65 base pair long. One of PCR primers is labeled at its 5′ end with a biotin molecule, allowing the PCR product to be captured on a streptavidin coated micro well plate. The biotinylated strand anchored to the plate is then freed of its complementary stand by rinsing with NaOH. It is then interrogated to establish which allele of the SNP is present. This is done by hybridizing with a short (15-17 bp) oligonucleotide probe at low temperature. This probe is perfectly complementary to only one of the two SNP alleles and so it will anneal stably if it matches sequences or less stably if it is mismatched. This differential stability is exploited in the detection step. Wells containing

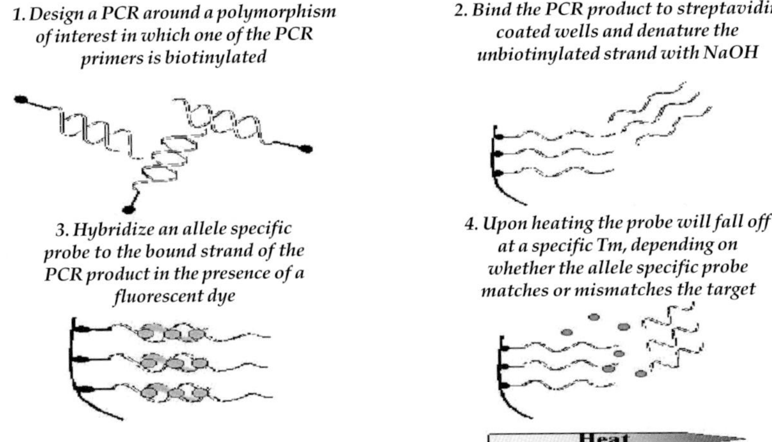

Fig. 1. Diagrammatic representation of principle of dynamic allele specific hybridization (Source : Brokes, 2002)

probe-target duplexes are heated beyond their melting temperature in presence of fluorescent dye, which binds to double standard DNA molecules. In presence of DNA duplex, the dye will fluoresce, emitting a detectable light signal that can be measured in real time. As the duplex destabilizes due to heating, it release the dye which thus ceases the fluorescing. This drop in fluorescence, which occurs earlier for mismatched structures than for matched duplexes, allows one to infer the SNP allele contained with in amplified DNA. To make observed denaturation profile more readily interpretable, the negative first derivative of the fluorescence versus time is plotted. Peaks that occur at lower temperature are scored as homozygous mismatch, peaks occurring at higher temperatures as homozygous matches and cases where both peaks are present are scored as heterozygous. Software that makes these assignments automatically is provided with commercial DASH devices. Thus DASH is a convenient method for SNP (insertion and detection) genotyping, which is highly applicable to both basic research and clinical diagnostics. Commercial DASH devices are now available, making the technology affordably accessible to all laboratories (Jonathan and Anthony, 2001). Russon *et al.* (2003) have developed a bead based DASH procedure which resulted in dramatic reduction of the working volume and is a step towards developing cost-effective high through put DASH method on array of single beads.

2.1.2 Denaturing High Performance Liquid Chromatography (dHPLC)

The presence of mutations and polymorphism can be detected with very high sensitivity and accuracy using temperature modulated heteroduplex analysis by denaturing high performance liquid chromatography (dHPLC) (Oefner and Underhill, 1998). In this method the DNA containing polymorphism or mutations is hybridized with wild type DNA which leads to formation of homoduplices or heteroduplices depending upon match or mismatch (Fig. 2). The homoduplices and heteroduplices are then can be separated by reverse phase high performance liquid chromatography in combinations with partial heat denaturation. Heteroduplices have lower denaturation temperatures and are generally retained less than homoduplices. Transgenomics Inc. (San Jose, USA) has developed the fully automated dHPLC WAVE™ system for temperature modulated heteroduplex analysis. This system has an autosampler, cartridge oven, analytical cartridge and a UV detector. The use of UV instead of fluorescence or radio isotopes for detection makes WAVE™ system extremely cost-effective. This method does not require prior knowledge of DNA sequence for detection of new SNPs and mutations. dHPLC can be used for finding new mutations, screening clones to identify candidate fragments for sequencing, evaluating fidelity of amplified fragments and diagnosing for presence of known mutations.

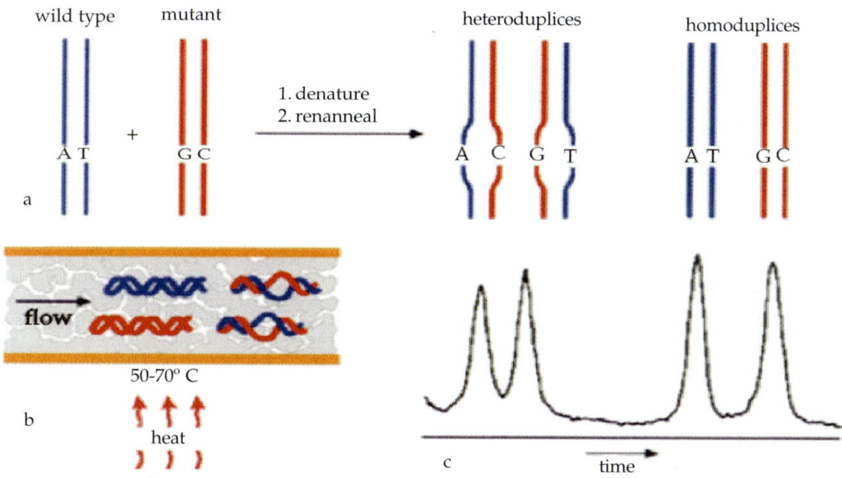

Fig. 2. Principle of mutation detection by dHPLC. ***a***, dHPLC compares typically two chromosomes as a mixture of PCR products denatured at 95°C for 3 min, and re-annealed over 30 min by gradual cooling from 95 to 65°C prior to analysis. In the presence of mismatch, not only are the original homoduplices formed, but also the sense and antisense strands of either homoduplex form heteroduplices; ***b***, Heteroduplices denature more extensively at the analysis temperature (ranges from 50 to 70°C) and are eluted earlier than the homoduplices in the DNA Sep column; ***c***, Corresponding chromatographic pattern shows four different peaks belonging to four different species of DNA (Source : Sivakumoran *et al.*, 2003)

2.1.3 Real-Time PCR (RT-PCR)

Real-Time PCR has made it possible to quantify accurately the staring amount of nucleic acid during PCR, without the need for post–PCR analysis. In real-time PCR, a fluorescent reporter is used to monitor the PCR product as it occurs (*i.e.*, in real time). The fluorescence of the reporter molecule increases as products accumulate with each successive round of amplification. The reporter can be a nonspecific intercalating double stranded DNA binding dye or a sequence specific fluorescent labeled oligonucleotide probe. The Oligonucleotide probe is labeled with both a reporter fluorescent dye and a quencher dye. While the probe is intact, the proximity of the quencher greatly reduces the fluorescence emitted by the reporter dye by Forster Resonance Energy Transfer (FRET) through space. Adequate quenching is observed for probes with the reporter at 5′ end and the quencher at 3′ end. When a probe molecule is incorporated into an amplicon, quenching is disrupted and the probe fluoresces (Dieffenbach and Dveksler, 2003).

The ability to monitor the real time progress of the PCR has completely revolutionized the approach to PCR-based quantification of DNA and RNA. It is now possible to quantify PCR products reliably by eliminating the variability associated with conventional quantitative techniques. Because reporter fluotrescence is monitored externally, the reaction tubes do not need to be opened after the PCR is complete; this prevents the aerosol contamination by PCR products and reduces the number of false positives. TaqMan assay and molecular beacons are two popular real time PCR probe study.

2.1.3.1 *TaqMan assay*

One popular real time PCR probe strategy is the TaqMan assay (Applied Biosystems). TaqMan assay detects specific sequences in target DNA is based on the 5′-3′ exonuclease activity of the *Taq* DNA polymerase which degrades a fluorescent labeled oligonucleotide TaqMan probe during polymerase chain reaction (Holland *et al.*, 1991). The TaqMan probe (20-30 bp), disabled from extension at the 3′ end consist of site specific sequence. It has two fluorescent tags attached to it. One is a reporter dye such as 6-carboxyfluorescein (FAM) at the 5′ end and another quencher dye such as 6-carboxy-tetramethyl-rhodamine (TAMRA) at the 3′ end, or any T position. When both dyes are attached to the probe, reporter dye's emission spectrum is quenched. The TaqMan probes generally have melting temperature 10°C higher than primers. TaqMan probe hybridizes to its complementary single stand target DNA sequence located between two PCR primers. When amplification occurs the TaqMan probe is degraded due to its intrinsic 5′-3′ exonuclease activity of Taq DNA polymerase, thereby separating the reporter from quencher. Due to the release of the reporter dye from quencher dye the fluorescence intensity of the reporter dye increases, which gives a detectable signal (Fig. 3). The increase in fluorescence of reporter dye is proportional to amount of PCR product formed and is measured in real time by a laser spectrophotometer which monitors the position of the 96 well microlitre plate 8 times per minute. Amplicon size should range between 50 and 150 bp and should not exceed 400 bp. Smaller amplicons give more consistent results because PCR is more efficient and more tolerant of reaction conditions.

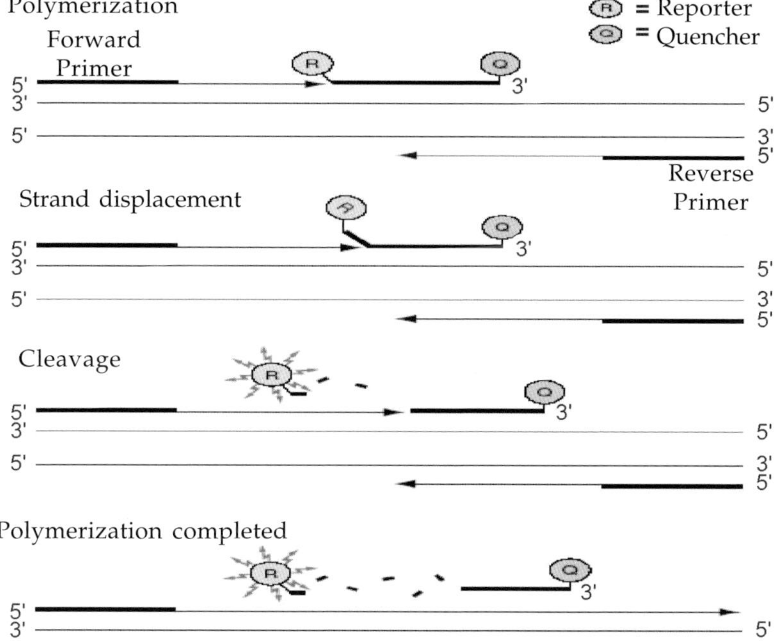

Fig. 3. TaqMan probe, labeled with a reporter dye, FAM, (R) and a quencher dye, TAMRA, (Q), bind to the DNA template. The 3' phosphate group (P) prevents extension of the TaqMan probe., Taq polymerase during extension of the primer degerades the TaqMan probeseparating reporter from quencher resulting in an increase in relative fluorescence of the reporter (Source : TaqMan, 2006)

If a TaqMan probe is being designed for allelic discrimination, it is advisable to place the mismatching nucleotide (the polymorphic site) in the middle of the probe rather than at the ends. These probes should be as short as possible to provide the greatest discrimination by the mismatch. The TaqMan assays are developed after sequencing the alleles at both loci, and by designing suitable primer and probes based on SNP or insertion/deletion polymorphism. However, when due to presence of SNP or insertion/deletion probe mismatched the target, no such degradation of probe is possible and there is no rise in fluorescence signal. Thus TaqMan procedure allows fast and highly reproducible analysis directly in PCR vials. The complete automation of the process makes this technique highly valuable for large scale screening required for marker aided selection (MAS), map-based cloning and any other diagnostic application such as screening of samples for presence or incorporation of favourable traits and detection of pathogen and diseases. Different combinations of reporters with different probes also permit multiplexing so that as many as six SNPs can be scored in a single PCR reaction (Lee *et al.*, 1999). TaqMan probes for polymorphism linked to soybean cyst nematode resistance gene, *Rhg4* has been standardized (Meksem *et al.*, 2001).

2.1.3.2 *Molecular Beacons*

Molecular beacons are allele specific single stranded oligonucleotide hybridization real time probes. The molecular beacon enable dynamic, real-time detection of nucleic acid hybridization events both *in-vivo* and *in-vitro* (Tyagi and Kramer, 1996; Kostrikis *et al.*, 1998; Tyagi *et al.*, 1998). A molecular beacon is a dual-labeled oligonucleotide (25-40 nt) probes that form stem and loop structure and become fluorescent upon target binding (Tyagi and Kramer, 1996). The loop contains a probe sequence that is complementary to a target sequence, and stem is formed by the annealing of complementary arm sequences that are located on either side of the probe sequence. A fluorophore is covalently linked to the end of one arm and a quencher is covalently linked to the end of the other arm. In the absence of targets, molecular beacons do not fluoresce, because the stem places the fluorophore so close to the non-fluorescent quencher that they covalently share electrons, such that emission spectrum of fluorophore is observed by quencher and converted in to heat energy. When probe encounters a target molecule, it forms a probe target hybrid that is longer and more stable than the stem hybrid. Consequently, the molecular beacon undergoes a spontaneous conformational reorganization that forces the stem hybrid to dissociate, the fluorophore and quencher move away from each other, which enable them to fluoresce brightly (Fig. 4). Molecular beacons are added to the assay mixture before carrying out gene amplification and fluorescence is measured in real time. The assay tubes remains sealed. Consequently, the amplification can not escape to contaminate untested samples. Tm of the probe region should be 7-10 °C higher than the PCR annealing temperature. For preferential hybridization the probe must be complenmentary to a region of minimal secondary structure. The beacon should bind at or near the centre of the amplicon and the distance between the 3′ end of the upstream primer and 5′ end of the beacon stem (5-7 bp; GC 70-80%) should be greater than 6 nucleotides.

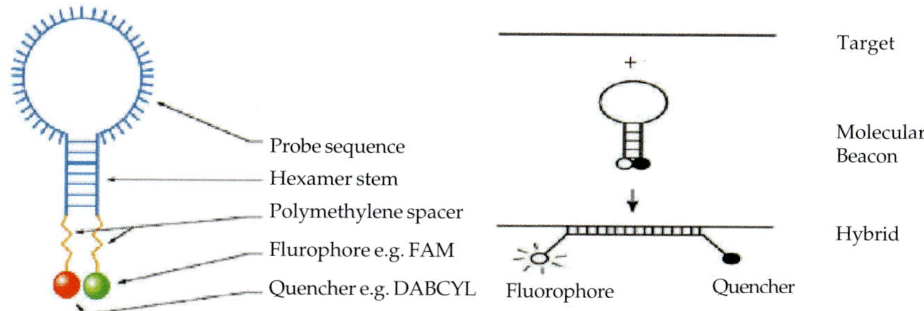

Fig. 4. Molecular beacons. On their own, these molecules are non-fluorescent, because the stem hybrid keeps the fluorophore close to the quencher. When the probe sequence in the loop hybridizes to its target, forming a rigid double helix, a conformational reorganization occurs that separates the quencher from the fluorophore, restoring fluorescence (Source : Yongsheng *et al.*, 2008)

Molecular beacons can be synthesized that possess differently coloured probes, enabling assay to be carried out that simultaneously detect different targets in same reaction. For example, multiplex assay can contain a number of different primer sets, each set enabling the amplification of unique gene sequence and a corresponding number of molecular beacons can be present each containing probe sequence for one of the amplicon and each labeled with a fluorophore of a different colour. The colour of resulting fluorescence, identify the specific amplicon in sample, and the number of amplification cycles required to generate detectable fluorescence can give the quantitative measure of target *e.g.* number of pathogenic agent in a sample. Molecular beacons are extraordinary specific. They easily discriminate target sequences that differ from one another by a single nucleotide substitution. Molecular beacons are thus ideal probes for use in diagnostic assays, designed for genetic screening, detection of single nucleotide polymorphism, and pharmacogentic application.

Molecular beacons are finding a multitude of uses in many quantitative and qualitative target detection assays. They are used for real time monitoring of DNA amplification during PCR (Tyagi and Kramer, 1996); they can be multiplexed using multiple dye labels and used for real time fluorescence genotyping (Kostrikis *et al.*, 1998; Tyagi *et al.*, 1998) and in the simultaneous detection of different pathogens in clinical samples (Vet *et al.*, 1999). They have also been used to detect RNA transcripts in living cells (Sokol *et al.*, 1998) and to detect DNA binding proteins (Heyduk and Heyduk, 2002). Kota *et al.*(1999) demonstrated that molecular beacons can be used to detect the transgenic barley containing *bar* gene. The ability of molecular beacons to analyze single nucleotide differences has been exploited to develop rapid test for SNP analysis. Piatek *et al.* (1998) employed this assay for detecting drug resistance in *Mycobacterium tuberculosis*. Although limited information of the many agronomical important plants is presently available it is envisaged that the molecular beacon assay will have the potential to rapidly analyze SNPs in plants, and these probes will find then way into nucleic acid research and diagnostics.

2.1.4 Scorpion Primers

Scorpion primers are flourogenic PCR primers with a probe attached at the 5′end *via* a PCR stopper which can be used in real time Amplicon-specific detection of PCR products in homogenous solution. Scorpion primers are bi-functional molecules in which a primer is covalently linked to a probe. The basic elements of scorpion primers are a PCR primer, a PCR stopper to prevent PCR-read through probe sequences and fluorescence detection system containing at least one flourophore and quencher. Two different formats are possible, the stem and loop format and the duplex format. In stem and loop format, the probe sequence form a loop followed by a stem which is formed by self complementary sequences on either side of the probe. This stem and loop structure is attached to the primer *via* a PCR blocker. The primer hybridizes to the target and extension occurs due to the action of polymerase. In absence of the target, the quencher nearly absorbs the fluorescence emitted by the flourophore. After PCR extension of the scorpion primer, if resultant amplicons contains a sequence that is complementary to the probe, the probe binds to its complementary sequence which is rendered single stranded during the denaturation stage of PCR cycle. This results in increase in florescence as quencher is no longer in the vicinity of flourophore. The scorpion primer pair should be designed to give an amplicon of approximately 100-200 bp and probe sequences should be ideally 17-27 bases.The melting temperature of stem structure should be lower than probe target hybrid so that hairpin loop require less energy than the new DNA duplex product. Scorpion technology can be used in allelic discrimination and in SNP genotyping. The intramolecular probing mechanism of the scorpion primers offer a significant advantage over other genotyping systems such as TaqMan, molecular beacons and hybridization probes that they rely on bimolecular probing. Unlike TaqMan probes, scorpions do not depend upon enzymatic cleavage, therefore rapid PCR cycling is possible (Thelwell *et al.*, 2000). The scorpion primers can also be used to detect multiple targets, SNPs simultaneously by using different flourophore tagged to different probes which result in increase throughput.

2.2. Flap Endonuclease Discrimnation

Flap endonuclease (FEN) discrimination assay is based on flap endonuclease which recognize and cleave three dimensional structure formed by hybridization of overlapping oligonucleotide probes, near the SNP site. Invader assay is based on the following chemistry :

2.2.1 Invader Assay

The invader assay enables simultaneous detection of two different alleles in a non-PCR-based assay. The invader assay employs endonuclease from flap endonuclease family, to cleave structure-specific rather than sequence-specific sites. The flap endonuclease-1 (FEN-1) is obtained from the bacteria *Pyrococcus furiosus* which has its role in DNA repair and replication. The invader technology relies on the specificity of recognition and cleavage by a flap endonuclease of the three dimensional structure formed, when two overlapping oligonucleotide, an invader oligonucleotide, and a signal oligonucleotide with reporter arm hybridize to target DNA containing a polymorphic site (Lyamichev *et al.,* 1999). The invasive cleavage assay involves hybridization of genomic DNA with two sequence specific oligonucleotides, one termed the *'invader oligonucleotide'* and other as *'probe oligonucleotide'*. The invader oligonucleotide has a sequence homology with a segment of genomic DNA upstream of the SNP site. The probe oligonucleotide on the other hand has a segment at its 3′ end that is homologous to the target allele and another segment at its 5′ end has no homology with target DNA. On hybridization, duplex is formed between the homologous segment of the probe oligonucleotide and the target DNA. The invader oligonucleotide now invades into the duplex forming a overlap at this point. The flap endonuclease recognizes the overlapping structure and cleaves the 5′ flap on the primary probe at the base of the overlap releasing it as target specific product. If the probe does not hybridize perfectly at the site of interest no overlapping structure is formed and no cleavage occurs and target specific product is not released. The target specific oligos are involved in a secondary reaction where they act as invader probes on a fluorescent resonance energy transfer (FRET™) cassette leading to the formation of an overlapping structure that is recognized by cleavage enzyme. When FRET™ cassette is cleaved a fluorophore is released from quencher on the FRET™ generating a fluorescence signal (Fig. 5). There are two signal fluorophore attached to two different FRET™ cassettes (FRET™1 and FRET™2) that are spectrally distinct and specific to either allele of the bi-alleic system. The ratio of two fluorescent signals then allows a genotype to be assigned.

2.3. Primer Extension

Primer extension also known as mini-sequencing and forms the basis of a number of methods for allelic discrimination. An extension primer is annealed at 5′ end of SNP, either adjacent but not including the SNP or several bases upstream of the SNP. The primer is then extended for one or several nucleotides to include SNP site by DNA polymerase. In case of single base extension, a primer is annealed adjacent to SNP and extended to incorporate a di-deoxynucleotide (ddNTP) at polymorphic site. The

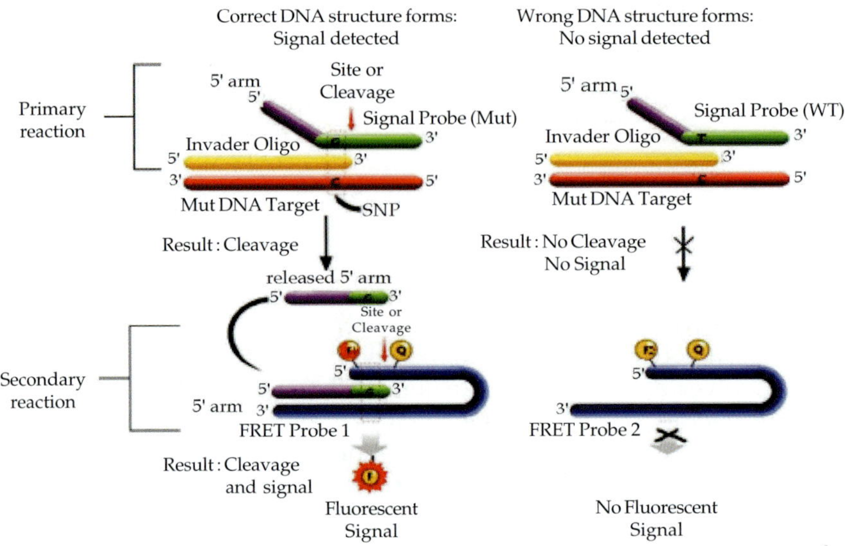

Fig. 5. Schematic representation of the invader assay. During the primary phase, an Invader® oligo and a primary probe are annealed to target DNA, overlapping at the SNP position. The arrow indicates the site of cleavage by the cleavage enzyme. The released 5' flap anneals to the FRET cassette during the secondary phase and initiates a second cleavage reaction that releases the fluorescent dye. The signal is only released when the invasive structure is formed on the target DNA ('Perfect Match', left reaction). If the primary probe does not match the nucleotide at the SNP position, cleavage will not act (reaction on right) (Source : Bunyan, 2008)

ddNTP stops further extension and detected through fluorescence attached to it.

2.3.1 Genetic bit analysis

Genetic bit analysis is a primer guided genotyping method for high throughout detection of polymorphism at known SNP site (Nikiforv *et al.*, 1994). The sequence information surrounding the site of variation in target DNA is used to design an oligonucleotide primer that is complementary to the region immediately adjacent to, but not including, the variable oligonucleotide in the target DNA. Specific fragment of genomic DNA containing query base are first amplified by the PCR using one regular and other diphosphorothioate modified primers. The double stranded PCR product is rendered single stranded by treatment with exonuclease. One of the stranded of DNA is protected from exonuclease digestion due to use of diphosphorothioate modified primer. The resulting single standard DNA is captured onto individual wells of polystyrene plate by hybridization to an immobilized oligonucleotide primer which immediately flanks the query base at its 3′ end. Now DNA polymerase extends the 3′ end of oligonuceotide primer is by one base using mixture of one biotin labeled, one fluorescein labeled and two unlabeled dideoxynucleotides. The success of single base extension is dependent upon the base sequence of single DNA template. The incorporation of a single ddNTP at SNP site will terminate the rection and will allow the detection of SNP in form of labeled ddNTP. The detection of labeled dideoxynucleotide is achieved using either horseradish peroxidase, antibiotin or alkaline phosphate, antinfluorescein conjugates in combination with enzyme specific colorimetric determination using a plate reader. Enzyme reaction is performed sequentially with each enzyme detection system. A positive signal with both enzyme detection systems indicates a heterozygous sample. The technique allows the detection of two possible alleles in the same well of a microlitre plate which result in both operational and biochemical advantages. The technique has been used as diagnostic tool in human paternity tests as well as in a pedigree analysis. Among plants systems the successful use of the technique has been demonstrated in onion for distinguishing plastomes of cytoplasmic male sterile (CMS) and fertile lines, which differ by single SNP (Alcala *et al.,* 1997).

2.3.2 Masscode™ system

Qiagen Genomics Inc. has developed the Masscode tagging system for DNA labeling and detection. The labeling system is based on a small molecular weight tag that is covalently attached through a photo cleavable linker to a DNA oligonucleotide (Fig. 6). The tagged oligonucleotide is designed to be used as primer for PCR amplification of specific allele discriminated by a SNP. The SNP specific amplicons are then purified to remove any unincorporated tagged oligonucleotide primer. The tags are then cleaved from purified products by exposure to a 254-nm mercury lamp, and isolated tags are subjected to mass spectrometry analysis using a single quadruplole mass spectrometer. The system has a lower limit of detection in the fentomolar range. This system claims to offer the highest throughput; at present 30 different masscode tag can be used in a multiplex fashion which could provide more than 40,000 SNP genotyping measurements daily on a single masscode system.

2.3.3 Pyrosequencing

Pyrosequencing, a non-electrophoretic method for sequencing of short stretches of DNA and is emerging as a popular platform for analysis of SNPs. With help of PSQ™96, sequences of 96 samples can be obtained simultaneously within 15 minutes. In this method, a sequencing primer is hybridized to a single, PCR amplified DNA template and incubated with four enzymes; DNA polymerase, ATP sulfurylase (yeast), luciferase (fire fly), apyrase (potato) and the substrates adenosine 5′ phosphosulphate (APS) and luciferin. The single standard DNA is used as template for

Masscode Tag Technology

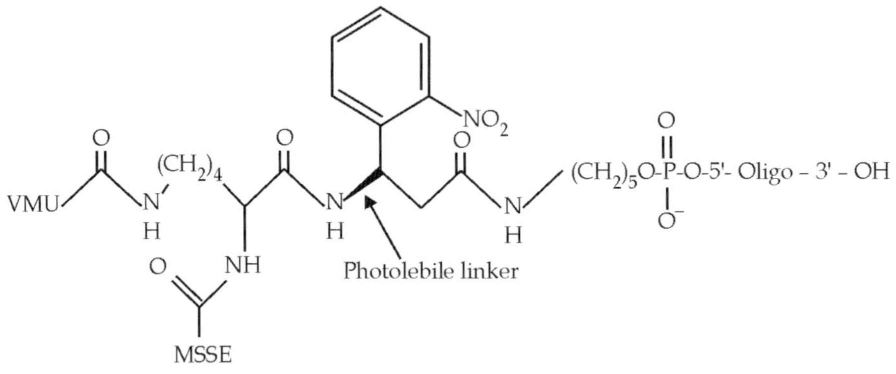

Fig. 6. Molecular weight tag covalently attached through a photo cleavable linker to a DNA oligonucleotide used in masscode system (Source : Qiagen Mews, 2001).

synthesizing its complementary strand. The synthesis proceeds by incorporating one nucleotide at a time. When first of four deoxynucleotide triphosphate (dNTP) is added to the reaction, DNA polymerase catalyzes its incorporation into DNA strand if it is complementary to the base in the template strand. Each incorporation event is accompanied by release of pyrophosphate (PPi) in a quantity equimolar to the amount of incorporated nucleotide. ATP sulfurylase quantitatively converts PPi to ATP in the presence of adenosine 5′ phosphosulfate. This ATP drives the luciferase mediated conversion of luciferin for oxyluciferin that generate visible light. The light produced in luciferase catalyzed reaction is detected by a charged coupled device (CCD) camera and seen as a peak in pyrogram. Each light signal is proportional to number of nucleotides incorporated. Apyrase, a nucleotide degrading enzyme, continuously degrades unincorporated dNTPs and excess ATP. When degradation is complete, another dNTP is added (Fig. 7). Addition of dNTPs is performed one at a time. The following reaction is involved in the emission of light:

$$(DNA)_n + dNTP \xrightarrow{\text{DNA polymerase}} (DNA)_{n+1} + PPi$$

$$APS + PPI \xrightarrow{\text{sulfurylase}} ATP$$

$$\text{Luciferin} + ATP \xrightarrow{\text{Luciferase}} \text{oxyluciferin} + \text{light}$$

If the added dNTP does not match, it is not incorporated into complementary strand and no light is emitted. As the process continues, the

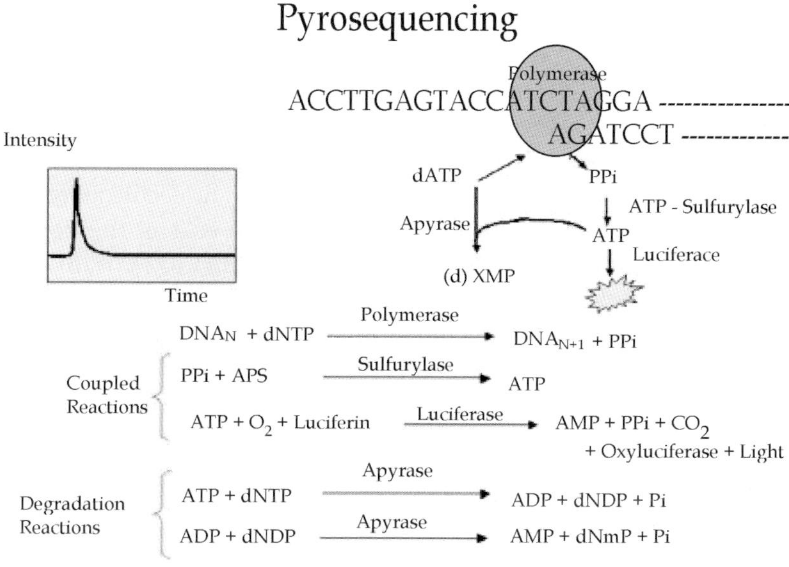

Fig. 7. Diagrammatic presentation of Pyrosequencing (Source : Wijesuriya, 2006).

complementary strand is built up and nucleotide sequence is determined from the signal peak in pyrogram. It should be noted that deoxyadenosine alfa thiotriphosphate (dATP) is used as substitute for the natural deoxyadenosine triphosphate (dNTP) since it is efficiently used by the DNA polymerase, but not recognized by the luciferase. Pyrosequencing is particularly suitable for SNP genotyping, since genotyping of previously identified SNP by this method require sequencing of only a few nucleotides. Pyrosequencing is applied in diverse clinical research areas such as cancer mutation analysis, DNA methylation analysis, microbial identification and antibiotic resistance studies, clinical genetic and pharmacogenetics. Plant genetic studies are also well served by pyrosequencing. These applications benefit from the rapid, qualitative sequence data with built in quality controls that pyrosequencing offers. Pyrosequencing is being used for SNP genotyping in wheat and corn by DuPont and Pioneer Hibred in USA.

2.4. Oligonucleotide Ligation

Oligonucleotide ligation chemistry is based on DNA ligase enzyme which closes down the nick in the double stranded DNA molecule by formation of phosphodiester bond between two adjacent nucleotides. Oligonucleotide ligation assay is one of most well established examples based on DNA ligase chemistry.

2.4.1 Oligonucleotide ligation assay

Oligonucleotide ligation assay uses two independent oligonucleotide probes that hybridize with PCR product adjacent to each other. The probe on 5′ end is an allele specific oligonucleotide (ASO) to one allele of the target. The last base at 3′ end of this ASO is positioned at the site of the target DNA polymorphism. The ASO also has a biotin molecule at its 5′ end that functions as chemical hook. The probe on 3′ end is the common oligonucleotide probe for both the alleles. It hybridizes with target immediately next to the query base. It is fluorescently labeled at 3′ end and is called as reporter probe. The gene fragment containing the polymorphic site is amplified by PCR and incubated with these probes in presence of DNA ligase. The ligation of allele specific probe and reported probe occur when there is a complete match of allele specific probe to the target sequence. The ligation product, which is fluorescent labeled at 3′end captured on a solid streptavidin coated matrix due to biotinylation at its 5′ end and fluorescence signal can be detected. However when there is a mismatch due to presence of SNP, the biotinylated oligonucleotide probe and fluorescent labeled reporter probe are unable to ligate, so that biotinylated oligonucleotide captured by slreptavidin, carry no signal. Oligonucleotide ligation assay can also be suitably modified to develop and multiplex systems. Edelstein *et al.* (1998) reported the detection of mutations in the *pol* gene of human immuno deficiency virus type1 by genotyping by an oligonucleotide ligation assay.

3.0 CONCLUSION

The pharmaceutical industry, biotechnology companies and academic group alike aspire to carry out an increasing volume of genotyping with high throughput without increasing the cost of genetic research. An ideal system for high throughput testing would be one in which process of amplification and detection are performed simultaneously or at least gel electrophoresis step could be replaced by simpler step. To meet this demand, the search is on to find a genotyping technology able to deliver accurate genotyping at very high throughput and at low cost per genotype. As discussed in the chapter there are various convenient methods for SNP (insertion and detection) genotyping, which is highly applicable to both basic research and clinical diagnostics. Commercial devices are also available, making the technology affordably accessible to all laboratories. The current technologies have resulted in dramatic reduction of the working volume and are a step towards developing cost-effective high throughput detection of PCR products and SNPs for molecular breeding.

4.0 REFERENCES

Alcala, J., Giovannoni, J.J., Pike, L.M. and Reddy, A.S. 1997. Application of genetic bit analysis for allelic selection in plant breeding. Molecular Breeding, 3: 495–502.

Brookes, A.J. 2002. DASH (Dynamic Allele Specific Hybridization) for SNP Genotyping. Atelier 136, La Roche Posay Méthodologie pour l'étude de la variation génomique Methods for the elucidation of genomic variation. (http://ist.inserm.fr/basisateliers/atel136/brok.pdf).

Bunyan, D. 2008. Evaluation of the Invader® assay platform for the detection of allelic variants. National Genetics Reference Laboratory (Wessex), Salisbury District Hospital, Salisbury. (http://www.ngrl.co.uk).

Dieffenbach, C.W. and Dveksler, G.S. 2003. PCR Primer: A laboratory manual. Sec. Ed. Cold Spring Harbor Laboratory Press, New York, 67-72.

Edelstein, R.E., Nickerson, D.A., Tobe, V.O., Manns-Arcuino, L.A. and Frenkel, L.M. 1998. Oligonucleotide ligation assay in the human immunodeficiency virus Type I *pol* gene that are associated with resistance to Zidovudine, Didanosine and lamivudine. Journal of Clinical Microscopy, 36:559-572.

Heyduk, T. and Heyduk, H. 2002. Molecular beacons for detectimg DNA binding proteins. Nature Biotechnology, 20:171:176.

Higuchi, R., Dollinger, G., Walsh, P.S. and Grifth, R. 1992. Simultaneous amplification and detection of specific DNA sequences. Bio-Technology, 10: 413-417.

Higuchi, R., Fockler, C., Dollinger, G. and Watson, R. 1993. Kinetic PCR analysis: real time monitoring of DNA amplification reactions. Bio-Technology, 11:1026-1030.

Holland, P., Abramson, R.D., Watson, R. and Gelfand, D.H. 1991. Detection of specific PCR reaction product by utilizing the 5'-3' exonuclease activity of *Thermus aquaticus* DNA polymerase. Proceedings of the National Academy of Sciences USA, 88:7276-7280.

Jonathan, A.P. and Anthony, J.B. 2001. Towards high-throughput genotyping of SNPs by dynamic allele specific hybridization. Expert Review of Molecular Diagnostics, 1(3):89-94.

Kostrikis, L.G., Tyagi, S., Mhlanga, M.M., Ho, D.D. and Kramer, F.R. 1998. Spectral genotyping of human alleles. Science, 279:1228-1229.

Kota, R., Holton, T.A. and Henry, R.J. 1999. Detection of transgenes in crop plants using molecular beacon assays, Plant Molecular Biology Reporter, 17, 363-370.

Lee, L.G., Livak, K.J., Mullah, B., Graham, R.J., Vinayak, R.S. and Wondenberg. T.M. 1999. Seven-Color, Homogeneous Detection of Six PCR Products. Biotechniques, 27(2):342-349.

Lyamichev, V., Mast, A. L., Hall, J.G., *et al.* 1999. Polymorphism identification and quantitative detection of genomic DNA by invasive cleavage of oligonucleotide probes. Nature Biotechnology, 17:292-296.

Meksem, K., Hyten, D., Ruben, E. and Lightfoot, D.A. 2001. High throughput genotyping for a polymorphism linked to soybean cyst nematode resistance gene *Rhg 4* by using TaqMan probes. Molecular Breeding, 77:63-71.

Nikiforv, T.T., Rendle, R.B., Goelet, P., Rogers, Y.H., Kotewicz, M.L., Anderson, S., Trainor, G.L. and Knapp, M.R. 1994. Genetic bit analysis: a solid phase nucleotide polymorphisms. Nucleic Acids Research, 22: 4167-4175.

Oefner, P.J. and Underhill, P.A. 1998. DNA mutation dilution using denaturing high performance liquid chromatography (DHPLC). Current protocols in Human Genetics, 1998 Supplement 19, 7.10, 1-7, 10.12 (Willey and Sons, N.Y.).

Piatek, A.S., Tyagi, S., Pol, A.C., Telente, A., Miller, L.P., Kramer, F.R. and Alland, D. 1998. Molecular beacon sequence analysis for detecting drug resistance in *Mycobacterium Tuberclosis*. Nature Biotechnology, 16: 359-363.

QIAGEN News. 2001. High-throughput SNP genotyping using Masscode™ technology. Issue No. 2, pp 11-12.

Russon, A., Gustav, A., Torsten, M., Patrik, M., Anthony, J.B., Anderson, H. and Stemme. G. 2003. SNP analysis by dynamic allele specific hybridization on patterned monolayers of beads. 7th international conference on miniaturized chemical and biochemical analysis systems, October 5-9, Squan Valley, California USA.

Sivakumaran, T.A., Kucheria, K., and Oefner, P.J. 2003. Denaturing high performance liquid chromatography in the molecular diagnosis of genetic disorders. Current Science, 84 (3): 291-296.

Sokol, D.L., Zhang, X., Lu, P. and Gewirtz, A.M. 1998. Real time detection of DNA , RNA hybridization in living cells. Proceedings of the National Academy of Sciences USA, 95:11538-11543.

TaqMan® 2006. White Paper on TaqMan® Gene Expression Assays, TaqMan® Gene Expression Assays for Validating Hits from Fluorescent Microarrays. (www.appliedbiosystems.com).

Thelwell, N., Millington, S., Solinas, A., Booth, J. and Brown, T. 2000. Mode of action and application of scorpion primers to mutation detection. Nucleic Acids Research, 28:3752–3761.

Tyagi, S. and Kramer, F.R. 1996. Molecular beacon: probes that fluoresce upon hybridization. Nature Biotechnology, 14: 303-308.

Tyagi, S., Bratu, D.P. and Kramer, F.R. 1998. Multicolor molecular beacons for allelic discrimination. Nature Biotechnology, 16:49-53.

Vet, J.A., Majithia, A.R., Marras, S.A.E., Tyagi, S., Dube, S., Poiesz, B.J. and Kramer, F.R. 1999. Multiplex detection of fouir pathogenic retroviruses using molecular beacons. Proceedings of the National Academy of Sciences USA, 96:6394-6399.

Wijesuriya, H. 2008. Overview of pyrosequencing. GENOSEQ, ULCA, Genotyping and sequencing. (http://www.genoseq.ucla.edu).

Yongsheng, L., Xiaoyan, Z., Duyun Y. 2008. Molecular beacons: An optimal multifunctional biological probe. Biochemical and Biophysical Research Communications, 373: 457–461.

Molecular Plant Breeding: Principle, Method and Application
Eds : R.K. Singh, Rajesh Singh, Guoyou Ye, A. Selvi and G.P. Rao
Studium Press LLC, Texas, USA, 2009, pp. 101-134

CHAPTER 4

Molecular Mapping and Mapping Populations in Plants

GYAN P. MISHRA[1], SHASHI BALA SINGH and R.K. SINGH[2]*

ABSTRACT

Molecular makers are useful tools for developing detailed linkage maps of various plant species. These maps have obvious utility for the identification of markers linked to genes of interest. Molecular markers are equally good for mapping and tagging of both qualitative and quantitative trait loci (QTLs). There are different types of mapping populations like $F_{2,}$ backcross, doubled haploids, recombinant inbred lines etc. which are used extensively for mapping of trait of interest. Beyond their practical utility, molecular markers and molecular maps can provide detailed information regarding: the structure of the genome of a species, evolutionary relatedness of species within a family, sequence rearrangements during evolution, relationship between physical and genetic distances linkage drag.

Key Words: Molecular mapping, Mapping population, Linkage, QTL, BSA

[1]*Defence Institute of High Altitude Research, DRDO, C/o 56 APO, Leh, India,*
[2]*Indian Institute of Sugarcane Research, Lucknow, UP, India*
* *Corresponding author e-mail: gyan.gene@gmail.com*

1.0 INTRODUCTION

1.1 Plant Genome Structure

The genomes of all eukaryotic species consist of single-copy, middle repetitive and high copy number sequences. To gain an understanding of the percentages of each of the classes within any particular species, a group of random clones can be hybridized to blots of plant DNA. One such study was performed by Zamir and Tanksley (1988). They hybridized 50 random genomic clones to tomato DNA. Washing was performed at two stringencies. The following are the results with regards to copy number.

Table 1. Hybridization stringency and its effect upon the number of sequences to which it hybridizes (Zamir and Tanskley, 1998)

Clone Class	Low Stringency Wash[1]	High Stringency Wash[2]
Single copy clones	44%	78%
Multiple copy clones	46%	18%
Repetitive clones	10%	4%

[1]1.0X SSC, 65 C;<80% homology
[2]0.05X SSC, 65 C;>98% homology

The table shows that hybridization stringency has a significant effect upon the number of sequences to which a sequence hybridizes. At the higher stringencies, most of the clone recognized only a single copy within the genome. What this also shows is that the tomato genome contains many sequences that are about 80% homologous, but fewer sequences that are highly homologous. The principal conclusion that can be drawn from these hybridizations is that the random sequences (as represented by the random genomic clones) are evolving faster than the single-copy sequences. Why? At the higher stringency, homology to distant related species was reduced. Therefore these sequences have undergone greater divergence.

1.2 Evolutionary Relatedness

Many papers have been published in which molecular data are used to assess the evolutionary relatedness of both cultivated (Hoey *et al.*, 1996; Kaga *et al.*, 1996; Santalla *et al.*, 1998; García *et al.*, 1998; Blair *et al.*, 1999; Nebauer *et al.*,1999; Sharma *et al.*, 1995; Ratnaparkhe *et al.*, 1995; Sonnante *et al.*, 1997; Moeller and Schaal, 1999) and wild plant species (Yeh *et al.*, 1995; Khasa and Dancik, 1996; Owuor *et al.*, 1997; Fornari *et al.*, 1999; Nebauer *et al.*, 1999) using molecular markers. The typical experiment begins with a large collection of cultivars or species. These samples are then analyzed by hybridization with RFLP clones or more recently by the PCR amplification (RAPD, ISSR, SSR, STMS, etc). The similarity of each pair of samples is measured by calculating the number of common bands or amplification products. One estimator was developed by Nei and Li (1979). The formula is: F = 2nXY/(nX + nY).

Where, nX and nY are the total number of fragments for sample X and Y and nXY is the number of fragments shared by the samples. Data are normally collected for a relatively large number of hybridizations or PCR-based primers amplifications. The similarity data are then used in a cluster analysis to develop dendrograms which show the molecular relatedness of the species.

These types of analysis can provide independent support to previous phylogenetic and evolution hypotheses, it can also identify gene pools within a genus, it can estimate genetic diversity within a genus and provide the necessary data to select appropriate parents for a molecular mapping project.

2.0 MAPPING WITH MOLECULAR MARKERS

2.1 Genetic Linkage and Linkage Mapping

Genetic linkage occurs when particular genetic loci or alleles for genes are inherently joined. Genetic loci on the same chromosome are physically connected and tend to stay together during meiosis, and are thus genetically linked. This is called autosomal linkage. Alleles for genes on different chromosomes are usually not linked, due to independent assortment of chromosomes during meiosis. Because there is some crossing over of DNA when the chromosomes segregate, alleles on the same chromosome can be separated and go to different daughter cells. There is a greater probability of this happening if the alleles are far apart on the chromosome, as it is more likely that a cross-over will occur between them. The relative distance between two genes can be calculated using the offspring of an organism showing two linked genetic traits, and finding the percentage of the offspring where the two traits do not run together. The higher the percentage of descendants that does not show both traits, the further apart on the chromosome they are.

The observations by Thomas Hunt Morgan showed that the amount of crossing over between linked genes differs led to the idea that crossover frequency might indicate the distance separating genes on the chromosome. Morgan's student Alfred Sturtevant developed the first genetic map, also called a linkage map. Sturtevant proposed that the greater the distance between linked genes, the greater the chance that non-sister chromatids would cross over in the region between the genes. By working out the number of recombinants it is possible to obtain a measure for the distance between the genes. This distance is called a genetic map unit (m.u.), or a centimorgan and is defined as the distance between genes for which one product of meiosis in 100 is recombinant. A recombinant frequency (RF) of 1 % is equivalent to 1 m.u. A linkage map is created by finding the map distances between a number of traits that are present on the same chromosome, ideally avoiding having significant gaps between traits to avoid the inaccuracies that will occur due to the possibility of multiple recombination events.

2.2 Calculating Linkage Using Molecular Markers

A. The % recombinants in F_2 analysis doesn't directly give us the recombination frequency (Fig. 1.).

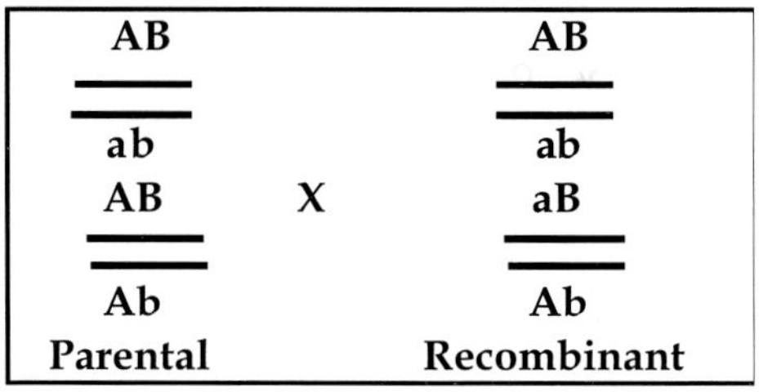

AB / ab		AB / ab
AB / Ab	X	aB / Ab
Parental		Recombinant

Fig. 1. Progenies from the F_1 selfing (F_2 population) are both parental and recombinant types (Source : Fristensky, 2004)

The progenies from the F_1 selfing (F_2) would be heterozygous at both loci (Recombinant types, Fig. 1), which might be natively scored as non-recombinant types (If the trait under study are dominant nature). This is why geneticists have traditionally gone to great extremes to construct tester stocks for mapping by testcrosses. However, that is not always possible to do, especially if our objective is to map hundreds or thousands of markers to construct a genomic map. Consequently analytical methods were used which are appropriate for F_2 data. Note: The examples in this section apply to dominant markers, such as RAPDs or AFLPs. Linkage calculations for co-dominant markers such as STMS and RFLPs are slightly different.

Note: The examples in this section apply to dominant markers, such as RAPDs or AFLPs. Linkage calculations for co-dominant markers such as STMS and RFLPs are slightly different.

B. Calculating Linkage between Two Loci

Although we have seen that many recombination events are masked by dominant phenotypes, it is straight forward to calculate the expected number of progeny in each of 4 possible phenotypic classes in a segregating population. A chromosome is said to be in coupling when it carries both dominant or both recessive alleles, for two loci. A chromosome is in repulsion when it carries one dominant and one recessive allele, with the complementary set of alleles on the other chromosome (Fig. 2.).

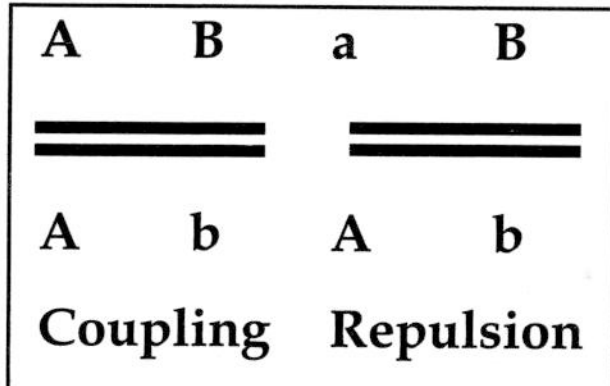

Fig. 2. Coupling and repulsion phase linkage of genes in a chromosome

As implied above, when markers with complete dominance, such as RAPDs, or AFLPs are scored, there are only 4 phenotypic classses: AB; Ab; aB; ab. The expected number of progeny in each of four possible phenotypic classes is determined by the phase of the two gametes in each mating, and frequency of recombination. In coupling phase, therefore, parental gametes AB or ab, and recombinants are Ab and aB (Table 2.). In repulsion, parental gametes carry Ab or aB, while recombinant gametes carry AB or ab (Table 3.).

Table 2. A two-locus model for two loci linked in coupling and recombination fraction θ in F_2 progeny (Lui, 1998)

F1 Gamete frequency	AB 0.5 (1-θ)	Ab 0.5θ	aB 0.5θ	Ab 0.5 (1-θ)
AB 0.5 (1-θ)	A_B_ $0.25 (1-\theta)^2$	A_B_ 0.25 θ (1- θ)	A_B_ 0.25 θ (1- θ)	A_B_ $0.25(1-\theta)^2$
Ab 0.5θ	A_B_ 0.25 θ (1- θ)	A_bb $0.25\ \theta^2$	A_B_ $0.25\ \theta^2$	A_bb 0.25 θ (1- θ)
aB 0.5θ	A_B_ 0.25 θ (1- θ)	A_B_ $0.25\ \theta^2$	aaB_ $0.25\ \theta^2$	aaB_ 0.25 θ (1- θ)
ab 0.5 (1-θ)	A_B_ $0.25\ \theta (1-\theta)^2$	A_bb 0.25 θ (1- θ)	aaB_ 0.25 θ (1- θ)	aabb $0.25\ \theta (1-\theta)^2$

Table 3. A two locus model for two loci linked in repulsion and recombination fraction θ in F_2 progeny (Lui, 1998)

F1 Gamete frequency	Ab 0.5 (1-θ)	AB 0.5θ	ab 0.5θ	aB 0.5 (1-θ)
Ab 0.5 (1-θ)	A_bb $0.25 (1-\theta)^2$	A_B_ 0.25 θ (1- θ)	A_bb 0.25 θ (1- θ)	A_B_ $0.25(1-\theta)^2$
AB 0.5θ	A_B_ 0.25 θ (1- θ)	A_B_ $0.25\ \theta^2$	A_B_ $0.25\ \theta^2$	A_B_ 0.25 θ (1- θ)
ab 0.5θ	A_bb 0.25 θ (1- θ)	A_B_ $0.25\ \theta^2$	aabb $0.25\ \theta^2$	aaB_ 0.25 θ (1- θ)
aB 0.5 (1-θ)	A_B_ $0.25\ \theta (1-\theta)^2$	A_B_ 0.25 θ (1- θ)	aaB_ 0.25 θ (1- θ)	aaB_ $0.25\ \theta (1-\theta)^2$

The expected frequencies of progeny from each type of mating can be calculated directly by adding the recombination fractions for each mating class. These are summarized for coupling phase (Table 4) and repulsion phase below (Table 5).

Table 4. Expected genotypic frequency for an F_2 population using dominant markers in coupling linkage phase (Lui, 1998)

Genotype	Observed count	Expected frequency	P_i (R|G)
A_B_	f_1	$0.25\ (3 - 2\theta + \theta^2)$	$2\theta / (3 - 2\theta + \theta^2)$
A_bb	f_2	$0.25\theta\ (2 - \theta)$	$1/ (2 - \theta)$
aaB_	f_3	$0.25\theta\ (2 - \theta)$	$1 / (2 - \theta)$
Aabb	f_4	$0.25\ (1 - \theta)^2$	0.0

Table 5. Expected genotypic frequency for an F_2 population using dominant markers in repulsion linkage phase (Lui, 1998)

Genotype	Observed count	Expected frequency	P_i (R|G)
A_B_	f_1	$0.25\ (2+\theta^2)$	$\theta\ (2+ \theta)/(2+ \theta^2)$
A_bb	f_2	$0.25\ (1-\theta^2)$	$\theta / (1+\theta)$
aaB_	f_3	$0.25\ (1- \theta^2)$	$\theta / (1+\theta)$
Aabb	f_4	$0.25\theta^2$	1.0

The observed count f is the actual number of progeny in each phenotypic class. The expected frequency is simply the sum of the frequencies of each type of mating, listed in Tables 4&5. The frequencies for each mating can also be used to calculate P_i(R|G) the probability for any given gamete in phenotypic class **i**, that that gamete was recombinant. To cite a trivial example, if A and B are in repulsion, and we see an aabb individual, all gametes must be recombinant, so P_i(R|G) = 1.0. There is no chance (ignoring double crossovers) that aabb can be generated by non-recombinant chromosomes. In another example, if A and B are in repulsion, and we see an aaB_ individual, the probability a given gamete in the aaB_ phenotypic class in recombinant is $\theta/(1+\theta)$.

2.3 How Do We Get From Numbers of Progeny to Recombination Frequencies?

Suppose that out of progeny from 2 crosses examined, we get the following offspring (Table 6)

Table 6. Progenies from two cross *i.e.* F_2 and Backcross (from Liu, 1998)

Genotype	Observed Count F_2	Observed Count Backcross
A_B_	140	162
A_bb	10	40
aaB_	10	40
aabb	40	158

First, just by looking at the data it is obvious that A and B are in coupling phase, since the two major classes of progeny are AB and ab. It is also apparent that these loci must be linked. While it is not strictly impossible that such data could result when $\theta = 0.5$, that is, when the loci are not linked, the likelihood is extremely low. If these loci were linked at something like 25 cM (*i.e.* 25% recombination, or 25 map units), still expect the recombinant classes to have about 50 total progeny, out of 200 in F_2. (Actually, it is more correct to say that we see the number of progeny that would result if 25% of the gametes were recombinant.) In any case, it still seems unlikely that two loci linked at 25 cM would give this set of progeny. However, it is more likely for a 25 cM linkage to give this result than if the loci were unlinked.

The point is that as we get closer to the true recombination frequency, the likelihood of the data generating the observed result increases. For the dataset in Table 6, the likelihood of seeing the data, as a function of θ is shown graphically (Fig. 3).

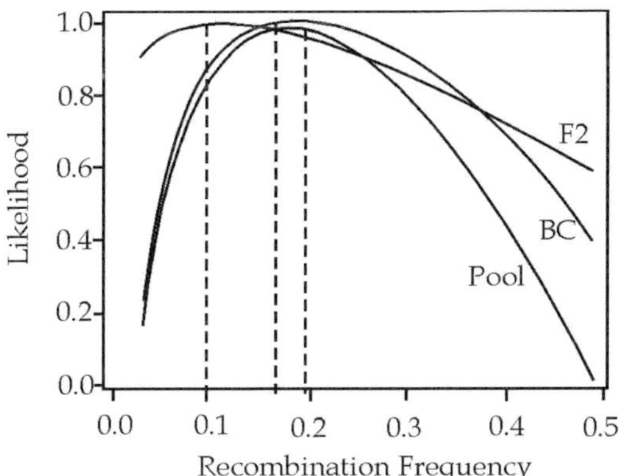

Fig. 3. Graphical representation of relationship between the likelihood of seeing the data, as a function of θ (Source : Fristensky, 2004)

There is no general formula that will directly calculate recombination frequency from phenotypic data. However, if we knew the precise value of θ, the following relation should be true:

$$\theta = 1/N \ \Sigma fi \ P_i \ (R \mid G)$$

All this equation says is that if we multiply the number of progeny in each phenotypic class by the probability that it represents a recombinant gamete we get the number of recombinant gametes. Summing all of these and dividing by the total number of progeny is, by definition, the true recombination frequency θ. However, we don't know the true value of θ. Nonetheless, if we were to calculate $P_i(R \mid G)$ using some other value, θ^{old}, the new value of θ^{new}, must lie somewhere closer to θ^{new} than θ^{old}. Therefore, $\theta^{new} = 1/N \ \Sigma \ fiP_i \ (R|G)$; gives a better estimate of θ than θ^{old}. The common maximum likelihood EM algorithm applies this principle iteratively.

2.4 Expectation-Maximization (EM)

An Expectation-Maximization (EM) algorithm is used in statistics for finding maximum likelihood estimates of parameters in probabilistic models, where the model depends on unobserved latent variables. EM alternates between performing an expectation (E) step, which computes an expectation of the likelihood by including the latent variables as if they were observed, and maximization (M) step, which computes the maximum likelihood estimates of the parameters by maximizing the expected likelihood found on the E step. The parameters found on the M step are then used to begin another E step, and the process is repeated (http://en.wikipedia.org/wiki/Expectation-maximization_algorithm)

2.4.1 EM algorithm to calculate recombination fraction θ

1. Make an initial guess: θ^{old}
2. Expectation step: Using θ^{old} as if it were the true recombination fraction, compute the expected number of recombinants.
3. Maximization step: Using the expected value, compute the maximum likelihood estimate θ^{new} for the recombination fraction.
4. Iterate E and M steps until the likelihood reaches its maximum, or the estimate converges:

$|\theta^{new} - \theta^{old}| <= tolerance$; (Liu, 1998).

Table 7. EM estimation of recombination fractions for data (Liu, 1998)

Iteration	F_2	Backcross	Pooled
0	0.25		
1	0.19373	0.25	0.25
2	0.15771	0.2	0.19686
3	0.13577		0.17983
4	0.12284		0.17457
5	0.11537		0.17296
6	0.11111		0.17248
7	0.10869		0.17233
8	0.10733		0.17228
9	0.10656		0.17227
...	...		0.17227
18	0.10558		

In this example, the initial estimate for the recombination fraction θ^{old} is arbitrarily set to 0.25. Because θ^{new} is always a better estimate than θ^{old}, the EM algorithm will always converge to θ (Table 7).

2.5 Inspection of mapping data illustrates the principle of co-segregation of closely-linked markers.

Neighboring loci in any region will always have the most similar segregation patterns. This is nothing new, it is simply a restatement of the idea of co-segregation of closely-linked loci. By definition, the more closely-linked two loci are, the more they will co-segregate (http://www.arabidopsis.org).

2.6 Calculation of linkage for multiple loci is done by iterative application of maximum likelihood methods.

For any possible order and spacing of loci, it is possible to calculate the probability that "the given map would exactly give rise to the observed data". Note that the liklihood is necessarily very small, because it is the probability that each meiosis under study would come out exactly the same if the experiment were repeated. Thus, likelihoods are useful only for comparative purposes. For example, if an alternative map had a 1000-fold lower chance of giving rise to the data, one might choose to reject it."

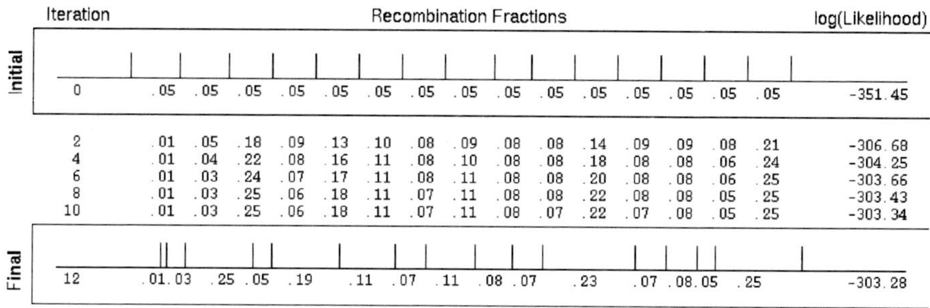

Fig. 4. Example of multipoint linkage analysis using EM algorithm, showing convergence to maximum-likelihood genetic map for 16 RFLPs on human chromosome 7, studied in CEPH families. The initial assumption of 5% recombination between consecutive RFLPs corresponded to a $\log_{10}$ likelihood -351.45. After 12 iterations, the recombination fractions converged to a map that was ~10^{48} times more likely to have produced the observed data (Fristensky, 2004)

In Fig. 4, we see that MAPMAKER begins considering a particluar map in the following way: Arbitrarily space all loci under consideration at a distance of 5cM. Calulate the likelihood of seeing the observed data, given the current map. Next, try another guess for map distances between markers.

Calculate the likelihood of seeing the data, given the new guess. If the new guess has a higher likelihood, choose it as the current working map. Repeat the process until the difference in log likelihood between the current map and the previous map is less than some threshold (*e.g.* 0.1). In this way, the program quickly converges on the spacing that is most consistent with the data.

3.0 DEVELOPMENT AND CHARACTERIZATION OF MAPPING POPULATIONS

The development of molecular marker technology has caused renewed interest in genetic mapping. Genetic map construction requires that the researchers to select the most appropriate mapping population (s), calculate pairwise recombination frequencies using these population, establish linkage groups and estimate map distances and determine the map order.

Since large mapping populations are often characterized by different marker systems, map construction has been computerized. Computer software package, such as LINKAGE-1, GMendel, MAPMAKER, Map Manager and Join Map have been developed to aid in the analysis of genetic data for map construction. These programmes use data obtained from the segregating populations to estimate recombination frequency that are then used to determine the linear arrangement of genetic markers.

3.1 Mapping Populations

A population used for gene mapping is commonly called a mapping population. Commonly used mapping populations are usually obtained from controlled crosses. Selection of a population for genome mapping involves choosing of parents and determining a mating scheme. Decisions on selection of parents, mating designs and the type of markers should depend upon the objectives of experiments.

Parents of mapping populations must have sufficient variation for the traits of interest at both the DNA sequence and the phenotype level. The variation at DNA level is essential to trace the recombination events. The more DNA sequence variation exists, the easier it is to find polymorphic informative makers. When the objective is to search for genes controlling a particular trait, genetic variation of trait between parents is important. If the parents are greatly different at phenotypic level for a trait, there is a reasonable chance that genetic variation exists between the parents, although uncontrolled environmental effects could create large phenotypic variation without any genetic basis for the effects. However, lack of phenotypic variation between parents does not mean that there is no genetic variation, as different sets of genes could result in same phenotype.

3.2 Selection of Parents for Development of Mapping Population

Selection of parents for developing mapping population is critical to successful map construction. Since a map's economic significance will depend upon marker-trait association, as many qualitatively inherited morphological traits as possible should be included in the genetic stocks chosen as parents for generating mapping population. Consideration must be given to the source of parents (adapted *vs* exotic) used in developing mapping population. Chromosome pairing and recombination rates can be severely disturbed (suppressed) in wide crosses and generally yield greatly reduced linkage distances. Wide crosses will usually provide segregating populations with a relatively large array of polymorphism when compared to progeny segregating in a narrow cross (adapted x adapted). To have significant value in crop improvement programme, a map made from a wide cross must be collinear (*i.e.* order of loci similar) with map constructed using adapted parents.

3.3 TYPES OF MAPPING POPULATIONS

Different types of mapping populations that are often used in linkage mapping are F_2, F_2 derived F_3 ($F_{2:3}$), backcrosses, doubled haploids (DHs), recombinant inbred lines (RILs), and near-isogenic Lines (NILs). The characteristic features, merits and demerits of each of these populations are briefly presented below:

3.3.1 F_2 Population

An F_2 population is developed by selfing (or intermating for cross pollinated species) among F_1 individuals. These F_1 individuals are developed by crossing two parents that show significant polymorphism for whichever type of loci we are going to score. F_2 individuals are products of single meiotic cycle. Ratio expected for dominant marker is 3:1 and for co-dominant marker is 1:2:1.

Merits: This population is considered best population for preliminary mapping. It requires less time for development and can be developed with minimum efforts, when compared to other populations.

Demerits: Linkage established using F_2 population is based on one cycle of meiosis. These populations are of limited use for fine mapping. Quantitative traits cannot be precisely mapped using F_2 population as each individual is genetically different and cannot be evaluated in replicated trials over locations and years. Thus, the effect the G x E interaction on the expression of quantitative traits cannot be precisely estimated. It is not a long-term population *i.e.* not immortal and is impossible to construct exact replica or increase seed amount.

3.3.2 F_2 derived F_3 ($F_{2:3}$) Population

$F_{2:3}$ Population is obtained by selfing the F_2 individuals for a single generation.

Merits: It is suitable for specific situations like mapping quantitative traits and recessive genes. This population is used for reconstituting the genotype of respective F_2 plants, if needed, by pooling the DNA from plants in the family.

Demerit: Like F_2 population, it is not 'immortal'.

3.3.3 Backcross Population

Backcross populations are developed by crossing the F_1 with one of the two parents used in the initial cross. Usually in genetic analysis, backcross with recessive parent (testcross) is used. With respect to molecular markers, the backcross with dominant parent (B_1) would segregate in ratio 1:0 and 1:1 for dominant and co-dominant markers, respectively. However, backcross with recessive parent (B_2) or test cross would segregate in a ratio of 1:1 irrespective of the nature of marker.

Merits: Like an F_2 population, the backcross populations require less time to be developed. However, the recombination information of backcrosses are based on only one parent (the F_1). The specific advantage of backcross populations is, the further utilization of populations for marker-assisted backcross breeding.

Demerits: The major drawback of using F_2 or backcross populations is that the populations are not eternal. Therefore, limited source of tissue to isolate DNA or protein will be exhausted at some point in time. Then, an other mapping population will be taken up for tagging and mapping of desired gene(s).

3.3.4 Doubled Haploids (DHs)

Chromosome doubling of anther culture derived haploid plants from F_1 generates DHs. The suitability of doubled-haploid progenies for mapping project has been demonstrated by Lefebvre *et al.* (1995) in pepper. DHs are also products of one meiotic cycle, and hence comparable to F_2 in terms of recombination information. The expected ratio for the marker is 1:1, irrespective of genetic nature of marker (whether dominant or co-dominant).

Merits: DHs are permanent mapping population and hence can be replicated and evaluated over locations and years and maintained without any genotypic change. They are useful for mapping both qualitative and quantitative characters. It can be used for instant production of homozygous lines, thus saving time.

Demerits: Recombination from the male side alone is accounted. Since, it involves *in vitro* techniques relatively more technical skills are required in comparison with the development of other mapping populations. It often happens that suitable culturing methods/haploid production methods are not available for number of crops and different crops differ significantly for their tissue culture response. Further, anther culture induced variability should be taken care of.

3.3.5 Near-Isogenic Lines (NILs)

NILs are generated either by repeated selfing or backcrossing the F_1 plants to the recurrent parents. NILs developed through backcrossing are similar to recurrent parent but for the gene of interest, while NILs developed though selfing are similar in pair but for the gene of interest (however, differ a lot with respect to the recurrent parent). Expected segregation ratio of the markers is 1:1 irrespective of the nature of marker.

Merits: Like DHs and RILs, NILs are also 'immortal mapping population'. They are suitable population for tagging the trait, and quite useful in functional genomics.

Demerits: Require many generations for development. It is directly useful only for molecular tagging of the gene concerned, but not for linkage mapping. Linkage drag is a potential problem in constructing NILs, which has to be taken care of.

3.3.6 Recombinant Inbred Lines (RILs)

Recombinant Inbred Lines (RILs) are developed by single-seed selections from individual plants of an F_2 population. Because of this procedure, these lines are also called F_2-derived lines. RILs are produced by continuous selfing or sib mating the progeny of individual members of an F_2 population until complete homozygosity is achieved. Single-seed Descent (SSD) method is best suited for developing RILs. Bulk method and pedigree methods without selection can also be used. (SSD) is repeated for several generations. At this point, all of the seed from an individual plant is bulked. For example, a $F_{3:4}$ RI populations underwent single-seed descent through the F_3 generation, and were bulked to develop the F_4. This population of seed can then be grown to obtain a large quantity of seed of each individual line. Importantly, each of the lines is fixed for many recombination events. RILs also equalize marker types like DHs, the genetic segregation ratio for both dominant and co-dominant marker would be 1:1. RILs developed though brother-sister mating require more time than those developed through selfing. The number of inbred lines required is twice, in case they are developed through brother-sister mating compared to selfing particularly, when linkage is not very tight.

Merits: Once homozygosity is achieved, RILs can be propagated indefinitely without further segregation (Table 8). Since RILs are immortal population, they can by replicated over locations and years and therefore are of immense value in mapping QTLs. RILs being obtained after several cycles of meiosis, are very useful in identifying tightly linked makers. RIL populations obtained by selfing have twice the amount of observed recombination between very closely linked markers as compared to population derived from a single cycle of meiosis.

Table 8. Relationship between recombinant inbred population (level of inbreeding) and within-line homozygosity at each locus (%)

RI population level of inbreeding	% within-line homozygosity at each locus
$F_{3:4}$	75.0
$F_{4:5}$	87.5
$F_{5:6}$	92.25
$F_{6:7}$	96.875
$F_{7:8}$	98.4375
$F_{8:9}$	99.21875

Demerits: Requires many seasons/generations to develop and, developing RILs are relatively difficult in crops with high inbreeding depression.

These lines have several uses. First, they can be used to derive a map because it essentially is an eternal F_2 population with unlimited mapping possibilities. Additionally, these lines can be scored for morphological traits (such as disease resistance or flower color) or quantitative traits (such as yield or maturity). This morphological trait data can then be compiled and those traits can be placed on the developing molecular map. These lines are especially powerful for analyzing quantitative traits because replicated trials can be analyzed using identical genetic material. The quantitative trait data can then be used to determine if any molecular markers are closely associated with those traits.

3.4 Genetics of Mapping Molecular Loci

Each of the mapping populations will give a specific segregation ratio at each locus. The knowledge of these ratios is important to determine if the population is expressing a skewed segregation ratio at any locus. The following are the ratios that we would expect at each locus for co-dominant and dominant makers segregating in the three types of populations (Table 9).

Table 9. Ratios expected at each locus for co-dominant and dominant makers segregating in the three types of populations

Population	Co-dominant loci	Dominant loci
F_2 population	1:2:1	3:1
Backcross population	1:1	1:1*
RI population	1:1	1:1
DH	1:1	1:1

*To score a dominant marker in a backcross population, we must cross the recessive parent with the F_1 plant. Therefore to score RAPD loci we would need to create two populations, each one developed by backcrossing to one of the two parents. For this reason, backcross populations have not been used for mapping RAPD loci.

Once the segregating population was analysed using RFLP, RAPD or isozyme markers and it was determined that the segregation ratio of each locus does not deviate from the expected ratio, the next step was to begin developing the map. (It should be noted here that scientists which develop molecular maps normally include those loci with skewed segregation ratios in their mapping analysis). All of the segregation data are then compiled and used to derive the linkage relationship among the markers. This analysis is performed using computers and one program widely used is called MAPMAKER. This procedure is based on the maximum likelihood method. The output from this program is a linear relationship among the markers and the distance between the markers is measured in centi-Morgans (CM).

4.0 SPECIALIZED MAPPING TOPICS

4.1 Bulk Segregant Analysis (BSA)

Often a geneticist is not interested in developing a molecular map, but would rather find a few markers that are closely linked to a specific trait. The identification of these markers is often achieved by a procedure called bulk segregant analysis. The essence of this procedure is the creation of a bulk sample of DNA for analysis by pooling DNA from individuals with similar phenotypes. Besides the above-mentioned populations, bulked segregant analysis (BSA) approach, using any one of the above-mentioned populations (except NILs) is frequently used in gene tagging. BSA is based on the principle of isogenic lines. In BSA, two parents (say a resistant and susceptible), showing high degree of molecular polymorphism and contrast for the target trait are crossed and F_1 is selfed to generate F_2 population. For example, we may be interested in finding a molecular locus linked to a disease resistance locus. In F_2 individual plants are phenotyped for resistance and susceptibility. Usually, the DNA isolated from 10 plants in each group is pooled to constitute resistant and susceptible bulks. The resistant parent, susceptible parent, resistant bulk and susceptible bulk, are surveyed for polymorphism using polymorphic markers. A marker showing polymorphism between parents as well as bulks is considered putatively linked to the target trait, and is further used for mapping using individual F_2 plants. Conceptually, the genetic constitution of the two bulks is similar, but for the genomic region associated with the target trait. Hence, they serve the purpose of isogenic lines in principle.

It has been observed over experiments that when 10 plants are sampled in each group for constituting the bulk, the probability of a polymorphic marker (between parents as well as bulks) not being linked to the target trait is extremely low (10^{-19}). Hence, usually 10 plants are used for constituting the bulks. However, this number may vary depending upon the types of mapping populations used. In absence of isogenic lines, the BSA approach provides a very useful alternative for gene tagging (Michelmore *et al.*, 1991).

The two DNA bulks were constituted in the polymorphism survey to eliminate from consideration those polymorphic markers that were less likely to be associated with disease resistance / susceptibility. Each of these bulk DNA samples will contain a random sample of all the loci in the genome, except for those that are in the region of the gene upon which the bulking occurred. Markers that showed a similar pattern of polymorphism between the parents and between the two bulks were considered as putative resistance / susceptibility related markers. These markers were used to genotype around 50-100 F_2 or >100 $F_{2:3}$ lines to test for their association with disease resistance / susceptibility. According to the published wheat and barley genetic linkage maps (http://wheat.pw.usda.gov). Therefore, any difference in RFLP or RAPD pattern between these two bulks should be linked to the locus upon which the bulk was developed. This is a powerful technique that has gained wide acceptance in the few years since it was first described.

4.2 Combining Markers and Populations

The genetic segregation ratio at maker locus is jointly determined by the nature of marker (dominant / co-dominant) and types of mapping populations (Table 10). Therefore, a thorough understanding of the nature of markers and mapping populations is crucial for any mapping projects. Markers such RFLPs, SSR /microsatellites and CAPS etc. are co-dominant in nature, while AFLP, RAPD, ISSR are often scored as dominant markers. Mapping populations such as RILs and DHs equalize marker type because of fixation of parental alleles at marker locus in homozygous condition. These population results in 1:1 segregation ratio at marker locus irrespective of genetic nature of markers, while an F_2 population segregates in 1:2:1 ratio for a co-dominant marker and in 3:1 ratio for dominant marker. Depending upon the segregation pattern, statistical analysis of marker data will vary.

Table 10. Genetic segregation ratio at marker locus in different marker-population combinations

Marker	Nature	Genetic segregation ratio					
		F_2	RILs	DHs	NILs	Backcross population	
						B_1	B_2
RFLP	Co-dominant	1:2:1	1:1	1:1	1:1	1:1	1:1
RAPD	Dominant	3:1	1:1	1:1	1:1	1:0	1:1
AFLP	Dominant	3:1	1:1	1:1	1:1	1:0	1:1
SSR	Co-dominant	1:2:1	1:1	1:1	1:1	1:1	1:1

4.3 Characterization of Mapping Populations

Precise molecular and phenotypic characterization of mapping population is vital for success of any mapping project. Since the molecular genotype of any individual is independent of environment, it is not influenced by G x E interaction. However, trait phenotype could be influenced by the environment, particularly in case of quantitative characters. Therefore, it becomes important to precisely estimate the trait value by evaluating the genotypes in multi-location testing over years using immortal mapping populations to have a valid marker-trait association.

4.4 Segregation Distortion in Linkage Mapping

Significant deviation from expected segregation ratio in a given marker-population combination is referred to as segregation distortion. There are several reasons for segregation distortion, including: gamete/zygote lethality, meiotic drive/preferential segregation, sampling/selection during population development and differential responses parental lines to tissue culture in otherwise normal mapping population. It is therefore important that the 'goodness of fit' of segregation ratio must be tested for individual marker locus and if necessary, the marker(s) showing high degree of segregation ratio must be tested for individual marker locus and those marker(s) showing high degree of segregation distortion be eliminated from the analysis.

4.5 Sequence Tagged Sites

For many research labs, RFLPs are not an attractive molecular marker system because of the labor involved and the requirement of radioisotopes. Sequence tagged sites (or STS) may become a popular alternative to RFLP markers because they are PCR-based and do not require radioactive probing. STSs are developed by first sequencing the ends of a RAPD product or a clone used as the RFLP probe. From the sequence information, oligo-nucleotide primers 18-20 nucleotides long are synthesized that are complementary to each end of the RAPD product or the clone. These new primers are then used to amplify DNA by PCR. Two results could occur. First the size of amplification products among different DNAs (for example, two parents differing for a disease resistance locus) could be polymorphic. Alternatively, the amplification products could be monomorphic (of the same size). If this is the case, then it will be necessary to cut the products with various restriction enzymes to identify polymorphisms.

Because the primers are longer than those used for RAPD mapping, the PCR reaction can be run at a higher annealing temperature. This simple change in reaction temperature results in specific and simple amplification pattern that is very reproducible from laboratory to laboratory. (This is often not the case with RAPD technology.) Because they are portable from lab to lab, it is then possible to develop STS markers that can allow the quick location of any gene on a molecular map. First, it is necessary to define from two to three evenly dispersed STS sites for each chromosome of the species with which we are working. Then whenever a new gene of interest is identified, the linkage relationship between that gene and each of the STS loci can be established. The new gene should map within 25 cM of one of the STS loci. Once the new gene is located in relationship to an STS, then we can go to a RFLP or RAPD map and select probes or primers that will allow us to identify markers more closely linked to the new gene.

4.6 Comparative Genome Mapping

Often the clones used to identify RFLPs or any other PCR based marker(s) in one species can be used in a second species. These heterologous probes/PCR amplified products can then be used to develop a RFLP/PCR marker based map in the second species. Once the two species have been mapped, the relative evolution of the two species can be compared. Comparative mapping can identify inversions, translocation and duplications that have occurred. Genetic factors can also be assessed by comparing map distances of genes with conserved gene order in the two species.

A comparison of the related grass species sorghum and maize was made using maize clones (Whitkus *et al.*, 1992). The diploid chromosome number of the two species is twelve. Many of the sorghum chromosomes contained regions from two of the maize chromosomes. This result may represent the ancestral duplication of chromosomal material. Furthermore, during maize evolution duplicate genes have also occurred to a greater extent than that seen in sorghum. Only nine inversions of gene order have been observed between the species. Maize and sorghum were next compared with regard to genetic distance. Conserved gene orders were compared, and these linkages measured 862 cM in maize and 835 cM in sorghum. Therefore, it can be concluded that since the divergence of sorghum and maize, large chromosomal changes have not occurred and much of the recombination distance has been maintained.

The same type of analysis was performed with two more distant species, tomato and pepper (Tanksley *et al.*, 1988). Little linkage conservation appears to have been maintained since the divergence of these two species, even though the chromosome number has been conserved. The largest conserved linkage block is a 63 cM block of tomato chromosome two that was located on chromosome one of pepper. Some chromosomes of pepper contained six distinct regions of the tomato genome (pepper chromosome X). These chromosomal breakages and rearrangements are not centromeric in nature, suggesting that evolution involved breakage throughout the genome. Because duplicated regions were found in pepper, it was concluded that the gene duplications events within pepper occurred after divergence from tomato. Linkage distances of conserved gene orders were quite similar, though (229 cM in pepper vs. 172 cM in tomato) (McClean, 1998). The DNA sequence of 106 BAC/PAC clones in the minimum tiling path (MTP) of the long arm of rice chromosome 11, between map positions 57.3 and 116.2 cM, has been predicted to contain 2,932 genes. The largest number of genes, about one-third, mapped to the homoeologous group 4 chromosomes of wheat, suggesting a common evolutionary origin. The remaining genes were located on wheat chromosomes of different groups with significantly higher numbers for groups 3 and 5 (Singh *et al.*, 2004). The degree of synteny between rice and other cereals is very high. Rice (*Oryza sativa* L.) chromosome 3 is evolutionarily conserved across the cultivated cereals and shares large blocks of synteny with maize and sorghum, which diverged from rice more than 50 million years ago (Buell *et al.*, 2005). In fact, the degree of synteny between any rice chromosome and maize is astounding (Fig. 5). This principle holds true for all the cereals, which means that the genes present in one cereal will almost certainly be present in the same order in another. Virtually every part of the maize genome finds a homologue in rice, where the sequence identity is greater than 80%.

4.7 Physical and Genetic Distances

All distances found on a linkage maps are the product of genetic recombination and therefore are considered to be genetic distances. Two pairs of loci that are genetically the same distance apart may not be physically the same distance apart because of suppressed recombination in the region between two of the loci. Those loci where suppression of recombination does occur will physically be farther apart. The Table 11 gives the relationship between the physical and genetic distance in three species.

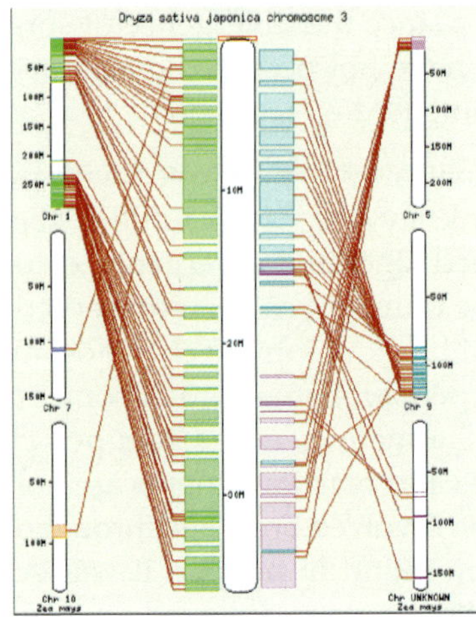

Fig. 5. Comparison of chromosome 3 of the rice japonica species with the maize species *Zea mays* (Source: http://www.gramene.org/).

Table 11. Relationship between the physical and genetic distance in three species

Species	Kilobases / CentiMorgan
Arabidopsis	139
Tomato	510
Corn	2140

But remember, these values are estimates of a single region of the genome of each species, and it is quite common to see that two regions of the same species have different amount of DNA per genetic distance.

Correlations between physical and genetic distances have also been performed using deletion stocks in wheat. A series of deletions stocks of a specific wheat chromosome were analyzed with probes known to hybridize to that chromosome. "Each locus was assigned to the chromosome region between the breakpoint of the largest deletion where the band was present and the next larger deletion where the band was absent" (Werner *et al.*,1992). The authors noted large discrepancies between the physical and genetic maps of chromosomes 7B and 7D. Two loci that are located near the centromere of 7B are 7 cM apart genetically, yet the distance between the two loci spans 25% of the chromosome. Another region at the distal region of long arm of 7B which accounts for about 15 % of the chromosome is 91 cM long genetically. This points out the difference of recombination that can occur within a single chromosome.

4.8 Linkage Drag

One goal of plant breeding is to introduce a gene from a donor parent to improve a cultivar for a specific trait. For example, a wild germplasm could be used as a source of disease resistance or male sterility or fertility restoration etc. At the same time the breeder does not want to carry any of the other genes from the wild germplasm that might reduce the agronomic fitness of the cultivar. The backcross method of plant breeding is one approach in which the introduction of a specific gene is accomplished. One major drawback of backcross breeding is 'linkage drag'. This refers to the reduction in fitness in a cultivar due to deleterious genes introduced along with the beneficial gene during backcrossing. Molecular markers offer a tool in which the amount of wild or alien DNA can be monitored during each backcross generation.

First, let's look at the amount of alien DNA that can be maintained after a backcrossing program. Young and Tanksley (1989) studied the chromosomal region around the Tm-22 allele in a number of tomato cultivars. This allele was introduced from the wild tomato species *L. peruvianum*, and it provides resistance to tobacco mosaic virus. Large variation in the amount of introgressed DNA was observed. For example, Craigella-Tm-22 contains 51 cM of wild DNA, whereas Vendor-Tm-22 and Nova-Tm- 22 each contain about 8 cM of introgressed DNA.

Backcrossed lines could be developed rapidly in which only a small amount of foreign DNA is linked to the gene of interest. After the first backcross all lines with the gene of interest could be screened with an RFLP or any other linked marker that is 1 cM away. Those lines in which a crossover occurred at this marker would be selected. Then those lines would be crossed to the recurrent parent. The progeny from these crosses would then be scored for a second marker 1 cM away on the other side of the gene. Again crossovers progeny could be selected. Thus, in only two generations the amount of wild DNA could be reduced to 2 cM. The key to this procedure is to have markers that are polymorphic between the two parents that are closely linked to the gene of interest.

Young and Tanklsley (1989) applied these principles to the reduction of *L. peruvianum* DNA in Craigella-Tm-22. The authors were able to reduce the amount of DNA on one side of the Tm-22 allele from 47 to 7 cM in one generation by selecting for crossovers at the CD32A locus.

4.9 Log of Odds (LOD) Ratio

The best hypothesis for a given locus (marker) is that hypothesis which maximizes the LOD score, where LOD is the log of the odds ratio between the two hypotheses:

$LOD = \log_{10} [P(H_1) / P(H_0)]$

Used a cutoff of LOD > 2.4

LOD scores were plotted for each locus (marker) on a chromosome. A curve-fitting algorithm was used to make a continuous curve. Remember, each unit LOD is an order of magnitude.

Kosambi mapping function:

$r = 1/2 \tanh 2m$

Where,

r = recombination frequency

m = map distance in Morgans

r itself is not really additive over long distances. The Kosambi function, like other mapping functions, is meant to give additive map distances (Paterson *et al.*, 1988).

LOD simply means the probability of linkage over probability of no linkage of any trait of interest with any marker type. It means if LOD= 4 then the probability of linkage of the trait of interest is 10^4 with the marker type over probability of no linkage. In other words trait of interest is highly linked.

Beavis *et al.* (1991) analyzed four populations of maize and found molecular markers linked to plant height. No marker was consistently associated as a QTL with plant height in all four populations. Each of the ten maize chromosomes contained a marker linked to a QTL for at least one of the four populations. The statistical analysis of quantitative traits provided valuable information for the plant breeder.

5.0 QUANTITATIVE TRAIT LOCUS (QTL)

Many traits are controlled by a single gene and fall into a few distinct phenotypic classes. These classes can be used to predict the genotypes of the individuals. For example, if we cross a round and wrinkled seeded pea plant and look at F_2 plants, we know the genotype of wrinkled plants, and we can give a generalized genotype for the round plant phenotype. Furthermore, if we know the genotype we could predict the phenotype of the plant. These types of phenotypes are called discontinuous or qualitative traits.

Some traits do not fall into discrete classes. Rather, when a segregating population is analyzed for these traits, a continuous distribution is found. Kernel color in wheat is an interesting polygenic trait to study. There are two genes that works together to determine kernel color. Dark red kernel plants are AABB while white kernel plants are aabb. When we cross a dark red with a white, the F_1 offspring are having genotype AaBb and kernels are with intermediate color *i.e.* medium pink. In case of F_2 progenies there are five different colors possible. The darkest red occurs when all four alleles are represented by upper case letters. With three upper case alleles, the color is dark pink while two upper case alleles give a medium pink, and one upper case allele results in light pink. The fifth possible color is white, represented by no upper case alleles.

We can find a whole range of a trait if it is controlled by many genes exhibiting a bell-shaped distribution. These types of traits are called continuous traits and cannot be analyzed in the same manner as discontinuous traits. Because continuous traits are often given a quantitative value, they are often referred to as quantitative traits, and the area of genetics that studies their mode of inheritance is called quantitative genetics. Furthermore, the loci controlling these traits are called quantitative trait loci or QTL.

Because many important agricultural traits such as crop yield, weight gain in animals, and fat content of meat are quantitative traits, much of the pioneering research into the modes of inheritance of these traits was performed by agricultural geneticists. Many human phenotypes such as IQ, learning ability and blood pressure also are quantitative traits. These traits are controlled by multiple genes, each segregating according to Mendel's laws. These traits can also be affected by the environment to varying degrees (McClean, 1998).

5.1 Mapping QTL with Molecular Markers

The improvement of quantitative traits has been an important goal for many plant breeding programs. With a pedigree breeding program, the breeder will cross two parents and practice selection until advanced-generation lines with the best phenotype for the quantitative trait under selection are identified. These lines will then be entered into a series of replicated trials to further evaluate the material with the goal of releasing the best lines as a cultivar. It is assumed that those lines which performed best in these trials have a combination of alleles most favorable for the fullest expression of the trait.

This type of program, though, requires a large input of labor, land, and money. Therefore plant breeders are interested in identifying the most promising lines as early as possible in the selection process. Another way to state this point is that the breeder would like to identify as early as possible those lines which contain those QTL alleles that contribute to a high value of the trait under selection. Plant breeders and molecular geneticists have joined efforts to develop the theory and technique for the application of molecular genetics to the identification of QTLs.

Traditional quantitative genetic research defined a quantitative trait in terms of variances. The total phenotypic variance was first partitioned into genetic and environmental variances. The genetic variance could then be further divided into additive, dominance and epistatic effects. From this information it was then possible to estimate the heritability of the trait and predict the response of the trait to selection. It was also possible to estimate the minimum number of genes which controlled the trait.

Mapping markers linked to QTLs identifies regions of the genome that may contain genes involved in the expression of the quantitative trait. But what functions could these genes be encoding? To answer this question we should consider a trait such as yield. What types of qualitative genes (genes inherited as simple genetic factors) could be involved in the expression of yield? The first event required for yield is meiosis. Therefore any gene that is involved in gamete formation could potentially be considered a QTL. Any of the genes involved in the protein and carbohydrate biosynthetic pathways could also affect the final yield of a plant and could also be considered to be QTLs. Each of these genes of known function may only account for a portion of the final yield. An important question that can now be posed is whether any known genes map as QTLs?

Molecular makers associated with QTLs are identified by first scoring members of a random segregating population for a quantitative trait. The molecular genotype (homozygous Parent A, heterozygous, or homozygous parent B) of each member of the population is then determined. The next step is to determine if an association exists between any of the markers and the quantitative trait.

Several methods have been described for estimating means of QTL genotypes using unreplicated progeny (Weller, 1986 and 1987; Jensen, 1989; Lander and Botstein, 1989; Simpson, 1989; Knapp *et al.*, 1990). These methods use marker genotypes or QTL genotypes estimated from marker genotypes as independent variables, genetic models for progeny derived from crosses between inbred lines, *e.g.*, BC, RI, DH, and F_2 progeny, and statistical models suitable for unreplicated individuals or lines.

The most common method of determining the association is by analyzing phenotypic and genotypic data by one-way analysis of variance and regression analysis. For each marker, each of the genotypes is considered a class, and all of the members of the population with that genotype are considered an observation for that class. (Data are typically pooled over locations and replications to obtain a single quantitative trait value for the line.) If the variance for the genotype class is significant, then the molecular marker used to define the genotype class is considered to be associated with a QTL. For those loci that are significant, the quantitative trait values are regressed onto the genotype. The R^2 value for the line is considered to be the amount of total genetic variation that is explained by the specific molecular marker. The final step is to take those molecular marker loci that are associated the quantitative trait and perform a multiple regression analysis. From this analysis, we will obtain an R^2 value which gives the percentage of the total genetic variance explained by all of the markers.

The two types of populations that are often used to identify markers linked to QTLs are $F_{2:3}$ families (or F_3 families from F_2 plants) and recombinant inbred lines (RILs). As described, above each population type has advantages and disadvantages. The primary advantage of $F_{2:3}$ families is the ability to measure the effects of additive and dominance gene actions at specific loci. Because RI lines are essentially homozygous, only additive gene action can be measured. The advantage, though, of the RI lines is the ability to perform larger experiments at several locations and even in multiple years. For many crops, it is not possible to generate enough seed to perform a multi-location experiment with population of $F_{2:3}$ families.

Cowenn (1988) and Lander and Botstein (1989) have proposed statistical models and QTL parameter estimation methods using replicated progeny to increase power for mapping quantitative trait loci. The power of tests of hypotheses about means of QTL genotypes are partly determined by the genetic model (Lander and Botstein, 1989) and the parameter estimation method (Simpson, 1989). These factors do not determine how experiment design factors affect power; however, they do determine the sample size needed to achieve a given power. The power of the test of a difference between means of QTL genotypes is increased by increasing the number of replications of lines *(r)* (Lander and Botstein, 1989). Lander and Botstein (1989) proposed using "progeny testing" $(r > 1)$ to increase the "power of QTL mapping" and compared using replicated recombinant inbred (RI) and un-replicated backcross (BC) progeny. They proposed, "recombinant inbred strains will thus typically be more efficient for QTL mapping than an equal number of backcross progeny" by "reducing the environmental variance through replicate phenotypic measurements within each recombinant inbred line" (Lander and Botstein, 1989). Power is obviously increased by increasing *r*, but the effect of *r* has not been quantified. Furthermore, power for replicated progeny is determined by *n* and other factors in addition to *r*, and how these factors affect power has not been quantified. The standard error of a QTL genotype mean is a function of the environmental variance and unexplaines genetic variance. Increasing *r* decreases the standard error of a QTL genotype mean, and consequently increases power for testing hypotheses about means of QTL genotypes, but this standard error is determined by *r* and *n*.

6.0 MARKER ASSISTED SELECTION: MAPPING AND TAGGING OF GENES

Plant improvement, either by natural selection or through the efforts of breeders, has always relied upon creating, evaluating and selecting the right combination of alleles. The manipulation of a large number of genes is often required for improvement of even the simplest of characteristics (Flavell, 1995). With the use of molecular markers it is now a routine to trace valuable alleles in a segregating population and mapping them. These markers once mapped enable dissection of the complex traits into component genetic units more precisely (Hayes, 1993), thus providing breeders with new tools to manage these complex units more efficiently in a breeding programme.

The very first genome map in plants was reported in maize (Helentjaris *et al.*, 1986; Gardiner *et al.*, 1993), followed by rice (Mc Couch *et al.*, 1988), Arabidopsis (Chang *et al.*, 1988; Nam *et al.*, 1989) etc. using RFLP markers. Maps have since then been constructed for several other crops like potato, barley, banana, members of Brassicaceae, etc. (Winter and Khal, 1995). Once the framework maps are generated, a large number of markers derived from various techniques are used to saturate the maps as much as is possible. Microsatellite markers, especially STMS markers, have been found to be extremely useful in this regard. Owing to their quality of following clear Mendelian inheritance, they can be easily used in the construction of index maps, which can provide an anchor or reference point for specific regions of the genome. The very first attempt to map microsatellites in plants was made by Zhao and Kochert (1992) in rice using (GGC)*n*, followed by mapping of (GA)*n* and (GT)*n* by Tanksley *et al.* (1995) and (GA/AG)*n*, (ATC) 10 and (ATT) 14, by Panaud *et al.* (1995) in rice. The potato microsatellite map has been generated by Milbourne *et al.* (1998). Similar to microsatellites, looking at the pattern of variation, generated by retrotransposons, it is now proposed that apart from genetic variability, these markers are ideal for integrating genetic maps (Ellis *et al.*, 1998).

The very first reports on gene tagging were from tomato (Paterson *et al.*, 1988; Weller *et al.*, 1988; Williamson *et al.*, 1994), followed by identification of markers linked to genes involved in several traits like water use efficiency (Martin *et al.*, 1989), resistance to *Fusarium oxysporum* (the 12 gene) (Sarfatti *et al.*, 1989), leaf rust resistance genes *LR9* and *LR24* (Schachermayr *et al.*, 1994, 1995), and root knot nematodes (*Meliodogyne* sp.) (the *mi* gene) (Klein *et al.*, 1991; Messeguer *et al.*, 1991). Xiao *et al.* (1998) have shown the utility of RFLP markers in identifying the trait improving QTL alleles from wild rice relative *O. rufipogon*.

Allele-specific associated primers have also exhibited their utility in genotyping of allelic variants of loci that result from both size differences and point mutations. Some of the genuine examples of this are the *waxy* gene locus in maize (Shattuck *et al.,* 1991), the *Glu D1* complex locus associated with bread making quality in wheat (D'Ovidio and Anderson, 1994), the *Lr1* leaf rust resistance locus in wheat (Feuillet *et al.*, 1995), the *Gro1* and *H1* alleles conferring resistance to the root cyst nematode *Globodera rostochiensis* in potato (Niewohner *et al.,* 1995), and allele-specific amplification of polymorphic sites for detection of powdery mildew resistance loci in cereals (Mohler and Jahoor, 1996). A number of other traits have been tagged using ASAPs in tomato, lettuce, etc. (Olson *et al.*, 1989; Klein *et al.*, 1991; Paran *et al.*, 1991). Besides ASAPs, AFLP and SSR markers have been identified to be associated with quantitative resistance to *Globodera pallida* (stone) in tetraploid potato, which can be very well employed in marker-assisted selection (Bradshaw *et al.,* 1998).

STMS markers have displayed a potential use as diagnostic markers for important traits in plant breeding programmes, *e.g.* $(AT)_{15}$ repeat has been located within a soybean heat shock protein gene which is about 0.5 cM from *Rsv* a gene conferring resistance to soybean mosaic virus (Yu *et al.,* 1994). Several resistance genes including peanut mottle virus (*Rpv*), phytophthora (*Rps3*) and Javanese root-knot nematode are clustered in this region of the soybean genome.

Similar to specific markers like RFLPs, STMS and ASAPs, arbitrary markers like RAPDs have also played important role in saturation of the genetic linkage maps and gene tagging. Their use in mapping has been especially important in systems, where RFLPs have failed to reveal much polymorphism. One of the first uses of RAPD markers in saturation of genetic maps was reported by Williams *et al.* (1991). They have proven utility in construction of linkage maps among species where there has been inherent difficulty in producing F_2 segregating populations and have large genome size, *e.g.* conifers (Carlson *et al.*, 1991; Chaparoo *et al.*, 1992). RAPD markers in near isogenic lines can be converted into SCARs and used as diagnostic markers. SCAR/STS marker linked to the translocated segment on 4 AL of bread wheat carrying the *Lr28* gene has been tagged by Naik *et al.* (1998). Recently, ISSRs, which too belong to the arbitrary marker category, but are found to be devoid of many of the drawbacks shared by RAPD class of markers, have been employed as a reliable tool for gene tagging. An ISSR marker $(AG)_8$YC has been found to be linked closely (3.7 ± 1.1 cM) to the rice nuclear restorer gene, *RF1* for fertility. *RF1* is essential for hybrid rice production and this marker would be useful not only for breeding both restorer and maintainer lines, but also for the purity management of hybrid rice seeds (Akagi *et al.,* 1996). Similarly ISSR marker $(AC)_8$ YT has been found to be linked to the gene for resistance to fusarium wilt race 4 in repulsion at a distance of 5.2 cM in chickpea (Ratnaparkhe *et al.,* 1998).

Apart from mapping and tagging of genes, an important utility of markers like RFLP has been observed in detecting gene introgression in a backcross breeding programme (Jena and Khush, 1990), and synteny mapping among closely related species (Gale and Devos, 1998). Similar utility of STMS markers has been observed for reliable pre-selection in a marker assisted selection backcross scheme (Ribaut *et al.*, 1997). Apart from specific markers, DAMD-based DNA fingerprinting in wheat has also been useful for monitoring backcross-mediated genome introgression in hexaploid wheat (Somers *et al.*, 1996).

7.0 CONCLUSION

It is evident from the foregoing discussion that the short-term mapping populations such as F_2, backcross and conceptual isogenic lines developed through BSA approach can be a good starting point in molecular mapping, while long-term mapping populations such as RILs, NILs for global mapping projects. As a matter of fact, the development and phenotypic characterization of mapping populations should become an integral part of the ongoing breeding programmes in important crops. At this point, the role of geneticists and plant breeders becomes crucial to reap the benefits of molecular plant breeding.

With the use of molecular markers it is now a routine to trace valuable alleles in a segregating population and mapping them like yield, disease resistance, stress tolerance, seed quality, etc. These markers once mapped enable dissection of the complex traits into component genetic units more precisely thus providing breeders with new tools to manage these complex units more efficiently in a breeding programme. A large number of monogenic and polygenic loci for various traits have been identified in a number of plants, which are currently being exploited by breeders and molecular biologists together, so as to make the dream of marker-assisted selection come true. Tagging of useful genes like the ones responsible for conferring resistance to plant pathogen, synthesis of plant hormones, drought tolerance and a variety of other important developmental pathway genes, is a major target. Such tagged genes can also be used for detecting the presence of useful genes in the new genotypes generated in a hybrid programme or by other methods like transformation, etc. DNA based markers have proved their importance as markers for gene tagging and are very useful in locating and manipulating quantitative trait loci (QTL) in a number of crops. Molecular analysis of quantitative traits now provides new tools, not only as selection tools for plant breeding, but as starting points for the cloning of these genes. These objectives could not have been realized without molecular markers.

8.0 REFERENCES

Akagi, H., Yokozeki, Y., Inagaki, A., Nakamura, A. and Fujimura, T. 1996. A codominant DNA marker closely linked to the rice nuclear restorer gene, *Rf-1*, identified with inter-SSR fingerprinting. Genome, 39:1205–1209.

Beavis, W.D., Grant, D., Albertsen, M. and Fincher, R. 1991. Quantitative trait loci for plant height in four populations. Theoretical and Applied Genetics, 83:141-145.

Blair M.W., Panaud O. and McCouch S.R. 1999. Inter-simple sequence repeat (ISSR) amplification for analysis of microsatellite motif frequency and fingerprinting in rice (*Oryza sativa* L.) Theoretical and Applied Genetics, 98:780–792.

Bradshaw, J.E., Hackett, C.A., Meyer, R.C., Milbourne, D., McNicol, J.W., Phillips, M. S. and Waugh, R. 1998. Identification of AFLP and SSR markers associated with quantitative resistance to *Globodera pallida* (Stone) in tetraploid potato (*Solanum tuberosum* subsp. *tuberosum*) with a view to marker-assisted selection. Theoretical and Applied Genetics, 97:202–210.

Buell, C.R., Yuan, Q., Ouyang, S., Liu, J., Zhu, W., Wang, A., Maiti, R., Haas, B., Wortman, J., Pertea, M., Jones, K.M., Kim, M., Overton, L., Tsitrin, T., Fadrosh, D., Bera, J., Weaver, B., Jin, S., Johri, S., Reardon, M., Webb, K., Hill, J., Moffat, K., Tallon, L., Van, Aken S., Lewis, M., Utterback, T., Feldblyum, T., Zismann, V., Iobst, S., Hsiao, J., de Vazeille, A.R., Salzberg, S.L., White, O., Fraser, C., Yu, Y., Kim, H., Rambo, T., Currie, J., Collura, K., Kernodle-Thompson, S., Wei, F., Kudrna, K., Ammiraju, J.S., Luo, M., Goicoechea, J.L., Wing, R.A., Henry, D., Oates, R., Palmer, M., Pries, G., Saski, C., Simmons, J., Soderlund, C., Nelson, W., de la Bastide, M., Spiegel, L., Nascimento, L., Huang, E., Preston, R., Zutavern, T., Palmer, L., O'Shaughnessy, A., Dike, S., McCombie, W.R., Minx, P., Cordum, H., Wilson, R., Jin, W., Lee, H.R., Jiang, J. and Jackson, S. 2005. Rice Chromosome 3 sequencing consortium. sequence annotation and analysis of synteny between rice chromosome 3 and diverged grass species. Genome Research, 15(9):1284-91.

Carlson, J.E., Tulsieram, L.K., Glaubitz, J.G., Luk, V.W.K., Kauffeldt, C. and Rutledge, R. 1991. Segregation of random amplified DNA markers in F_1 progeny of conifers. Theoretical and Applied Genetics, 83:194–200.

Chang, C., Bowman, A.W., Lander, E.S. and Meyerowitz, E.W. 1988. Restriction fragment length polymorphism linkage map for *Arabidopsis thaliana.* Proceedings of the National Academy of Sciences USA, 85:6856–6860.

Chaparoo, J., Werner, D., O'Malley, D. and Sederoff, R. 1992. Plant Genome I, San Diego, Nov. 9–11, 1992, p. 21.

Cowenn, M. 1988. The use of replicated progenies in marker based mapping of QTL's. Theoretical and Applied Genetics, 75:857-862.

D'Ovidio, R. and Anderson, O.D. 1994. PCR analysis to distinguish between alleles of a member of a multigene family correlated with wheat bread making quality. Theoretical and Applied Genetics, 88:759–763.

Ellis, T.H., Poyser, S.J., Knox, M.R., Vershinin, A.V. and Ambrose, M.J. 1998. Polymorphism of insertion sites of *Tyl-copia* class and its use for linkage and diversity analysis in pea. Molecular and General Genetics, 260:9–91.

Feuillet, C., Messmer, M., Schachermayr, G. and Keller, B. 1995. Genetic and physical characterisation of the *Lrl* in wheat. Molecular and General Genetics, 248:553–562.

Flavell, R. B. 1995. Plant biotechnology R&D -The next ten years. Tibtech, 13:313–319.

Fornari B, Taurchini D, and Villani F. 1999. Genetic structure and diversity of two Turkish *Castanea sativa* Mill. populations investigated with isozyme and RAPD polymorphisms. Journal of Genetics and Breeding, 53:315–325.

Fristensky, B. 2004. Plant 7680, Plant Molecular Genetics, University of Manitoba. (http://www.umanitoba.ca)

Gale, M. D. and Devos, K. M. 1998. Comparative genetics in the grasses. Proceedings of the National Academy of Sciences USA, 95:1971–1974.

García E., Jamilena M. and Alvarez J.I. 1998. Genetic relationships among melon breeding lines revealed by RAPD markers and agronomic traits. Theoretical and Applied Genetics, 96:878–885.

Gardiner, J. M., Coe, E. H., Melia-Hancock, S., Hoisington, D. A. and Chao, S. 1993. Development of a core RFLP map in maize using and immortalized F_2 population. Genetics, 134:917–930.

Hayes, P. M. 1993. Quantitative trait locus effects and environmental interaction in a sample of North American barley germplasm. Theoretical and Applied Genetics, 87:392–401.

Helentjaris, T., Slocum, M., Wright, S., Schaefer, A. and Nienhuis, J. 1986. Construction of genetic linkage maps in maize and tomato using restriction fragment length polymorphisms. Theoretical and Applied Genetics, 72:761–769.

Hoey, B.K., Crowe, K.R. and Jones, V.M. 1996. A phylogenetic analysis of *Pisum sativum* based on morphological characters and allozyme and RAPD markers. Theoretical and Applied Genetics, 92:92–100.

http://en.wikipedia.org/wiki/Expectation-maximization_algorithm

http://wheat.pw.usda.gov

http://www.arabidopsis.org

http://www.gramene.org

http://www.ndsu.edu/instruct/mcclean/plsc731/index.htm

Jena, K.K. and Khush, G.S. 1990. Introgression of genes from *Oryza officinalis* Well ex Watt to cultivated rice, *O. sativa* L. Theoretical and Applied Genetics, 80:737–745.

Jensen.J. 1989. Estimation of recombination parameters between a quantitative trait locus (QTL) and two marker gene loci. Theoretical and Applied Genetics, 78 (6):13-18.

Kaga, A.N., Tomooka, N., Egawa, Y., Hosaka, K. and Kamijima, O. 1996. Species relationships in subgenus *Ceratotropis* (genus *Vigna*) as revealed by RAPD analysis. Euphytica, 88:17–24.

Khasa, P.D. and Dancik, B.P. 1996. Rapid identification of white-Engelman spruce species by RAPD markers. Theoretical and Applied Genetics, 92:46–52.

Klein-Lankhorst, R.M., Vermunt, A., Weide, R., Liharska, T. and Zabel, P. 1991. Isolation of molecular markers for tomato (*L. esculentum*) using random amplified polymorphic DNA (RAPD). Theoretical and Applied Genetics, 83:108–114.

Klein-Lankhorst, R.M., Rietveld, P., Machiels, B., Verkerk, R., Weide, R., Gebhardt, C., Kornneef, M. and Zabel, P. 1991. RFLP markers linked to the root knot nematode resistance gene *Mi* in tomato. Theoretical and Applied Genetics, 81:661–667.

Knapp, S.J., Bridges, W.C. and Birkes, D. 1990. Mapping quantitative trait loci using molecular marker linkage maps. Theoretical and Applied Genetics 79: 583-592.

Lander. S., and Botstein, D. 1989. Mapping Mendelian factors underlying quantitative traits using RFLP linkage maps. Genetics, 121: 185-199.

Lefebvre, V., Palloix, A., Caranata, C. Pochard, E. 1995. Construction of an intraspecific linkage map of pepper using molecular markers and doubled-haploid progenies. Genome 38:112-121.

Liu, B.M. 1998. Statistical Genomics. Genomics 1:174-181.

Martin, B., Nienhuis, J., King, G. and Schaefer, A.1989. Restriction fragment length polymorphisms associated with water use efficiency in tomato. Science, 243:1725–1728.

McClean, P.E. 1998. Plant Molecular Genetics. PLSC 731. (http://www.ndsu.nodak.edu).

Mc Couch, S.R., Kochert, G., Yu, Z.H., Wang, Z.Y., Khush, G.S., Coffman, W.R. and Tanksley, S.D. 1998. QTL and Molecular mapping of rice chromosomes. Theoretical and Applied Genetics, 76:815–829.

Messeguer, R., Ganal, M., DeVincente, M.C., Young, N.D., Bolkan, H. and Tanksley, S.D. 1991. High resolution RFLP map around the root-knot nematode resistance gene (Mi) in tomato. Theoretical and Applied Genetics, 82:529–536.

Michelmore, R.W., Paran, I. and Kesseli, R.V. 1991. Identification of markers linked to disease resistance genes by bulked segregant analysis: A rapid method to detect markers in specific genomic regions by using segregating populations. Proceedings of the National Academy of Sciences USA 88:9829-9832.

Milbourne, D., Meyer, R.C., Collins, A.J., Ramsay, L.D., Gebhardt, C. and Waugh, R. 1998. Isolation, characterization and mapping of simple sequence repeat loci in potato. Molecular and General Genetics, 259:233–245.

Moeller D.A. and Schaal, B.A. 1999. Genetic relationships among native American maize accessions of Great Plains assessed by RAPDs. Theoretical and Applied Genetics 99:1061–1067.

Mohler, V. and Jahoor, A. 1996. Allele-specific amplification of polymorphic sites for the detection of powdery mildew resistance loci in cereals. Theoretical and Applied Genetics, 93:1078–1082.

Naik, S., Gill, K.S., Prakasa Rao, V.S., Gupta, V.S., Tamhankar, S.A., Pujar, S., Gill, B.S. and Ranjekar, P.K. 1998. Identification of a STS marker linked to the *Aegilops speltoides*-derived leaf rust resistance gene *Lr28* in wheat. Theoretical and Applied Genetics, 97:535–540.

Nam, H.G., Giraudat, J., Den Boer, B., Moonan, F., Loos, W.D.B., Hauge, B.M. and Goodman, H.M. 1989. Restriction fragment length polymorphism linkage map of *Arabidopsis thaliana.* Plant Cell, 1:699–705.

Nebauer, S.G., Del CastilloAgudo, L. and Segura, J. 1999. RAPD variation within and among natural populations of outcrossing willow- leaved foxglove (*Digitalis obscura* L.). Theoretical and Applied Genetics, 98:985–994.

Nei and Li. 1979. Mathematical Model for Studying Genetic Variation in Terms of Restriction Endonucleases. Proceedings of the National Academy of Sciences USA, 76:5269-5273.

Niewohner, J., Salamini, F. and Gebhardt, C. 1995. Development of PCR assays diagnostic for RFLP marker alleles closely linked to alleles *Gro1* and *H1*, conferring resistance to the root cyst-nematode *Globodera rostochiensis* in potato. Molecular Breeding, 1:65–78.

Olson, M., Hood, L., Cantor, C. and Botstein, D. 1989. A common language for physical mapping of the human genome. Science, 245:1434–1435.

Owuor, E.D., Fahima, T., Beiles, A. and Korol, A. 1997. Population genetic response of microsite ecological stress in wild barley *Hordeum spontaneum*. Molecular Ecology, 6:1177–1187.

Panaud, O., Chen, X. and McCouch, S.R. 1995. Frequency of microsatellite sequences in rice (*Oryza sativa* L.). Genome, 38:1170–1176.

Paran, I., Kesseli, R. and Michelmore, R. 1991. Identification of RFLP and RAPD markers linked to downy mildew resistance genes in lettuce using near isogenic lines. Genome, 34:1021–1027.

Paterson, A.H., Lander, E.S., Hewitt, J.D., Peterson, S.E., Lincoln, S.E. and Tanskley, S.D. 1988. Resolution of quantitative traits into Mendalian factors by using a complete linkage map of restriction fragment length polymorphisms. Nature, 335:721-726.

Phillip McClean. 1998. http://www.ndsu.edu/instruct/mcclean/plsc731/index.htm

Podlich, D.W., Winkler, C.R. and Cooper, M. 2004. Mapping As you Go: An Effective Approach for Marker-Assisted Selection of Complex Traits. Crop Science, 44:1560-1571.

Ratnaparkhe, M.B., Gupta, V.S., Ven Murthy, M.R. and Ranjekar, P.K. 1995. Genetic fingerprinting of pigeonpea (*Cajanus cajan*) and its wild relatives using RAPD markers. Theoretical and Applied Genetics 91:893-898.

Ratnaparkhe, M.B., Santra, D.K., Tullu, A. and Muehlbauer, F.J. 1998. Inheritance of inter-simple sequence repeat polymorphisms and linkage with a fusarium wilt resistance gene in chickpea. Theoretical and Applied Genetics, 96:348–353.

Ribaut, J.M., Hu, X., Hoisington, D. and Gonzalez-de-leon, D. 1997. Genome mapping, molecular markers and marker-assisted selection in crop plants. Plant Molecular Biology, 15:154–162.

Santalla, M., Power, J.B. and Davey, M.R. 1998. Genetic diversity in mungbean germplasm revealed by RAPD markers. Plant Breeding, 117:473–478.

Sarfatti, M., Katan, J., Fluhr, R. and Zamir, D. 1989. An RFLP marker in tomato linked to the *Fusarium oxysporum* resistance gene I-2. Theoretical and Applied Genetics, 78:755–759.

Schachermayr, G., Messmer, M.M., Feuillet, C., Winzeler, H. and Keller, B. 1995. Identification of molecular markers linked to *Agropyron elongatum* derived leaf rust resistance gene *Lr24* in wheat. Theoretical and Applied Genetics, 90:982–990.

Schachermayr, G., Siedler, H., Winzeler, H. and Keller, B. 1994.Identification and localization of molecular markers linked to *Lr9* leaf rust resistance gene of wheat. Theoretical and Applied Genetics, 88:110–115.

Sharma, S.K., Dowsons, I.K. and Waugh, R. 1995. Relationships among cultivated and wild lentils revealed by RAPD analysis. Theoretical and Applied Genetics, 91:647–654.

Shattuck-Eidens, D., Bell, R.N., Mitchell, J.T. and McWorther, V.C. 1991. Rapid detection of maize DNA sequence variation. Genetic Analysis Techniques and Applications, 8:240–245.

Simpson, S.P. 1989. Detection of linkage between quantitative trait loci and restriction fragment length polymorphisms using inbred lines. Theoretical and Applied Genetics, 77:815-819.

Singh, N.K., Raghuvanshi, S., Srivastava, S.K., Gaur, A., Pal, A.K., Dalal, V., Singh, A., Ghazi, I.A., Bhargav, A., Yadav, M., Dixit, A., Batra, K., Gaikwad, K., Sharma, T.R., Mohanty, A., Bharti, A.K., Kapur, A., Gupta, V., Kumar, D., Vij, S., Vydianathan, R., Khurana, P., Sharma, S., McCombie, W.R., Messing, J., Wing, R., Sasaki, T., Khurana, P., Mohapatra, T., Khurana, J.P. and Tyagi, A.K. 2004. Sequence analysis of long arm of rice chromosome 11 for rice-wheat synteny. Functional and Integrative Genomics. 4(2):102-17.

Somers, D.J., Zhou, Z., Bebeli, P. and Gustafson, J.P. 1996. Repetitive, genome-specific probes in wheat (*Triticum aestivum* L.) amplified with minisatellite core sequences. Theoretical and Applied Genetics, 93:983–989.

Sonnante, G., Marangi, A., Venora, G. and Pignone, D. 1997. Using RAPD markers to investigate genetic variation in chickpea. Journal of Genetics and Breeding, 51:303–307.

Tanksley, S.D., Bernatzky, R., Lapitan, N.J. and Prince. J.P. 1988. Conservation of gene repertoire but not gene order in pepper and tomato. Proceedings of the National Academy of Sciences USA, 85:6419.

Tanksley, S.D., Ganal, M.W. and Martin, G.B. 1995. Chromosome landing: A paradigm for map-based cloning in species with large genomes. Trends in Genetics, 11:63–68.

Weller J.I. 1986. Maximum likelihood techniques for the mapping and analysis of quantitative trait loci with the aid of genetic markers. Biometrics, 42:627-640.

Weller, J.I., Soller, M. and Brody, T. 1988. Linkage analysis of quantitative traits in an inter-specific cross of tomato (*Lycopersicon esculentum* x *Lycopersicon pimpinellifolium*) by means of genetic markers. Genetics, 118:329–339.

Weller, J.I. 1987. Mapping and analysis of quantitative trait loci in *Lycopersicon* (tomato) with the aid of genetic markers using approximate maximum likelihood methods. Heredity, 59:413-421.

Werner, J.E., Endo, T.R. and Gill, B.S. 1992. Toward a cytogenetically based physical map of the wheat genome. Proceedings of the National Academy of Sciences USA, 89:11307-11311.

Whitkus, R., Doebley, J. and Lee, M. 1992. Comparative Genome Mapping of Sorghum and Maize. Genetics, 132:1119.

Williams, M.N.V., Pande, N., Nair, S., Mohan, M. and Bennett, J. 1991. Restriction fragment length polymorphism analysis of polymerase chain reaction products amplified from mapped loci of rice (*Oryza sativa* L.) genomic DNA. Theoretical and Applied Genetics, 82:489–498.

Williamson, V.M., Ho, J-Y., Wu, F.F., Miller, N. and Kaloshian, I. 1994. PCR-based marker tightly linked to the nematode resistance gene, *Mi*, in tomato. Theoretical and Applied Genetics, 87:757–763.

Winter, P. and Kahl, G. 1995. Molecular marker technologies for plant improvement. World Journal of Microbiology and Biotechnology, 11:438–448.

Xiao, J., Li, J., Grandillo, S., Ahn, S.N., Yuan, L., Tanksley, S.D. and McCouch, S.R. 1998. Identification of trait-improving quantitative trait loci alleles from a wild rice relative, *Oryza rufipogon*. Genetics, 150:899–909.

Yeh, F.C., Chong, D.K.X. and Yang, R.C. 1995. RAPD variation within and among natural populations of Trembling Aspen (*Populus tremuloides* Michx.) from Alberta. Journal of Heredity, 86:454–460.

Young, N.D and Tanksley, S.D. 1989. RPLP analysis of the size of chromosomal segments retained around the Tm-2 locus of tomato during backcross breeding. Theoretical and Applied Genetics, 77: 353-359.

Yu, Y.G., Saghai, Maroof, M.A., Buss, G.R., Maughan, P.J. and Tolin, S.A. 1994. RFLP and microsatellite mapping of a gene for soybean mosaic virus resistance. Phytopathology, 84:60–64.

Zamir and Tanksley. 1988. Tomato genome is comprised largely of fast-evolving, low copy-number sequences. Molecular and General Genetics,, 213:254-261.

Molecular Plant Breeding: Principle, Method and Application
Eds : R.K. Singh, Rajesh Singh, Guoyou Ye, A. Selvi and G.P. Rao
Studium Press LLC, Texas, USA, 2009, pp. 135-159

CHAPTER

5

Mapping and Tagging of Qualitative Traits in Crop Plants

J. RAMALINGAM[1] *and A. SELVI*[2]

ABSTRACT

Qualitative traits are simple in inheritance and follow Mendelian principles. Developing new faster tools are needed to make such simply inherited gene hunt cheaper and practical which stays the primary goal of any genome assisted selection programme. Genetic mapping is one such tool to isolate gene of interest causing discrete phenotypic variation. Genetic mapping offers firm evidence that a disease/insect/ quality characters are transmitted from parent to offspring is linked to one or more genes. It also provides information on chromosomal location of the gene and more precisely where it exactly lies on the chromosome. Genetic maps have been used successfully to find the single gene responsible for most of biotic stresses in plants like disease and insect resistance, quality characters etc. This chapter deals with the difference between qualitative and quantitative traits, linkage map construction, efficient marker assisted selection strategies and case studies with reference to disease resistance in crop plants. It also deals with the introgression of qualitative traits which are typically controlled by single, dominant genes, by backcross breeding and the significance of foreground, recombinant and background selection for precise transfer of specific genomic regions and quick recovery of recurrent parent genotype.

Key Words: Single gene, Qualitative traits, Linkage map, MAS.

[1]*Department of Plant Molecular Biology and Biotechnology,Centre for Plant Molecular Biology,Tamil Nadu Agricultural University,Coimbatore-641007 ,Tamil Nadu, India*
[2]*Sugarcane Breeding Institute,Coimbatore-641007, Tamil Nadu, India*
Corresponding author, e-mail: jrjagadish@tnau.ac.in

1.0 INTRODUCTION

The main objective of any plant breeding programme is to introgress one or a few genes from donor into highly adopted variety and to recover most of the recipient parental genome as rapidly as possible. This can achieve by either traditional back cross or through transgenic approach. Molecular markers turn handy when the system of screening for a pest or disease or any simple inherited character (qualitative) is laborious and difficult (recessive genes control). The main use of molecular markers has been the construction of linkage map and to locate the chromosomal location of the gene of interest. With the use of molecular markers, it is now a routine to trace valuable alleles in a segregating population and mapping them. These markers once mapped enable easy identification of both dominant and recessive alleles more precisely, thus providing breeders with new tools to manage the traits more efficiently in a breeding programme.

Molecular marker assisted (MAS) selection is powerful tool for indirect selection of simply inherited traits at early stage of crop thereby speedup the conventional plant breeding programme. MAS have been particularly effective for selection of simply inherited traits which are qualitative in nature. Qualitative traits are those in which the phenotype of the plants in segregating populations can be classified easily into discrete classes as opposed to quantitative traits which show continuous variation and are influenced by environment. Qualitative traits are mostly controlled by one or two major genes and are generally not influenced by environment. For a better understanding of the qualitative and quantitative traits, Mackill

Table 1. Classification of major and minor gene traits

Trait	Segregation	Percentage of phenotypic variation explained	Example	Classification
Qualitative	Discrete	100	Purple leaf, Blast resistance	Major gene
Semi-Quantitative	Discrete	100	Semi-dwarfism in rice *sd 1*	Major gene
Quantitative	Continuous	> 50	Submergence tolerance gene *Sub1*	Major gene
Quantitative	Continuous	25-50	Stem rot resistance in rice	Major QTL
Quantitative	Continuous	< 25	QTL for most agronomic and physiological traits	QTL

and Junjian (2001) have classed them based on the mechanism of gene control (Table 1). DNA markers have enormous potential to improve the efficiency and precision of the screening process *via* marker-assisted selection (MAS). A number of varieties/genotypes had been developed in different crops using MAS. This chapter deals with genetic markers, linkage map construction and marker assisted selection for qualitative traits in several crops.

2.0 MOLECULAR MARKERS

Number of phenotypically neutral, random DNA markers can be generated for any species and have been successfully used in many studies to map gene of interest. However genetic linkage between a specific marker and target locus allele can be broken by genetic recombination. This limits the use of random markers as genetic tools (Rafalski and Tingsey, 1993). In recent times a number of third generation markers such as EST, SNPs etc are also used to saturate linkage maps. These markers are already discribed in detail in the chapter 2.

3.0 LINKAGE MAP

A linkage map is a graphical representation of the position of genes along with linked markers within a linkage group or it is a genetic map of a species or experimental population that shows the position of its known genes and/or genetic markers relative to each other in terms of recombination frequency, rather than as specific physical distance along each chromosome. The recombination frequency is a measure of genetic linkage and is used in the creation of a genetic linkage map. A centimorgan (cM) named for the pioneering geneticist Thomas Hunt Morgan is a unit that describes a recombination frequency of 1% (recombination distance between two loci that recombine in 1 out of 100 gametes). Genetic mapping is based on the principle that gene and markers cosegregate *via* chromosomal recombination (crossing over) during meiosis, thus allowing their analysis in the progeny (Paterson, 1996). The tightly linked markers or genes will inherit together from parent to progeny more frequently than to those of located apart (distantly linked). The percentage of recombinants formed by F_1 individuals can range from a fraction of 1% up to the 50% always seen with gene loci on separate chromosomes. The farther two markers apart, the higher the frequency of recombination should be. Recombination frequency is an indirect indication of map distance. The relative distance between two genes/markers can be calculated using the segregation pattern of gene/markers in the offspring of an organism showing two linked genetic traits, and finding the percentage of the offspring where the two traits/markers do not run together. The higher the percentage of descendants that does not show both traits/markers, the further apart on the chromosome they are. In other words, higher percentage of parental types for a pair of traits/markers, closer the distance between them. There is number of parental and recombinants types in segregating population and the recombinant frequency can be calculated from this. The important usage is to identify the chromosomal locations containing genes associated with trait of interest.

4.0 MAPPING POPULATIONS

The segregating generations like F_2, back cross, doubled haploid lines or recombinant inbred lines are prerequisites for molecular mapping (Fig. 1). An F_2 population is developed by selfing (intermating for cross pollinated species) among F_1 individuals. These F_1 individuals are developed by crossing two parents that show significant polymorphism for whichever type of loci you are going to score. Backcross populations are developed by crossing the F_1 with one of the two parents used in the initial cross. The major drawback to using F_2 or backcross populations is that the populations are not eternal. Therefore, the source of tissue to isolate DNA or protein will be exhausted at some point in time.

Populations of recombinant inbred lines (RI) can be a powerful solution to this problem. Recombinant inbred lines are developed by single-seed descent method from individual plants of an F_2 population. Single-seed descent is repeated for several generations. At this point, all of the seed from an individual plant is bulked. For example, a $F_{3:4}$ RI populations underwent single-seed descent through the F_3 generation, and were bulked to develop the F_4. This population of seed can then be grown to obtain a large quantity of seed of each individual line. Importantly, each of the lines is fixed for many recombination events (Table 2).

Table 2. Percentage of homozygosity at cash locus within line in different generations

RI population level of inbreeding	Percentage within-line homozygosity
$F_{3:4}$	75.0
$F_{4:5}$	87.5
$F_{5:6}$	92.25
$F_{6:7}$	96.875
$F_{7:8}$	98.4375
$F_{8:9}$	99.21875

The recombinant inbred lines has several uses such as, to derive a map because essentially it is an eternal F_2 population with unlimited mapping possibilities and scoring for qualitative traits (such as disease or insect resistance or flower color) or quantitative traits (such as yield or maturity). This morphological trait data can then be compiled and can be placed on the developing molecular map. The qualitative trait data can then be used to determine if any molecular markers are closely associated with those traits. The main problem associated with RI is time taken to develop the population as it require nearly six to eight generations.

Doubled haploid (DH) population is also used for mapping within given plant species. It is produced from pollen grains or egg cells to produce a haploid cell line and the haploid cells undergo chromosome doubling. It is standard resource in genetic mapping and can be produced within two years of initial cross regardless of the species but limited to tissue culture amenable species. The doubled haploid lines increase the genetic advance per unit of time or even of means. Furthermore, it allows efficient use of marker-assisted selection. Indeed the combination of the use of doubled haploids and molecular markers is a powerful tool at the disposal of the plant breeder to bring together in the same genotype the favorable alleles of detected quantitative trait loci in a population of crosses (Gallais and Bordes, 2007). DH populations have been used to produce genetic maps in number of crop species including rice (Ramalingam *et al.*, 2003), wheat (Zhang *et al.*, 2008), oat (Tanhuanpää, 2008), barley (Dahleen *et al.*, 2003) etc. DH populations played a major role in facilitating the generation of the molecular marker maps and eternal source for linkage analysis.

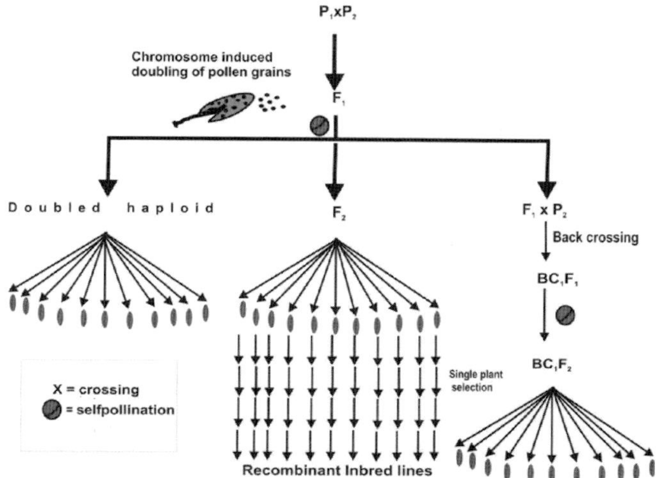

Fig. 1. Development of different mapping population for self pollinated crops (Source : Collard *et al.*, 2005)

5.0 LINKAGE MAP CONSTRUCTION

Linkage maps have been constructed from the analysis of many polymorphic segregating markers. The range of polymorphisms that are available is increasing and the advent of large-scale cDNA and genomic sequencing is a source of an ever-increasing set of available markers. Sufficient polymorphism between parental lines is critical for construction of linkage map. It is mostly observed when the chosen parents are genetically divergent (Collard *et al.*, 2003; Yu and Nguyen, 1994). Depending upon the types of species and availability of markers, the DNA markers are selected. The selected polymorphic markers will be screened over entire mapping population and parents to get the genotypic data for map construction.

6.0 GENETICS OF MAPPING MOLECULAR LOCI

Each of the mapping populations (RI, backcross or F_2) will give a specific segregation ratio at each locus (Table 3). The knowledge of these ratios is important to determine if the population is expressing a skewed segregation ratio at any locus. The following are the ratios that you would expect at each locus for co-dominant and dominant makers segregating in these populations.

Table 3. Segregation ratio in different mapping populations

Population	Co-dominant	Dominant
Recombinant inbred or doubled haploid	1:1	1:1
Backcross	1:1	1:1
F_2	1:2:1	3:1

Chi-square test is performed to test the significant deviations of observed ratios from above expected ratios.

According to the formula,

$$\chi^2 = \frac{\Sigma (O - E)^2}{E}$$

Where 'O' denotes for Observed frequency, 'E' for Expected frequency. The goodness of fit was tested using the c^2 value for the segregation data.

Classical linkage analysis is used to determine the arrangement of genes on the chromosomes of an organism. To explain mapping, one should understand the genetic concepts of segregation and recombination, and it is explained with classical Mendelian markers showing full dominance. Dominant and recessive alleles are given as upper and lower case letters respectively. Let consider A and B genes that are unlinked (located on the different chromosomes). As a result of meiosis, two alleles of a locus will segregate (separate from one another) with equal frequencies into the gametes. If a and A are two such alleles, then a diploid individual heterozygous at this locus (genotype Aa) will give gametes half of which are A and half of which are a. Similarly alleles b and B at a separate locus will segregate fifty -fifty into the gametes. The alleles will undergo independent segregation, giving four possible combinations in the gametes: AaBb = AB, Ab, aB, ab. In a test cross (F_1 x recessive parent) AaBb x aabb will produce the following progenies

AB ab : Parental type

Ab ab : Recombinant

aB ab : Recombinant

ab ab : Parental type

It will give the expected 1:1:1:1 ratio (50 % parental type and 50% recombinant type). Suppose "A" and "B" are linked and that a parent carries the dominant alleles A and B on one chromosome of a homologous pair and the alleles a and b on the other chromosome. The offspring usually co-inherit either 'A' with 'B' or 'a with 'b', the recombinants will only arise when crossing over occurs between them, and then their frequency will be less than 50%, in this case, the law of independent assortment is not valid.

The recombination value for a pair of loci from a segregating backcross population is:

Number of recombinants/ Total number of progeny x 100

21/300 x 100 = 7%

Suppose the recombination between loci 1 and 2 is 7 %, that between loci 2 and 3 is 21%, and that between 1 and 3 is 25%, the order of the loci along the chromosome is

It should be noted that genetic distances in the map are not additive. The underestimate based on the recombination between 1 and 3 is due to double (or multiple) crossovers.

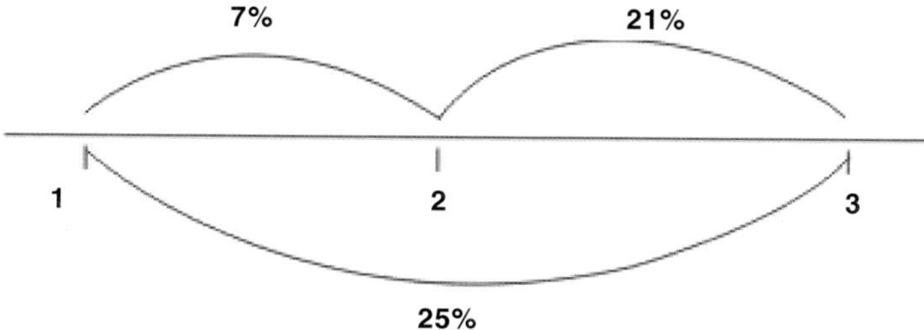

The construction of linkage map involves coding genotypic data (marker data on each individual of the population) and conducting linkage analysis. Largely linkage analysis is performed using computer programs as the manual analysis is laborious and time consuming for large number of markers. Linkage between markers or markers/ phenotypes is calculated using odds ratio. It is expressed as logarithm of the ratio and is called as logarithm of odds (LOD) value or LOD score (Risch,1992). By convention, a LOD score greater than 3.0 is considered evidence for linkage. A LOD value of three between two markers indicated that linkage of 1000 times more likely than no linkage (1000:1). LOD values may be lowered in order to detect a greater level of linkage or to place additional markers within maps constructed at higher LOD values.

The following steps will be considered while generating a genetic linkage map using a mapping population.

1. Generate marker data
2. Calculate all pairwise distances
3. Group makers into linkage group
4. Order markers in each linkage group

Diagrammatic presentation of construction of linkage map is given in Fig. 2. The marker data are generated as outlined in the diagram. Coding of marker data varies depending on the type of population used. Linked markers are grouped together into linkage groups. The ordering of marker is a crucial point in linkage map construction. The common criterion is finding the lowest sum of adjacent recombination.

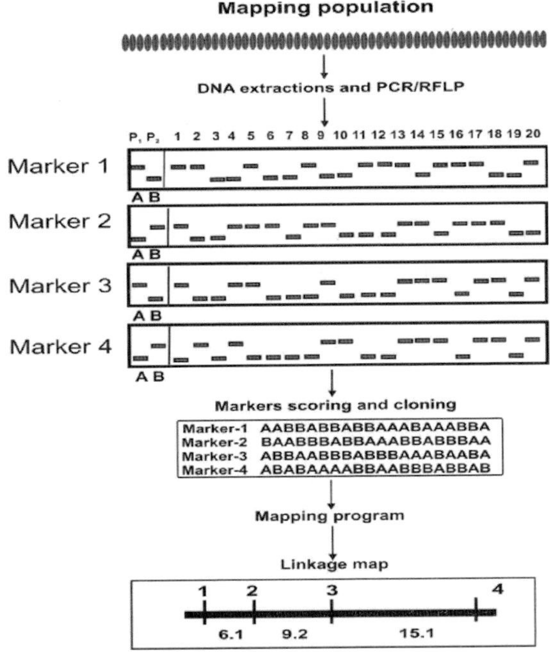

Fig. 2. Diagrammatic view of linkage map construction. The linkage values are hypothetical (Source : Collard *et al.*, 2005)

7.0 SOFTWARE USED FOR MAPPING

The commonly used software programs include Mapmaker/ EXP (Lander *et al.*, 1987; Lincoln *et al.*, 1993) and MapManager QTX (Manly *et al.*, 2001), which are freely available on the internet. JoinMap is another commonly-used program for constructing linkage maps (Stam, 1993).

Mapping functions are required to convert recombination fractions into centimorgans (cM) because recombination frequency and the frequency of crossing-over are not linearly related (Hartl and Jones, 2001;Kearsey and Pooni, 1996).When map distances are small (<10 cM), the map distance equals the recombination frequency. However, this relationship does not apply for map distances that are greater than 10 cM (Hartl and Jones, 2001). Two commonly used mapping functions are the Kosambi mapping function, which assumes that recombination events influence the occurrence of adjacent recombination events, and the Haldane mapping function, which assumes no interference between crossover events (Hartl and Jones, 2001; Kearsey and Pooni, 1996). It should be noted that distance on a linkage map is not directly related to the physical distance of DNA between genetic markers, but depends on the genome size of the plant species (Paterson, 1996). Furthermore, the relationship between genetic and physical distance varies along a chromosome (Kunzel *et al.*, 2000; Tanksley *et al.*, 1992; Young, 1994). For example, there are recombination 'hot spots' and 'cold spots,' which are chromosomal regions in which recombination occurs more frequently or less frequently, respectively (Faris *et al.*, 2000; Ma *et al.*, 2001; Yao *et al.*, 2002).

8.0 MARKER ASSISTED SELECTION

Marker assisted selection is the selection of plants that carry the desired genomic region that is involved in the expression of the trait of interest. It is a powerful tool for indirect selection of traits at an early stage of crop growth. Initially molecular markers were adopted for selection of qualitative traits for which phenotypic assessment was slow or difficult. Selection for nematode resistance in a range of crop species is often used as an example of a qualitative trait particularly amenable to marker assisted selection. The next step in the progression of marker application has been in the selection of "qualitative traits" under more complex genetic control known as the "pseudo-quantitative" traits for which genotype x environment interactions are generally not significant. Examples include abiotic stress tolerances and disease resistance based on pyramided genes. Applications of molecular markers are now moving towards the selection of "real" quantitative traits, for which breeding progress using traditional selection methodologies is often limited by genotype x environment interactions. Traits falling into this category include grain yield and grain protein concentration. With the development of high-resolution, easy-to-use genetic maps, coupled with the successful sequencing of model plant species *Arabidopsis* and rice and physical mapping of the entire genome, the efficiency of MAS has tremendously increased. Dense molecular genetic maps are now available in major cereal crops which offer indirect selection of many agronomically important traits thus helping the plant breeders in reaching their goals within a shorter period.

8.1 Considerations for Efficient MAS

The efficiency of MAS depends on several factors including the selection of the type of marker to be used and its linkage with the trait of interest. The main considerations for selecting a suitable marker system are 1. Reliability of the marker, 2. Quantity and the quality of DNA that is required for the assay of the marker, 3. Ease with which the assay can be carried out 4. Level of polymorphisms exhibited by the marker and 5. Cost effectiveness of the marker (Mackill and Ni, 2000).

Reliability of the marker depends on the close linkage of the marker with the major gene. Markers that are linked at distances less than 5 cM are largely preferred for use in selections. The use of flanking markers instead of a single marker increases the reliability of screening since a double cross over between the flanking makers is less likely than the single crossover (Figure 3). Markers that are developed from the target gene itself are more precise and are also referred to as genic markers that allow gene assisted selections for the trait.

Quantity and quality of the DNA that is required by various marker systems largely differ. Restriction based markers require large quantity of good quality DNA than the amplification based markers. It is important to choose a marker system that requires nominal quantities of DNA thereby reducing the costs of extraction and purification of DNA.

Ease with which the assay can be carried out is of utmost importance and a marker system that allows the screening of a large number of progenies at a given period of time is highly preferred. High-throughput DNA markers that are simple to perform and are amenable for automation are desirable. Sequence tagged site (STS), sequence characterized amplified region (SCAR) or single nucleotide polymorphism (SNP) markers that are derived from specific DNA sequences of markers that are linked to a gene are extremely useful for MAS.

Level of polymorphism exhibited by the marker is another important consideration. The markers must be highly polymorphic and should ideally be able to distinguish between different genotypes that form the breeding population.

Cost effectiveness of a marker greatly determines the utility of the markers especially when dealing with large population sizes.

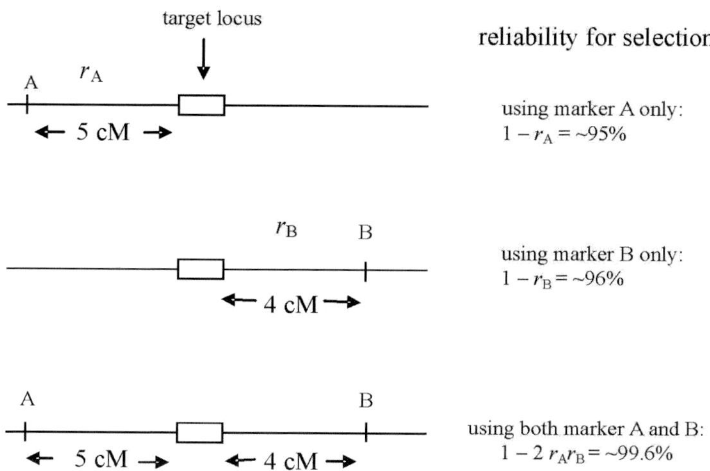

Fig. 3. Reliability of selection using single and flanking markers (adapted from Tanksley (1983), assuming no crossover interference). The recombination frequency between the target locus and marker A is approximately 5% (5 cM). Therefore, recombination may occur between the target locus and marker in approximately 5% of the progeny. The recombination frequency between the target locus and marker B is approximately 4% (4 cM). The chance of recombination occurring between both marker A and marker B (*i.e.* double crossover) is much lower than for single markers (approx. 0.4%). Therefore, the reliability of selection is much greater when flanking markers are used (Source : Liu, 1998)

8.2 MAS for Untrogression of Qualitative Traits

Transferring a single gene controlled trait from a donor parent into the genetic background of a recurrent parent is carried out by repeated backcrossing. Backcrossing is a method that is adopted mostly to incorporate highly heritable qualitative traits such as pathotype-specific disease resistances governed by one or a few genes into an adapted or elite variety. The adapted variety is used as a recurrent parent and has a large number of desirable traits but is deficient in one or few important traits (Allard, 1999).

8.2.1 Backcrossing for recessive genes

Depending on the gene of interest whether recessive or dominant, two types of backcrossing is carried out. In the case of a trait that is controlled by a recessive gene the recurrent parent and the donor parent is crossed and the resulting F_1 is selfed to generate F_2 that segregates for the trait. The F_2 plants that possess the trait are selected and backcrossed to the recurrent parent to generate BC_1 population. The individuals that possess the trait in BC_1 are self pollinated to produce segregating progenies that are selected for the trait and used for backcrossing to the recurrent parent. This is repeated until the BC_6 generation and improved inbreds are generated with the target gene introduced.

8.2.2 Backcrossing for dominant genes

In the case of traits under dominant gene control, the recurrent parent is crossed with the donor parent and the F_1 backcrossed to the recurrent parent. The individuals that possess the trait in the BC_1 generation are backcrossed to the recurrent parent until BC_6 and the best individuals are selected and self-pollinated for homozygous expression of the trait.

8.3 Linkage Drag

Backcrossing is characterized by genetic linkage of undesirable traits with the desirable target traits that result in the fragments of the donor genome surrounding the target gene to be 'dragged' along. Young and Tanksley (1989) demonstrated linkage drag in Tomato cultivars that were developed by introgression of the *Tm2* disease resistance gene from *Lycopersicon peruvianum* a wild relative of tomato. It was seen that cultivars developed after eleven backcrosses had an entire chromosome arm along with the gene from the donor parent and cultivars developed after twenty backcrosses had introgressed segments as large as 4 cM. Often, six generations of selection and backcrossing to the recurrent parent are carried out to recover the 99% of the genotype of the recurrent parent along with the target gene (Babu *et al.*, 2004).

8.4 Marker Assisted Backcrossing

The use of DNA markers in backcrossing greatly increases the efficiency of selection. Marker assisted back crossing (MAB) can be carried out in three general levels as described by Holland (2004) (Fig. 4).

8.4.1 Foreground selection

The first level of selection is the marker assisted foreground selection where markers that are closely linked to the target gene or markers that are flanking the target gene in the donor genotype are used for selections (Hospital and Charcosset, 1997). This is particularly important for traits that have laborious or time-consuming phenotypic screening procedures. Marker assisted foreground selection is very effective for the transfer of traits controlled by recessive genes since their classical transfer requires additional recurrent selfing generations, which is very time consuming in commercial breeding programs. It is also advantageous for selection of reproductive stage traits in the seedling stage.

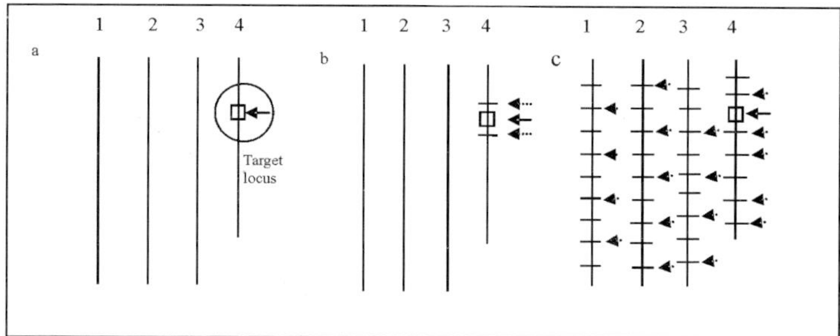

Fig. 4. Levels of selection during marker-assisted backcrossing. A hypothetical target locus is indicated on chromosome 4. (a) Foreground selection, (b) Recombinant selection and (c) Background selection (Source: Holland, 2004)

8.4.2 Recombinant selection

The second level of selection used for selecting backcross progenies with the target gene and recombination events between the target locus and linked flanking markers is referred to as 'recombinant selection'. By using recombinant selection the size of the donor chromosome segment that is linked with the gene of interest that is introgressed can be reduced. This method of selection drastically reduces the linkage drag thereby preventing the transfer of many undesirable genes that can negatively affect the performance of the crop. In conventional breeding methods donar chromosomal segments as large as 10 cM have been introgressed with the target gene even after several back cross generations (Ribaut and Hoisington 1998; Salina *et al.*, 2003). Flanking markers that are linked at less than 5 cM distance on either side of the target gene have found to be very useful in minimizing the linkage drag. Since double recombination events occurring on both sides of a target locus are extremely rare, recombinant selection is usually performed using at least two BC generations (Frisch *et al.*, 1999b).

8.4.3 Background selection

The third level of marker assisted backcross involves selecting backcross progeny with the maximum genome of the recurrent parent. This is done by using markers that are unlinked to the target gene and are distributed throughout the genome of the recurrent parent. Markers that distinguish between donor and recurrent parent genome will accelerate the recovery of recurrent parent genome remarkably. While in conventional backcrossing it takes a minimum of six BC generations to recover the recurrent parent using marker assisted background selection it can be achieved in BC_4, BC_3 or sometimes even in BC_2 generations (Frisch *et al.*, 1999a,b) thereby saving 2-4 backcross generations. The use of background selection during MAB to accelerate the recovery of the recurrent parent with the addition of the target gene has been referred to as 'complete line conversion' (Ribaut *et al.*, 2002).

9.0 CASE STUDIES

Many economically important traits have been identified and tagged using molecular markers (Li and Gill, 2004). Marker-assisted background selection has been extensively used in commercial hybrid maize breeding programs. Conversion of normal maize lines into quality protein maize (QPM) through marker assisted transfer of a recessive mutant allele, *opaque2*, using allele-specific molecular markers was reported at CIMMYT (International Center for Wheat and Maize Improvement), Mexico at http://www.maizegdb.org. QPM are maize genotypes that were developed by introgressing the *opaque2* gene along with endosperm modifiers that provide the hard kernel texture. The QPM genotypes are characterised with twice the amount of lysine and tryptophan along with kernel vitreousness that is comparable to the normal maize genotypes. Introgression of *opaque2*, a recessive allele along with endosperm modifiers which are polygenic in nature, into elite inbreds by conventional methods was complicated because of three major constraints: (i) each BC generation has to be selfed to identify the *opaque2* recessive gene, and a minimum of six such BC generations are required to recover satisfactory levels of recovery of the recurrent parent genome; (ii) Along with the homozygous *opaque2* gene, multiple endosperm modifiers must be selected; and (iii) laborious and time consuming biochemical tests are required to ensure enhanced lysine and tryptophan levels in the selected materials in each breeding generation (Babu *et al.*, 2004). Scientists at CIMMYT, Mexico used a combination of SSR markers linked to the *opaque2* allele and phenotypic selection for kernel vitreousness for converting normal maize lines into QPM. SSR markers that were identified within the *opaque 2* gene viz., *umc1066*, *phi057* and *phi112* were used to select individuals carrying the gene in the backcross generations, and the time required was reduced to half. In addition, marker-aided background selection helped to reduce to three generations to recover the recurrent parent genome as opposed to six generations in conventional selection programs. Similarly Ragot *et al.* (1995) described the transfer of CryIA(b) Bt gene from a Lancaster to a Stiff Stalk maize inbred line. Since the Bt gene was fused to the *pat* gene, encoding glufosinate (herbicide) tolerance, foreground selection for Bt gene was followed by spraying each BC generation with Bastaä (glufosinate). RFLP markers were successfully used for background selection resulting in the saving of two backcross generations. Another excellent example of successful MAS is the development of an improved version of the pearl millet hybrid HHB 67 with resistance to downy mildew, described at http://www.dfid-psp.org/AtAGlance/HotTopic.html.

9.1 MAS for Disease Resistance

Disease resistance is often considered to be controlled by major genes or oligogenes. The term "qualitative resistance" denotes Mendelian genes of large effect that interact on a gene-for-gene basis with the pathogen. Quantitative resistance refers to resistance that shows continuous variation and is usually incomplete in expression. Qualitative resistance is characterized by race-specificity, whereas the race specificity of quantitative resistance is still not very clear. Introgression of qualitative resistance is risky owing to its specificity to a particular race of the pathogen and hence pyramiding qualitative resistance genes with different race specificities ensures durable resistance. Gene pyramiding for resistance overcomes the possibility of the resistance being broken down with a single mutation in the pathogen that could give rise to a new race.

9.2 Marker-Assisted Pyramiding

Pyramiding is the process of combining several genes together into a single genotype. The most widespread application for pyramiding has been for combining qualitative resistance genes together into a single genotype. Pyramiding is possible through conventional breeding but the identification of plants that carry more than one gene is often difficult. Since the multiple resistant genes show the same phenotype, a progeny test has to be performed in order to identify plants that carry more than one gene. DNA markers that are linked to multiple genes have proved to be very useful for selections during pyramiding of several genes. With linked DNA markers, the number of resistance genes in any plant can be easily determined. Marker assisted selection is very promising for pyramiding resistance genes of both dominant and recessive nature. But still large scale application of this technique in actual plant breeding is not encouraging (Koebner, 2004).

Using functional markers, several pest and disease resistance genes have been identified in a segregating population. Cloning of some of the BLB resistant genes (*Xa1, xa5, xa13, Xa21, Xa26* and *Xa27*) (Yoshimura *et al.*, 1998; Iyer and McCouch ,2004; Jiang *et al.*, 2006; Chu *et al.*, 2006; Song *et al.*, 1995; Sun *et al.*, 2004; Gu *et al.*, 2005) made it possible to develop functional markers. Recently, a functional marker for recessive resistant gene *xa5* was developed (Iyer and McCouch ,2007). In case of *xa13* gene, markers producing promoter level variations between resistant and susceptible parents were identified (Chu *et al.*, 2007). Perumalsamy *et al.* (2008) designed the functional marker for *Xa21* based on coding sequence of both the alleles (*Xa21* and *xa21*) reported by Song *et al.* (1995, 1997). These functional markers are expected to enhance the reliability of MAS since it helps in direct selection of resistant genes involved in BLB resistance. The genes conferring bacterial blight resistance in rice were pyramided (*xa5*

+ *xa13* + *Xa21*) using this approach (Bharani *et al.*, 2008, Perumalsamy *et al.*, 2008). These pyramided genotypes with two and three resistant genes found to be highly resistant to *Xanthomonas oryzae* pv. *oryzae* isolates. Among 30 pyramided genotypes, 17 were found to be high yielding with good agronomic characteristics. These staked lines showed wider spectrum of resistance against BLB pathogen resistance (Fig. 5). Rice blast is another important disease where pyramiding of several resistant genes resulted in durable resistance. Three genes viz., *Pil*, *Piz5* and *Pita* have been pyramided in a susceptible rice variety Co39 using RFLP and PCR based markers to impart durable blast resistance (Hittalmani *et al.*, 2000). The list of some important achievements made using MAS in several crops is presented in Table 4.

The success of MAS depends upon several critical factors, including the number of target genes to be transferred, the distance between the flanking markers and the target gene, the number of genotypes selected in

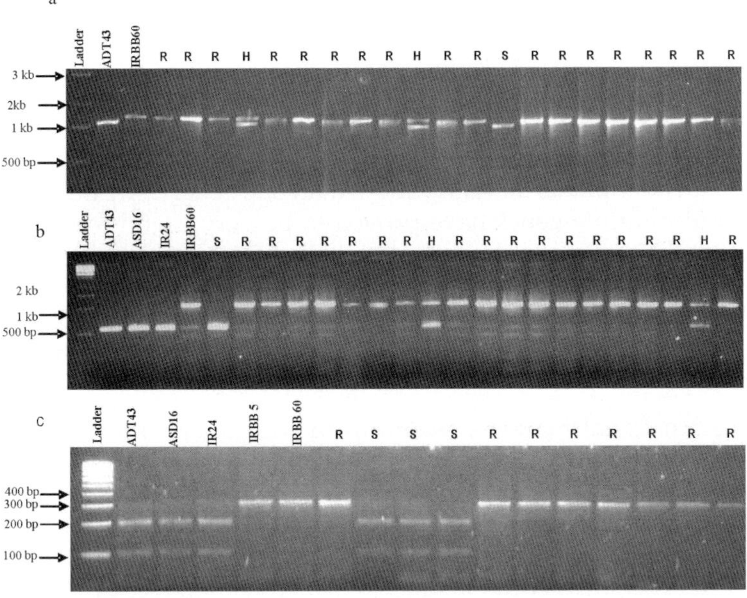

Fig. 5. PCR amplification of F_2 plants of ADT 43 x IRBB 60 cross (a) with Xa21F, Xa21R (b) with xa13F, xa13R (c) with xa5_1F, xa5_1R. R, H, S indicates resistant, heterozygote and susceptible lines, respectively. ADT43, ASD16, IR24 were susceptible genotypes and IRBB60 was resistant source (Source : Perumalsarry *et al.*, 2008)

each breeding generation, the nature of germplasm and the technical options available at the marker level. MAS is currently used more widely for simply inherited traits than for polygenic traits. Marker-assisted foreground selection is already gaining rapid momentum as allele-specific markers are now becoming available, through crop genomics research, on a number of

Table 4. Notable achievements made through marker assisted selection

Crop	Examples of MAS (Trait improved)	References
Rice	Gene pyramiding for BLB –*xa*5 and *xa*4 Two Indonesian rice cultivars Angke and Conde were developed	Toenniessen *et al* (2003)
	Gene pyramiding for bacterial leaf blight	Yoshimura *et al.* (1995) Huang *et al.* (1997) Sanchez *et al.* (2000) Ramalingam *et al.* (2002) Singh *et al.* (2001) Joseph*et al.* (2004) Gopalakrishnan*et al.* (2008) Sundaram*et al.* (2008)
	Gene pyramiding for blast	Chen *et al.* (2004) Hittalmani*et al.*(2000) Wang *et al.*(2004)
	Gene pyramiding for bacterial leaf blight and blast	Toojinda *et al.* (2005)
	Gene pyramiding for bacterial leaf blight and bacterial leaf streak	Zhou *et al.* (2009)
	Pyramiding of brown plant hopper resistance genes, *Bph1* and *Bph2*	Sharma *et al.* (2004)
Wheat	Pyramiding of cereal cyst nematode genes, *CreX* and *CreY*	Barloy *et al.* (2007)
	Powdery mildew resistance genes	Wang *et al.* (2001) Zhang*et al.*(2002)
	Rust resistance	Singh*et al.*(2004)
Barley	Yellow dwarf virus resistance gene Yd2	Jefferies *et al.* (2003)
	Tango variety with stripe rust resistance gene	Toojinda *et al.* (1998)
Soybean	Mosaic virus resistance	Saghai Maroof *et al.* (2008)
Common bean	Anthracnose resistance	Garzon *et al.* (2008)
	Common bacterial blight	Yu *et al.* (2000)

agronomically important traits. The cost of MAS is one of the important factors that limit commercial application. Nearly less than one third of cost for MAS was reported than the conventional screening method. Even though marker-assisted selection now plays a prominent role in the field of plant breeding, examples of successful, practical outcomes are limited and a cautious optimistic vision is needed for MAS as suggested by Young (1999). As molecular breeders adopt more rigorous experimental guidelines and ambitious goals, they also need to integrate the growing body of knowledge from genomics and bioinformatics. The optimization of marker genotyping methods in terms of cost-effectiveness and a greater level of integration between molecular and conventional breeding represent the main challenges for the greater adoption and impact of MAS in the near future.

10.0 CONCLUSION

Qualitative traits that are under simple genetic control including one or two major genes have been traits of choice for initial mapping and tagging studies. Permanent populations like double haploids and recombinant inbred lines have been more often used for tagging of qualitative traits though F_2s and NILs have also been used. With the concept of synthesizing DNA pools from individuals of the same phenotype by Michelmore (1991) several qualitative traits have been tagged using molecular markers. Most of the major genes that have been mapped and tagged are disease and insect resistant genes. On the other hand major genes for certain abiotic stresses like submergence tolerance and salt tolerance and for traits like photoperiod sensitivity, semi- dwarfism, shattering resistance, grain quality, seed color, aroma, amylose content, cooked kernel elongation, heterosis and wide compatibility have also been mapped and tagged. The markers identified to be linked to such qualitative traits have been widely used for selections for traits that are difficult to measure, for pyramiding of several genes into a particular background and in foreground and background selections in backcross breeding. Marker assisted selection programme would greatly help the plant breeders in reaching their goal within shorter period of time although currently the number of varieties developed are relatively low. Uncoupling and recombination are main problem for indirect selection of number of traits. As the number of genes cloned and sequence information are rapidly increasing it will circumvent the bottleneck. Development of functional markers and pyramiding of genes based on these markers will play a major role as it specifies a gene of interest and overcome linkage problems. Development of new markers such as gene specific markers for important traits will potentially reduce the problems associated with uncoupling of markers from the gene of interest.

11.0 REFERENCES

Allard, R.W. 1999. Principles of plant breeding, 2nd edn. New York, NY: Wiley.

Andersen, J.R. and Lubberstedt, T. 2003. Functional markers in plants. Trends in Plant Science, 8:554–560.

Babu, R., Sudha K Nair, Prasanna, B.M. and Gupta, H.S. 2004. Integrating marker assisted selection in crop breeding – Prospects and challenges. Current Science 87: 607-619.

Barloy, D., Lemoine , J., Abelard, P., Tanguy, A.M., Rivoal, R. and Jahier, J. 2007. Marker-assisted pyramiding of two cereal cyst nematode resistance genes from *Aegilops variabilis* in wheat. Molecular Breeding, 20:31–40.

Bharani, M., Nagarajan, P., Rabindran, R., Saraswathi, R., Balasubramanian, P. and Ramalingam, J. 2009. Bacterial leaf blight resistance genes (Xa21, xa13 and xa5) pyramiding through molecular marker assisted selection into rice cultivars. Archives of Phytopathology and Plant Protection, (In press).

Chen, X.W., Li, S.G., Ma, Y.Q., Li, H.Y., Zhou, K.D. and Zhu, L.H. 2004. Marker-assisted selection and pyramiding for three blast resistance genes, Pi-d(t)1, Pi-b, Pi-ta2, in rice. Sheng Wu Gong Cheng Xue Bao, 20(5):708-714.

Choudhary, S., Sethv, N.K., Shokeen, B. and Bhatia, S. 2008. Development of chickpea EST-SSR markers and analysis of allelic variation across related species. Theoretical and Applied Genetics.(online first).

Chu, Z., Yang, B., Fu, H., Xu, C., Li, Z., Sanchez, A., Park, Y.J., Bennetzen, J.L., Zhang, Q. and Wang, S. 2006. Targeting *xa13*, a recessive gene for bacterial blight resistance in rice. Theoretical and Applied Genetics, 112:455–461.

Chu, Z., Yuan, M., Yao, J., Ge, X., Yuan, B., Xu, C., Li, X., Fu, B., Li, Z., Bennetzen, J.L., Zhang, Q. and Wang, S. 2007: Promoter mutations of an essential gene for pollen development result in disease resistance in rice. Genes and Development, 20:1250-1255.

Collard, B.C.Y., and David, J.M. 2008: Marker assisted selection: an approach for precision plant breeding in the twenty first century. Philosophical Transactions of the Royal Society Biological Sciences, 363:557- 572.

Collard, B.C.Y., Jahufer, M.Z.Z., Brouwer, J.B. and Pang, E.C.K. 2005. An introduction to markers, quantitative trait loci (QTL) mapping and marker-assisted selection for crop improvement: The basic concepts. Euphytica, 142: 169-196.

Collard, B.C.Y., Pang, E.C.K. and Taylor, P.W.J. 2003. Selection of wild *Cicer* accessions for the generation of mapping populations segregating for resistance to ascochyta blight. Euphytica, 130:1–9.

Collins, A.J., Milbourne, D., Meyer, R.C., Ramsay, L.D., Gebhardt, C. and Waugh, R., 1998. Molecular Genetics and Genomics, 259:233–245.

Dahleen, L.S., Agrama, H.A., Horsley, R.D., Steffenson, B.J., Schwarz, P.B., Mesfin, A. and Franckowiak, J.D. 2003. Identification of QTLs associated with Fusarium head blight resistance in Zhedar 2 barley. Theoretical and Applied Genetics,108(1):95-104.

Faris, J.D., Haen, K.M. and Gill, B.S. 2000. Saturation mapping of a gene-rich recombination hot spot region in wheat. Genetics, 154:823–835.

Feltus, F.A., Wan, J., Schulze, R.S., Estill, J.C., Jiang, N. and Paterson,A.H. 2004. An SNP resource for rice genetics and breeding based on subspecies *Indica* and *Japonica* genome alignments, Genome Reseach,14:1812-1819.

Frisch, M., Bohn, M. and Melchinger, A.E. 1999a. Comparison of selection strategies for marker assisted backcrossing of a gene. Crop Science 39:1295-1301.

Frisch, M., Bohn, M. and Melchinger, A.E. 1999b. Minimum sample size and optimal positioning of flanking markers in marker-assisted backcrossing for transfer of a target gene. Crop Science 39:967-975.

Gallais, A. and Bordes, J. 2007. The use of doubled haploids in recurrent selection and hybrid development in Maize. Crop Science 47:190-201.

Garzón, L.N., Gustavo A. Ligarreto and Matthew W. Blair.2008. Molecular Marker-assisted backcrossing of *Anthracnose* resistance into Andean climbing beans (*Phaseolus vulgaris L.*). Crop Science, 48:562-570.

Gopalakrishnan, S., Sharma, R.K., Anand Rajkumar, K., Joseph, M., Singh, V.P., Singh, A.K., Bhat, K.V., Singh, N.K. and Mohapatra, T. 2008. Integrating marker assisted background analysis with foreground selection for identification of superior bacterial blight resistant recombinants in Basmati rice. Plant Breeding, 127:131-139.

Gu, K., Yang, B., Tian, D.,Wu, L., Wang, D. and Sreekala, C. 2005. R gene expression induced by a type-III effector triggers disease resistance in rice. Nature, 435: 1122-1125.

Gupta, P.K., Rustgi, S. and Mir, R.R. 2008. Array-based high-throughput DNA markers for crop improvement. Heredity, 101(1):5-18.

Hartl, D.L. and Jones, E.W. 2001. Genetics. Analysis of genes and genomes. Seventh edition. pp. 763. Jones and Bartlett Publishers Inc., Sudbury, Massachusetts.

Hittalmani, S., Parco, A., Mew, T.V., Ziegler, R.S. and Huang, N. 2000. Fine mapping and DNA marker assisted pyramiding of the three major genes for blast resistance in rice. Theoretical and Applied Genetics, 100:1121–1128.

Holland, J.B. 2004 Implementation of molecular markers for quantitative traits in breeding programs—challenges and opportunities. *In:* Proc. 4th Int. Crop Sci. Congress., Brisbane, Australia, 26 September—1 October.

Hospital, F. and Charcosset, A. 1997 Marker-assisted introgression of quantitative trait loci. Genetics 147:1469–1485.

Huang, N., Angeles, E.R., Domingo, J., Magpantay, G., Singh, S., Zhang, G., Kumaravadivel, N., Bennett, J. and Khush, G.S. 1997. Pyramiding bacterial blight resistance genes in rice: Marker assisted selection using RFLP and PCR. Theoretical and Applied Genetics, 95 313-320.

Iyer, A.S. and McCouch, S.R. 2007. Functional markers for *xa5* mediated resistance in rice (*Oryza sativa* L.). Molecular Breeding, 19:291-296.

Iyer, A.S. and McCouch, S.R. 2004: The rice bacterial blight resistance gene *xa5* encodes a novel form of disease resistance. Molecular Plant Microbe Interactions, 17:1348-1354.

Jander, G., Norris, S.R., Rounsley, S.D., Bush, D.F., Levin, I.M., and Last, R.L. 2002. *Arabidopsis* map-based cloning in the post-genome era. Plant Physiology, 129: 440–450.

Jefferies, S.P., King, B. J., Barr, A. R., Warner, P., Logue, S. J. and Langridge, P. 2003. Marker-assisted backcross introgression of the Yd2 gene conferring resistance to barley yellow dwarf virus in barley. Plant Breeding, 122 (1):52-56.

Jiang, G.H., Xia, Z.H., Zhou, Y.L., Wan, J., Li, D.Y., Chen. R.S., Zhai, W.X. and Zhu, L.H. 2006: Testifying the rice bacterial blight resistance gene *xa5* by genetic complementation and further analyzing *xa5* (*Xa5*) in comparison with its homolog TFIIA. Molecular Genetics and Genomics, 275:354-366.

Joseph, M., Gopalakrishnan, S., Sharma Singh, R.K., Singh, A.K., Singh, N.K. and Mohapatra, T. 2004: Combining bacterial blight resistance and basmati quality characteristics by phenotypic and molecular marker assisted selection in rice. Molecular Breeding, 13:377-387.

Kearsey, M. and Pooni, H. 1996. The genetical analysis of quantitative traits. Journal of Genetics, 76:93-95.

Koebner, R.M.D. 2004. Marker assisted selection the cereals: The dream and the reality, Cereal Genomics, 317-329.

Kunzel, G., Korzun, L. and Meister, A. 2000. Cytologically integrated physical restriction fragment length polymorphism maps for the barley genome based on translocation breakpoints. Genetics, 154:397–412.

Lander, E.S., Green, P., Abrahamson, J., Barlow, A., Daly, M.J., Lincoln, S.E. and Newburg, L. 1987. Mapmaker an interactive computer package for constructing primary genetic linkage maps of experimental and natural populations. Genomics, 1:174–181.

Li, W. and Gill, B.S. 2004. Genomics for cereal improvement pp. 585-634. *In*: Cereal Genomics (*eds*: Gupta, P.K. and Varashney, R.K.,). Klluwer Academic Publishers.

Lincoln, S.E., Daly, M.J. and Lander, E.S. 1993. Constructing genetic linkage map with MAPMAKER/EXP v3.0: a tutorial and reference manual. Whitehead Institute Technical Report, Cambridge, MA.

Liu, B. 1998. Statistical genomics: linkage, mapping and QTL analysis. Boca Raton, FL: CRC Press.

Lübberstedt, T., Zein, I., Andersen, J.R., Wenzel, G., Krützfeldt, B., Eder, J., Ouzunova, M. and Chun, S. 2005. Development and application of functional markers in maize. Euphytica, 146:101–108

Ma, X.F., Ross, K. and Gustafson, J.P. 2001. Physical mapping of restriction fragment length polymorphism (RFLP) markers in homoeologous groups 1 and 3 chromosomes of wheat by *in situ* hybridization. Genome, 44:401–412.

Mackill, D. J. and Junjian, N. 2001. Molecular mapping and marker assisted selection for major gene traits in rice. *In* Rice Genetics, IV, Science Publishers Inc., IRRI, Philippines, pp. 137–151.

Mackill, D.J, and Ni, J. 2000. Molecular mapping and marker assisted selection for major-gene traits in rice. *In: Proc. Fourth Int. Rice Genetics Symp.* (*eds* Khush, G.S. , Brar, D.S. and Hardy, B.) 137-151.

Manly, K.F., Cudmore, H. Robert, Jr. and Meer, J.M. 2001. Map Manager QTX, cross-platform software for genetic mapping, Mamm Genome, 12:930–932.

Paterson, A.H., Lan, T.H., Reischmann, K.P., Chang, C., Lin, Y.R., Liu, S.C., Burow, M.D., Kowalski, S.P., Katsar, C.S., DelMonte, T.A., Feldmann,K.A., Schertz,

K.F., and Wendel, J.F. 1996. Toward a unified genetic map of higher plants, transcending the monocot-dicot divergence. Nature Genetics 14:380-382.

Perumalsamy, S., Bharani, M., Sudha, M., Nagarajan, P., Arul, L., Saraswathi, R., Balasubramanian, P. and Ramalingam, J. 2008. Functional markers assisted selection for bacterial leaf blight resistant genes in rice (*Oryza sativa* L.). Plant Breeding (In press).

Ragot, M., Biasiolli, M., Delbut, M.F. and Gay, G. 1995. Marker-assisted backcrossing: a practical example. Pages 45-56 in: Techniques et utilisations de marqueurs moleculaires. Les Colloques No.72, INRA, Paris, France.

Rafalski, J. and Tingsey, S. 1993. Genetic diagnostics in Plant Breed: RAPDs, microsatellites and machines. Trends in Genetics, 9:275–280.

Ramalingam, J., Basharat, H.S. and Zhang. G. 2002. STS and microsatellite marker-assisted selection for bacterial blight resistance and waxy genes in rice, *Oryza sativa* L. Euphytica, 1 (27): 255-260.

Ramalingam, J., Vera Cruz, C.M., Kukreja. K., Chittoor, J.M., Wu, J.-L., Lee, S.W., Baraoidan, M., George, M.L., Cohen, M.B., Hulbert, S.H., Leach, J.E. and Leung, H. 2003. Defense genes from rice, barley, and maize and their association with qualitative and quantitative resistance in rice. Molecular Plant-Microbe Interactions, 16:14-24.

Ribaut, J.-M. and Hoisington, D.A., 1998. Marker assisted selection: new tools and strategies. Trends in Plant Science, 1998, 3, 236–239.

Ribaut, J.-M., Jiang, C. and Hoisington, D. 2002. Simulation experiments on efficiencies of gene introgression by backcrossing. Crop Science, 42, 557–565.

Risch, N. 1992. Genetic linkage: Interpreting LOD scores. Science, 255: 803–804.

Sachidanandam, R., Weissman, D., Schmidt, S.C., Kakol, J.M., Stein, L.D., Marth, G., Sherry, S., Mullikin, J.C., Mortimore, B.J., Willey, D.L., *et al.* 2001. A map of human genome sequence variation containing 1.42 million single nucleotide polymorphisms. Nature, 409:928–933.

Saghai Maroof, M.A., Jeong, S.C., Gunduz, I. ,Tucker, D.M., Buss, G.R. and S.A. Tolin, 2008: Pyramiding of soybean mosaic virus resistance genes by marker-assisted selection. Crop Science 48:517–526.

Salina, E., Dobrovolskaya, O., Efremova, T., Leonova, I. and Roder, M.S. 2003 Microsatellite monitoring of recombination around the Vrn-B1 locus of wheat during early backcross breeding. Plant Breed. 122, 116–119. (doi:10.1046/j.1439-0523.2003.00817.x)

Sanchez, A.C., Brar, D.S., Huang, N., Li, Z. and Khush, G.S. 2000. Sequence tagged site marker-assisted selection for three bacterial blight resistance genes in rice. Crop Scence, 40:792-797.

Schmid, K.J., Sorensen, T.R., Stracke, R., Torjek, O., Altmann, T., Mitchell-Olds, T., and Weisshaar, B. 2003. Large-scale identification and analysis of genome-wide single-nucleotide polymorphisms for mapping in *Arabidopsis thaliana*. Genome Research, 13:1250–1257.

Sharma, P.N., Torll, A., Takumi, S., Mori, N., and Nakamura, C.U. 2004. Marker assisted pyramiding of brown plant hopper (*Nilaparvata lugens* S.) resistance genes *Bph1* and *Bph2* on rice chromosome 12. Hereditas, 140(1): 61-69.

Singh, R., Dibendu Datta, Priyamvada, Somvir Singh, and Ratan Tiwari. 2004. Marker-assisted selection for leaf rust resistance genes *Lr19* and *Lr24* in wheat (*Triticum aestivum* L.). Journal of Applied Genetics, 45(4):399-403.

Singh, S., Sidhu, J.S., Huang, N., Vikal, Y., Li, Z., Brar, D.S., Dhaliwal, H.S., and Khush, G. S. 2001. Pyramiding three bacterial blight resistance genes (*xa5, xa13* and *Xa21*) using marker assisted selection into *indica* rice cultivar PR106. Theoretical and Applied Genetics, 102:1011-1015.

Song, W.Y., Pi, L.Y., Wang, G.L., Gardener, J., Holsten, T. and Ronald, P.C. 1997. Evolution of the rice *Xa21* disease resistance gene family. Plant Cell, 8:429 - 445.

Song, W.Y., Wang, G.L., Chen, L.L., Kim, H.S., Pi, L., Hosten, T., Gardener, J., Wang, B., Zhai, W.X., Fauquet, C. and Ronald, P. 1995. A receptor kinase like protein encoded by the rice disease resistance gene, *Xa21*. Science, 270:1804-1806.

Stam, P. 1993. Construction of integrated genetic linkage maps by means of a new computer package: JoinMap. Plant Journal, 3:739–744.

Sun, X., Cao, Y., Yang, Z., Xu, C., Li, X., Wang , S. and Zhang, Q. 2004. *Xa26*, a gene conferring resistance to *Xanthomonas oryzae pv. oryzae* in rice encodes an LRR receptor kinase-like protein. Plant Journal, 37:517-527.

Sundaram, M., Vishnupriya, R., Biradar, K.S., Laha, S., Reddy, G.A., Rani, N.S., Sarma, P. and Sonti, R.V. 2008. Marker assisted introgression of bacterial blight resistance in Samba Mahsuri, an elite *indica* rice variety. Euphytica, 160:411-422.

Tanhuanpaa, P., Kalendar. R., Schulman, A.H. and Kiviharju, E. 2008. The first doubled haploid linkage map for cultivated oat. Genome, 51(8):560-569.

Tanksley, S.D., Ganal, M.W., Prince, J.P., De Vicente, M.C., Bonierbale, M.W., Broun, P., Fulton, T.M., Giovannoni, J.J. and Grandillo, S. 1992. High density molecular linkage maps of the tomato and potato genomes. Genetics, 132: 1141-1160.

Toenniessen, G.H., John C O'Tooley and Joseph DeVriesz. 2003. Advances in plant biotechnology and its adoption in developing countries. Current Opinion in Plant Biology, 6:191-198.

Toojinda, T., Baird, E., Booth, A., Broers, l., Hayes, D., Powel, W., Thomas, W., Vivar, H. and Young, G. 1998. Introgression of quantitative trait loci (QTLs) determining stripe rust resistance in barley: an example of marker assisted line development. Theoretical and Applied Genetics, 96:123-131.

Toojinda, T., Tragoonrung, S., Vanavichit, A., Siangliw, J.L., In, N.P., Jantaboon, J., Siangliw, M. and Fukai, S. 2005. Molecular breeding for rainfed lowland rice in the Mekong region. Plant Production Science, 8:330-333.

Tanksley, S.D. 1983. Molecular markers in plant breeding. Plant Molecular Biology Reports, 1, 3–8.

Torjek, O., Berger, D., Meyer, R.C., Mussig, C., Schmid, K.J., Rosleff Sorensen, T., Weisshaar, B., Mitchell-Olds, T., and Altmann, T. 2003. Establishment of a high-efficiency SNP-based framework marker set for *Arabidopsis*. Plant Journal, 36:122-140.

Wang, X.Y., Chen, P.D. and Zhang, S.Z. 2001. Pyramiding and marker assisted selection for powdery mildew resistance genes in common wheat. Acta Genetica Sinica, 28(7):640-646.

Wang, Z.H., Jia, Y.L., Wu, D.X. and Xia, Y.W. 2004. Molecular marker assisted selection of *Pi-ta* gene resistant to rice bacterial blight, Acta Agronomica Sinica, 30(12):1259-1265.

Wolters, A.M. and Visser, R.G.F. 2000. Gene silencing in potato: allelic differences and effect of ploidy. Plant Molecular Biology, 43:377–386.

Yamanaka, S., Nakamura, I., Watanabe, K.N. and Sato, Y. 2004. Identification of SNPs in the waxy gene among glutinous rice cultivars and their evolutionary significance during the domestication process of rice. Theroetical and applied Genetics, 108(7):1200-1204.

Yao, H., Zhou, Q., Li, J., Smith, H., Yandeau, M., Nikolau, B.J. and Schnable, P.S. 2002. Molecular characterization of meiotic recombination across the 140-kb multigenic *a1-sh2* interval of maize. Proceedings of the National Academy Science, U.S.A, 99:6157–6162.

Yoshimura S, Yoshimura, A., Iwata, N., McCouch, S.R., Abenes, M.N., Baraoidan, M.R., Mew, T.W. and Nelson, R.J. 1995. Tagging and combining bacterial blight resistance genes using RAPD and RFLP markers. Molecular Breeding, 1:375-387.

Yoshimura, S., Yamanouchi, U., Katayose, Y., Toki, S., Wang, Z.-X., Kono, I., Yano, M., Iwata, N., and Sasaki, T. 1998. Expression of *Xa1*, a bacterial blight-resistance gene in rice, is induced by bacterial inoculation. Proceedings of the National Academy Science. U.S.A., 95:1663–1668.

Young, N.D. 1999. A cautiously optimistic vision for marker-assisted breeding. Molecular Breeding, 5:505–510.

Young, N.D. 1994. Constructing a plant genetic linkage map with DNA markers,pp.39-57. *In:* I. (*eds*: K.V. Ronald and L. Phillips). DNA-based markers in plants.

Young, N.D. and Tanksley, S.D. 1989. RFLP analysis of the size of chromosomal segments retained around Tm-2 locus of tomato during back cross breeding. Theoretical and Applied Genetics, 77:353–359.

Yu, K., Park, S. and Poysa, V. 2000. Marker-assisted selection of common beans for resistance to common bacterial blight: Efficacy and economics. Plant Breeding 119:411–415.

Yu, L.-X. and Nguyen, H. 1994. Genetic variation detected with RAPD markers among upland and lowland rice cultivars (*Oryza sativa L.*). Theoretical and Applied Genetics, 87:668-672.

Zhang, K.P., Zhao, L., Tian, J.C., Chen, G.F., Jiang, X.L. and Liu, B. 2008. A genetic map constructed using a doubled haploid population derived from two elite Chinese common wheat varieties. Journal of Integrative Plant Biology, 50(8):941-950.

Zhang. Z.Y., Chen, X., Zhang, C., Xin, Z.Y. and Chen, X.M. 2002. Selecting the pyramids of powdery mildew resistance genes *Pm4b*, *Pm13* and *Pm21* in wheat assisted by molecular marker. Scientia Agricultura Sinica, 35(7):789-793.

Zhou, Y.L., Xu, J.L., Zhou, S.C., Yu, J., Xie, X.W., Xu, M.R., Sun,Y., Zhu, L.H., Fu, B.Y., Gao, Y.M. and Li, Z.K. 2009. Pyramiding Xa23 and Rxo1 for resistance to two bacterial diseases into an elite indica rice variety using molecular approaches. Molecular Breeding, 23:279–287.

Molecular Plant Breeding: Principle, Method and Application
Eds : R.K. Singh, Rajesh Singh, Guoyou Ye, A. Selvi and G.P. Rao
Studium Press LLC, Texas, USA, 2009, pp. 161-188

QTL Mapping in Crop Plants: Principle and Methodology

GYAN P. MISHRA[1], R.K. SINGH[2], MEETUL KUMAR[1] and SHASHI BALA SINGH[1]

ABSTRACT

Some of the most difficult tasks of plant breeders relate to the improvement of traits that show a continuous range of values i.e. quantitative traits. The environmental variance resulting from differences in growing conditions further obscures the relation between phenotype and genotype. In practice, this problem is typically dealt with by evaluating large, replicated trials, which allow identification of genotypic differences through statistical analysis. Plant breeders would like to get a better grip on quantitative traits by direct selection for the genetic factors that are responsible for the observed variability in quantitative traits. This can be achieved through indirect selection: selection for other, well recognizable factors that are linked to the target genes. Since the development of molecular biology, the potential of molecular breeding contributing significantly to crop improvement has been controversial. With the identification of first QTL in 1980s, the ability of molecular markers to streamline the selection of complex traits has been oversold because scientist have largely underestimated the impact of gene networks and their interaction on plant phenotype. Some of these limitations have now been overcome, due to the development of more sophisticated statistical approaches, which allow categorizing both QTL and QTL by environment interaction (QEI), as well as the contributions made by plant models. Today, the use of markers to track transgenes or stack favorable alleles determining a significant proportion of the phenotypic variance is routine for many crops. QTL, functional genomic, and association studies are complementary approaches, which can quantify the genetic effects of specific alleles at target loci. Once the genetic gain of favorable alleles has been validated in a suitable biological context and environment, allele based markers can be easily developed and employed.

Key Words: QTL, Mapping, MAS, MAP MAKER, QTL detection methods

[1] *Defence Institute of High Altitude Research, DRDO, C/o 56 APO, Leh, 901205, India*
[2] *Indian Institute of Sugarcane Research, Lucknow-226 002, UP, India*
Corresponding author e-mail: gyan.gene@gmail.com

1.0 INTRODUCTION

1.1 Plant Breeding

Plant Breeding is a dynamic area of applied science. It relies on genetic variations and uses selection to gradually improve plants for traits and characteristics that are of interest for the grower and the consumer. Practical breeding of many economic important crops is performed by commercial companies that strive in a fierce competition for the favour of agricultural producers and consumers (Zuurbier, 1994). This competition has ensured a continued improvement of cultivated varieties (cultivars) over the past decades. These improvements were partly realised through an efficient use of existing variability, present within the available material. Another important way of improvement is the introduction of new genetic material (*e.g.* genes for disease resistance) from other sources, such as gene bank accessions and related plant species. Although current breeding practises have been very successful in producing a continuous range of improved varieties, recent developments in the field of biotechnology and molecular biology can be employed to enhance plant breeding efforts and to speed up the creation of cultivars by utilization of both qualitative and quantitative traits.

1.2 Quantitative Traits

Many traits of agronomic and horticultural interest are controlled by a single gene and fall into a few distinct phenotypic classes. These classes can be used to predict the genotypes of the individuals. Furthermore, if we know the genotype we could predict the phenotype of the plant. These types of phenotypes are called discontinuous traits *e.g.* round and wrinkled seed shape in pea. Other traits do not fall into discrete classes. Rather, when a segregating population is analyzed for these traits, a continuous distribution is observed. An example is yield of a crop, which does not fall into a tight distribution, but exhibits a bell-shaped distribution or a normal distribution. These types of traits are called continuous traits and cannot be analyzed in the same manner as discontinuous traits. Because continuous traits are often given a quantitative value, they are often referred to as quantitative traits, and the area of genetics that studies their mode of inheritance is called quantitative genetics. Furthermore, the loci controlling these traits are called quantitative trait loci or QTL. In other words genetic factors that are responsible for a part of the observed phenotypic variation for a quantitative trait can be called quantitative trait loci (QTLs). Although similar to a gene, a QTL merely indicates a region on the genome, and could be comprised of one or more functional genes (Falconer and Mackay, 1996).

Quantitative character have been a major area of study in genetics for over a century, as they are a common feature of natural variation in populations of all eukaryotes, besides being the typical of several important traits in both crop plants and animals. Because many agriculturally important traits such as crop yield, milk yield in cattle, fat content of meat etc., are quantitative traits, much of the pioneering research into the modes of inheritance of these traits was performed by agricultural geneticists. These traits are controlled by multiple genes, each segregating according to Mendel's laws. These traits can also be affected by the environment to varying degrees (McClean, 1998).

1.3 QTLs as Hypotheses

We may say that the Quantitative Trait Loci (QTLs) are hypotheses, that specific chromosomal regions contain genes that make a significant contribution to the expression of a complex trait. QTLs are generally identified by comparing the linkage (degree of covariation) of polymorphic molecular markers and phenotypic trait measurements. The ultimate goal of complex trait dissection is to identify the actual genes involved in the trait and to understand the cellular roles and functions of these genes. The accuracy and precision of locating QTLs depends, in part, on the density of the linkage map created. The higher the density of the map, the more precise the location of the putative QTL. When QTLs can be mapped to a relatively small chromosomal region or regions other methods, such as positional cloning, can be used effectively to isolate specific genes. Unfortunately, the denser the map, the more likely that false positive QTLs will be detected. Most, but not all, complex traits are conditioned by more than one locus. QTLs often interact in complex ways and their expression can also be influenced by non-genetic factors.

Because QTLs are hypotheses, they are subject to reinterpretation and revision. Because the location of QTLs are provisional, their nomenclature is likely to be fluid and temporary. Some of the terms associated with QTLs (phenotype, trait, etc.) are semantically "fuzzy."

1.4 What are the Most Important Things to Know about QTLs?

A useful description of a QTL includes at least the following elements (with indications of the certainty associated with each point):

- What traits were used to define a complex phenotype
- What lines / progenies were used in a study
- The type and size of the mapping population
- Methods used for QTL mapping
- Chromosomal location(s) of the QTL
- Candidate genes in the QTL region

- Magnitude of effect of the QTL on phenotype (% phenotypic/genetic variance accounted for)
- Model(s) of gene action (additive, dominance, etc.)
- Epistatic interactions, if any
- Comparison of QTL information to studies of the genetic basis of similar complex trait phenotypes in other organisms

2.0 QTL MAPPING

2.1 Historical Background

For most of the period up to 1980, the study of quantitative traits has involved statistical techniques based on means, variances and covariances of relatives. These studies provided a conceptual base for partitioning the total phenotypic variance into genetic and environmental variances, and further analyzing the genetic variance in terms of additive, dominance and epistatic effects. From this information, it became feasible to estimate the heritability of the trait and predict the response of the trait to selection. It was also possible to estimate the minimum number of genes that controlled the trait of interest. However, little was known about what these genes were, where they are located, and how they controlled the trait(s), apart from the fact that for any given trait, there were several such genes segregating in a Mendelian fashion in any given population, and in most cases their effects were approximately additive (Kearsey and Pooni, 1996). These genes were termed 'polygenes' by Mather (1949). During this phase, mapping of polygenes was based on methods involving a limited number of major marker loci (Sax, 1923; Thoday, 1961). Sax's (1923) experiment with beans demonstrated that the effect of an individual locus affecting a quantitative trait could be isolated through a series of crosses resulting in randomization of the genetic background with respect to all genes not linked to the genetic markers under observation. Even though all of the markers used by Sax were morphological seed markers with complete dominance, he was able to show a significant effect on seed weight associated with some of his markers. Despite this demonstration, there were extremely few successful examples were reported with some of his markers-QTL linkage in crop plants during 1930-80s, and of these even fewer were repeated. The major limitation was the lack of availability of adequate polymorphic markers.

Two major developments during the 1980s changed the scenario:

(i) Discovery of extensive, yet easily visualized, variablility at the DNA level that could be used as markers.

(ii) Development of statistical packages that can help in analyzing variation in a quantitative trait in congruence with molecular marker data generated in a segregating population.

2.2 Mapping Studies

In a process called QTL-mapping, association between observed trait values and presence/absence of alleles of markers that have been mapped onto a linkage map is analyzed. QTL mapping is the statistical study of the alleles that occur in a locus and the phenotypes (physical forms or traits) that they produce. When it is significantly clear that the correlation that is observed did not result from some random process, it is proclaimed that a QTL is detected. A breeder can analyze QTL occurrences and use this knowledge to his advantage, for instance by using indirect selection. When selection is (partly) based on genetic information retrieved through the application of molecular markers this is called marker-assisted selection (MAS).

Statistical analysis is required to demonstrate that different genes interact with one another and to determine whether they produce a significant effect on the phenotype. QTLs identify a particular region of the genome as containing a gene that is associated with the trait being assayed or measured. They are shown as intervals across a chromosome, where the probability of association is plotted for each marker used in the mapping experiment. The QTL techniques were developed in the late 1980s and can be performed on inbred strains of any species. To begin, a set of genetic markers must be developed for the species in question.

2.3 Advent of Marker Technologies

The advent of molecular marker techniques has had a large impact on quantitative genetics. A marker is an identifiable region of variable DNA. Biologists are interested in understanding the genetic basis of phenotypes. The aim is to find a marker that is significantly more likely to co-occur with the trait than expected by chance, that is, a marker that has a statistical association with the trait. Marker-based methods applied to segregating populations have provided us with a means to locate Quantitative Trait Loci (QTLs) to chromosomal regions and to estimate the effects of QTL allele substitution (Lander and Botstein, 1989). When a QTL is found, it is often not the actual gene underlying the phenotypic trait, but rather a region of DNA that is closely linked with the gene. Marker and QTL information obtained from a segregating population, the ability to estimate gene effects and locations for quantitative traits can be very useful for the design and application of new, efficient, breeding strategies. In recent years major advances in marker availability and statistical methods for assessing marker-trait correlations have been achieved (Lander and Botstein 1989; Haley and Knott, 1992; Van Ooijen, 1992; Jansen and Stam 1994; Falconer and Mackay 1996). MAS has been advocated as a useful tool for rapid genetic advance in the case of quantitative traits (Lande and Thompson, 1990; Knapp, 1994; Knapp, 1998).

With phenomenal improvements in molecular marker technology in the last two decades, identification and utilization of polymorphic DNA markers as a fame work around which the polygenes could be located, has improved multiple-fold. It is now clear that a genetic map saturated with polymorphic co-dominant Mendelian markers can be generated for almost any species. Nearly saturated genetic maps have already been produced for most species of economic or scientific interest. We now refer the polygenes by a more catchy acronym, QTL (Quantitative Trait Loci), a term first coined by Gelderman in 1975. A QTL is defined as "a region of the genome that is associated with an effect on a quantitative trait". Conceptually, a QTL can be a single gene, or it may be a cluster of linked genes that affect the trait.

QTL mapping studies have been reported in most crop plants for diverse traits including yield, quality, disease and insect resistance, abiotic stress tolerance, and environmental adaptation. Here, a brief overviews of the principle of QTL mapping, salient requirements for QTL mapping, common statistical tools and techniques employed in QTL analysis, and the strengths, constraints and applications of QTL mapping for crop improvement shall be provided.

2.4 Principle of QTL Mapping

It is not difficult in population of most crop plants to identify and map a good number of segregating markers (10 to 15) per chromosome. However, most of these markers would be in non-coding regions of the genome and might not affect the trait of interest directly; but, a few of these markers might be linked to genomic regions (QTLs) that do influence the trait of interest. Where such linkage occurs, the marker locus and the QTL will co-segregate. Therefore, the basic principle of determining whether a QTL is linked to a marker is to partition the mapping population into different genotypic classes based on genotypes at the marker locus, and then apply correlative statistics to determine whether the individuals of one genotype differ significantly with the individuals of other genotype with respect to the trait being measured. Situations where genes fail to segregate independently are said to display "linkage disequilibrium". QTL analysis, thus, depends on linkage disequilibrium.

With natural populations, consistent association between QTL and marker genotype will not usually exists, except in a very rare situation where the marker is completely linked to the QTL. Therefore, QTL analysis is undertaken in segregating mapping populations, such as F_2-derived populations, recombinant inbred lines (RILs), near-isogenic lines (NILs), doubled haploid lines (DHs), and backcross populations.

2.5 Objectives of QTL Mapping

The vast majority of molecular marker research in quantitative traits has been devoted to mapping QTL. These experiments basically have the following major objectives:

1. To identify the regions of the genome that affects the trait of interest.
2. To analyze the effect of the QTL on the trait.

Some other relevant questions must be answered during QTL mapping:

- How much of the variation for the trait is caused by a specific region?
- What is the gene action associated with the QTL (additive effect, dominant effect)?
- Which allele is associated with the favorable effect?

3.0 SALIENT REQUIREMENTS FOR QTL MAPPING

- A suitable mapping population generated from phenotypically contrasting parents
- A saturated linkage map based on molecular markers
- Reliable phenotypic screening of mapping population
- Appropriate statistical packages to analyze the genotypic information in combination with phenotypic information for QTL detection

3.1 Types and Size of Mapping Population

Linkage disequilibrium is a key to detecting QTLs with markers. It is therefore essential to develop a suitable experimental mapping population using parental lines that are highly contrasting phenotypically for the target trait (*e.g.* highly resistant and susceptible lines). Another important requirement is that these parental lines should be genetically divergent; this is important to enhance the possibility of identifying a large set of polymorphic markers that are well-distributed across the genome. To fulfill the second criterion, one may have to carry out molecular polymorphism survey across a set of potentially useful lines so as to identify the most suitable ones for generation of mapping population.

3.1.1 Various Types of Mapping Populations

F_2-derived populations, RILs, backcross populations, NILs, DHs etc. are employed for QTL mapping in various crop plants. The choice of a mapping population could vary based upon the objectives of the experiment, the time frame as well as resources available for undertaking QTL analysis. But, the ability to detect QTLs or the information contained in F_2 or F_2-derived populations and RILs are relatively higher than other. The primary advantage of $F_{2:3}$ families is the ability to measure the effects of additive and dominance gene actions at specific loci. Because RILs are essentially homozygous, only additive gene action can be measured. The advantage, though, of the RILs is the ability to perform larger experiments at several locations and even in multiple years. For many crops, it is not possible to generate enough seed to perform a multiplications experiment with population of $F_{2:3}$ families. Modification of the genetic model is necessary to accommodate different types of populations.

3.2 Size of the Mapping Population

This depends on several factors, including the type of mapping population employed for analysis, genetic nature of the target trait, objectives of the experiment, and the resources available for handling a sizable mapping population in terms of phenotyping and genotyping. Usually the possible number of QTLs that can be detected with a population size of 25, 50, 100, 500, 1000 and infinity are 0, 1, 2, 3, 4 and 5, respectively. While analysis of a large number of individuals (~500 or more) would enable detection of even QTLs having small effects on the target trait, from the practical point of view (MAS), the basic purpose of QTL mapping would be largely served if one can detect the QTLs with major effects. This would require, in general, a mapping population of a size of 200-300 individuals (Prasanna, 2007).

3.2.1 Need for Generating a Reasonably Saturated Linkage Map

By screening the mapping population using polymorphic molecular markers (popularly called as 'genotyping'), we can analyze the segregation patterns for each of the markers. The segregation patterns are usually in consonance with the type of mapping population used. The genotypic data are then analyzed using a statistical package such as MAPMAKER (Lander *et al.*, 1987) or JOINMAP (Stam, 1993), for construction of a linkage map of the molecular markers analyzed in the study. Mapping means placing the markers in order, indicating the relative genetic distances between them, and assigning them to their linkage groups on the basis of recombination values from all pair-wise combinations between the markers.

To perform a whole-genome QTL scan, it is desirable to have a saturated marker map. In such a map, markers are available for each chromosome from one end to the other, and adjacent markers are spaced sufficiently close so that the recombination events only rarely occur between them. For practical purposes, this is generally considered to be less than 10 recombinations per 100 meioses, or a map distance of less than 10 centi-Morgan (cM). In the model plant *Arabidopsis thaliana*, which has a particularly small genome, this requires as few as 50 markers. Several-fold more markers are needed for plant genomes such as wheat and maize. In crops like maize, a broad 'rule-of-the-thumb' is to cover each of the chromosomal (bin) locations with at least one or two polymorphic molecular markers.

3.3 Phenotyping of Mapping Population and Sample Size

The target quantitative traits have to be measured as precisely as possible and limited amounts of missing data can be tolerated. The power to resolve the QTL location is limited first by sample size, and then by genetic marker coverage of the genome. Often, the number of individuals in sample might appear to be large, but missing data or skewed allele frequencies in the population cause the effective sample size to diminish, thus sacrificing statistical power. Sometimes, it may be necessary to sacrifice population size in favour of data quality, and this trade-off means that only major QTL (with relatively large effect) can be detected. Data are typically pooled over locations and replications to obtain a single quantitative trait value for the line. It is also preferable to measure the target traits(s) in experiments conducted in multiple (and appropriate) locations to have a better understanding of the QTL x environment interaction, if any.

3.3.1 QTL Detection

Many studies on simultaneous detection of QTLs for multiple traits were described (Korol *et al.*, 1998; Ronin *et al.*, 1999). Such an approach may result in an increased power of QTL-detection. Multiple-trait QTL detection might become a natural partner of multiple-trait marker-assisted selection procedures. Of course, new problems arise when these kinds of procedures are to be applied. Not all traits can be measured with the same accuracy; hence data of unequal quality are used for QTL detection. The reliability of the data should in some way be reflected in the QTL-analysis. When data from a variety of sources are used, the inclusion of the experimental design into QTL detection methods could become important. Several authors reported conservation of QTL locations over populations that were derived from different progenitors. This indicates that the usefulness of a detected QTL may transcend the population in which it was found. More general QTL detection methods, which are not limited to the use of mapping populations and assumptions on the normality of trait distributions, could enhance the applicability of QTL-based selection strategies. Such detection methods are already being developed for analysis of human and animal populations, and could also be employed fruitfully in plant populations that are derived from a diversity of sources, as is common practice in plant breeding.

3.3.2 Analysis of Variance

The simplest method for QTL mapping is analysis of variance (ANOVA, sometimes called "marker regression") at the marker loci. In this method, in a backcross, one may calculate a t-statistic to compare the averages of the two marker genotype groups. For other types of crosses (such as the intercross), where there are more than two possible genotypes, one uses a more general form of ANOVA, which provides F-statistic. The ANOVA approach for QTL mapping has three important weaknesses:

1. We do not receive separate estimates of QTL location and QTL effect. QTL location is indicated only by looking at which markers give the greatest differences between genotype group averages, and the apparent QTL effect at a marker will be smaller than the true QTL effect as a result of recombination between the marker and the QTL.
2. We must discard individuals whose genotypes are missing at the marker.
3. When the markers are widely spaced, the QTL may be quite far from all markers, and so the power for QTL detection will decrease.

3.3.3 Methods for QTL Detection

The beauty of QTL mapping in plants is facility of experimental crosses in contrast to humans. The basic objective in QTL mapping studies is to detect QTL, while minimizing the occurrence of false positives (Type 1 errors, that is, declaring an association between a marker and QTL when in fact one does not exist). Tests for QTL/trait association are often performed by the following approaches:

3.3.3.1 Single Marker Analysis

The single marker analysis approach, sometimes referred to as the single factor analysis of variance (SF-ANOVA) or single point analysis, has been used extensively, especially with isozymes (Edwards *et al.*, 1987). SF-ANOVA is done for each marker locus (*e.g.* M_1 and M_2) and the putative QTLs (Q_1 & Q_2) which are independent from other loci (Fig. 1). F-tests provide evidence whether differences between marker locus genotype classes are significant or not. Although computationally simple, this approach suffers from some major limitations:

- The likelihood of QTL detection significantly decreases as the distance between the marker and QTL increases
- The method can not determine whether the markers are associated with one or more QTLs

- The effects of QTL are likely to be underestimated because they are confounded with recombination frequencies.

However problems due to the recombination can be minimized by using a large number of markers so that most part of the genome is covered with minimum distance between a marker and the target QTL.

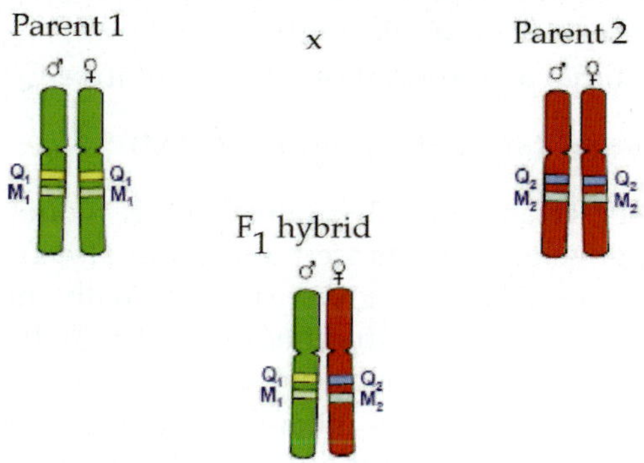

Fig. 1. Single Marker Analysis approach for QTL mapping (Source : Myberg, 2005)

3.3.3.2 Interval mapping

Lander and Botstein (1989) developed interval mapping, which overcomes the three disadvantages of analysis of variance at marker loci and takes full advantage of a complete linkage map. In this procedure, the intervals formed by the pairs of adjacent markers (r_1 & r_2) are taken as the unit of analysis and tested for the presence of a single QTL (Q) by using genetic information from the flanking markers (Fig. 2). Presence of a putative

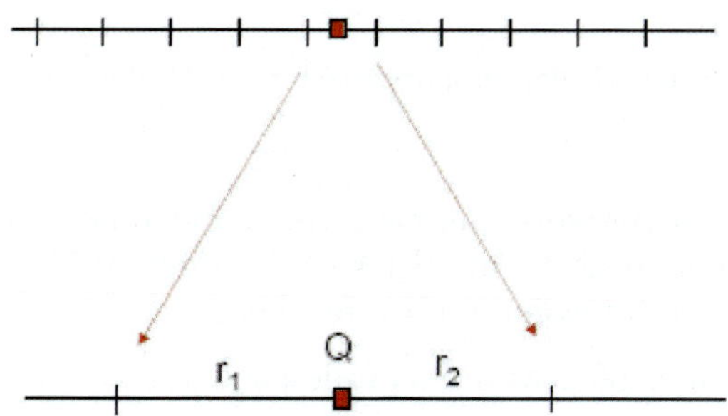

Fig. 2. Interval Mapping approach for QTL detection (Source : Myberg, 2005)

QTL is assumed if the log of odds ratio (LOD) exceeds a critical value. This method compensates for the recombination between marker and the QTL, thereby increasing the probability of statistically detecting the QTL. However when multiple QTLs are segregating in a cross, which is usually the case, it fails to eliminate the genetic variance caused by other QTLs, similar to single marker approach. Interval mapping is currently the most popular approach for QTL mapping in experimental crosses. The method makes use of a genetic map of the typed markers, and, like analysis of variance, assumes the presence of a single QTL. Each location in the genome is marked, one at a time, as the location of the putative QTL.

3.3.3.3 *Composite Interval Mapping (CIM)*

CIM (Zeng, 1994) and MQM (multiple-QTL model or marker-QTL-marker analysis) developed by Jansen and Stam (1994) combine interval mapping for a single QTL in a given interval with multiple regression analysis on marker associated with other QTL (Fig. 3). In this method, one performs interval mapping using a subset of marker loci as covariates. These markers serve as proxies for other QTLs to increase the resolution of interval mapping, by accounting for linked QTLs and reducing the residual variation.

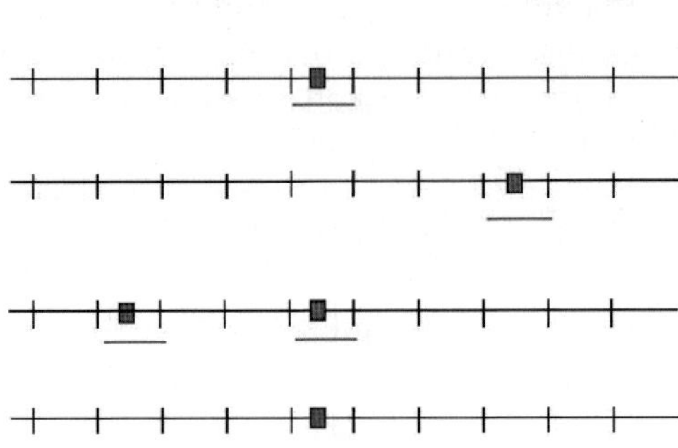

Fig. 3. Composite Interval Mapping approach for QTL detection (Source : Myberg, 2005)

It considers a marker interval plus a few other well-chosen single markers in each analysis, so the n-1 tests for interval-QTL associations are performed on a chromosome with n markers.

The advantages of CIM are as follows:

(i) Mapping of multiple QTLs can be accomplished by the search in one dimension.

(ii) By using linked markers as co-factors, the test is not affected by QTL outside the region, thereby increasing the precision of QTL mapping.

(iii) By eliminating much of the genetic variance by other QTL, the residual variance is reduced, thereby increasing the power of detection of QTL. CIM is more powerful than SIM, but is yet to be used extensively in QTL mapping.

The key problem with CIM concerns the choice of suitable marker loci to serve as covariates; once these have been chosen, CIM turns the model selection problem into a single-dimensional scan. The choice of marker covariates has not been solved, however. Not surprisingly, the appropriate markers are those closest to the true QTLs, and so if one could find these, the QTL mapping problem would be complete anyway.

3.3.3.4 *Multiple Interval Mapping (MIM)*

A new statistical method for mapping quantitative trait loci (QTL), called multiple interval mapping (MIM) (Kao *et al.*, 1999). It uses multiple marker intervals simultaneously to fit multiple putative QTL directly in the model for mapping QTL (Fig. 4). The MIM model is based on Cockerham's

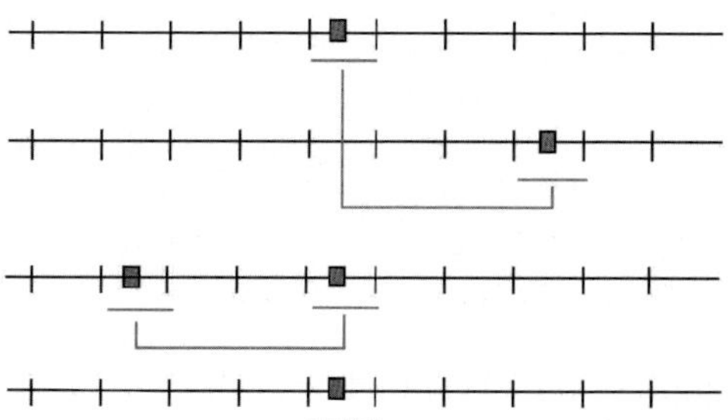

Fig. 4. Multiple Interval Mapping approach for QTL detection (Source : Myberg, 2005)

model for interpreting genetic parameters and the method of maximum likelihood for estimating genetic parameters. With the MIM approach, the precision and power of QTL mapping could be improved. Also, epistasis between QTL, genotypic values of individuals, and heritabilities of quantitative traits can be readily estimated and analyzed.

Using the MIM model, a stepwise selection procedure with likelihood ratio test statistic as a criterion is proposed to identify QTL. With the estimated QTL effects and positions, the best strategy of marker-assisted selection for trait improvement for a specific purpose and requirement can be explored.

3.3.3.5 *Non-traditional methods*

***Family-pedigree based mapping*:** The human genetic approach has taken attention to plant scientist to fit in to plant breeders families (Jannink *et al.*, 2001). There are some successful attempts to do so. One of quick method of QTL mapping is recently discussed (Rosyara *et al.*, 2007).

3.3.3.6 *Introgression of QTL through backcrossing*

Current developments in molecular and statistical genetics allow estimation of positions and effects of genome fragments responsible for variation in quantitative traits. Such quantitative trait loci (QTLs) are an important and essential source for crop improvement. This improvement can be achieved by making selected crosses within existing elite material in which QTLs have been assessed, followed by marker based selection of superior genotypes. Another option is to aim for the introgression of favorable genome fragments from unadapted material in order to obtain superior trait values, by making selected backcrosses (Tanksley and McCouch, 1997). In the case of backcrosses for introgression of QTLs a marker-assisted approach is required to verify, in each generation, the presence of the favorable allele at the QTL.

The presence or absence of a favorable QTL allele cannot be determined by screening the plant phenotype. In most cases no replicated trials are possible due to the small number of plants that make up a backcross population. Also, these backcross populations may differ with respect to their genetic background, and other factors that also influence the trait may still segregate. Screening backcross populations with molecular markers with known positions on a genetic map can provide valuable information. Not only can the origin of the QTL allele be determined, also the remainder of the genome, both linked and unlinked to the QTL, can be monitored. Using a theoretical approach, Stam and Zeven (1981) considered the amount of unwanted donor genome on the same chromosome as the gene of interest in a regular backcross program, without the use of molecular markers to control unwanted linkage.

3.3.3.7 *Advanced backcross QTL (AB-QTL) analysis*

This strategy was proposed by Tanksley and Nelson (1996) for combining QTL analysis with variety development, particularly for the transfer of valuable QTLs from unadapted germplasm (such as landraces and sexually-compatible wild relatives of crop plants) into elite breeding lines. In this strategy, QTL analysis is delayed until the BC_2 or BC_3 generation and, during the development of these populations, negative selection is exercised to reduce the frequency of deleterious donor alleles. QTL-NILs can be derived from advanced backcross populations in one or two additional generations and utilized to verify QTL activity. Such lines also represent improved commercial inbreds (over the original parental line) for one or more quantitative traits. The utility of this approach has been demonstrated for quantitative trait improvement in tomato and rice.

Once QTL are detected, the next step is to estimate the genotypic effect of the QTL and to localize the QTL to a precise genomic region. The interval mapping approach is superior to the ANOVA approach in terms of both localization and estimation of the effects of QTL. However, the reliability in terms of estimation of the QTL effect depends on the linkage between marker (s) and QTL, the number and type of progeny evaluated, and the heritability of the trait. From multiple regression analysis, one can also obtain an R^2 value which gives the percentage of the total genetic variance explained by all of the markers. The R^2 value for the line is considered to be the amount of total genetic variation that is explained by the specific molecular marker. Recently developed statistical packages also offer the means to analyze the QTL x environment interaction.

3.3.4 Interaction factors and QTL mapping

Studies on genetic improvement mainly focus on main effects. Interaction factors such as genotype by environment (GxE and QTLxE) interaction or interacting genes (epistasis) are difficult to handle and unpredictable. Improved algorithms that enable detection of epistasis provide new options to steer selection in these cases. Selection decisions could take advantage of knowledge on interacting genes and, through the use of marker assisted selection, favorable sets of genes could be assembled or undesired combinations of alleles prevented.

In all simulations it could be assumed that all QTLs affect a single trait. This is, of course, a simplification but not a limitation; one can easily imagine the case where the QTLs of the model are divided into subsets, each set affecting a different trait. The 'final trait' could then be an index value, composed of a linear combination of traits. This will not change our general results, as long as the traits involved are comparable in their importance to the breeder. When many QTLs are to be accumulated the chance of getting them all with just one pair of lines is small. In this case, one may think of an extension of the procedure to three way crosses or four way crosses. A possible way to compensate for QTLs or polygenes that have gone unnoticed in the mapping procedure is to combine marker information and phenotypic values into the index, in a way similar to the index proposed by Lande and Thompson (1990). This method basically assigns weights to markers and phenotype relative to the proportion of variance explained by the markers. Such an approach will be subject of a future study.

3.4 Software Tools for QTL Analysis

The increasing supplies of large molecular data sets demand the availability of a robust set of tools for analysis. The capacities of modern computers (the hardware) seem to keep up with the growing supply of data; therefore, the real demand is for intelligent software that is able to use the available data to provide answers to scientific and applied questions.

The high speed at which developments in marker and computer technology continue to advance induce a need for standardization and a more automated processing and analysis of molecular marker data. The large size and multidimensional character of marker data sets invite novel approaches to data visualization (Nelson, 1997). User friendly 'smart' software packages are therefore a prerequisite for practical use of marker derived information on a large scale. Although there are efforts to provide users with software that is easier to use (Korol *et al.* 1998; Van Berloo, 1999), it would be a good idea if professional software developers were to be involved in the development and introduction of standardized, robust and user-friendly software. A wide acceptation of such a suite of programs would not only keep the software affordable, but would also permit easier transfer, sharing and combining of data, which could help to increase experimental resolution (Beavis, 1999).

3.4.1 Software used for mapping of QTLs

For QTL mapping experiments there are different computer programs, allowing simulation of crossing and selection. A locus (marker or QTL) is the smallest unit that is present in the computer model. Loci are linked together in linkage groups or chromosomes and Mendelian rules apply to the simulation of recombination during meiosis. QTLs and allelic effects remain visible, but are not used for selection. Selection is based only on marker loci and intervals of marker loci. Within the model, indices are calculated for pairs of lines. Based on these index values, pairs of lines are either selected or disregarded of the selected fraction. In conventional selection, phenotypic values are used as the criterion to select RIL-pairs. Few important types of software which are used for QTL mapping are given below.

3.4.1.1 MAPMAKER/QTL

Over the last half century, a number of methods have been developed to map genes controlling quantitatively measured phenotypes segregating in experimental populations. In general, these techniques work by finding correlations between the inheritance of particular genetic markers and variation in the phenotype for each individual in the population.

MAPMAKER/QTL extends these methods to (1) provide support for "interval mapping", allowing one to fully exploit the information provided by a genetic linkage map, and (2) to calculate LOD scores for putative QTLs, providing a measure of the support for any particular hypothesis. MAPMAKER/QTL implements these features with a user-friendly interface which is similar to that used by the multipoint genetic linkage analysis package, MAPMAKER/EXP, and actually shares data files with the new versions of that program. In fact, MAPMAKER/QTL and MAPMAKER/EXP together form a complete linkage analysis package (which we call MAPMAKER) for mapping work. However, MAPMAKER is not a closed package: it is fairly easy to use other methods along with the MAPMAKER analyses. (http://www.broad.mit.edu/genome_software/other/qtl.html)

3.4.1.2 QTL Cartographer

It is a suite of programs to map quantitative traits using a map of molecular markers. QTL Cartographer is WinQTLCart's command-line sibling. QTLCart is a suite of command-line programs to map quantitative traits using a map of molecular markers. The programs are available *via* an anonymous ftp server. (http://statgen.ncsu.edu/qtlcart/index.php).

3.4.1.3 R/qtl

R/qtl is an extensible, interactive environment for mapping quantitative trait loci (QTLs) in experimental crosses. It is implemented as an add-on package for the freely available and widely used statistical language/software R. The development of this software as an add-on to R allows us to take advantage of the basic mathematical and statistical functions, and powerful graphics capabilities, that are provided with R. Further, the user will benefit by the seamless integration of the QTL mapping software into a general statistical analysis program. A key component of computational methods for QTL mapping is the hidden Markov model (HMM) technology for dealing with missing genotype data. It includes the main HMM algorithms, with allowance for the presence of genotyping errors, for backcrosses, intercrosses, and phase-known four-way crosses. The current version of R/qtl have the facilities for estimating genetic maps, identifying genotyping errors, and performing single-QTL genome scans and two-QTL, two-dimensional genome scans, by interval mapping (with the EM algorithm), Haley-Knott regression, and multiple imputation. All of this may be done in the presence of covariates (such as sex, age or treatment). One may also fit higher-order QTL models by multiple imputation.

3.4.1.4 Other QTL mapping softwares

Most of these methods of analysis used by these packages yield essentially similar QTL locations and gene effects on a given data set, while there could be slight variation in the confidence intervals. Recent advances

Table 1. Other software for QTL mapping

R/qtlbim (runs in R)	QTLNetwork
Pseudomarker (runs in Matlab)	WebQTL
Qgene	Multimapper
Map Manager QTL	MultiQTL
QTL Cartographer	PLABQTL
HAPPY	The QTL Cafe
MapQTL	QTL Express
BQTL (runs in R)	Kajsa Ljungberg's software

using nonparametric statistics or association-based approaches to identify QTLs and to calculate empirical critical (threshold) values for declaring significant QTLs will further refine QTL mapping. As with the improvement in marker technologies, the statistical tools needed for QTL mapping have evolved from a rudimentary to a very sophisticated level (Borevitz, 2004). Current approaches are based on multiple regression methods, using least square estimation methods, mixed model approaches such as maximum likelihood, and Markov Chain Monte-Carlo algorithms (MCMC), which use Bayesian statistics to estimate posterior probabilities by sampling from the data. In parallel, with progress in the characterization of genetic effects at QTL and refinement of QTL peak position through meta-analysis (Chardon *et al.*, 2004), advances have also been made in understanding the impact of the environment on plant phenotype. The mapping of QTL for multiple traits has allowed for the quantification of QEI (Jiang and Zeng, 1995), and more recently, approaches using factorial regression models have been applied to model both QEI and genotype by environmental interaction (GEI), using genetic and environmental co-variables in the same model (Varges *et al.*, 2006).

The theoretical work on the principles of QTL mapping has now achieved a solid background (Lander and Botstein, 1989; Haley and Knott, 1992; Van Ooijen, 1992; Jansen and Stam, 1994; Jansen, 1995; Doerge and Rebaï, 1996; and others) and is still subject of further study and improvement. Implementations of the developed methods, in a diversity of 'flavours' are now available (Lander *et al.*, 1987; Basten *et al.*, 1994; Holloway and Knapp, 1994; Tinker and Mather, 1995; Van Ooijen and Maliepaard, 1996a,b; Nelson, 1997). Many of the currently available software packages were created by enthusiastic scientists, sometimes on an ad-hoc basis directed at solving an emerging problem. The available software packages each incorporate different (sets of) solutions to tackle genetic problems (Li, 1999). Most scientific programmers have given emphasis to a sound methodology and functionality of their software, but paid less attention to the user-friendliness and standardization of data used for in- and output. Furthermore, since many of these software packages were created and are freely available, user support is rarely provided and maintenance is irregular or absent. This diversity has not made the practical use of QTL mapping very accessible to the community of scientists, working in related fields, and plant breeders.

4.0 FACTORS AFFECTING MARKER ASSISTED SELECTION IN QTL BREEDING

Most simulation studies show good results for application of marker-assisted selection but, as indicated in other studies (Whittaker *et al.* 1995; Knapp, 1998), simplifications and assumptions favouring MAS are often made. However, marker-assisted selection should not always be preferred over phenotypic selection. Many factors play a role in the decision which selection strategy to apply.

Heritability: Trait heritability is the most important factor influencing the effectiveness of MAS. MAS seems to be most promising for traits with low heritability. But trait heritability is also of major importance for accuracy in the mapping of QTLs. For instance, for traits with a high heritability MAS may still outperform phenotypic selection, but the high costs of obtaining genetic fingerprints, necessary for performing MAS, may render the procedure cost ineffective. Traits with low heritability reduces the power of detecting QTLs, which is based on correlation between phenotype and marker genotype. This could mean that for well-mapped QTLs MAS may add little to phenotypic selection. Traits with a *very* low heritability the underlying QTLs cannot be identified. It is the area in between these two extreme cases that looks most promising for application of MAS. If QTLs can be mapped for a trait having a low heritability the accuracy of the QTL position may not be very high, which is reflected in a large QTL support interval on the genetic map (Lee, 1995). In most cases MAS will give better selection results than phenotypic selection, for a low heritability trait. The breeder can decide if the increased selection results are worth the extra cost involved in obtaining the marker data. Index based selection opens new ways to quantify performance with regard to several traits into one index value, and use markers to select for those plants that give an optimization of this index in the current or a future generation. This may facilitate breeding for several traits simultaneously.

QTL Information and generation of selection: When MAS is applied in a case with incomplete QTL information the efficiency may actually be worse than phenotypic selection, since some undetected factors remain unselected by MAS. Furthermore, long term objectives should be considered. In population improvement, the high efficiency that is observed when MAS is applied is seen mostly during the first generations of selection. It has been reported (Gimelfarb and Lande, 1994a; Hospital *et al.* 1997; Dekkers, 1999) that continued marker-assisted selection may yield lower selection efficiency in the long term, compared with conventional selection procedures. This is mainly seen when stringent MAS is applied in early generations and 'minor QTLs', which remain undetected until all 'major QTLs' have become fixed in later generations, are lost.

Population Size: Another important parameter is population size (Gimelfarb and Lande, 1994a; Moreau *et al.*, 1998). When larger populations are used it may be expected that MAS will be able to extract, in a more efficient way than phenotypic selection, the superior genotypes or parents from this population. Also, large mapping populations allow a more reliable detection of QTLs. However, in most cases practical and economic considerations limit the population sizes that can be used. In a situation where budgets are fixed and the costs of genotyping plants in order to be able to perform MAS come at the expense of fewer plants that can be grown (*i.e.* a smaller population size), it remains to be seen if MAS will end up as the superior selection strategy.

5.0 FACTORS AFFECTING THE POWER OF QTL MAPPING

QTLs are statistically inferred from the data generated in an experiment. However, statistical influence does not always indicate biological significance due to multiple test problems associated with QTL mapping (Liu, 2002). The following factors affect the power of QTL mapping:

- Number of genes controlling the target trait(s) and their genome positions
- Distribution of genetic effects and existence of genetic interactions
- Heritability of the trait
- Number of genes segregating in a mapping population
- Type and size of mapping population
- Density and coverage of markers in the linkage map
- Statistical methodology employed and significance level used for QTL mapping

Excellent discussions on the above factors and the possible means to improve the power of QTL mapping are available in literature (Kearsey and Farquhar, 1998; Liu, 2002). Replicate progeny analysis, selective genotyping, sample pooling and sequential sampling are some of the suggested approaches for optimization of experimental designs, so as to enhance the power of QTL detection and estimation of QTL effects.

6.0 UNCERTAINTY ASSOCIATED WITH QTLS

There is a fair amount of uncertainty associated with QTLs, including:

- Uncertainty in whether a QTL is really associated with the complex trait in question.
- Uncertainty as to the chromosomal location of the QTL.
- Uncertainty in whether the QTL contains one or more linked genetic factors.
- Uncertainty in nomenclature.
- Uncertainty as to which genes/alleles in the mapped QTL region are likely to be directly related to the complex trait in question.

7.0 UTILITY OF QTL INFORMATION

While the methods for QTL analysis described above provide us with information about the putative locations and effects of QTLs influencing a quantitative trait, these do not tell us anything about the molecular nature of the QTL. In other words:

- Are these QTL coding for specific enzymes involved in a particular pathway?
- Do they act as regulators of gene expression?
- Are they non-coding regions that have some influence in the expression of the trait of interest?

The fact is, we do not know the precise nature of QTL (with a handful of exceptions), despite hundreds of studies on QTL mapping in diverse plants over the last two decades. However, information about the structure and function of a QTL is possible to obtain after cloning and characterization of the QTL using recombinant DNA technology. Identification of putative QTL locations and DNA markers linked to QTLs have opened up opportunities for isolation and molecular characterization of QTLs *via* map-based cloning. There are important caveats, however. Only the QTLs of largest effect and those closest to a marker locus, will show statistically reliable associations. It may be difficult to estimate even the presence of QTL if they interact strongly in their effects. And the regions to which a QTL is localized can be quite large (Several cM, where 1 cM can range from hundreds to thousands of kilobases). Such regions may contain many genes, and there is no guarantee that a QTL will correspond to only one gene. Thus, with QTLs in hand, much further work is necessary to truly dissect quantitative variation at the mechanistic level. Particularly important is fine-mapping or high-resolution mapping of the QTL, if the QTL information has to be effectively applied in basic/applied research. Once fine-mapped, QTLs can also serve as useful tools for comparative genomics, functional genomics and evolutionary studies.

8.0 FROM QTL TO FUNCTIONAL GENES

In the past decade, quantitative trait locus (QTL) mapping has identified hundreds of chromosomal regions containing genes affecting a range of traits. The ultimate goal of QTL mapping is to identify the genes underlying these polygenic traits and to gain a better understanding of their physiology and biochemistry. But identifying the QTL genes has been slow and difficult. It is believe that QTL mapping and mutagenesis will lead to unique insights about complex phenotypes, although it is thought that each approach will yield different outcomes. Mutagenesis will find all genes in a pathway, many of which will resemble rare mutations. In contrast, QTL mapping is more likely to identify genes that encode regulatory proteins or rate-limiting enzymes, which will provide important targets (Fig. 5). The mapping of QTL will be particularly important in the future, because QTL found in animal models may predict their location in the human genome (Sandberg *et al.*, 2000). In future the QTL mapping will help identify several genes underlying the major complex traits.

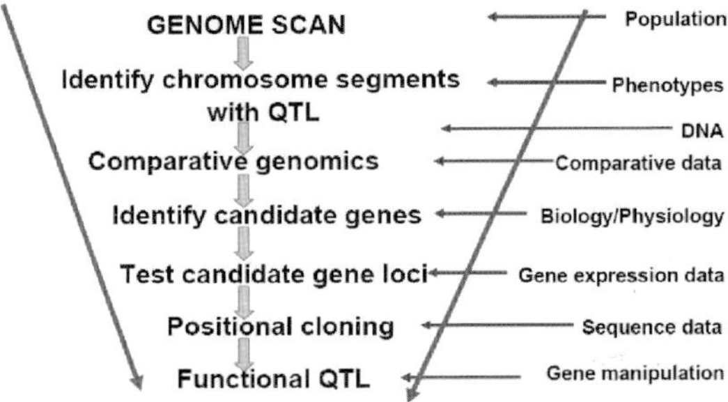

Fig. 5. Schematic display of sequence of events from genome scans to functional QTL (Source : http://www.angelfire.com).

9.0 ERRORS IN QTL MAPPING

To study the effect of uncertainty in QTL number and positions, Van Berloo and Stam (1998) have run simulations for the following situations:

9.1 QTLs Mapped to Incorrect Marker Intervals

It is assumed that the mapped positions of some QTLs do not correspond with the true positions on the genome. Instead these QTLs are mapped to intervals adjacent to the true intervals, leading to selection of incorrect marker intervals by the MAS procedure. The performance of MAS is affected when QTLs are not mapped at their true position. A reduction in selection response will be observed, as the number of incorrectly located QTLs increased, but the effect will be small. We believe this is because using a neighbouring marker interval for calculation of the index will in most cases still result in the same index. Only when recombination has occurred within either or both of the correct and incorrect intervals will the resulting index be affected, and thus the performance of a RIL pair inaccurately predicted.

9.2 Undetected QTLs (Type II errors)

In most QTL mapping studies, some QTLs are detected, but even when the same populations are used, different QTLs may be found in replicated trials. As a result, these QTLs can not be selected by the MAS procedure. A reduction in selection response was observed as the proportion of undetected QTLs increased. Here it was assumed that the QTL mapping procedure failed to locate one or more QTLs, causing reduced selection opportunities for MAS. Beavis (1999) found up to 60 QTLs in a very large experiment. When using a subset of the data, representative in size to a commonly used mapping population, only about 15 QTLs were detected. This example illustrates the common knowledge among quantitative geneticists that any single QTL study will usually not be conclusive. Some QTLs, also of larger effects, will remain undetected due to the limited detection power available in common mapping populations. This limitation is mainly due to the population size. Accurate mapping of many QTLs depends on the occurrence of rare crossovers. Since this is a process of chance, only very large populations are likely to contain individuals in which several rare crossover events did occur. Beavis (1999) therefore suggested to pool available experimental results to obtain a better power of QTL detection.

9.3 False Positive QTL Detection (Type I errors)

When QTL detection is conducted, there is always the risk that the QTL mapping procedure falsely indicates the presence of one or more QTLs at positions where none in fact exist. These 'false QTLs' were included in the index used by MAS, introducing errors in the overall combination-index. The introduction of false QTL *i.e.* QTL that are not actually present, but were falsely identified by the QTL mapping procedure showed no effect on the MAS selection results. Even when the number of false QTLs equalled the number of true QTLs no significant decrease in selection response was found. Apparently the MAS procedure does not suffer much from extra information. This may be due to the configuration we tested. All QTLs were linked in coupling phase, so adding QTLs to the map will inflate the index value, but the order of index values and the line pairs that will be selected will not change dramatically (Van Berloo and Stam, 1998).

10.0 CONCLUSION

Molecular markers provide plant breeding with an important and valuable new source of information not only for qualitative but for quantitative traits too. Selection of parents is an important issue in plant breeding. Basing selection on QTL information, *i.e.* applying marker-assisted selection, can result in increased selection efficiency. This is especially true for quantitative traits with a low heritability. Knowledge that results from such analyses, *i.e.* the location on the genome of important genetic factors (quantitative trait loci or QTLs), can and should be applied when making selection and breeding decisions. For efficient application of marker-assisted selection reliable and fairly complete QTL-mapping results are required. When QTLs were mapped for several traits a multiple trait-selection can be devised, through the use of a suitable index. In this case an ideal target genotype, containing favorable alleles for QTLs that affect the traits of interest, can be constructed and crosses can be made between selected parents in such a way that the probability of obtaining the target genotype is maximized. A more reliable and complete mapping of QTLs, including mapping of interaction between QTLs, mapping of QTLs with a higher reliability, for instance resulting from a combined mapping of several traits, and mapping of QTLs in more diverse non-mapping types of populations could greatly contribute to an increased application of marker-assisted selection, and hence a more efficient selection in plant breeding.

One of the achievements of plant biotechnology revolution for the last two decades has been the development of molecular genetics and associated technologies, which have led to the development of an improved understanding of the basis of inheritance of quantitative traits. The genomic segments or QTL involved in the determination of phenotype can be identified from the analysis of phenotypic data in conjunction with allelic segregation at loci distributed throughout the genome. Because of this the mode of inheritance, as well as the gene action underlying the QTL can be deduced (Lander and Botstein, 1989) a large volume of QTL data has been generated since 1980, and thousands of QTL now populate the various crop databases. About 150 original research papers reporting plant QTL data have been published annually from 2000-2004, covering Arabidopsis, soybean, rice, sorghum, maize, barley and wheat. Over 2,200 maize QTLs have been documented in MaizeGDB (Lawrence *et al.*, 2004), underlining the effort to identify phenotype/genotype associations in experimental crosses. Favorable QTL alleles can be used to transfer one or more discrete segments from a donor to an elite cultivar by backcrossing, or to conduct marker assisted population improvement by the stacking of favorable alleles into individuals selected on the basis of marker genotype.

Comparison of the map position of functional sequences with detected QTLs could provide information on the metabolic function of genes located at QTLs, which would increase the understanding of genetics and genetic regulation and provide new options for selection and controlled genetic improvement. The search for genes that interact with other genes, environment and the detection of QTLs through the analysis of correlated traits all require extensive calculations, due to the large number of combinations that need to be evaluated. Nevertheless, it is expected that these options, together with the introduction of desired alleles from related gene-pools will receive the most attention in the time that lies ahead.

11.0 REFERENCES

Basten, C.J., Weir, B.C. and Zeng, Z.B. 1994. Zmap-a QTL cartographer. *In* Smith, Gavora, Chesnais, Fairfull, Gibson, Kennedy, and Burnside (*Eds.*), Proc 5th World Congress on Genetics Applied to Livestock Production: Computing Strategies and Software, Vol 22, Guelph, Ontario, Canada, pp. 65-66.

Beavis, W.D. 1999. Integrating the central dogma with plant breeding. Abstracts Plant & Animal Genome Conference VII: Scherago Intl. Inc. New York p.28.

Borevitz, J. 2004. Genomic approaches to identifying QTL: lessons from *Arabidopsis thaliana. In* Molecular Genetics and Ecology of Plant Adaptation. Proceedings of an International Workshop held in December 13, 2002, in Vancouver, British Columbia, Canada. *Edited* by Cronk, Q.C.B., Whitton, J., Ree, R.H. and Taylor, I.E.P. NCR Research Press, Ottawa, Ontario, pp. 53-60.

Chardon, F., Virlon, B., Moreau, L., Joets, J., Decousset, L., Murigneux, A. and Charcosset, A. 2004. Genetic architecture of flowering time in maize as inferred from QTL meta analysis and synteny conservation with rice genome. Genetics, 168:2169-2185.

Dekkers, J.M. 1999. Optimizing strategies for marker-assisted selection. Abstracts Plant Animal Genome Conference VII. Scherago Intl. Inc. New York. p.29.

Doerge, R.W. and Rebaï, A. 1996. Significance thresholds for QTL interval mapping tests. Heredity, 76:459-464.

Edwards, M.D., Stubwer, C.W. and Wendel, J.F.C. 1987. Molecular marker facilitated investigations of quantitative trait loci in maize. Genetics, 116: 113-125.

Falconer and MacKay, T.F.C. 1996. Introduction to Quantitative Genetics, 4th ed. Longman Press. pp. 356-378.

Gelderman, H. 1975. Investigations on inheritance of quantitative characters in animals by gene markers methods. Theoretical and Applied Genetics, 46:319-330.

Gimelfarb, A. and Lande, R. 1994. Simulation of marker assisted selection in hybrid populations. Genetical Research Cambridge, 63:39-47.

Haley, C.S. and Knott, S.A. 1992. A simple regression method for mapping quantitative trait loci in line crosses using flanking markers. Heredity, 69:315-324.

Holloway, J.L. and Knapp, S.J. 1994. G-Mendel Users guide. Oregon State University; Corvallis, Oregon, USA.

Hospital, F., Moreau, L., Lacoudre, F., Charcosset, A. and Gallais, A. 1997. More on the efficiency of marker-assisted selection. Theoretical and Applied Genetics, 95:1181-1189.

http://famprevmed.ucsd.edu:16080/faculty/cberry/bqtl/

http://research.jax.org/faculty/churchill/software/pseudomarker/index.html

http://statgen.ncsu.edu/qtlcart/cartographer.html

http://statgen.ncsu.edu/qtlcart/index.php

http://www.angelfire.com/mn2/nath/BP/QTL.pdf

http://www.biosciences.bham.ac.uk/labs/kearsey/

http://www.broad.mit.edu/genome_software/other/qtl.html

http://www.genenetwork.org/home.html

http://www.kyazma.nl/index.php/mc.MapQTL

http://www.mapmanager.org/qtsoftware.html

http://www.multiqtl.com/

http://www.qgene.org/

http://www.qtlbim.org/

http://www.well.ox.ac.uk/~rmott/happy.html

Jannink, J., Bink, M.C.A.M., and Jansen, R.C. 2001. Using complex plant pedigrees to map valuable genes. Trends in Plant Science, 6: 337-342.

Jansen, R.C. 1995. Genetic mapping of quantitative trait loci in plants - a novel statistical approach. Thesis Wageningen Agricultural University, 109pp. Wageningen, The Netherlands.

Jansen, R.C. and Stam, P. 1994. High resolution of quantitative traits into multiple loci *via* interval mapping. Genetics, 136:1447-1455.

Jiang, C. and Zeng, Z.B. 1995. Multiple trait analysis of genetic mapping for QTL. Genetics, 140:1111-1127.

Kao, C.H., Zeng, Z.B. and Teasdale, R.D. 1999. Multiple Interval Mapping for Quantitative Trait Loci. Genetics, 152: 1203-1216.

Kearsey, M.J. and Farquhar, A.G.L. 1998. QTL analysis in plants; where are we now? Heredity, 80: 137-142.

Kearsey, M.J. and Pooni, H.S. 1996. The genetical analysis of quantitative traits. Chapman & Hall, London. 381 pp.

Knapp, S.J. 1994. Selection using molecular marker indexes. *In* Proc second plant breeding symposium of the Crop Science Society America And American Society of Horticulture Sciences, Corvallis, Oregon. pp 1-11. American Society of Horticulture Science Alexandria, Virginia, USA.

Knapp, S.J. 1998. Marker-assisted selection as a strategy for increasing the probability of selecting superior genotypes. Crop Science, 38:1164-1174.

Korol, A.B., Ronin, Y.I., Nevo, E. and Hayes, P.M. 1998. Multi-interval mapping of correlated trait complexes. Heredity, 80:273-284.

Lande, R. and Thompson, R. 1990. Efficiency of marker assisted selection in the improvement of quantitative traits. Genetics, 124:743-756.

Lander, E.S. and Botstein, D. 1989. Mapping Mendelian factors underlying quantitative traits using RFLP linkage maps. Genetics, 121:185-199.

Lander, E.S., Green, P., Abrahamson, J., Barlow, A., Daly, M.J., Lincoln, S.E. and Newburg, L. 1987. Mapmaker: an interactive computer package for constructing primary genetic linkage maps of experimental and natural populations. Genomics, 1:174-181.

Lawrence, C.J., Dong, Q., Polacco, M.L., Seigfried, T.E. and Brendel, V. 2004. MaizeGDB, the community database for genetics and genomics. Nucleic Acid Research, 32:393-397.

Lee, M. 1995. DNA markers in plant breeding programs. Advances in Agronomy, 55: 265-344

Li, W. 1999. An alphabetic list of genetic analysis software. (http://linkage.rockefeller.edu/soft/list.html)

Liu, B.H. 2002. Statistical Genomics: Linkage, mapping and QTL Analysis. 2nd Edition. CRC Press, Boca Raton, Florida.

Ljungberg, K., Holmgren S. and Carlborg, Ö. 2002. Efficient algorithms for quantitative trait loci mapping problems. Journal of Computational Biology, Vol 9, pp. 793-804.

Manly, K.F., Cudmore, Jr, R.H. and Meer, J.M. 2001. Map Manager QTX, cross-platform software for genetic mapping. Mammalian Genome, 12: 930-932.

Martinez, V., Thorgaard, G., Robinson, B. and Silanappa, M.J. 2005. An application of Bayesian QTL mapping to early development in double haploid lines of rainbow trout including environmental effects. Genetical Research, 86: 209-221.

Mather, K., 1949. Biometrical Genetics. London, Methuen & Co.

McClean, P. 1998. (http://www.ndsu.nodak.edu/)

Moreau, L., Charcosset, A., Hospital, F. and Gallais, A. 1998. Marker-assisted selection efficiency in populations of finite size. Genetics, 148:1353-1365.

Myberg, Z. 2005. How to map QTL. Generation Challenge workshop, 27 May, 2005, Department of Genetics and FABI. Room No. 6-21, Agricultural Science Building, University of Pretoria .

Nelson, J.C. 1997. QGENE: software for marker-based genomic analysis and breeding. Molecular Breeding, 3:239-245.

Prasanna, B.M. 2007. QTL mapping and its applications in crop plants. e-Book, Statistical methods for agricultural research, Compiled and Edited by Jaggi, S., Varghese, C., Batra, P.K. and Sharma, V.K., IASRI, ICAR, Pusa, New Delhi, http://www.iasri.res.in/ebook/EB_SMAR/

Ronin, Y.I., Korol, A.B. and Nevo, E. 1999. Single- and Multiple-trait mapping analysis of linked quantitative trait loci: some asymptotic analytical approximations. Genetics, 151:387-396.

Rosyara, U.R., Maxson-Stein, K.L., Glover, K.D., Stein, J.M. and Gonzalez-Hernandez, J.L. 2007. Family-based mapping of FHB resistance QTLs in hexaploid wheat. Proceedings of National Fusarium head blight forum, 2007, Dec 2-4, Kansas City, MO.

Sandberg, R., Yasuda, R., Pankratz, D.G., Carter, T.A., Del Rio, J.A., Wodicka, L., Mayford, M., Lockhart, D.J. and Barlow, C. 2000. Regional and strain-specific gene expression mapping in the adult mouse brain. Proceedings of the National Academic of Sciences, 97:11038?11043.

Sax, K. 1923. The association of size differences with seed-coat pattern and pigmentation in *Phaseolus vulgaris*. Genetics, 75:709-726.

Seaton, G., Haley, C.S., Knott, S.A., Kearsey, M. and Visscher, P.M. 2002. QTL Express: mapping quantitative trait loci in simple and complex pedigrees. Bioinformatics, 18, 339-340.

Stam, P. 1993. Construction of integrated genetic linkage maps by means of a new computer package: Joinmap. The Plant Journal, 3:739-744.

Stam, P. and Zeven, A.C. 1981. The theoretical proportion of the donor genome in near isogenic lines of self-fertilisers bred by backcrossing. Euphytica, 30 :227-238.

Tanksley, S.D. and McCouch, S.R. 1997. Seed banks and molecular maps: unlocking genetic potential from the wild. Science, 277:1063-1066.

Tanksley, S.D. and Nelson, J.C. 1996. Advanced backcross QTL analysis: a method for the simultaneous discovery and transfer of valuable QTLs from unadapted germplasm into elite breeding lines. Theoretical and Applied Genetics, 92:191-203.

Thoday, J.M. 1961. Location of polygenes. Nature, 191:368-370.

Tinker, N.A. and Mather, D.E. 1995. MQTL: software for simplified composite interval mapping of QTL in multiple environments. Journal of Agricultural Genomic, Vol 1: 2 (http://www.ncgr.org/ag/jag/papers95/paper295/jqtl16r2.html).

Utz and Melchinger 1996. http://www.uni-hohenheim.de/~ipspwww/soft.html

Van Berloo, R. 1999. MapQTL Assistant: A user-friendly shell round MapQTL®. Journal of Agricultural Genomics, Vol 4: 2 (www.ncgr.org/ag/jag/papers99/paper199/indexp199.html).

Van Berloo, R. and Stam, P. 1998. Marker assisted selection in autogamous RIL populations: a simulation study. Theoretical and Applied Genetics, 96: 147-154.

Van Ooijen, J.W. 1992. Accuracy of mapping quantitative trait loci in autogamous species. Theoretical and Applied Genetics, 84:803-811.

Van Ooijen, J.W. and Maliepaard, C. 1996a. MapQTL ™ version 3.0: Software for the calculation of QTL positions on genetic maps. Plant Genome IV Conference abstracts, p.316.

Van Ooijen, J.W. and Maliepaard, C. 1996b. MapQTL ™ version 3.0: Software for the calculation of QTL positions on genetic maps. CPRO-DLO, Wageningen.

Varges, M., vanEeuwijk, F., Crosa, J. and Ribaut, J.M., 2006. Mapping QTLs and QTLenvironment interaction for CIMMYT maize draught stress programme using factorial regression and partial least square methods. Theoretical and Applied Genetics, 112:1009-1023.

Whittaker, J.C., Curnow, R.N., Haley, C.S. and Thompson, R. 1995. Using marker-maps in marker-assisted selection. Genetical Research Cambridge, 66:255-265

Yang, J., Hu, C., Ye, X., and Zhu, J. 2006. Software for Mapping QTL with epistatic and QE interaction effects in experimental populations. http://ibi.zju.edu.cn/software/qtlnetwork/index.htm.

Zeng, Z.B. 1994. Precision mapping of quantitative trait loci. Genetics, 136:1457-1468.

Zuurbier, P.J.P. 1994. Diagnosing the seed industry. Prophyta, 4:10-27.

Molecular Plant Breeding: Principle, Method and Application
Eds : R.K. Singh, Rajesh Singh, Guoyou Ye, A. Selvi and G.P. Rao
Studium Press LLC, Texas, USA, 2009, pp. 189-208

Association Mapping in Crop Improvement

M. RAVEENDRAN, V.G. SHOBHANA and N. SENTHIL*

ABSTRACT

Advancements in the field of QTL mapping have revolutionized the field of molecular breeding. Identification of DNA markers linked to QTLs controlling target phenotype was mainly done based on classical linkage analysis involving segregating progenies between parents of known ancestry. Recent advances in genomic technology intended to exploit natural diversity and development of robust statistical analysis methods have led to the development of a strategy called "association mapping" affordable to plant research programs. Association mapping identifies quantitative trait loci (QTLs) by examining the marker-trait associations that can be attributed to the strength of linkage disequilibrium between markers and functional polymorphisms across a set of diverse germplasm. In this review, differences between classical linkage mapping and association mapping procedures are explained and procedures involved in genome wide association mapping and candidate gene based association mapping are discussed in detail. We have also described the methodologies involved in association mapping including both phenotyping and genotyping, current status of association mapping in plants and opportunities and challenges in complex trait dissection and genomics-assisted crop improvement.

Key Words: Crop improvement, Natural diversity, Association mapping, Linkage analysis vs association mapping, Applications in plant breeding

---

Department of Plant Molecular Biology and Biotechnology, Centre for Plant Molecular Biology, Tamil Nadu Agricultural University, Coimbatore – 641 003, India.
**Corresponding author e-mail: raveendrantnau@gmail.com*

1.0 INTRODUCTION

The practice of agriculture was invented by humans approximately 10,000 years ago with an aim of cultivating and harvesting specific plants to produce their food. The improved plant traits selected by early agriculturists transmitted to succeeding generations of plants. Further improvement in terms of agricultural production, quality and other agronomic traits has been made through domestication of wild species, hybridization, mutation etc. The domesticated plants have been resulted from the genetic modification of wild plants through thousands of years of natural and artificial selection. Improved understanding of genetic principles based on Mendelian theory and their application to plant breeding technology has greatly accelerated the rate of improvement of crop plants. Breeders were looking for continual search for novel genetic combinations from which plants with superior traits, such as crop quality, yield, regional performance, and tolerance to pests and diseases were selected. This was achieved by making sexual crosses between diverse genotypes to produce new combinations of genetic traits, which then resulted in diverse phenotypes, or observable morphological or quality traits in the progeny plants.

Primary source of genetic variation is the wide array of germplasm within each crop species and closely related wild species that are crossable. Some of the primitive crop cultivars, also known as land races, were adapted to local growing conditions and preferences, and therefore remained genetically diverse for traits such as product qualities, stress tolerance, disease resistance, and yield stability. For many crops, breeders were practicing introgression of genes from closely related wild plants to increase genetic variation in the crops. Hybridization between a crop plant and a related wild species (a wide cross) enabled transfer of valuable genes from the wild species. For example, most of the modern crop varieties possessing resistances against fungal, bacterial, and viral diseases have been developed through wide crosses between domesticated varieties and related wild species (introgression).

Many agriculturally important traits such as yield, quality and some forms of disease resistance are controlled by many genes and are known as quantitative traits (also 'polygenic,' 'multifactorial' or 'complex' traits). Genomic regions controlling these quantitative traits are called as "Quantitative Trait Loci" (QTL). Most traits of agricultural or evolutionary importance are controlled by multiple quantitative trait loci. When transferring a specific trait from a wild plant into a crop plant *via* wide crosses, the simultaneous transfer of undesired genes from the wild plant is often a problem. In repeated crosses to the cultivated type, the percentage of wild type genes in each generation would be 50% in the first generation, 25% in the second, 12.5% in the third, 6.25% in the fourth, etc. Even after many generations, wild-type genes located close to the desired gene on the chromosome may still be present (*i.e.*, are closely linked to the desired gene). Identifying markers closely linked to the QTL will help us to introgress the target region with only few undesirable genes along with it into the desired background. In addition to its use in targeted introgression, DNA markers like SSRs and AFLPs are useful in assessing the recovery of recurrent genome. Identifying such DNA markers linked to target traits/QTLs is possible by two commonly used methods called linkage analysis and association mapping.

1.1 Linkage Analysis Vs Association Mapping

Linkage analysis is involved in detecting QTLs controlling the target traits in a mapping population developed using parents of known ancestry. Linkage analysis in plants has been typically conducted with experimental populations that are derived from a bi-parental cross. Although based on the same fundamental principles of genetic recombination as linkage analysis, association mapping examines this shared inheritance for a collection of individuals often with unobserved ancestry (Thornsberry *et al.*, 2001; Remington *et al.*, 2001). By exploring deeper population genealogy rather than family pedigree, association mapping offers three advantages over linkage analysis: 1)much higher mapping resolution, 2) greater allele number and broader reference population and 3) less research time in establishing an association (Buckler *et al.*, 2002; Flint-Garcia *et al.*, 2003). Linkage analysis and association mapping, however, are complimentary to each other in terms of providing prior knowledge, cross-validation, and statistical power (Wilson *et al.*, 2004). Systematic comparisons of these two different approaches have been reviewed in detail by Hirschhorn *et al.*,

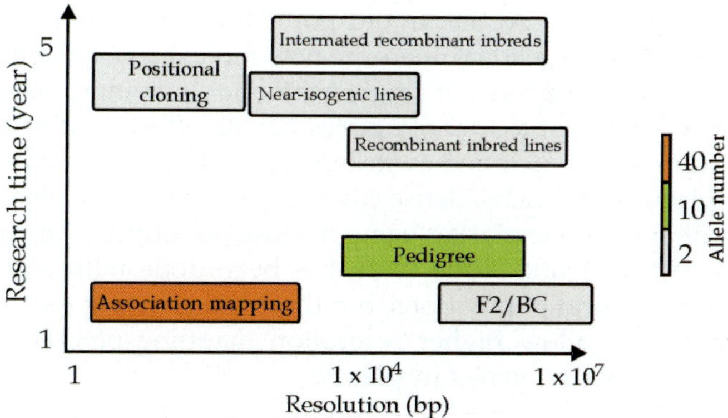

Fig. 1. Schematic comparison of various methods for identifying nucleotide polymorphism trait association in terms of resolution, research time and allele number. BC, backcross. (Source : Yu and Buckler, 2006)

(2005) and Wilson *et al.*, (2004). Procedures for conducting an association mapping study in plants have also been well documented (Wilson *et al.*, 2004; Whitt *et al.*, 2003).

Both linkage analysis and association studies rely on co-inheritance of neighboring DNA variants with the phenotype. In both analyses, recombination is the main force eliminating linkage over generations. The difference is that in linkage analysis, there are only few opportunities allowed for recombination to occur (one or two generations), resulting in comparatively low resolution of the map. In association analysis which exploits natural variation in the population, recombination events over many generations are expected to eliminate linkage between a mutated (variant) gene and molecular markers, except the markers that are located close to the gene. Only very close markers are in Linkage Disequilibrium (LD) with the mutated gene, and form the basis for association mapping. LD measures the closeness of the genetic association between markers and a particular trait, and may be used to identify markers in close proximity to the gene(s) responsible for the trait.

1.2 Association Analysis

Genetic linkage between two loci (genes or molecular markers) on a genome is measured by the fact that they are in association and the involved loci are very closely associated and are located very close to each other. A complementary approach is to analyze Linkage Disequilibrium (LD) between the marker and target gene in a natural population. This has been done successfully in humans where large mapping populations(s) (families) do not exist. LD between two loci in natural populations is affected by all the recombination events that have happened. Thus, LD declines as the number of generations increases, so that in old populations LD is limited to small distances. In the human genome, there are now more than 2 million molecular markers available in the form of single nucleotide polymorphisms (SNPs). This density is sufficient to identify genes having effect on traits governed by a large number of, each one contributing a relatively small percentage to the phenotype. With such a dense coverage of the genome, there is a high probability of at least one marker being in close proximity to any particular gene of interest. In plants, little work has been done with regard to LD mapping using natural populations, but there is a growing realization that this approach may allow higher resolution mapping of QTL, and even identification of specific genes in plants.

Efficiency and power of association mapping was well demonstrated at Institute of Grassland and Environmental Research, UK, Europe in finding markers associated with flowering time in ryegrass germplasm. Association studies resulted in the identification of four markers occurring at high frequency in early flowering ryegrass populations and very infrequently in late flowering populations. Two of them were in highly significant LD with each other and by linkage analysis it was shown that they are located very close to each other on Linkage Group (chromosome) 7 of perennial ryegrass. This location coincides with a region of that chromosome already known to contain genes of importance for flowering time (a QTL). The other two markers are also in significant LD with each other, and they have been mapped to Linkage Group 2, near a second QTL for flowering time.

There are two types of association mapping procedures in plants.

1. **Genome wide association mapping**: It is a comprehensive approach to search the genome for causal genetic variation. A large number of markers are tested for association with various complex traits and it doesn't require any prior information on the candidate genes.
2. **Candidate gene association mapping**: It is a hypothesis driven approach to dissect out the genetic control of complex traits, with biologically relevant candidates selected and ranked based on the available results from genetic, biochemical, or physiology studies in model and non-model plant species (Mackay, 2001; Risch and Merikangas, 1996). Candidate-gene association mapping requires the identification of SNPs between lines and within specific genes. Identifying candidate gene SNPs relies on the resequencing of amplicons from several genetically distinct individuals of a larger association population. Because SNPs offer the highest resolution for mapping QTL and are potentially in LD with the causative polymorphism they are the preferential candidate-gene variant to genotype in association studies (Rafalski, 2004).

2.0 ASSOCIATION MAPPING PROCEDURES

Association analysis involves assessment of linkage disequilibrium (LD) based association. LD refers to a historically reduced (non equilibrium) level of the recombination of specific alleles at different loci controlling genetic variation in a population which can be detected statistically. This method has been widely applied to map and clone a number of genes underlying the complex genetic traits in humans.

Although the overall approach of population-based association mapping in plants varies based on the methodology chosen, the performance of association mapping includes the following basic steps.

1. Selection of a group of individuals from a natural population or germplasm collection with wide coverage of genetic diversity for the target trait
2. Recording or measuring the target phenotypic characteristics

3. Obtain diversity estimates and evaluate patterns of selection
4. Genotyping of selected lines by molecular marker analysis
5. Statistically evaluate associations between genotypes and phenotypes
6. Identify markers that are associated with the target trait; Marker alleles in strong LD with the trait can be used in MAS.
7. If the purpose is to find the gene controlling the trait, then further LD analysis can be performed with more markers to narrow down the length of DNA containing the gene, using the markers already identified as starting points.

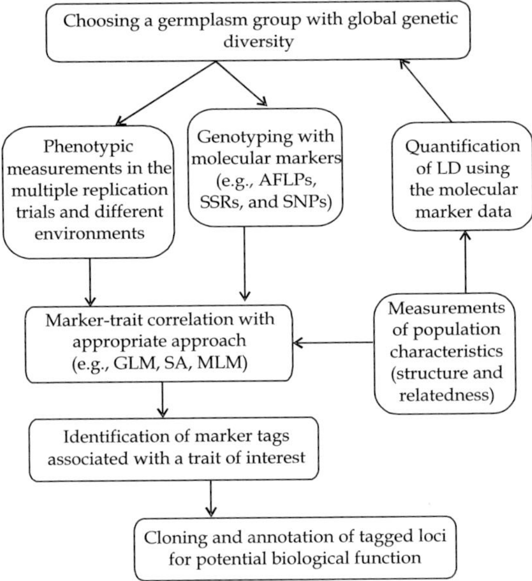

Fig. 2. The scheme of genome wide association mapping for tagging gene of interest using germplasm accessions (Source : Abdurakhmonov and Abdukarimov, 2008)

Choice of germplasm is critical to the success of association analysis (Yu *et al.*, 2006). Genetic diversity, extent of genome-wide LD and relatedness within the population determine the mapping resolution, marker density, statistical methods, and mapping power. For genotyping a set of unlinked, selectively neutral background markers scaled to achieve genome-wide coverage are employed to broadly characterize the genetic composition of individuals. Background genetic markers are useful in assigning individuals to populations (Pritchard and Rosenberg, 1999), preventing spurious associations. Random amplified polymorphic DNA (RAPD) (Williams *et al.*, 1990) and amplified fragment length polymorphism (AFLP) (Vos *et al.*, 1995) markers can serve as background markers, but almost all RAPD and

AFLP markers are dominantly inherited and thus demand special statistical methods if used to estimate population genetic parameters (Falush *et al.*, 2007; Ritland, 2005). Conversely, codominant microsatellites, or simple sequence repeats (SSRs), and SNPs are more revealing (*i.e.*, no allelic ambiguity) than their dominant counterparts. However, due to higher genome density, lower mutation rate and better amenability to high-throughput detection systems, SNPs are rapidly becoming the marker of choice for complex trait dissection studies.

A standard procedure for carrying out association analysis on candidate genes is as follows:

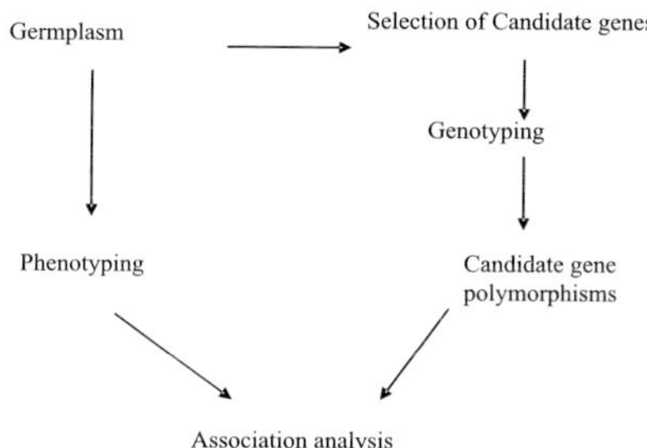

1. Choosing germplasm exhibiting bulk of diversity for the target trait.
2. Make selection positional candidate genes using existing QTL and positional cloning studies.
3. Scoring of phenotypic traits in replicated trials.
4. Amplification and sequencing of candidate genes in the population.
5. Manipulating sequence into valid alignments to identify polymorphisms.
6. Obtain diversity estimates and evaluate patterns of selection
7. Statistically evaluate associations between genotypes and phenotypes taking population structure into account.

Candidate-gene association mapping involves the identification of SNPs between lines and within specific genes. Candidate-gene selection is straightforward for relatively simple biochemical pathways (*e.g.*, starch synthesis in maize) or well characterized pathways (*e.g.*, flowering time in *Arabidopsis*) that have been resolved mainly through genetic analysis of mutant loci (natural or induced). But for complex traits such as grain or biomass yield, the entire genome could potentially serve as a candidate (Yu and Buckler, 2006). Therefore, the most straightforward method of identifying candidate gene SNPs relies on the resequencing of amplicons from several genetically distinct individuals of a larger association population. Fewer diverse individuals in the SNP discovery panel are needed to identify common SNPs, whereas many more are needed to identify rarer SNPs. Promoter, intron, exon, and 5'/3'-untranslated regions are all reasonable targets for identifying candidate gene SNPs, with non-coding regions expected to have higher levels of nucleotide diversity than coding regions.

3.0 FACTORS TO BE CONSIDERED FOR ASSOCIATION MAPPING ANALYSIS

3.1 Choice of Germplasm

The choice of germplasm is crucial to the discovery of useful alleles. In order to have enough statistical power to find an association, it is critical that the samples span the full range of phenotypic variation. To maximize the range of alleles tested, a genotypically diverse set of germplasm should be chosen. When available marker or phenotypic surveys can be used to choose a subset of the germplasm that is most diverse. From a practical level, it is found that a sample of 100 diverse inbred lines has enough statistical power to identify associations that control 10% of the phenotypic variation. Larger samples and/or more replications of phenotypic evaluation could be used to identify associations with smaller effects. Inbred samples allow direct identification of haplotypes throughout the genome, generally have more consistent phenotypes than segregating populations, and provide evaluation of phenotype without the complications of dominance. In some species, core sets of germplasm have been defined and characterized, and these are excellent starting points for association studies.

3.2 Population Structure

The final consideration in selecting the sample population is whether to use randomly or non-randomly mated germplasm. Randomly mated populations represent a rather narrow group of germplasm, likely to lower resolution and harbor only a narrow range of alleles. However, if non-randomly mated germplasm is used, population structure needs to be controlled in the statistical analyses. In addition, the genome for each sample population should be genotyped with SSRs, SNPs, RFLPs, RAPDs or AFLPs to provide an estimate of population structure. The ideal markers are either a modest number of SSRs or large numbers of SNPs, while if resources are limited AFLP may provide a good compromise. Usually, that 50 to 150 markers generally provide good estimates of population structure.

3.3 Phenotyping

3.3.1 Field Design

The importance of phenotyping has not received as much attention as genotyping. While accuracy and throughput of genotyping have dramatically improved, obtaining robust phenotypic data remains a hurdle for large-scale association mapping projects. Because association mapping often involves a relatively large number of diverse accessions, phenotypic data collection with adequate replications across multiple years and multiple locations is challenging. Efficient field design with incomplete block design (*e.g.*, α-lattice), appropriate statistical methods (*e.g.*, nearest neighbor analysis and spatial models), and consideration of QTL × environmental interaction should be explored to increase the mapping power, particularly if the field conditions are not homogenous (Eskridge, 2003). The increase in power of detecting QTLs with repeated measurements is well known and also has been demonstrated by simulation studies in mapping with pedigree-based breeding germplasm (Arbelbide *et al.*, 2006).

It is also important to consider the influence of flowering time on the expression of other correlated traits. It is worthwhile to block a field by flowering time if traits of interest are dependent on developmental transitions. Other issues that need be considered in phenotyping include photoperiod sensitivity, lodging, and susceptibility to prevalent pathogens because these traits affect the measurement of other morphological or agronomic traits at field condition.

3.3.2 Data Collection

Collection of high quality phenotypic data is essential for genetic mapping research. Association mapping studies often are long-term projects, with phenotyping being conducted over years in multiple locations (Flint-Garcia *et al.*, 2005). To ensure that high quality data are obtained from a wide range of conducted experiments, each researcher should assess the quality of the experiment for which they are responsible. Specific information about the experiment, such as check performance and environmental growth conditions (field or greenhouse), should be included as an annotation to the experiment in the trait database. In association studies for candidate genes, any newly discovered candidate gene polymorphism need to be tested for association with existing phenotypic data.

3.4 Genotyping

In association studies, a set of unlinked, selectively neutral background markers scaled to achieve genome-wide coverage are employed to broadly characterize the genetic composition of individuals. RAPD, AFLP codominant microsatellites, or SSRs, and SNPs can serve as background markers. If whole-genome association scans are to be conducted in crops, an important first step is to use high-capacity DNA sequencing instruments or high-density oligonucleotide (oligo) arrays to efficiently identify SNPs at a density that accurately reflects genome-wide LD structure and haplotype diversity.

3.5 Number of Background Markers

The number of background markers required to accurately estimate genetic relationships is a common issue that needs to be addressed in candidate-gene association mapping studies. The number of required markers is much higher for biallelic SNPs than for multiallelic SSRs. Generally it is advised to choose markers numbering about four times the chromosome number of that species, which translates to two markers per chromosome arm. Of course, length of the chromosome, diversity of the species, diversity of the particular sample, and cost and availability of different marker systems also will impact the number of background markers used in a study.

3.6 Candidate Gene Selection and Amplification

In case of association analysis for candidate genes, selection of candidate genes, *i.e.* those genes most likely to contain the polymorphism responsible for the phenotype, is one of the most critical steps. Candidate genes that fall within QTL intervals are referred as "positional candidate genes". QTL mapping often has limited resolution, but is an excellent way to narrow the search for candidates to specific chromosomal regions. Focusing on positional candidate genes will maximize the opportunity to find associations. The major aspects of choosing genes are:

1. Collect a list of genes that affect the phenotype of interest. Mutagenesis, biochemistry, various profiling technologies, comparative genomics, and positional cloning techniques/studies can help in the identification of genes.

2. Collect a list of map positions of QTL for the trait of interest over all previous experiments. Various databases such as MaizeDB (http://www.agron.missouri.edu) and Gramene (http://www.gramene.org) provide a good starting point for determining these positions.

3. Compare the two lists, and generate a list of all known genes with potential phenotypic effects in QTL confidence intervals. These positional candidate genes with the most neighboring QTL are most likely to have segregating variation at the locus.

Once a candidate gene has been identified, the following set of standard procedures is to be adopted for amplification including various molecular techniques.

1. Design compatible primer pairs from candidate gene sequence.
2. Employ PCR to amplify the target.
3. Verify product from PCR by agarose gel electrophoresis and purify the DNA.
4. Determine nucleotide sequence by DNA sequencing
5. Sequence manipulation and aligning sequence data.
6. Evaluation of diversity and selection

3.7 Statistical Analysis

Under ideal situations, the basic statistics for association analysis would be linear regression, analysis of variance (ANOVA), *t* test or chi-square test. However, as population structure can generate spurious genotype – phenotype associations, different statistical approaches have been designed. For family-based samples, the transmission disequilibrium test (TDT) (Spielman *et al.*, 1993) is used to study the genetic basis for human disease, whereas the quantitative transmission disequilibrium test (QTDT) is employed in the dissection of quantitative traits (Abecasis *et al.*, 2000; Allison, 1997). To address the issue of population structure in population-based samples, genomic control (GC) and structured association (SA) are the two most common methods utilized in both human and plant association studies. With GC, a set of random markers is used to estimate the degree that test statistics are inflated by population structure, assuming such structure has a similar effect on all loci (Devlin and Roeder, 1999). By contrast, SA analysis first uses a set of random markers to estimate population structure (Q) and then incorporates this estimate into further statistical analysis (Falush *et al.*, 2003). Modification of SA with logistic regression can also be used for association studies (Wilson *et al.*, 2004). A unified mixed-model approach for association mapping that accounts for multiple levels of relatedness has been developed recently (Yu *et al.*, 2006). In this method, random markers are used to estimate Q and a relative kinship matrix (K), which are then fit into a mixed model framework to test for marker-trait associations. Principal component analysis (PCA) has long been used in genetic diversity analysis and was recently proposed as a fast and effective way to diagnose population structure (Patterson *et al.*, 2007).

3.8 Softwares

A variety of software packages are available for data analysis in association mapping. TASSEL is the most commonly used software for association mapping in plants and is frequently updated (Bradbury *et al.*, 2007). In addition to association analysis methods (*i.e.*, logistic regression, linear model, and mixed model), TASSEL is also used for calculation and graphical display of linkage disequilibrium statistics and browsing and importation of genotypic and phenotypic data. STRUCTURE soft ware is used to estimate Q (Pritchard *et al.*, 2000). The Q is an $n \times p$ matrix, where n is the number of individuals and p is the number of defined subpopulations. SPAGeDi soft ware is used to estimate K among individuals (Hardy and Vekemans, 2002). K is an $n \times n$ matrix with off –diagonal elements being *Fij*, a marker-based estimate of probability of identity by descent. EINGENSTRAT soft ware is used to estimate PCs of the marker data and correct test statistics resulting from population stratification (Price *et al.*, 2006). SAS soft ware (SAS Institute, 1999) or R (Ihaka and Gentleman, 1996) often is used by advanced researchers with programming skills as the platform to develop various methods.

4.0 EVALUATION OF GENOTYPE - PHENOTYPE ASSOCIATIONS

4.1 Estimation of Population Structure

Samples used in association mapping studies can be grouped by the level of population structure and within group familial relatedness. The concern about population structure is that LD can be caused by admixture of subpopulation, which leads to false-positive results if not correctly controlled in statistical analysis. To reduce this risk, estimates of population structure must be included in association analysis. If the samples are not randomly mated, it is critical that population structure be included in the association analysis.

Different statistical approaches have been designed to deal with the population structure issue for different association samples. For family-based samples, the transmission disequilibrium test (TDT) has long been used to study the genetic basis for human disease, whereas the quantitative TDT (QTDT) has been employed in the dissection of quantitative traits. To address the issue of population structure in population based samples, genomic control (GC) and structured association (SA) are the two most common methods utilized in both human and plant studies. The STRUCTURE software is used to estimate population structure for association approaches.

4.2 Evaluation of Linkage Disequilibrium

Understanding the structure of linkage disequilibrium (LD) for a specific locus will, in turn reveal the association resolution possible at that locus. For example if LD decays within 1000 bp, then 1 or 2 markers per 1000 bp will be needed to identify associations. DnaSP, Arlequin, or TASSEL will calculate linkage disequilibrium between pairs of polymorphisms (r^2 or D′).

5.0 EVALUATION OF ASSOCIATION

5.1 Polymorphism Filtering

The segregating sites need to be extracted from the sequence alignments either by hand or by programs such as TASSEL and DnaSP. Normally polymorphisms that are present in less than three samples or with a frequency less than 5% are not included in the analyses. These low frequency polymorphisms may be the product of PCR or sequencing error. Insertions and deletions also need to be identified and coded for analysis.

When samples are truly randomly mated, no correction for population structure is required. If the trait is binary (*e.g.*, yellow versus white kernels), then a series of chi-square tests (X^2) can be used to evaluate whether the segregating polymorphisms associate. If the trait is quantitative, then a series of t-tests or ANOVA can be used to evaluate the associations. In case of structured samples, statistical analysis must account for it. If the trait is binary, the STRAT program can be used to evaluate the associations. If the trait is quantitative, either SAS or TASSEL can be used.

5.2 Determining the Significance of Association

The above said statistical tests will result in a P-value associated with each polymorphism - trait pair. For many association tests there will be 10s or 100s of polymorphisms to test. The trait values should be permuted relative to the fixed haplotypes, and then associations recalculated for 100 to 1000 permutations. The permuted P-value should be compared to the distribution of P-values for random markers across the genome. In some cases, the estimates of population structure do not explain all of the structure. Subsequently, the random markers used for estimating population structure could be used, as could data from unrelated candidate genes. The candidate gene P-value could be rescaled based on the P-values for the random markers.

5.3 Interpretation of Genotype - Phenotype Associations

Once an association is empirically determined, the validity of the association must be ascertained as follows:

1. Association studies will often find multiple polymorphisms that significantly associate. The LD structure surrounding the association can help to identify the candidate polymorphism. Although the most significant site is the most likely cause for the association, many of the

slightly less significant sites could actually be the functional cause of the phenotypic variation. Breaking the polymorphisms into likely functional (biologically significant) versus likely silent is useful in developing lists of sites for future evaluation.

2. The most straightforward way to prove an association is to evaluate the candidate polymorphisms in an entirely different population sample. Only polymorphisms that are closely linked to the cause of a phenotype should be significant in (Confirmation Population-CP). It is important that the population structure of the CP is truly independent of the first sample. Only the candidate polymorphisms need to be retested in the CP.
3. In some cases associations will suggest a molecular or biochemical mechanism of action. Following up hypotheses generated by association analysis with molecular biology and biochemistry will be very productive.
4. Final proof of the association can be obtained through marker assisted selection and production of near isogenic lines (NIL).

6.0 POWER OF ASSOCIATION MAPPING

The power of association mapping is the probability of detecting the true associations within the mapping population size that really depends on:

1. Extent and evolution of the LD in a population,
2. Complexity and mode of gene action of the trait of interest and
3. Sample size and experimental design.

The power can be increased utilizing the better data (knowledgeable experimental design and accurate measurements) and increasing the sample size. In QTL mapping studies, there are specific statistical approaches to estimate the false-positive level of the obtained strong (p-value) associations (control for Type I error) such as a permutation test or false discovery rate (FDR). A statistical approach within the Bayesian framework is used to test the reliability of obtained significance (p-values) in association mapping because of possibility of getting unreliable values due to-

1. Overestimation of effects (selection bias),
2. Association coming from neglecting confounding effects of a sample,
3. Poor experimental design, and
4. Instability of genetic effects across different environments.

At some point, requirement for larger sample size might make association mapping disadvantageous over a traditional QTL-mapping. However, the sample size for association mapping can be decreased keeping the high power with-

1. Pre-selecting a known QTL regions or candidate genes (from QTL-mapping and expression analyses),
2. Using the large populations with samples longer LD block that require a less number of markers to find useful associations,
3. An alternative experimental design (*i.e.*, TDT), and
4. Choosing the single marker from the haplotypes

7.0 EXAMPLES OF ASSOCIATION MAPPING STUDIES

First attempt on candidate-gene association mapping study in plants resulted in the identification of DNA sequence polymorphisms within the *D8* locus associated with flowering time (Thornsberry *et al.*, 2001). This research marked the first empirical association study in any organism for which background molecular markers were used to control for population structure (Pritchard *et al.*, 2001). Later studies of the same population associated the candidate gene *su1* with sweetness taste (Whitt *et al.*, 2002), *bt2, sh1* and *sh2* with kernel composition, and *ae1* and *sh2* with starch pasting properties (Wilson *et al.*, 2004). In a separate study, candidate genes *a1* and *whp1* were associated with maysin synthesis (Szalma *et al.*, 2005). Association mapping has also been used to successfully associate candidate gene *Y1* with maize endosperm color (Palaisa *et al.*, 2003). Recent studies have also shown that gene discovery can be initiated by analyzing existing data for pedigreed maize inbred lines or hybrids (Parisseaux *et al.*, 2004– Yu *et al.*, 2005). Few examples where association mapping strategy is successfully employed to map QTLs controlling different traits in plants are given in Table 1.

8.0 CONCLUSION

Compared to the conventional linkage mapping, linkage disequilibrium (LD)-mapping using the nonrandom associations of loci in haplotypes, is a powerful high-resolution mapping tool for complex quantitative traits. Association mapping methodology, initially developed by the human geneticists, has found its successive application in plant germplasm resources, in particular after recent improvements in minimization of spurious associations. The examples of association mapping studies performed in various plant germplasm resources are good indicatives of the potential utilization of this technology in both genome wide and candidate gene association studies. The near-future completion of genome sequencing

TABLE 1 : Linkage disequilibrium and association mapping studies in plants.

Species	Mating system	LD extent	Mapped traits	*Approach used
Arabidopsis	Selfing	10-250 kb and 50-100 cM [20, 21, 64, 66, 67]	Flowering time, growth response, pathogen resistance, and branching architecture [66, 129, 145-148]	One way ANOVA, simple regression, SA, MLM
Maize	Outcrossing	200-2000 bp [43, 68]. 3-500 kb [43, 69-71], 4-41cM [9, 22]	Plant height, flowering time, endosperm color,starch production, maysin and chlorogenic acid accumulation, cell wall digestibility, forage quality, and oleic acid level [43, 69, 71, 87, 88, 149-154]	GLM, SA, MLM, WGA
Rice (indica, japonica ad rufipogon)	Selfing	5-500 kb [73, 75, 76] 50-225 cM [74], 20-30 cM [155]	Multiple agronomic traits such as plant height, heading date, flag leaf length and width, grain length/width ratio, grain thickness, 1000-grain weight, width and length of milled rice grains [58, 155, 156]	DA, MLM, mixed model with mltip0le QTL effect
Barley	Selfing	10-50 cM [16, 77] 98-500kb [51], 300 bp [78]	Yield, yield stability, heading date, flowering time, plant height, rachilla length, resistance to mildew, ad leaf rust were associated with many different types of molecular markers [17, 18, 157, 158]	Pearson correlation; regression, ANOVA
Tetraaploid wheat	Selfing	10 and 20 cM [50]	N/A	N/A
Hexaploid wheat	Selfing	<1-10 cM [52, 56, 72]	Kernel size and milling, a high molecular weight glutenin and blotch resistance [52, 56, 159]	GLM-Q, LMM

contd...

Species	Mating system	LD extent	Mapped traits	*Approach used
Potato	Selfing	0.3-1 cM [25, 60], 3 cM [160]	Resistance to wilt disease, bacterial blight, phytophtora, and potato quality (tuber shape, flesh color, under water weight, maturity, and etc,)[59, 60, 138, 160]	Nonparametric Mann-Whitney U test, standard two sample t-test, GMM
Soybean	Selfing	10-50 cM [79, 80]	Seed protein content [80]	WGA
Sorghum	Outcrossing	50 cM [44]	N/A	N/A
Grape	Vegatative Propagation Outcrossing	5-10 cM [53]	N/A	N/A
Sugarcane	Vegatative Propagation	10 cM [10]	N/A	N/A
Sugar beet	Outcrossing	3 cM [81]	N/A	N/A
Forage grasses (silage maize and ryegrass)	Outcrossing	200-2000 bp [87-91]	Cold tolerance, flowering time and forage quality, water-soluble carbohydrate content [87, 88, 161, 162]	Multiple linear regressing; ANOVA
Forest trees (Norway spruce, Loblolly pine, Poplar, European aspen, Douglass-fir)	Outcrossing	100-200 bp [86], 500-2000 bp [83-85]	Early-wood microfibril angle trait, wood density and wood growth rate [141, 163]	ANOVA; combination of LD and QTL mapping

*MLM: mixed linear model [133]: GLM: general linear model without population structure [71]: GLM-Q general linear model using population structure matrix (Q) or the least square solution to the fixed effects GLM [56]: DA: discriminant analysis [156]: SA-structured association [47]: LMM: linear mixed model [52]: WGA: whole genome association [154, 164, 165]: GMM: general mixed model [59]: ANOVA: analysis of variance test: N/A-not available (search of known major online library database as of December 2007).

projects of crop species, powered with more cost-effective sequencing technologies, will certainly create a basis for application of whole genome-association studies. In summary, association mapping platforms are being developed for multiple plant species. Genetic diversity and phenotyping are expected to gain further attention, as researchers become more aware of their importance. Superior allele mining for trait improvement will be greatly facilitated by synergy among various research groups involved in different aspects of association mapping. This will provide with more powerful association mapping tool(s) for crop breeding and genomics programs in tagging true functional associations conditioning genetic diversities, and consequently, its effective utilization.

9.0 REFERENCES

Abdurakhmonov, I.Y. and Abdukarimov, A. 2008. Application of Association Mapping to Understanding the Genetic Diversity of Plant Germplasm Resources. International Journal of Plant Genomics, Article ID 574927.

Abecasis, G.R. and Cookson, W.O.C. 2000. GOLD—graphical overview of linkage disequilibrium. Bioinformatics, 16:182–183.

Allison, D.B. 1997. Transmission-disequilibrium tests for quantitative traits. American Journal of Human Genetics, 60:676-690.

Arbelbide, M., Yu, J. and Bernardo, R. 2006. Power of mixed-model QTL mapping from phenotypic, pedigree and marker data in self-pollinated crops. Theoretical and Applied Genetics, 112:876–884.

Bradbury, P.J., Zhang, Z., Kroon, D.E., Casstevens, T.M., Ramdoss, Y. and Buckler, E.S. 2007. TASSEL: Soft ware for association mapping of complex traits in diverse samples. Bioinformatics, 23:2633–2635.

Buckler, E.S. and Thornsberry, J.M. 2002. Plant molecular diversity and applications to genomics. Current Opinion in Plant Biology, 5:107-111.

Ching, A., Caldwell, K.S., Jung, M., Dolan, M., Smith, O.S., Tingey, S., Morgante, M. and Rafalski, A.J. 2002. SNP frequency, haplotype structure and linkage disequilibrium in elite maize inbred lines. BMC Genetics, 3:19.

Devlin, B. and Roeder, K. 1999. Genomic control for association studies. Biometrics, 55:997–1004.

Eskridge, K.M. 2003. Field design and the search for quantitative trait loci in plants. Available at: http://www.stat.colostate.edu/graybillconference/Abstracts/Eskridge.html;

Falush, D., Stephens, M. and Pritchard, J.K. 2003. Inference of population structure using multilocus genotype data: linked loci and correlated allele frequencies. Genetics, 164:1567-1587.

Falush, D., Stephens, M. and Pritchard, J.K. 2007. Inference of population structure using multilocus genotype data: Dominant markers and null alleles. Molecular Ecology Notes, 7:574–578.

Flint-Garcia, S.A., Thornsberry, J.M. and Buckler, E.S. 2003. Structure of linkage disequilibrium in plants. Annual Review in Plant Biology, 54:357-374.

Flint-Garcia, S.A., Thuillet, A., Yu, J., Pressoir, G., Romero, S.M., Mitchell, S.E., Doebley, J.F., Kresovich, S., Goodman, M.M. and Buckler, E.S. 2005. Maize association population: A high resolution platform for QTL dissection. Plant Journal, 44:1054-1064.

Hardy, O.J. and Vekemans, X. 2002. SPAGeDi: a versatile computer program to analyze spatial genetic structure at the individual or population levels. Molecular Ecology Notes, 2:618–620.

Hirschhorn, J.N. and Daly, M.J. 2005. Genome-wide association studies for common diseases and complex traits. Nature Review in Genetics, 6:95-108.

Ihaka, R. and Gentleman, R. 1996. A language for data analysis and graphics. Journal of Computer Graphics and Statistics, 5:299–314.

Mackay, T.F. 2001. The genetic architecture of quantitative traits. Annual Review in Genetics, 35:303-339.

Palaisa, K.A., Morgante, M., Williams, M. and Rafalski, A. 2003. Contrasting effects of selection on sequence diversity and linkage disequilibrium at two phytoene synthase loci. Plant Cell, 15:1795-1806.

Parisseaux, B. and Bernardo, R. 2004. In silico mapping of quantitative trait loci in maize. Theoretical and Applied Genetics, 109:508-514.

Patterson, N., Price, A.L. and Reich, D. 2007. Population structure and eigen analysis. PLoS Genetics, 2:190-195.

Price, A.L., Patterson, N.J., Plenge, R.M., Weinblatt, M.E., Shadick, N.A. and Reich, D. 2006. Principal components analysis corrects for stratification in genome-wide association studies. Nature Genetics, 38:904–909.

Pritchard, J.K. and Rosenberg, N.A. 1999. Use of unlinked genetic markers to detect population stratification in association studies. American Journal of Human Genetics, 65:220-228.

Pritchard, J.K., Stephens, M., Rosenberg, N.A. and Donnelly, P. 2000. Association mapping in structured populations. American Journal of Human Genetics, 67:170–181.

Pritchard, J.K. 2001. Deconstructing maize population structure. Nat Genet, 28:203-204.

Rafalski, A. and Morgante, M. 2004. Corn and humans: recombination and linkage disequilibrium in two genomes of similar size. Trends in Genetics, 20:103–111.

Remington, D.L., Thornsberry, J.M., Matsuoka, Y., Wilson, L.M., Whitt, S.R., Doebley, J., Kresovich, S., Goodman, M.M. and Buckler, E.S. 2001. Structure of linkage disequilibrium and phenotypic associations in the maize genome. Proceedings of the National Academy of Sciences USA, 98:11479-11484.

Risch, N. and Merikangas, K. 1996. The future of genetic studies of complex human diseases. Science, 273:1516–1517.

Ritland, K. 2005. Multilocus estimation of pairwise relatedness with dominant markers. Molecular Ecology, 14:3157–3165.

SAS Institute. SAS/STAT user's guide. Version 8 SAS Institute,1999. Inc, Cary, NC.

Spielman, R.S., McGinnis, R.E. and Ewens, W.J. 1993. Transmission test for linage disequilibrium: the insulin gene region and insulin-dependent diabetes mellitus (IDDM). American Journal of Human Genetics, 52:506-516.

Szalma, S.J., Buckler, IV E.S., Snook, M.E. and McMullen, M.D. 2005. Association analysis of candidate genes for maysin and chlorogenic acid accumulation in maize silks. Theoretical and Applied Genetics, 110:1324–1333.

Thornsberry, J.M., Goodman, M.M., Doebley, J., Kresovich, S., Nielsen, D. and Buckler, IV E.S. 2001. Dwarf 8 polymorphisms associate with variation in flowering time. Nature Genetics, 28:286-289.

Vos, P., Hogers, R., Bleeker, M., Reijans, M., van de Lee, T., Hornes, M., Frijters, A., Pot, J., Peleman, J. and Kuiper, M. 1995. AFLP: A new technique for DNA fingerprinting. Nucleic Acids Research, 23:4407–4414.

Whitt, S.R., Wilson, L.M., Tenaillon, M.I., Gaut, B.S. and Buckler, E.S. 2002. Genetic diversity and selection in the maize starch pathway. Proceedings of the National Academy of Sciences USA, 99:12959-12962.

Whitt, S.R. and Buckler, E.S. 2003. Using natural allelic diversity to evaluate gene function. Methods in Molecular Biology, 236:123-140.

Williams, J.G.K., Kubelik, A.R., Livak, K.J., Rafalski, J.A. and Tingey, S.V. 1990. DNA polymorphisms amplified by arbitrary primers are useful as genetic markers. Nucleic Acids Research. 18:6531–6535.

Wilson, L.M., Whitt, S.R., Ibanez, A.M., Rocheford, T.R., Goodman, M.M. and Buckler, E.S. 2004. Dissection of maize kernel composition and starch production by candidate gene association. Plant Cell, 16:2719-2733.

Yu, J. and Buckler, E.S. 2006. Genetic association mapping and genome organization of maize. Current Opinion in Biotechnology, 17:155–160.

Yu, J., Arbelbide, M. and Bernardo, R. 2005. Power of in silico QTL mapping from phenotypic, pedigree, and marker data in a hybrid breeding program. Theoretical and Applied Genetics, 110:1061-1067.

Yu, J., Pressoir, G. and Briggs, W.H. 2006. A unifiedmixed-model method for association mapping that accounts for multiple levels of relatedness. Nature Genetics, 38:203– 208.

Molecular Plant Breeding: Principle, Method and Application
Eds : R.K. Singh, Rajesh Singh, Guoyou Ye, A. Selvi and G.P. Rao
Studium Press LLC, Texas, USA, 2009, pp. 209-239

Statistical Analysis of Molecular Data

R. BALAKRISHNAN

ABSTRACT

Statistical tools have played a major role in the development of biological sciences, especially the science of genetics right from the late 19th century when Mendel's theory of inheritance was laid on a strong scientific foundation. The application of statistical methods in the study of variation in quantitative traits of economic importance that began early in the 20th century, convinced many researchers that Mendelian principles apply to qualitative as well as quantitative traits. This has also shaped the general model that embraces the multiple-factor hypothesis for quantitative traits. More recently, the linking of molecular genetics and genomics with the study, evaluation and improvement of quantitative traits has become the central theme of much exciting research and statistical tools greatly help in the analysis of interpretation of the data generated on quantitative trait locus mapping, and molecular marker studies. This chapter deals with some of the essential statistical tools that are used in the study and interpretation of quantitative trait loci (QTL), analysis of molecular variation (AMOVA) and analysis of genetic diversity using molecular data. Some useful softwares for such analyses are discussed in detail.

Key Words: Statistical tools, Genetic diversity, QTL, AMOVA

Sugarcane Breeding Institute, Coimbatore–641007 (India)
Corresponding author e-mail: balakrishnan.ragupathy@gmail.com

1.0 INTRODUCTION

Even before the science of genetics was born in 1900, evolution and inheritance of quantitative characters were topics of intensive research in the hands of stalwarts like Francis Galton, Karl Pearson, and others. The theory of ancestral heredity propounded by Galton (1889, 1897) resulted in the invention of regression method and Pearson's Mathematical contributions to the theory of evolution in 1904 dealing with correlation between relatives in large populations laid the foundation for the branch of biometrical genetics or quantitative genetics. It was interest in phenotypic diversity that drove the development of modern synthesis in the first half of the 20th century, and the primary research of many evolutionary geneticists was mainly focused on the mechanisms responsible for morphological, physiological, and/or behavioral diversification within and among species. The vast majority of characters falling into these categories are quantitative, encoded by a large number (dozens to perhaps hundreds) of loci. It was Fisher (1918) who provided a sound and theoretical basis for the inheritance of quantitative characters. In doing so, Fisher for the first time introduced genetics in an otherwise statistical approach and assuming a very large number of unlinked loci, each with small additive effects gave what is known in the literature as infinitesimal model. Later on, Darwin's theory of evolution by natural selection was given a quantitative orientation by Fisher, Wright and Haldane in their works during 1920 to 1930 leading to the birth of science of statistical genetics. In subsequent years, the principles of statistical genetics greatly influenced the theory and practice of plant and animal breeding. In particular, the theoretical consequences of artificial selection in domesticated species of plants and animals for the improvement of economic characteristics, which are usually quantitative, were adequately worked out and proved to be quite effective in predicting the gains expected from selection.

1.1 Phenotypes versus Genetic Values

A major limitation of the classical theory of quantitative genetics is due to the multi-locus nature and environmental dependence of the expression of most quantitative traits- it is usually impossible to use phenotypic information to make precise statements about the underlying genotypes of individuals. As a consequence, most studies in quantitative genetics concentrate on variation at the population level and the partitioning of such variation into causal components (associated, for example, with additive effects, maternal effects, developmental noise, etc.) Sophisticated statistical methods (for example, BLUP or best linear unbiased prediction) do provide a basis for estimating the breeding values of individuals when phenotypic measures are available from multiple relatives, and these methods are employed extensively in animal breeding to identify elite individuals for selective breeding. In addition, for crop plants that can be propagated clonally or as inbred lines, general combining abilities can be estimated for individual lines and specific combining abilities for pairs of lines. However, all of these concepts have some limitations in that they do not yield much insight into the biological nature of quantitative variation. For example, a breeding value is simply an estimate of the expected phenotype of an offspring when the focal parent is randomly mated. As an estimate, a breeding value is never known with absolute certainty, but more significantly, such a composite measure tells us nothing about the actual allelic states of the loci involved in the expression of the phenotype or about modes of gene action. Similarly, the additive genetic variance, which is used to estimate heritability of quantitative traits, is formally related to the variance of breeding values and it need not convey any information on the physiological mode of gene action (Cheverud and Routman, 1995). Situations exist in which there can be substantial non-additive interactions among segregating genes and yet very little non-additive genetic variance; the outcome depends on the average effects of alleles in all genetic backgrounds. As a consequence of the hierarchical way in which genetic components of variance are defined in a statistical sense, higher-order components involving non-additive interactions are virtually always smaller in magnitude than the first-order (additive) component, and this appears to have led to widespread misconception that dominance and epistasis are of minor importance in the expression of quantitative traits. Despite these limitations, quantitative genetics theory has played a major role in applications of selective breeding and its ability to reveal the extent to which standing variation for complex traits is due to genes or environment.

1.2 Genomics

The path breaking advances in genetics, particularly molecular genetics, before the close of the 20th century are, however, influencing the field of quantitative genetics in a big way. Biotechnology - both plant and animal - have provided tools to produce genetically modified organisms which incorporate desirable genes from other species for enhancing yield and making them resistant to various diseases. Worldwide efforts are being made in human, plant and animal genome research on mapping the genes, generating DNA sequences, studying their homology and transferring them to other organisms. For the model plant *Arabidopsis thaliana,* which is a small mustard species, the entire genome has been sequenced. It has a highly compact genome of about 130 Mb with little interspersed repetitive DNA and is related to many food plants like rice, wheat, maize, sorghum, millets, etc, and can therefore provide a focus from which the genome content of other higher plants can be extrapolated. The gene order and the organization in this plant can be used to isolate and characterize genes in other cereals. Such studies have given birth to a new science of Genomics - study of an organism by looking at many genes at a time - that interfaces the traditional branches of genetics among themselves along with automated systems on computers. Genomics is primarily concerned with genetic markers, DNA sequences, linkage analysis, gene ordering and quantitative trait loci (or QTL) mapping. It has interfaces with almost all the areas of genetics, *viz.* Mendelian Genetics, Population Genetics, Quantitative Genetics with applications in Plant and Animal Breeding, Cytogenetics and Molecular Genetics. In almost all of these areas, statistics plays an important role in the elucidation of the concepts, planning of the programmes, data analysis and appropriate inferences drawn there from. With the generation of DNA sequence data at a very rapid rate and of enormous volume, issues of informatics and automation have also become important. A useful reference, in this regard, is a book by Liu (1998).

2.0 QUANTITATIVE TRAIT LOCI (QTL)

The advent of techniques to detect 'molecular variation', beginning with protein electrophoresis and culminating in DNA sequencing, has revealed huge amount of genetic variability in almost all species. This has allowed the construction of detailed genetic maps comprising hundreds of genetic markers, such as for instance restriction fragment length polymorphism (RFLP), evenly spaced throughout the genome, and has led to mapping of quantitative trait loci (QTLs). A newly emerging era of breeding and selection in plants and animals is opening up in which traditional quantitative genetic methods need to be reoriented to take into account the new information contained in genomic data accumulating so rapidly. With the help of the data on molecular markers one can now identify and locate quantitative trait loci as has been done for instance in tomatoes by Causse *et al.* (1988). The methods of molecular genetics are also being integrated with those of the artificial selection on individual and/or collateral basis by applying what has come to be known as marker-assisted selection (MAS). With information available on the character and the molecular score, statistical considerations involve developing optimum selection index that maximize genetic improvement in the character.

For quantitative traits, we can not identify the effect of individual genes and follow them individually through the generations as we do in the case of Mendelian genes having major effects in controlling the qualitative traits. Sax (1923) brought out the association between a qualitative trait like seed colour in *Phaseolus vulgaris* plants, controlled by a major gene, and a quantitative trait like seed weight. An attempt to resolve the underlying basis for variability in quantitative traits into the effects of individual loci using linkage to markers were put, for the first time, on a firm theoretical basis by Thoday (1961). Since that time several contributions to the theory of marker-based detection of quantitative trait loci (QTL) were made such as those by McMillan and Robertson (1974), Soller *et al.* (1976), Soller and Genizi (1978) and several others. Broman and Speed (1999) reviews the methods of identifying QTLs in experimental crosses.

An essential pre-requisite for the study of QTL is the construction of linkage maps. A linkage map is a chromosome map of a species or experimental population that shows the position of its known genes and/ or markers relative to each other in terms of recombination frequency, rather than as specific physical distance along each chromosome. Mapping functions are used to convert recombination fractions into map units called centi-Morgans (cM). Linkage maps are constructed from the analysis of many segregating markers. The three main steps of linkage map construction are: (1) production of a mapping population; (2) identification of polymorphism, and (3) linkage analysis of markers.

2.1 Mapping Populations

The construction of a linkage map requires a segregating plant population (*i.e.* a population derived from sexual reproduction). The parents selected for the mapping population will differ for one or more traits of interest. Population sizes used in preliminary genetic mapping studies generally range from 50 to 250 individuals. Larger populations are required for high-resolution mapping. If the map will be used for QTL studies (which is usually the case), then an important point to note is that the mapping population must be phenotypically evaluated (*i.e.* trait data must be collected) before subsequent QTL mapping. Generally in self-pollinating species, mapping populations originate from parents that are both highly homozygous (inbred). F_2 populations, derived from F_1 hybrids, and backcross (BC) populations, derived by crossing the F_1 hybrid to one of the parents, are the simplest types of mapping populations developed for self-pollinating species, while in cross-pollinating species mapping populations may be derived from a cross between a heterozygous parent and a haploid or homozygous parent (Wu *et al.*, 1992).

2.2 Identification of Polymorphism

The next step in the construction of a linkage map is to identify DNA markers that reveal differences between parents (*i.e.* polymorphic markers). It is critical that sufficient polymorphism exists between parents in order to construct a linkage map (Young, 1994). In general, cross pollinating species possess higher levels of DNA polymorphism compared to inbreeding species; mapping in inbreeding species generally requires the selection of parents that are distantly related. In many cases, parents that provide adequate polymorphism are selected on the basis of the level of genetic diversity between parents. Once polymorphic markers have been identified, they must be screened across the entire mapping population, including the parents (and F_1 hybrid, if possible). This is known as marker 'genotyping' of the population.

2.3 Linkage Analysis of Markers

The final step of the construction of a linkage map involves coding data for each DNA marker on each individual of a population and conducting linkage analysis using computer programs. Linkage between markers is usually calculated using odds ratios (*i.e.* the ratio of linkage versus no linkage). This ratio is more conveniently expressed as the logarithm of the ratio, and is called a logarithm of odds value or LOD score (LOD - logarithm (base 10) of odds, also called logit by mathematicians as it is given in detail below.

2.3.1 LOD score method for estimating recombination frequency

The lod score is a statistical test often used for linkage analysis. The test was developed by Newton E. Morton. Computerized lod score analysis is a simple way to analyze complex family pedigrees in order to determine the linkage between Mendelian traits (or between a trait and a marker, or two markers). The method is described in greater detail by Strachan and Read (1999). Briefly, it works as follows:

1. Establish a pedigree

$$LOD = Z = \log 10 \frac{\text{probability of birth sequence with a given linkage value}}{\text{probability of birth sequence with no linkage}}$$

$$= \log 10 \frac{(1-\theta)^{NR} \times \theta^{R}}{0.5^{(NR+R)}}$$

2. Make a number of estimates of recombination frequency
3. Calculate a lod score for each estimate
4. The estimate with the highest Lod score will be considered the best estimate

The Lod score is calculated as follows

NR denotes the number of non-recombinant offspring, and R denotes the number of recombinant offspring. The reason 0.5 is used in the denominator is that any alleles that are completely unlinked (alleles on separate chromosomes) have a 50% chance of recombination, due to independent assortment. In practice, lod scores are looked up in a table which lists lod scores for various standard pedigrees and various values of recombination frequency. By convention, a lod score greater than 3.0 is considered evidence for linkage. (A score of 3.0 means the likelihood of observing the given pedigree if the two loci are not linked is less than 1 in 1000). On the other hand, a lod score less than -2.0 is considered evidence to exclude linkage. Although it is very unlikely that a LOD score of 3 would be obtained from a single pedigree, the mathematical properties of the test allow data from a number of pedigrees to be combined by summing the LOD scores. Commonly used software programs for the computation of LOD scores and linkage maps include free softare like Mapmaker/EXP (Lander *et al.*, 1987; Lincoln *et al.*, 1993) and MapManager QTX (Manly *et al.*, 2001), or JoinMap (Stam, 1993).

2.3.2 Permutation test

Suppose we want to test the difference between two population means using one sample each from each of the populations. Let the samples be denoted by:

S^X = {x1, x2, x3,...........xn} and S^Y = {y1, y2, y3,.........yn}. The estimated difference between the two sample means is d = $\bar{x}$ - $\bar{y}$. If the distribution of differences is known, a standard test using the distribution (say, t-test) can be applied. However, if the distribution is unknown or complex, then a standard test can not be applied. Let us assume that there is no difference between the two population means – the two sampled are assumed to be drawn from the same population. Now, imagine putting the 2*n* observations into a bag and shuffle the bag well. We can draw *n* observations from the bag without replacement. Two samples, one containing the *n* drawn observations and the other containing the remaining *n* observations in the bag are obtained. These samples are called shuffling samples and the process is a permutation. The difference in the means of the two shuffling samples can be obtained. The difference obviously depends on the permutation rather than the difference between the population means. If we put all the drawn observations back into the bag, shake the bag well and carry out another draw again, another permutation is obtained. If this is done *p* times, a distribution of the difference can be obtained. This distribution is an empirical distribution of the test statistic under the null hypothesis. If the observed difference between the two sample means falis in the middle of 95% of the distribution, we can conclude that the difference between the two population means are not significant at 5% level. On the other hand, if the difference between the sample means falls outside the middle 95% of the empirical distribution, we can conclude that the two population means differ at 5% level. Thus a test based on the empirical distribution obtained through permutation, is called a *permutation test*. The permutation test is often used to determine the threshold for the significance of a likelihood ratio test for declaring a QTL. Normally a large number of permutations (like 10000) are carried out.

2.3.3 Haldane's Mapping Function.

This function is based on Poisson process – if the average number of crossovers is λ, then the probability of no crossovers occurring in an interval is:

$$P[\text{no crossover}] = e^{-\lambda}. \ldots\ldots\ldots\ldots\ldots..(a)$$

the probability of a crossover occurring in an interval is:

$$P[\text{crossover}] = 1 - e^{-\lambda}. \ldots\ldots\ldots\ldots\ldots(b).$$

The probability of a recombinant is half the probability of a crossover (because each pair of homologs with one crossover results in one-half recombinant gametes). If we define the expected number of recombinants as a mapping function ($m = 0.5\,\lambda$), then

$$r = 0.5\,(1-e^{-2m}). \ldots\ldots\ldots\ldots\ldots\ldots\ldots..(c).$$

This is the form of Haldane's mapping function. The inverse of equation (c) is

$$m = -0.5\log(1-2r)\ldots\ldots\ldots\ldots\ldots\ldots...(d),$$

Which would convert an estimated recombinant fraction ***r*** to Haldane's map distance m (Liu, 1990).

2.4 Mapping functions

Mapping functions are required to convert recombination fractions into centiMorgans (cM) because recombination frequency and the frequency of crossing-over are not linearly related (Hartl and Jones, 2001). When map distances are small (<10 cM), the map distance equals the recombination frequency. However, this relationship does not apply for map distances that are greater than 10 cM. Two commonly used mapping functions are the Kosambi mapping function, which assumes that recombination events influence the occurrence of adjacent recombination events, and the Haldane mapping function, which assumes that the crossovers occur randomly along the length of the chromosomes.

Kosambi's mapping function allows for interference (I), whereby one crossover tends to prevent other crossovers in the same region, where I is estimated using:

$$I = 1 - \frac{\text{Observed\_nuimber\_double\_recombinants}}{\text{Expected\_number\_double\_recombinants}}$$

If C is defined as the coefficient of coincidence, then C = 1 –I. The Haldane's mapping function defines that $r_{AC} = r_{AB} + r_{BC} - 2r_{AB}r_{BC}$, (with C = 0), whereas Kosambi's map function defines that $r_{AC} = r_{AB} + r_{BC} - 2Cr_{AB}r_{BC}$.

In Kosambi's map function C = 2r. When $r_{AB} = r_{BC} = 0.5$, then C = 1. This means that when the loci are located far apart, interference is absent. When the loci are close together, interference occurs and Kosambi's map function would be expected to give the best results. In the case of Kosambi's mapping function, the map distance is calculated as

$$m = \tfrac{1}{4} \log_e [(1+2r)/ (1-2r)] \ldots\ldots\ldots\ldots\ldots (e).$$

3.0 PRINCIPLE OF QTL ANALYSIS

In simple terms, QTL analysis is based on the principle of detecting an association between phenotype and the genotype of markers. Markers are used to partition the mapping population into different genotypic groups based on the presence or absence of a particular marker locus and to determine whether significant differences exist between groups with respect to the trait being measured (Young, 1996). A significant difference between phenotypic means of the groups, depending on the marker system and type of population, indicates that the marker locus being used to partition the mapping population is linked to a QTL controlling the trait.

3.1 Methods to Detect QTLs

Three widely-used methods for detecting QTLs are single-marker analysis, simple interval mapping and composite interval mapping (Liu, 1998).

3.1.1 Single Marker Analysis

Single-marker analysis (also 'single-point analysis') is the simplest method for detecting QTLs associated with single markers. For linkage effects to be detected and estimated, the population must come from a cross between two parents at least one of which should be heterozygous for both the loci. The statistical methods used for single-marker analysis include *t*-tests, analysis of variance (ANOVA) and linear regression. Linear regression is most commonly used because the coefficient of determination (R^2) from the marker explains the phenotypic variation arising from the QTL linked to the marker. This method does not require a complete linkage map and can be performed with basic statistical software programs. However, the major disadvantage with this method is that the farther a QTL is from a marker, the less likely it will be detected. This is because recombination may occur between the marker and the QTL. This causes the magnitude of the effect of a QTL to be underestimated. The use of a large number of segregating

DNA markers covering the entire genome (usually at intervals less than 15 cM) may minimize both problems (Tanksley, 1993). The results from single-marker analysis are usually presented in a Table 1, which indicates the chromosome (if known) or linkage group containing the markers, probability values, and the percentage of phenotypic variation explained by the QTL.

Let us consider the case of a double back cross. Let the two loci, corresponding to the marker locus and QTL, be A-a and Q-q, respectively. This gives rise to four marker-QTL genotypes viz. ***AAQQ, AAQq, AaQQ***, and ***AaQq***. Their frequencies would be (1-r)/2, r/2, r/2, and (1-r)/2 respectively where ***r*** is the recombination probability between the two loci and the linkage is in the coupling phase. Since the marginal frequency of the two possible marker genotype **AA** and **Aa** are one-half each, the QTL genotypic frequencies, conditional on the marker genotype, would be as given in Table 1.

Table 1. Conditional frequencies of QTL genotypes given the marker genotypes in a back cross

Marker Genotype	Frequency	QQ	Qq	qq
AA	1/2	(1-r)	r	*0*
Aa	1/2	r	(1-r)	*0*
aa	0	0	0	0

AA: $(1\text{-}r)d_a + rh_a$

Aa: $rd_a + (1\text{-}r)h_a$

aa: 0

We assume that the genotypic values of the QTL genotypes are $\boldsymbol{d_a}$, $\boldsymbol{h_a}$ and $\boldsymbol{-d_a}$, respectively for ***QQ, Qq*** and ***qq*** and that the marker locus has no pleiotropic effect on the quantitative trait. Then the expected values of the means for the marker classes are:

The contrast between the two marker genotype classes is then:

(AA-Aa) = $(1\text{-}2r)(d_a\text{-}h_a) = 2g_a\,(1\text{-}2r)$, where $g_a = (d_a\text{-}h_a)$ is the genetic effect in a back cross progeny in terms of additive and dominance effects. We can test the null hypothesis H_0: (difference in the true means of the two marker classes AA and Aa = 0) by the usual *t-test.* But it can be interpreted either as g_a=0 or as r = *0.5.* The power of the test is, however, low when the marker is loosely linked with the QTL and the genetic effect cannot be estimated without bias. Single marker analysis is therefore not commonly employed in practice.

3.1.2 Simple Interval Mapping

This approach, given by Lander and Botstein (1989), involves formation of intervals by pairing of adjacent markers and treating them as a single unit of analysis for detection and estimation purposes. It is based on the joint frequencies of a pair of adjacent markers and a putative QTL flanked by the two markers. Suppose markers A and B are linked with recombination fraction r and Q is located between them with r_1 as recombination fraction from A and r_2 from B.

$$\text{Then } r = r_1 + r_2 - 2\, r_1\, r_2$$

on the assumption of no crossover interference. When r is small, no double cross-over can be assumed and the relationship reduces to $r = r_1 +$

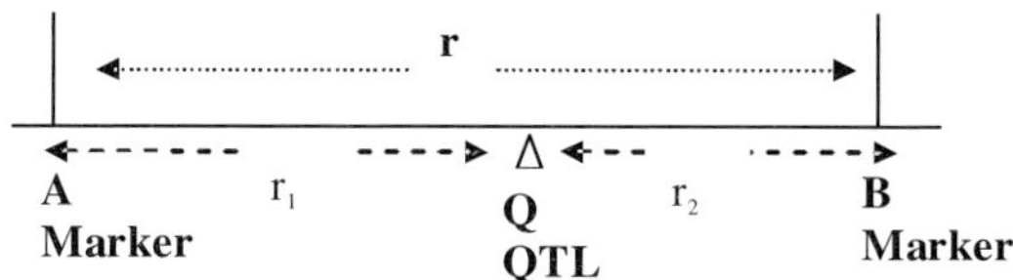

r_2. The QTL position is usually represented by a position relative to the interval between A and B as shown below:

Therefore the relative position of the putative QTL in the genome segment flanked by the two markers is defined as **ñ** = r_1/r and **1- ñ** = r_2/r. In the classical back cross design with three loci each with two alleles, A-a, *B-b,* and *Q-q,* the expected frequencies for the eight marker-QTL genotypes can be used to obtain the conditional probabilities of the QTL genotypes given the marker genotypes. These probabilities and the expected trait values for the observable marker genotypes are given in Table 2.

Let the trait value for the i-th individual in the population be y_i, ***ì*** be the population mean, g_i the indicator variable taking the value 1 if the QTL is *QQ* and -1 if it is *Qq,;* ***b*** *is* the allelic substitution effect for this QTL and ***e*** *is a* random variable with zero expectation and variance ó². Then the linear regression model is given by

$$Y_i = \mu + bg_i + e_i \quad \text{.............................. (e)}$$

Table 2. Expected QTL genotypic frequencies conditional on the marker genotypes with no double crosover:

Marker genotype	Frequency p_i	P (Q_j / M_i)		Expected value(g_i)
		QQ	*Qq*	
AABB	*0.5 (1-r)*	1	0	μ_1
AABb	*0.5 r*	1 - p = r_2/r	p	(1-p) μ_1+p μ_2
AaBB	*0.5 r*	p = r_1/r	1-p	p μ_1+(1-p) μ_2
AaBb	*0.5 (1-r)*	0	1	μ_2
Mean	025	μ_1	μ_2	0.5 (μ_1+ μ_2)

In this model the QTL genotype for the i-th individual is unknown. gi – random indicator variable with conditional probabilities of obtaining QQ or Qq at the QTL which depend on the marker genotype at the flanking markers and the three loci recombination frequencies between them and the putative QTL as we have seen in the Table 2.

In the above table μ_1 and μ_2 indicate respectively the mean values of genotypes QQ and Qq.

3.1.2.1 *Maximum likelihood method*

We assume that the character is normally distributed within each of the eight marker QTL classes with equal variance $\boldsymbol{\sigma}^2$. If $\mathbf{y}_i$ is the observed trait value for the i-th individual, *p(Qi/Mi)* is the conditional probability and $\boldsymbol{\mu}_j$ is the trait value for the j-th QTL, with i =1,2..._N; j =1,2; the likelihood function is given by

$$L = \frac{1}{\{\sigma\sqrt{2\pi}\}^N} * \prod_{i=1}^{N}\sum_{j=1}^{2} p(Q_j / M_i) * \exp\left[-\frac{(y_i - \mu_j)^2}{2\sigma^2}\right]$$

Under the null hypothesis H_o: $(\mu_1 - \mu_2) = 0$, the log likelihood is

$$\mathrm{Log}[L(\mu_1 = \mu_2 = \mu)] = -\frac{1}{2\sigma^2}\sum_{i-1}^{4}\sum_{j-1}^{n_i} (y_{ij} - \mu)^2 - \frac{N}{2}\mathrm{Log}(2\pi\sigma^2)$$

where $N = n_1+n_2+n_3+n_4$; n_i being the number of individuals in the i-th marker class.

The log likelihood ratio for testing the hypothesis that the QTL is not located in the interval is given by:

$A = 2\{[\mathrm{Log}[L(\mu1, \mu2, \sigma^2, r_1) - \mathrm{Log}[L(\mu_1 = \mu_2 = \mu)]\}$

Where the log likelihoods are evaluated using the maximum likelihood estimates of the genotypic values for the two QTL genotypes, the variance σ^2 and the recombination fraction between marker A and the putative QTL using iterative procedures based on EM (Expectation-Maximization) algorithm. This statistic is distributed as X^2 with 1 d.f. The associated lod score for the interval mapping is then lod score= $(1/2)(\log_{10}e) = 0.217$. This statistic is evaluated at regularly spaced points, say 1 or 2 cM distance, covering the interval as a function of the presumed QTL position. Repeating this procedure for each interval along the chromosome and plotting the lod score curve against the interval gives a QTL likelihood map that presents the evidence for the QTL at any position in the genome. Presence of a putative QTL is assumed if lod score exceeds the threshold T and the maximum of the lod score function in the map gives an estimate of the QTL position and the gene effects. The mapping of QTL by interval method is widely used in practice. The analysis is done through the software package MAPMAKER/QTL developed by Lincoln and Lander (1990) and Lincoln *et al.* (1990).

3.1.2.2 *Regression method*

An approximate method of analysis of data obtained in simple interval mapping (SIM), independently suggested by Haley and Knott (1992) and Martinez and Curnow (1992), uses regression approach wherein, in the linear model given by (e), the indicator variable g for the QTL is replaced by its *conditional expected* value given the observed marker genotype at the flanking markers. A regression based on such a model is fitted at regular positions within each interval along with the residual sum of squares (SSE) and the estimate of the QTL position is taken at the position where the SSE is minimum. The test statistic is now

$$[-N \log_e(1-R^2)]$$

where R^2 is the proportion of the total variance explained by the model and can be chosen to be equal to the likelihood ratio A approximately as

$$A = N \log_e(SSE_{reduced} / SSE_{full}) = - N \log_e(1-R^2)$$

where, full refers to the model with the QTL and reduced refers to the model without it. If we plot, this statistic along the chromosome, it gives a very close approximation to the QTL likelihood map provided the distance between the adjacent markers is sufficiently small (<20 cM).

3.1.3 Composite Interval Mapping(CIM):

Although SIM is the method for QTL mapping most widely used with advantage in several practical situations, it ignores the fact that most quantitative traits are influenced by numorous QTLs.This is overcome either by adopting a model of Multiple QTL Mapping (MQM) or by combining SIM with the method of multiple linear regression, a procedure known as Composite Interval Mapping (CIM) (Jansen, 1993; Jansen and Stam, 1994). In all these methods, one uses the approach of maximum likelihood whch produces only point estimates of the parameters such as the number of QTLs, their location, and effects. The corresponding confidence intervals are required to be determined separately by resampling-based methods. Further, the correct number of QTLs are difficult to determine using traditional methods. Their incorrect specification leads to distortion of the estimates of locations and effects of QTLs. To address these problems a Bayesian approach is adopted wherein the joint posterior distribution of all unknown parameters given their *prior* distributions and the observed data are computed. Of course this is done using iterative simulation procedures on high-speed computers. The main advantage of CIM is that it is more precise and effective at mapping QTLs compared to single-point analysis and interval mapping, especially when linked QTLs are involved. Many researchers have used QTL Cartographer (Basten *et al.*, 2001), MapManager QTX (Manly *et al.*, 2001) and PLABQTL (Utz and Melchinger, 1996) to perform CIM.

3.1.4 Computer-intensive Statistical Methods

The revolution in the field of electronic computation has led to a new orientation in the theory and methods of statistics. The research efforts are now directed towards the development of computer-intensive statistical methods in terms of simulation, algorithms, related issues like convergence diagnostics and *exact* statistical testing procedures. The applications of computer-intensive methods such as Jack Knife, Bootstrap, EM (Expectation-Maximization) algorithms, Markov Chain Monte Carlo (MCMC) methods, Gibbs Sampling etc. and permutation testing in several issues on QTL mapping have enabled a better and more realistic understanding of the problems at hand.

Bootstrap methods are basically simulation methods conducted on high-speed computers and are aimed at generating new data sets from the observed original data set to extract as much information as possible from the data on hand (Efron and Tibshirani, 1993). Jack Knife and Bootstrap methods are often used in genetic studies like estimation of recombination fraction involving markers for reducing bias, estimating standard errors and generating confidence intervals (Visscher *et al*, 1996). EM algorithms are iterative procedures programmed on computers to compute maximum likelihood estimates from incomplete observed data (Zehua Chen, 2005). MCMC integration methods and Gibbs Sampler are used in Bayesian approach to QTL mapping that involves evaluation of high-dimensional integrals to obtain *posterior* distributions of unobserved quantities (Sen and Churchil, 2001; Banerjee *et al.*, 2008).

As we know, classical methods of statistical tests of significance are based on large sample approximations which usually do not work with small sample sizes. Exact test procedures, also called *permutation tests*, are therefore called for but they involve enormous computation. The rapid advance of computer power has however made such procedures available for virtually all significance testing. Concerned agencies engaged in the practice of statistics are therefore emphasizing the use of exact tests as an insurance against reporting misleading results. The exact test procedures require the dataset to be permuted in *all* possible ways under the null model that is being tested, the value of the test statistic to be computed for each permutation, and how extreme the observed value of the test statistic is, compared to its permutation distribution, to be determined. In contrast, the earlier methods of statistical inference assumed, often inappropriately, that the dataset is large enough or well-behaved enough, for the test statistic to follow a normal or a chi-square distribution. When these assumptions are met, the permutation test and an appropriate asymptotic approximation will yield the same result. But when these assumptions fail, only the permutation test can be relied upon. Several statistical software have therefore been developed and have become indispensable tools for such state-of-the-art statistical inference (for example StatXact). In QTL mapping procedures also such an approach of permutation testing has been particularly useful for determining the critical values to declare significant QTL effects.

4.0 ANALYSIS OF MOLECULAR VARIATION (AMOVA)

Owing to the development in the application of molecular techniques, the knowledge of the genetic diversity of the plant or animal or human populations has increased considerably over the last two decades. Progress in molecular genetics has enabled the study of extremely high number of samples and has provided enough scope for the most complete description of the population under investigation. Hence quantitative resolution has improved as a large number of haplotypic markers are found within each sample and as a large number of samples can be simultaneously investigated. As a consequence, powerful statistical techniques are needed in order to describe the structure of the populations and to highlight the contribution of its components for interpretation of the results. Statistical techniques like ANOVA in general uncover the effects of categorical independent variables on an interval dependent variable under the assumption of normality of data but which is not the case of the molecular polymorphisms. The Analysis of Molecular Variance (AMOVA, Excoffier *et al.*, 1992) is a methodology for the analysis of variance that makes use of molecular data derived from the analysis of DNA. AMOVA is adaptable to different kind of assumption on the evolution of a genetic system. Starting from data of various origin (sequences, iso-enzymatic patterns, microsatellites, RFLP, RAPD, PCR-RFLP, RFLP, AFLP), AMOVA makes use of the molecular information gathered in the population study to investigate the genetic differentiation of the sampled populations. Moreover the significance in AMOVA procedure is tested *via* a permutational approach, eliminating the need of normal distribution that is required for the analysis of variance but inappropriate for molecular data. The effectiveness and the usefulness of AMOVA for the analysis of molecular data have been widely demonstrated by several authors.

4.1 Basic Principle of AMOVA

Population genetic structure has traditionally been studied using departures of allele frequencies from random expectations using estimation procedures related to Wright's *F*-statistics (Wright, 1951, 1965). AMOVA was proposed by Excoffier *et al.* (1992) to make use of the available molecular information provided in population studies. AMOVA differs from analysis of variance in that it can accommodate different evolutionary assumptions without modifying the basic structure of the analysis, and in that hypotheses are tested using permutational methods, so that normal distribution assumption is not required. Considering a large set of samples, they can be

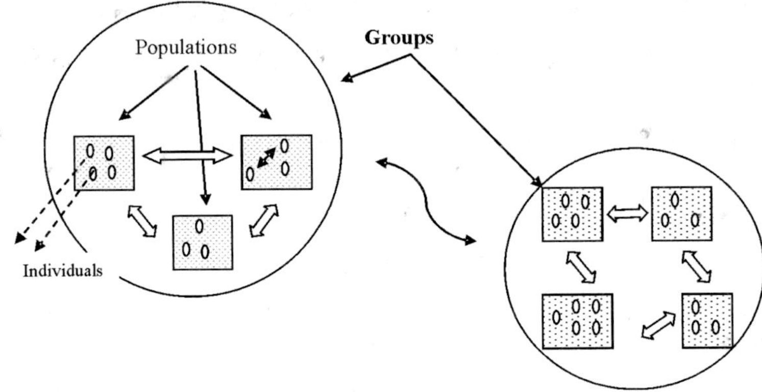

Fig.1. Representation of AMOVA - The circles indicate the groups; squares the populations; small dots indicate the strains (individuals). The double arrows show the comparisons made: among strains within the population, among populations, among groups (Source: Excoffier, 2000).

sorted among different groups any of which contains in turn different populations each formed by several individuals. The aim of the analysis is to find out and describe that hierarchical structure splitting the total variance into covariance components due to intra-populations, inter-populations and inter-groups differences (Excoffier, 2000). AMOVA computes the differences between groups, among populations within group and among strains (or individuals) within the population (Fig. 1).

Molecular biology techniques (*i.e.* RFLP, RAPD, PCR-RFLP, AFLP, etc) produces bands after electrophoresis that can be interpreted in a binary form, i.e. in the form of 0s and 1s, where 1 denotes the presence of a band and zero its absence. That is a haplotype data for an individual is represented by a vector $\boldsymbol{p}$. For instance, for RFLP data $\boldsymbol{p}$ has a 1 if a restriction site is present and 0 if absent; for RAPD data $\boldsymbol{p}$ has a 1 if an homologous band is present and 0 if absent. We are interested in the genetic distance between pairs of haplotypes. For binary data, distances can be defined as Euclidean distance (the number of allelic differences between two haplotypes, see Nei and Tajima, 1981). In the example below, the Euclidean distance between A and B is equal to 8, which is a simple count of differences among the haplotypes (indicated in bold letters).

A 0 1 0 0 0 0 0 0 1 0 0 1 0 **1** 0 **0** 0 **0** 0 **1** 1 **0** **0** 1 1 0 0 **0** **0** 0 0 1 1 0

B 0 1 0 0 0 0 0 0 1 0 0 1 0 **0** 0 **1** 0 **1** 0 **0** 1 **1** **1** 1 1 0 0 **1** **1** 0 0 1 1 0

This method is used when

- identities of restriction sites or RAPD markers are well defined;
- some haplotypes are clearly more different than others;
- no evolutionary network relating the haplotypes is available.

Using the haplotype data the inter-distances for all combinations is computed in the form of a matrix.

Let Xj be a Boolean vector, of length S, representing the jth haplotype where S is the number of restriction sites or marker bands.

The squared distance between haplotypes h_j and h_k is:

$$\delta_{jk}^2 = (X_j - X_k)'W(X_j - X_k)$$

can be defined in a variety of ways depending up on the weight-matrix **W** on the assumptions as stated below:

- W is an Identity Matrix (all elements equal to 1), indicating that all sites or bands are independent and given equal weight;
- W is Diagonal, indicating that all sites or bands are independent of one another but some are given more weight than others;
- W is Non-zero off-diagonal elements, indicating that sites are not independent.

Details can be viewed from: http://www.bioss.ac.uk/smart/unix/mamova/slides/frames.htm

The total sum of squares of the distances between all pairs of haplotypes is equivalent to the conventional sum of squares of deviations of multidimensional vectors from the centroid of a multidimensional space. The analysis of molecular variance is essentially an analysis of variance using squared distances between pairs of haplotypes as the data. To understand the working principle of AMOVA the equation, we may extend the notation to include population structure. If $X_{(jig)}$ indexes haplotype *j* in subpopulation *i* in group *g*, it can be represented by the linear model

$$X_{jig} = X + a_g + b_{ig} + c_{jig}$$

where,

- X is the unknown expectation
- *a* is the group effect with variance σ_a^2, $g = 1 \ldots G$

- *b* is the subpopulation effect within a group with variance σ_b^2, i = 1 ... I_g
- *c* is the individual effect within a subpopulation with variance σ_c^2, j = 1 ... N_{ig}

These effects are assumed to be additive, random and uncorrelated.

The basic steps to be performed in an AMOVA approach can be summarized as follows: i) binary scoring of the molecular fingerprint of each isolate/strain/individual; ii) computing of the distance matrix; iii) assuming an hypothesis of variance partition; iv) testing the hypothetical partition.

4.2 Software for AMOVA Computations

AMOVA is a computer program developed by Laurent Excoffier at the Genetics and Biometry Laboratory of the Department of Anthropology & Ecology in the University of Geneva. The program is a very useful tool for the analysis of population genetic structure at the molecular level. It runs under PC Windows and is currently available as freeware from the URL:: http://acasun1.unige.ch/LGB/software/win/amova/

To run the AMOVA program it is necessary to construct

- a group file
- a distance file
- several population files

There are settings for choosing various options in the analysis. The user manual provides the details of the data file formats. Using the input files and settings an output file is obtained. Since developing AMOVA, the Genetics & Biometry Laboratory has produced a much more sophisticated program - **ARLEQUIN** (Schneider *et al.*, 2000)- which can handle data from RFLPs, DNA sequences, microsatellites as well as standard multi-locus or allele frequency data. Arlequin can be freely downloaded from the URL: http://lgb.unige.ch/arlequin/.

The AMOVA approach used in ARLEQUIN is essentially similar to other approaches based on analysis of variance of gene frequencies, but it takes into account the number of mutations between molecular haplotypes (which first need to be evaluated). The significance of the fixation indices is tested using a non-parametric permutation approach described in Excoffier *et al.* (1992), consisting in permuting haplotypes, individuals, or populations, among individuals, populations, or groups of populations. After each permutation round, this package re-computes all statistics to get their null distribution. Depending on the tested statistic and the given hierarchical design, different types of permutations are performed. Under this procedure, the normality assumption usual in ANOVA tests is no longer necessary, nor is it necessary to assume equality of variance among populations or groups of populations. A large number of permutations (1,000 or more) is necessary to obtain some accuracy on the final probability.

ARLEQUIN provides 6 different types of hierarchical AMOVA. The number of hierarchical levels varies from two to four. In each of the situations, it describes the way the total sum of squares is partitioned, how the covariance components and the associated F-statistics are obtained, and which permutation schemes are used for the significance test. Before enumerating all the situations, let us introduce some notations:

SSD(T): Total sum of squared deviations.

SSD(AG): Sum of squared deviations among groups of populations

SSD(AP): Sum of squared deviations among populations

SSD(AI): Sum of squared deviations among individuals

SSD(WP): Sum of squared deviations within populations

SSD(WI): Sum of squared deviations within individuals

SSD(AP/WG): Sum of squared deviations among populations, within groups

SSD(AI/WP): Sum of squared deviations among individuals, within populations

G: Number of groups in the source

P: Total number of populations

N: Total number of individuals for genotype data or total number of gene copies for haplotypic data.

N_p: Number of individuals in population p for genotypic data or total number of gene copies in population p for haplotypic data.

N_g: Number of individuals in group g for genotypic data or total number of gene copies in group g for haplotypic data.

Now we consider two most common types of hierarchical data in the following sections.

I. Haplotypic data, one group of populations

Source of variation	D.F	S.S	E.M.S
Among populations	P-1	SSD (AP)	$n\,\sigma_a^2 + \sigma_b^2$
Within populations	N-P	SSD (WP)	σ_b^2
Total	N-1	SSD (T)	σ_T^2

In the case of a simple hierarchical genetic structure consisting of haploid individuals in populations, the implemented form of the algorithm leads to a fixation index F_{ST} which is identical to the weighted average F-statistic over loci, $\theta_{w,}$ defined by Weir and Cockerham (1984). In terms of inbreeding coefficients and coalescence times, this F_{ST} can be expressed as

$$F_{ST} = (f_0 - f_1)/(1-f_1) = (t_1-t_0)/t_1$$

Where $\mathbf{f_0}$ is the probability of identity by descent of two different genes drawn from the same population, $\mathbf{f_1}$ is the probability of identity by descent of two genes drawn from two different populations, $\mathbf{t_1}$ is the mean coalescence times of two genes drawn from two different populations, and $\mathbf{t_0}$ is the mean coalescence times of two genes drawn from the same population, where $\boldsymbol{n}$ and $\boldsymbol{F_{ST}}$ are defined by

$$F_{ST} = (\sigma_a^2 / \sigma_T^2); \text{ and } n = \frac{N - \sum_P \frac{N_P^2}{N}}{P-1}$$

Test σ_a^2 and F_{ST} by permuting haplotypes among populations.

II. Haplotypic data, several groups of populations

Source of variation	D.F	SSD	EMS
Among Groups	G-1	SSD (AG)	$n'' \sigma_a^2 + n' \sigma_b^2 + \sigma_c^2$
Among populations / Within Groups	P-G	SSD (AP/WG)	$n\sigma_b^2 + \sigma_c^2$
Within populations	N-P	SSD (WP)	σ_c^2
Total	N-1	SSD (T)	σ_T^2

The formulas for the average sample sizes and F statistics are given below.

$$S_G = \sum_{g \in G} \sum_{p \in g} \frac{N_P^2}{N_g}; \quad n = \frac{N - S_G}{P - G}; \quad n\times = \frac{S_G - \sum_{p \in P} \frac{N_P^2}{N}}{G-1}$$

$$n'' = \frac{N - \sum_{g \in G} \frac{N_g^2}{N}}{G-1}; \quad F_{CT} = \sigma_a^2 / \sigma_T^2; \quad F_{SC} = \sigma_b^2 / (\sigma_b^2 + \sigma_c^2)$$

$$F_{ST} = (\sigma_a^2 + \sigma_b^2) / \sigma_T^2$$

- Test σ_c^2 and F_{ST} by permuting haplotypes among populations among groups

- Test σ^2_b and F_{SC} by permuting haplotypes among populations within groups
- Test σ^2_a and F_{CT} by permuting populations among groups

4.3 AMOVA Illustration

In Table 3 an example of AMOVA result as reported by Mengoni and Bazzicalupo (2002) is given. The hypothesis tested is the existence of an uneven distribution of the genetic variation of nodulating bacteria (*Sinorhizobium meliloti*) linked to the difference among plant cultivars (*Medicago sativa*). Groups were formed joining together isolates from different plants belonging to the same cultivar. The total variance derived from the molecular data has been divided into the three hierarchical partitions aiming to test the hypothesis of a genetic differentiation based on the different plant cultivar. Columns show the percent of total variance attributed to the partition, the Φ statistics and its significance (P-value). In general, the differences within populations (plants) (3rd row) are the main component in the variance partition (highest percentage and Φ_{ST} value). P-value is highly significant in the first partition indicating that the plant cultivar is a factor influencing the shaping of populations. Moreover, the highly significant values in the second and third partition indicates that the different plants belonging to the same cultivar represent different bacterial populations and that in the same plant the different isolates can be distinguished.

Table 3. An example of Analysis of Molecular Variance (AMOVA) from populations of nodulating *Sinorhizobium meliloti* analysed with RAPD markers

Source of variation	% Total	Φ Statistics	P-value
Among cultivars	12.8	Φ_{CT} =0.128	<0.0001
Plants within cultivars	25.7	Φ_{SC} =0.295	<0.0001
Isolates within plant	61.5	Φ_{ST} =0.385	<0.0001

We have so far considered only two potential areas of statistical applications in molecular biology. However, there are other aspects like finding over and under-represented words in DNA sequences in molecular biology since exceptional words may be involved in the stability of DNA as well as in mechanisms like recombination, replication and repair. The concept of an exceptional word is based on a statistical comparison between the observed frequency of a word and the one expected under a probability model that reflects the composition of the DNA sequence in its small vocabulary. DNA sequences consist of one of four bases at each position (A, C, G or T) and the sequence is always 'read' in a specific direction: from the 5' to the 3' end in molecular biological language. Therefore DNA sequences take the form of a four-state series positioned in space. The most commonly used parsimonious model for discrete state series is a low order stationary Markov chain. Such models are used for predicting the occurrence of certain sequences, *e.g.* the sequence CTGAC in the DNA context (Avery, 1987 and Cowan, 1991). If the sequence occurs in particular positions much more often than the model predicts then this suggests that the sequence is probably there for a functional reason. Such findings are useful in suggesting research directions for molecular biologists. It is suggested that the reader refers the book by Ewens and Grant (2005) for a detailed treatment on the application of probability models in DNA sequence analysis.

5.0 MEASURING GENETIC DIVERSITY

Assessing genetic diversity is very essential for crop breeding for identifying diverse parental combinations to create progenies with maximum genetic variability and for incorporating desirable genes from diverse genetic resources into existing genetic stocks. In measuring genetic diversity we are attempting to quantify the extent of differences within or between populations. In analyzing genetic diversity we come across data sets ranging from pedigree information to DNA-based marker data. The data sets provide varying types of information, and the methods of measuring diversity depend on the objectives of the study and the level of the diversity required ranging from individuals to population level. Added to that, the measurement of genetic diversity is subjected to sampling variation due to varying number of individuals sampled per population and the number of loci sampled. Normally a larger portion of the sampling variance of the diversity measures is due to the variation of diversity levels among loci across the genome. Weir (1990) describes the statistical theory and formulae for estimating the sampling variance of some useful measures of genetic diversity. A review of the statistical tools for the analysis of genetic diversity in crop plants is provided by Mohammadi and Prasanna (2003).

Various genetic distance measures have been proposed for analysis of molecular marker data for the purpose of genetic diversity analysis among individuals. For molecular marker data, the amplification products may be equated to alleles and hence allelic frequencies can be calculated. A general formula for estimating the genetic distance between individuals *i* and *j* is:

$$D(i,j) = \text{Constant} * \sum_{a=1}^{n} |X_{ai} - X_{aj}|^{r})^{1/r},$$

where X_{ai} and X_{aj} are the frequencies of the allele *a* for individual *i* and *j*; *n* is the number of alleles per locus, and *r* is a constant based on the coefficient used. When *r*=2, $D(i,j) = 1/2 *(\sum^{n}_{a=1} |X_{ai} - X_{aj}|^{2})^{1/2}$, referred to as Rogers' (1972) measure of distance (RD). Other measures of genetic distance between any two individuals are described below.

Let N_{11} be the number of bands/ alleles present in both individuals; N_{00} be the number of bands/ alleles absent in both the individuals; N_{10} be the number of bands present only in individual *i*; N_{01} be the number of bands present only in individual *j* and N represent the total number of bands.

i. Nei and Li's Coefficient (GD_{NL}) = $1 - [2N_{11} / (2N_{11} + N_{10} + N_{01})]$. This measures the proportion of bands-alleles shares as the result of being inherited from a common ancestor. That is, it represents the proportion of bands-alleles present and shared in both individuals divided by the average of proportion of bands-alleles present in each individual.

ii. Jaccard's Coefficient (GD_J) = $1 - [N_{11} / (N_{11} + N_{10} + N_{01})]$. This measure considers only matches between bands-alleles that are present and ignores pairs in which a band-allele is absent in both the individuals.

iii. The simple matching Coefficient (GD_{SM}) = $1 - [(N_{11} + N_{00}) / (N_{11} + N_{10} + N_{01} + N_{00})]$. This measure takes into account mismatches and matches, and gives equal weight to both 0-0 and 1-1 matches in estimating the genetic distance. However, the 1-1 matches in reality indicate more similarity than the 0-0 matches because there are many reasons for lack of amplification or absence of bands (Mohammadi and Prasanna, 2003)

iv. Modified Rogers' Distance (GD_{MR}) = $[N_{10} + N_{01}) / 2N]^{0.5}$.

For the first two distance measures have unknown statistical distributions and need *bootstrap* estimates of their sampling variances. However, the GD_{SM} and GD_{MR} are Euclidean distances, and have binomial distributions (Tivang *et al.*, 1994). The distance measure GD_{MR} is widely preferred because of its good genetical and statistical properties and GD_{SM} is useful in hierarchical clustering methods and for the analysis of molecular variance (Mohammadi and Prasanna, 2003).

While the above distance measures are useful for estimating genetic diversity among individuals, for genetic differentiation of populations the F-statistics of Wright (Wright, 1951; Weir, 1996) are useful. The F-Statistics are computed based on genotypic frequencies. Let within-population gene diversity be H_I; mean within-population gene diversity be H_s and total diversity be H_T. Then the three different *F* coefficients are given as:

i. Correlation of genes within individuals over all populations (F_{IT})

ii. Correlation of genes of different individuals in the same population (F_{ST})

iii. Correlation of genes within individuals within populations (F_{IS}), where

$F_{IT} = 1 - (H_I / H_T)$; $F_{IS} = 1 - (H_I / H_S)$ and $F_{ST} = 1 - (H_S / H_T)$. FST, FIT and FIS are interrelated so that: $1 - F_{IT} = (1 - F_{ST})(1 - F_{IS})$ and $F_{ST} = (F_{IT} - F_{IS}) / (1 - F_{IS})$. The value of F_{ST} is always greater or equal to 0.Value of F_{ST}

in the range 0-0.05 is interpreted as little differentiation in the populations; value in the range 0.05-0.15 is interpreted as moderate differentiation; value in the range 0.15-0.25 is interpreted as large differentiation and greater than 0.25 as very large. Nei (1973) proposed a statistic G_{ST}, analogous to F_{ST} that is calculated from allelic frequencies rather than genotypic frequencies. Its estimate is given by the equation.

$G_{ST} = D_{ST} / H_{T}$, where,

H_T = total genic diversity = $H_S + D_{ST}$

H_S = intra-population genic diversity

D_{ST} = inter-population diversity. It may be noted that

$(H_T/H_T) = (H_S/H_T) + (D_{ST}/H_T) = 1.$

G_{ST} measures the proportion of gene diversity that is distributed among populations. It is required that a larger number of loci must be sampled and the equations are complex and may require specific computer software.

5.1 Classification Methods in Diversity Analysis

Multivariate statistical techniques which simultaneously analyze multiple measurements (morphological, biochemical or molecular marker data) on each individual are widely used for diversity analysis. Among them cluster analysis, principal component analysis, principal coordinate analysis and multi-dimension scaling are the commonly used procedures. In distance based clustering methods, pair-wise distance matrix is used as input for analysis. The clustering methods are broadly classified as hierarchical and non-hierarchical. The former methods are commonly employed in the analysis of genetic diversity of crop species. Among the algorithms for hierarchical clustering methods, UPGMA (Un-weighted Paired Group Method using Arithmetic averages) is the most widely used followed by Ward's Minimum Variance Method (Everitt, 1993). Statistical packages like SPSS and SAS provide a wide choice of methods of analysis using cluster analysis. In non-hierarchical cluster analysis, individuals are optimally allotted to different clusters, once the number of clusters is specified. An important aspect in cluster analysis is to determine optimum number of clusters such that the within-cluster genetic distance is less than the overall mean genetic distance and the between cluster distances are greater than their within cluster distance of the two clusters involved. Wallace and Boulton (1968) and Boulton and Wallace (1970) describe the algorithms of arriving at optimum cluster configurations using information theory concepts.

5.2 Resampling Techniques

Computer intensive resampling techniques such as "bootstrap" and "jackknife" methods are attracting considerable attention in analysis of diversity using molecular markers. It is obvious that use of a large number

Link	Description
NTSYS-PC (http://www.exetersoftware.com/cat/ntsyspc/ntsyspc.html)	Provides a wide range of multivariate statistical tools for genetic data analysis. A useful manual on the use of NTSYS with example data sets is provided by Warburton and Crossa (2002)
PHYLIP (http://evolution.genetics.washington.edu/phylip.html)	PHYLIP (the *PHYL*ogeny *I*nference *P*ackage) is a package of programs for inferring phylogenies (evolutionary trees). Methods that are available in the package include parsimony, distance matrix, and likelihood methods, including bootstrapping and consensus trees. Data types that can be handled include molecular sequences, gene frequencies, restriction sites and fragments, distance matrices, and discrete characters
Arlequin (http://lgb.unige.ch/arlequin/)	Arlequin is an exploratory population genetics software environment able to handle large samples of molecular data (RFLPs, DNA sequences, microsatellites), while retaining the capacity of analyzing conventional genetic data (standard multi-locus data or mere allele frequency data). A variety of population genetics methods have been implemented either at the intra-population or at the inter-population level, and they can be conveniently selected and parameterized through a graphical interface.
GenoType/GenoDive (http://staff.science.uva.nl/~meirmans/softindex.html)	GenoType and GenoDive are two programs for analysing the genotypic diversity in clonal/asexual organisms. GenoType assigns genotypes to individuals, based on molecular marker data. Possible input data comes from microsatellites, allozymes, AFLP's or RAPD's. GenoType can also handle haploid and polyploid data and has some features to take scoring errors or mutations into account. GenoDive reads genotypic data (*e.g.* from GenoType) and calculates a number of indices of genotypic diversity. It can also perform a bootstrap test to see whether these indices are different for pairs of populations. Finally it can also test for differentiation between populations in genotypic composition (this is not the same as the test for differences in genetic diversity).

contd....

Link	Description
POPGENE (http://www.ualberta.ca/~fyeh/)	POPGENE is a user-friendly computer freeware for the analysis of genetic variation among and within populations using co-dominant and dominant markers. This package provides the Windows graphical user interface that makes population genetics analysis more accessible for the casual computer user and more convenient for the experienced computer user. Simple menus and dialog box selections enable you to perform complex analyses and produce scientifically sound statistics, thereby assisting you to adequately analyze population genetic structure using the target markers.
TFPGA (http://www.marksgeneticsoftware.net/)	Tools for Population Genetic Analyses: A Windows program for the analysis of allozyme and molecular population genetic data. This program calculates descriptive statistics, genetic distances, and F-statistics. It also performs tests for Hardy-Weinberg equilibrium, exact tests for genetic differentiation, Mantel tests, and UPGMA cluster analyses. Additional features include the ability to analyze hierarchical data sets as well as data from either co-dominant markers such as allozymes or dominant markers such as AFLPs or RAPDs.

of polymorphic markers or bands will provide an increasingly more precise (with reduced variance) estimate of genetic relationship. Nevertheless there is a need to reduce the cost involved in assaying a large number of markers and bootstrap method are useful for finding the smallest set of markers that can provide an accurate assessment of genetic relationships among a set of genotypes or populations (Tivang *et al.*, 1994). Because molecular markers are capable of generating a large amount of data, they provide an excellent opportunity for bootstrap sampling using whole data sets as well as with smaller portions of the data set (Mohammadi and Prasanna, 2003). The basic principles of bootstrap method is described by Efron and Tibshirani. (1993). A simple tutorial on bootstrap method can be downloaded from the URL: http://people.revoledu.com/kard/ tutorial/bootstrap/.

5.3 Software for diversity analysis (A sample list)

Apart from the above list, there are a number of software packages available depending on the type of application, the type of computer and the type of input data for molecular data. The reader can refer to the following URL for a huge number of available software for molecular data analysis- http://evolution.genetics.washington.edu/phylip/software.html. The programs are classified under various categories of application and a brief description is given about each package.

6.0 CONCLUSION

The availability of molecular marker data has provided a new dimension to quantitative genetics methods, *viz.*, the QTL analysis. This method has become an important tool using which biologist can understand the genetics of complex characters by relating the molecular marker data with trait data. One of the key factors contributing to the success of a QTL mapping experiment is to estimate the strength of evidence for the presence of a QTL and the precision with which QTL positions can be estimated. The ANOVA method which is the simplest statistical method for QTL mapping has the advantage of simplicity, but may fail to identify the QTL's precisely when there is appreciable missing marker genotype data or when the markers are widely spaced. Single interval mapping and composite interval mapping estimation methods naturally increase the precision with which QTL positions can be identified but at the cost of increased computational needs that generally is taken care of by computer software. One major problem associated with QTL analysis is that the more QTLs are there in the population, the smaller their individual contribution and difficulties in detecting them. Nevertheless, QTL analysis of segregating populations have been found useful in identifying major QTLs and improve the chances of marker aided selection. The Analysis of Molecular Variance (AMOVA) provides a statistical framework using which the factors that influence the genetic structure of populations can be described using molecular data available at the population level. Statistical measures of genetic diversity help in quantifying the extent of differences within or between populations by using data ranging from phenotypic values to DNA-based marker data. Marker studies on the distribution of genetic variation within and between populations can also be used as a guide for the acquisition of new material for germplasm conservation and to identify germplasm that is under-represented in a gene bank. The genetic diversity estimated through a variety of distance measures are useful not only for estimating genetic diversity among individuals (thereby identifying diverse parental combinations to create progenies with maximum genetic variability), but also for genetic differentiation among populations. While structural and functional genomics help in the advancement of our knowledge about an organism, practical application of genomics would have to consider the interaction of genes with environmental factors such as drought. For example, a QTL analysis of the genetics of physiological traits associated with drought tolerance would help us to identify key pathways involved in drought tolerance and to know how the concerned genes interact. Such studies would be crucial in complementing the efforts of conventional breeding programmes. Integration of genomics into crop growth models can be a major strategy in evolving crop modeling that are robust over wide range of environments with accompanying diversity in actual and potential cultivar types (Hoogenboom *et al.*, 2004).

7.0 REFERENCES

Avery, P.J. 1987. The analysis of intron data and their use in the detection of short signals. Journal of Molecular Evolution, 26:335-340.

Banerjee, S., Yandell, B.S., and Yi, N. 2008. Bayesian quantitative trait loci mapping for multiple traits. Genetics, 179(4):2275-2289.

Basten, C., Weir, B. and Zeng, B. 2001. *QTL Cartographer*. Department of Statistics, North Carolina State University, Raleigh, North Carolina.

Boulton, D.M. and Wallace, C.S. 1970. A program for numerical classification. The Computer Journal, 13: 63-69.

Broman, K.W. and Speed, T.P. 1999. A review of methods for identifying QTLs in experimental crosses. *In*: Françoise Seillier-Moiseiwitsch, *ed*. Statistics in molecular biology and genetics: Selected proceedings of the Joint AMS-IMS-SIAM Summer Conference on Statistics in Molecular Biology held in Seattle, WA, June 22-26, 1997.

Causse, M., Saliba-Colombani, V., Lecomte, L., Duffé, P., Rousselle, P. and Buret M. 2002. QTL analysis of fruit quality in fresh market tomato: a few chromosome regions control the variation of sensory and instrumental traits. Journal of Experimental Botany, 53 (377):2089-2098.

Cheverud, J.M., and Routman, E.J. 1995. Epistasis and its contribution to genetic variance components. Genetics, 139:1455-1461.

Cowan, R. 1991. Expected frequencies of DNA patterns using Whittle's formula. Journal of Applied Probab, 28, 886892.

Efron, B. and Tibshirani, R.J. 1993. An introduction to the bootstrap, Chapman & Hall.

Everitt, B. 1993. Cluster Analysis. John Wiley & Sons Inc. (3rd Edition).

Ewens W.J. and Grant G.R. 2005. Statistical Methods in Bioinformatics: An introduction. Springer Science & Business Media, Inc.

Excoffier, L. 2000. Analysis of population subdivision. *In*: Balding, D., Bishop, M., Cannings, C., eds, Handbook of Statistical Genetics. Wiley and Sons, Ltd.

Excoffier, L., Smouse, P.E. and Quattro, J.M. 1992. Analysis of molecular variance inferred from metric distances among DNA haplotypes: application to human mitochondrial DNA restriction data. Genetics, 131:479-491.

Fisher, R.A. 1918. The correlation between relatives on the supposition of Medelian inheritance. Translation of Royal Society, 52:399-433.

Galton, F. 1889. Natural Inheritance. Macmillan, London.

Galton, F. 1897. The average contribution of each several ancestor to the total heritage of the offspring. Proceedings of Royal Society, 61:401–413.

Haley, C.S., and Knott, S.A. 1992. A simple regression method for mapping quantitative trait loci in line crosses using flanking markers. Heredity. 69: 315-324.

Hartl, D. and Jones, E. 2001. Genetics: Analysis of Genes and Genomes, Jones and Bartlett Publishers, Sudbury, MA.

Hoogenboom, G., White, J.W. and Messina, C.D. 2004. From genome to crop: integration through simulation modeling. Field Crops Research, 90:145-163.

Jansen, R. 1993. Interval mapping of multiple quantitative trait loci. Genetics, 135: 205-211.

Jansen, R. and Stam, P. 1994. High resolution of quantitative traits into multiple loci *via* interval mapping. Genetics, 136:1447–1455.

Lander, E. and Botstein, D. 1989. Mapping Mendelian factors underlying quantitative traits using RFLP linkage maps. Genetics, 121:185–199.

Lander, E.S., Green, P., Abrahamson, J., Barlow, A., Daly, M.J., Lincoln, S.E. and Newburg, L. 1987. Mapmaker an interactive computer package for constructing primary genetic linkage maps of experimental and natural populations. Genomics, 1:174–181.

Lincoln, S., Daly, M. and Lander, E. 1993. Constructing genetic linkage maps with MAPMAKER/EXP. Version 3.0. Whitehead Institute for Biomedical Research Technical Report, 3rd Edn.

Lincoln, S.E. and Lander, E.S. 1990. Mapping genes controlling quantitative traits using MAPMAKER/QTL. Technical Report. Cambridge, MA: Whitehead Institute for Biomedical Research.

Lincoln, S.E., Daly, M.J. and Lander, E.S. 1990. *Constructing genetic linkage maps with MAPMAKER: a tutorial and reference manual*. Technical Report. Cambridge, MA: Whitehead Institute for Biomedical Research.

Liu, Ben-Hui. 1998. Statistical genomics: Linkage, mapping, and QTL analysis. CRC Press.

Manly, K.F., Cudmore Robert, H. Jr. and Meer, J.M. 2001. Map Manager QTX, cross-platform software for genetic mapping. Mammalian Genome, 12:930–932.

Martinez,O., and Curnow, R.N. 1992. Estimation of the locations and the sizes of the effects of quantitative trait loci using flanking markers. Theoretical and Applied Genetics, 85:480—488.

Mcmillan, I. and Robertson, A. 1974. The power of methods for the detection of major genes affecting quantitative characters. Heredity, 32:349-356.

Mengoni, A. and Bazzicalupo, M. 2002. The statistical treatment of data and the analysis of molecular variance (AMOVA) in molecular microbial ecology. Annals of Microbiology, 52:95-101.

Mohammadi, S.A. and Prasanna, B.M. 2003. Analysis of genetic diversity in crop plants- Salient statistical tools and considerations. Crop Science, 43:1235-1248.

Nei, M. 1973. Analysis of gene diversity in subdivided populations. Proceedings of the National Academy of Sciences (USA). 70: 3321-3323.

Nei, M. and Tajima, F. 1981. DNA polymorphism detectable by restriction endonucleases. Genetics, 97:145-163.

Rogers, J.S. 1972. Measures of genetic similarity and genetic distance. Studies in genetics. VIII. University Texas. Publication, 2713:145-153.

Sax, K. 1923. The association of size differences with seed-coat pattern and pigmentation in *Phaseolus vulgaris*. Genetics, 8:552-560.

Schneider, S., Roessli, D. and Excoffier, L. 2000. Arlequin 2000: a Software for Population Genetic Data Analysis. Genetics and Biometry Laboratory, University of Geneva, Switzerland.

Sen. S. and Churchill, G.A. 2001. A statistical framework for quantitative trait loci. Genetics, 159: 371-387.

Soller, M., Brody, T. and Genizi, A. 1976. On the power of experimental design for the detection of linkage between marker loci and quantitative loci in crosses between inbred lines. Theoretical and Applied Genetics, 47:35-39.

Soller, M. and Genizi, A. 1978. The efficiency of experimental designs for the detection of linkage between a marker locus and a locus affecting a quantitative trait in segregating populations. Biometrics, 34:47-55.

Stam, P. 1993. Construction of integrated genetic linkage maps by means of a new computer package: JoinMap. Plant Journal, 3:739–744.

Strachan, T., and Read, A.P. 1999. Human Molecular Genetics. John Wiley & Sons, Inc.

Tanksley, S.D. 1993. Mapping polygenes. Annual Review Genetics, 27:205–233.

Thoday, J.M. 1961. Location of polygenes. Nature,191:368-370.

Tivang, G., Nienhuis, J. and Smith, O.S. 1994. Estimation of sampling variance of molecular marker data using bootstrap procedure. Theoretical and Applied Genetics, 89:259-264.

Visscher, P.M., Thompson, R. and Haley, C. S. 1996. Confidence Intervals in QTL Mapping by Bootstrapping. Genetics, 143:1013-1020.

Wallace, C.S. and Boulton, D.M. 1968. An information measure for classification. The Computer Journal, 11:185-194.

Warburton, M. and Crossa, J. 2002. Data Analysis in the CIMMYT Applied Biotechnology Center for Fingerprinting and Genetic Diversity Studies. CIMMYT (2nd Edition).

Weir, B.S. and Cockerham, C.C. 1984. Estimating F-statistics for the analysis of population structure. Evolution, 38:1359-1270.

Weir, B.S. 1990. Genetic data analysis. Sinauer Associates, Sunderland.

Weir, B.S. 1996. Intraspecific differentiation. pp. 385-403. *In*: D.M. Hillis *et al.* (ed.) Molecular systematics. 2nd Edition. Sinauer Associates, Sunderland.

Wright, S. 1951. The genetic structure of populations. Annals of Eugenics, 1:323-334.

Wright, S. 1965. The interpretation of population structure by *F*-statistics with special regards to systems of mating. Evolution, 19:395-420.

Wu, K., Burnquist, W., Sorrells, M.E., Tew, T., Moore, P. and Tanksley, S.D. 1992. The detection and estimation of linkage in polyploids using single-dose restriction fragments. Theoretical and Applied Genetics ,83:294–300.

Young, N.D. 1994. Constructing a plant genetic linkage map with DNA markers, p. 39–57, *In*: I. K.V. Ronald & L. Phillips (*Eds.*), DNA-based markers in plants. Kluwer, Dordrecht/Boston/London.

Young, N.D. 1996. QTLmapping and quantitative disease resistance in plants. Annual Review Phytopathology, 34:479–501.

Utz, H. and Melchinger,A. 1996. PLABQTL: A program for composite interval mapping of QTL. Journal of Quantitative traits, 2(1):15-23.

Zehua Chen. 2005. The full EM algorithm for the MLEs of QTL effects and positions and their estimated variances in multiple-interval mapping. Biometrics. 61: 474-480.

Molecular Plant Breeding: Principle, Method and Application
Eds : R.K. Singh, Rajesh Singh, Guoyou Ye, A. Selvi and G.P. Rao
Studium Press LLC, Texas, USA, 2009, pp. 241-268

In situ Hybridization – An Important Tool in Cyto-Molecular Mapping

ALKA SRIVASTAVA and MAHESH BASANTANI*

ABSTRACT

The information on the order of loci on chromosomes in most plant species is based almost entirely on genetic linkage maps that are generated by producing multi-hybrid crosses and determining the relative frequency of recombination between genes or molecular markers. The physical location of genes on chromosomes is important because chromosome structure has a significant bearing on gene activity but linkage maps cannot be superimposed on chromosomes because map distances are not proportional to physical distances . Negligible record of genes and subsequently crossing over in heterochromatin, and irregularly distributed crossing over in euchromatin is the reason for discrepancy in linkage and physical maps.

The classical approach to visualizing chromatin used DNA binding dyes such as Giemsa stain, acetocarmine and aceto-orcein, which had relatively low sensitivity and gave no information of the sequence. Linkage and recombination helped in constructing genetic maps of chromosomes which had no bearing either on the actual size of the chromosomes or the position and sequence of genes. The organization and behaviour of genome within the nuclei can now be studied with the optimization of fluorescence based hybridization technologies combined with advances in the instrumentation for microscopy. *In situ* hybridization, using the principle of complementarity and hybridization , initially done with radioactive probes , has evolved into a superior cytological tool for mapping of chromosomes in the form of fluorescence *in situ* hybridization (FISH), fibre FISH, Primed *in situ* variation, branched FISH etc. All such hybridization based methods have found application in mapping, transgene localization, satellite DNA localization and detection of chromosome aberrations. The rapid advance using these techniques has made cytomolecular mapping possible which correlates molecular maps to chromosome structure.

Key Words: ISH, FISH, Probe, Linkage, Cyto-molecular mapping

In vitro culture and Plant Genetics Unit, Department of Botany University of Lucknow, Lucknow 226 007, Uttar Pradesh, India
**Corresponding author e-mail: alkasrivastava@hotmail.com*

1.0 INTRODUCTION

The transition from the Mendelian concept of inheritance to the biometrical system based on the control of important traits in crops by many loci has almost necessitated the deciphering of the sequence and position of genes in a genome. Crop plants have been improved by various methods over the years. The selection done after any breeding programme has been on the basis of phenotype and not the number and sequence of genes involved. The success rate of breeding methods may probably be enhanced if the phenotypic basis is supplemented with information on genes affecting quantitative traits. A parallel line of investigation has developed based on several recent experimental approaches which help in mapping the genes present on chromosomes, to the last detail and also image the distribution and dynamics of DNA, mRNA, proteins and metabolites.

Chromosomes were first described cytologically in the 1870s, and later as the physical carriers of genetic information (Bridges, 1916). The first genetic map was created in 1913 by Sturtevant. He successfully developed a map of the *Drosophila* X chromosome, where the precise relationship between the physical features of the chromosome and its genes was elucidated. Cytogenetic maps integrate data from both genetic and cytological maps, and simultaneously report the cytological and genetic position of a marker. Mapping is putting markers (genes or QTL) in a precise order, indicating the relative distances among them, and assigning them to their linkage groups on the basis of their recombination values from all pairwise combinations. Knowledge about the genetic concepts of segregation and recombination is essential to the understanding of mapping. Genetic maps are produced on the basis of the percentage of recombination between markers, measured in centiMorgans (cM) (Haldane, 1919). These maps reveal the linear order of markers and the amount of recombination between linked markers. They do not contain information concerning the number of DNA base pairs between markers. The factors required to place a marker on a genetic map are: two different alleles of the marker, two different alleles of a second marker to map relative to, and a prospect for meiotic recombination (Harper and Cande, 2000). Cytological maps involve localization of visible structures on stained chromosomes, and are based both on meiotic and mitotic chromosomes. They do not contain information on recombination data, or number of DNA base pairs between two cytological structures. The positions of markers on a cytological map are reported as percentage of total chromosome arm length. The cytological features visible by chromosome staining are restricted to centromeres, the heterochromatin around the centromere (termed "centric heterochromatin"), the nucleolar organizing regions (NORs), regions of euchromatin and regions of non-centric heterochromatin such as knobs and chromomeres only. The unique and reproducible banding patterns known as C-banding (C=constitutive, stains heterochromatin, repetitive DNA) (Arrighi and Hsu, 1971), G-banding (G=giemsa, stains AT-rich heterochromatin) (Seabright, 1971), R-banding (R=reverse of giemsa) (Dutrillaux and Lejeune, 1971), and Q-banding (quinacrine mustard, stains AT-rich regions) (Caspersson *et al.*, 1970) may also serve as potential signposts on chromosomes for the creation of cytological maps. However, with the recent technological advancements and the availability of sophisticated techniques based on in situ hybridization (ISH), it is now possible to locate the position of specific DNA sequences on cytological maps.

In the course of evolution of angiosperms, spreading over several million years, the relative genome size has increased from 0.1 pg to more than 100 pg.The genome size variation reflects increase in ploidy and also repetitive DNA content. The analysis of genome evolution in higher plants indicates that speciation correlates with rapid changes in repetitive DNA, which cannot be elucidated by linkage mapping nor chromosome walking and therefore ISH is the method of choice for identifying repetitive and low copy/single copy sequences. The most direct means of determining the location of genes on chromosomes is through the use of fluorescence in situ hybridization (FISH). In this technique hapten-labeled DNA probes are hybridized to chromosomes that have been spread on glass microscope slides, and antibodies or other affinity reagents conjugated to fluorochromes are used to detect directly or indirectly sites of hybridization. FISH with repetitive sequences as probes has been widely reported for both animal and plant chromosomes, and single-copy FISH to mammalian chromosomes is fairly routine (Lichter *et al.*, 1991; Heng *et al.*, 1992). Fast and highly specific visualization of DNA sequences on chromosomes and in interphase nuclei can be done by primed *in situ* labeling (PRINS). Although the technique does not reach the sensitivity of fluorescence *in situ* hybridization that is needed for detection of single-copy targets, it is superior in its speed and simplicity and its applications include fluorescent labeling of repeated DNA sequences, such as ribosomal DNA and satellite repeats, which are used to discriminate individual chromosome types within karyotypes and to assess the purity of chromosome fractions separated by flow-sorting (Macas *et al.*, 2006).

Cytomolecular mapping unites molecular, genetic, genomic, and cytological data in a way that provide unique insight into how genomes function and evolve within the context of chromosomes. In cytomolecular mapping single-copy molecular markers are localized on synaptonemal complex (SC) spreads *via* fluorescence *in situ* hybridization. The chromosomal positions of the markers are then characterized with respect to each other and to cytological features such as centromeres, telomeres, heterochromatin, euchromatin, and chromomeres. Because the DNA markers used as probes in cytomolecular mapping are part of molecular maps (usually RFLP maps in plants), cytomolecular mapping allows molecular maps to be directly superimposed onto their corresponding chromosomes. In species where physical maps are being constructed (or in species in which genome sequencing is completed), cytomolecular mapping presumably can be used to position a complete (or nearly complete)

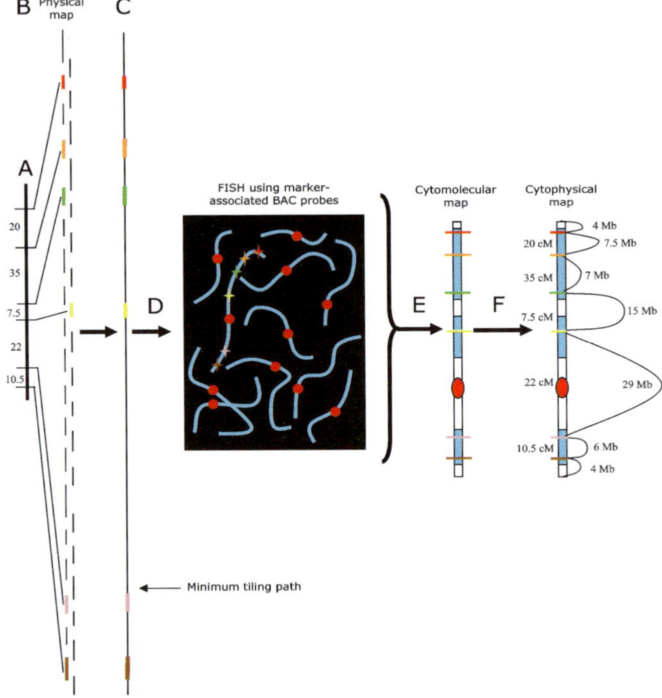

Fig. 1. Using cytomolecular markers to anchor the physical map for a particular linkage group onto the actual structure of its chromosome.(A) A molecular genetic map. Molecular markers from a particular linkage group are localized on BAC macroarrays.(B) A physical map is generated using physical mapping techniques such as BAC end sequencing, DNA fingerprinting, chromosome walking, and STS mapping. Marker-associated BAC clones (colored lines) are isolated.(C) Eventually, a minimum tiling path is assembled for the chromosome. (D) Insert DNA from marker-associated BAC clones is used in FISH to chromosome spreads in which individual chromosomes and heterochromatin/euchromatin can be differentiated. The precise location(s) of each locus is determined. (E) A cytomolecular map is constructed for the chromosome.In this example, heterochromatic regions are white, euchromatic regions are blue, and the kinetochoreis represented by a red circle. (F) The linkage map and physical map of the chromosome are superimposed directly onto the structure of the chromosome to produce a "cytophysical" map. Cytophysical mapping allows comparison of genetic linkage, chromatin configuration, and base pair distances. (Source : Draye *et al.*, 2001)

chromosomal DNA molecule directly onto its corresponding chromosome (*i.e.* cytophysical mapping) (Fig. 1).

The genome sequence data are not included on genetic, cytological or cytogenetic maps. DNA sequence maps are known as physical maps, and show the position of DNA motifs as the distance in base pairs between the two motifs in question. With many ongoing plant genome sequencing efforts, and the availability of whole genome sequences, it is now possible to include these data, creating a map that integrates cytogenetic data with DNA sequence information.

There are two methods to integrate genetic and cytological maps. The first one involves the localization of genetically mapped markers relative to chromosome breakpoints. The chromosomal rearrangements such as deficiencies, duplications, inversions and translocations contain breakpoints, and genes which are lost or gained on either side of a breakpoint are determined on the basis of gene order information from the genetic map (Harper and Cande, 2000). This ascertains the position of the breakpoint on the genetic map. This is then correlated with the physical position of the breakpoint as identified by the analysis of meiotic prophase or mitotic metaphase chromosomes. The second approach involves integration of the two maps by direct hybridization of genetically mapped sequences onto chromosomes by FISH.

2.0 GENETIC MAPPING

Genetic mapping, also known as linkage mapping or meiotic mapping, refers to the determination of the relative position and distances between markers along chromosomes, or the relative positions of genes on a DNA molecule and the distance between them (Semagn *et al.*, 2006a). It is based on the principle that genes (markers or loci) segregate through chromosomal recombination during meiosis (*i.e.* sexual reproduction), thus allowing their analysis in the progeny. Genetic map distances between two markers are defined as the mean number of recombination events in that region per meiosis. It indicates the position and relative genetic distances between markers along chromosomes (Paterson, 1996; Collard *et al.*, 2005). Genetic maps with considerable genome coverage are highly useful for many plant development and breeding programmes, and the importance of these maps is further implicit in their following functions:

a) They permit genetic analysis of qualitative and quantitative traits that enable localization of genes and quantitative trait loci (QTL), respectively (Doerge, 2002; Yim *et al.*, 2002; Long *et al.*, 2007; Torada *et al.*, 2008; Knoll *et al.*, 2008).

b) They facilitate the introgression of desirable genes or QTLs through marker-assisted selection.

c) They allow comparative mapping between different species so as to evaluate similarity between genes orders and function in the expression of a phenotype (Paterson *et al.*, 2000; Ohmido *et al.*, 2007).

d) They help create physical maps and assist in positional or map-based cloning of genes (Vuysteke *et al.*, 1999; Hass-Jacobus *et al.*, 2006; Xu *et al.*, 2008).

The pre-requisites for making a genetic map are: (a) an appropriate mapping population, (b) molecular markers for genotyping the mapping population, (c) screening of parents for marker polymorphism, and (d) linkage analysis to calculate recombination frequencies between markers, create linkage groups, estimate map distances, and determine map order.

2.1 Mapping Population

Two genetically divergent parents, which show clear genetic differences for one or more traits of interest and exhibit sufficient polymorphism, are selected to produce a mapping population. In self-pollinating species, mapping population consists of parents that are both highly homozygous (inbred) and progenies from the second filial generation (F_2), backcross (BC), recombinant inbred lines (RILs), double haploids (DHs), and near isogenic lines (NILs) (Burr *et al.*, 1988; He *et al.*, 2001; Doerge, 2002). However, in case of cross pollinating species the situation is quite complex as the two species do not tolerate inbreeding.

2.2 Molecular Markers

Molecular markers reveal sites of variation at the DNA sequence level. The most common DNA markers used for mapping in plants are restriction fragment length polymorphisms (RFLPs) (Grodzicker *et al.*, 1975), random amplified polymorphic DNAs (RAPDs) (Welsh and McClelland, 1990; Williams *et al.*, 1990), amplified fragment length polymorphisms (AFLPs) (Vos *et al.*, 1995), cleaved amplified polymorphic sequence (CAPS) (Konieczny and Ausubel, 1993), sequence characterized regions (SCARs) (Paran and Michelmore, 1993), sequence tag sites (STSs) (Olson *et al.*, 1989), diversity arrays technology (DArT) (Jaccoud *et al.*, 2001), single nucleotide polymorphism (SNP) (Jordan and Humphries, 1994) and microsatellites or simple sequence repeats (SSRs). Each marker system has its own advantages and disadvantages, and the factors involved in selecting one or more of these marker have been described by Semagn *et al.*, (2006b).The first large scale attempts aimed towards the generation of genetic maps involved the use of RFLP markers.

A molecular marker can be defined as a chromosomal landmark or allele that allows for the tracing of a specific region of DNA; alternatively it can also be referred to as a genomic region which is used to identify other traits. The information about the location of markers on the chromosome and the proximity of these markers to specific genes can be used to create genetic linkage map. Such genetic maps are used to analyze associations between economically important traits and genes or QTLs, which, in turn, may facilitate the introduction of desirable genes or QTLs through marker-assisted selection. Molecular markers usually do not have any biological effect, and they can be thought of as constant landmarks in the genome. They are identifiable DNA sequences, found at specific locations of the genome, and transmitted by the standard laws of inheritance from one generation to the next.

The various molecular markers can be classified into different groups on the basis of mode of transmission, mode of gene action (dominant or codominant markers) and method of analysis (hybridization-based or PCR-based).

2.2.1 Hybridisation-based markers

RFLP is the hybridization-based marker that exploits differences in restriction digestion pattern observed between two individuals when their DNA is digested with the same restriction enzyme.

2.2.2 PCR-based markers

Based on the types of primers used for amplification, PCR-based techniques are of two types:

1) Arbitrary or semi-arbitrary primed PCR techniques, which do not require prior sequence information (*e.g.*, AP-PCR, DAF, RAPD, AFLP, ISSR).

2) Site-targeted PCR techniques, which are developed from known DNA sequences (*e.g.*, EST, CAPS, SSR, SCAR, STS).

2.2.3 Nucleotide based marker : Single nucleotide polymorphisms (SNP)

SNP can be considered as 'third generation markers'. These are point mutations in which one nucleotide is subsituted for another at a particular locus. SNPs are the most common type of sequence differences between alleles, are codominant in nature, and represent an inexhaustible source of polymorphic markers for use in high-resolution genetic mapping of traits. Detection of the codominant SNPs is based on DNA assays can be carried out in plants, such as rice and maize, where genomics is either well advanced or is progressing at a rapid pace.

2.3 Screening of Parents for Marker Polymorphism

A population that exhibits sufficient number of different polymorphic markers between parents is chosen. Cross pollinating species show higher level of polymorphism as compared to inbreeding species. In case of dominant homozygous or recessive homozygous parents, the progeny inherits a marker from either of the parents; while in case of heterozygous parents, the progeny inherits a marker from both parents. The scoring methods between codominant and dominant markers, however, differ.

2.4 Linkage Analysis

Markers are allocated to different linkage groups on the basis of the odds ratio. This ratio is called logarithm of odds (LOD) value or LOD score, and refers to the ratio of the probability that two loci are linked with a given recombination value over a probability that the two are not linked (Risch, 1992; Stam, 1993a). The critical LOD scores used to establish linkage groups are called 'linklod' and the scores to calculate map distances are called 'maplod' (Stam, 1993b; Ortiz *et al.*, 2001). Marker pairs with a recombination LOD score above a critical 'linklod' are linked, whereas those with a LOD score less than 'linklod' are considered unlinked. A 'linklod' value of 3 is generally used as the minimum threshold value to decide whether or not loci are linked. A LOD value of 3 between two markers indicates that linkage is 1000 times more likely than no linkage (Stam, 1993a). After establishing linkage groups they are assigned to chromosomes. After the above information is gathered map distances and locus order are determined.

Linkage analysis and mapping are performed by specialized computer packages designed for these purposes. Some of the commonly used packages are MAPMAKER/EXP (Lander *et al.*, 1987; ftp://genome.wi.mit.edu/pub/mapmaker3/), GMendel (Echt *et al.*, 1992; http://gnome.agrenv.mcgill.ca/info/gmendel.htm), JoinMap (Stam and Van Ooijen, 1995; http://www.cpro.dlo.nl.cbw/), LINKAGE (Suiter *et al.*, 1983), and Map Manager QTX (Manly *et al.*, 2001).

3.0 CYTOLOGICAL MAPPING

Cytological mapping consists of localization of visible structures on stained chromosomes. The structures seen by chromosome staining include centromeres, NORs, knobs and chromomeres only, and this limited availability has discouraged the generation of cytological maps. However, modern techniques like FISH have proved to be very useful for cytological mapping.

3.1 *In situ* Hybridization

In situ hybridization (ISH), an assay of choice for localization of specific nucleic acid sequences in native context is a two decade old technique that has continuously been upgraded due to advancement in microscopy methods , simplification of probe recognition methods and automated data collection and analysis tools. The localization of DNA sequences to their respective chromosomal sites, is a powerful and useful tool for the molecular cytogeneticist and high sensitivity detection and simultaneous assay of many species by ISH and its modifications have made it the foremost biological assay. The technique was first described by Gall and Pardue (1969). They illustrated the use of this technique by hybridizing tritium-labeled rRNA to the extrachromosomal rDNA in the oocytes of *Xenopus* and since then the versatile technique has found diverse applications in detection of single-copy genes (Bhatt *et al.*, 1988), repetitive DNA, low-copy hybridization sites (Lawrence *et al.*, 1988; Spadoro *et al.*, 1990), transgene integration (Mouras *et al.*, 1987), chromosomal aberrations (Amiel *et al.*, 1996), etc. The technique has also been used to perform gene mapping studies and establish the relationship between genetic (recombinational) distances and physical distances, as significant differences occur between physical and genetic distances (Meagher *et al.*, 1988; Tel-Zur *et al.*, 2004).

Genomic *in situ* hybridization (GISH), established for plants by Schwarzacher *et al.* (1989) is a powerful tool to differentiate chromosomes of different parental genomes in allopolyploid species, interspecific hybrids, and their backcross progenies as well as to trace intergenomic chromosome rearrangements (Raina and Rani, 2001). However, GISH has found limited application in cases of plants with very small genomes, where the technique has failed to decorate entire chromosomes (Fahleson *et al.*, 1997; Snowdon *et al.*, 1997; Bennett 1995; Barre *et al.*, 1998; Raina *et al.*, 1998). Ali *et al.* (2004) used GISH with modifications to label entire mitotic and meiotic chromosomes of *Arabidopsis thaliana* ($2n = 10$) and closely related species with extremely small genomes. They used very high concentrations of genomic DNA probe and/or long hybridization times to achieve uniform labeling.

The shift from radioactive probes to fluorescence probes has been rapid. In the 1940s, antibodies were conjugated to fluorochromes without loss of their epitope binding specificity. The first antibody dependent fluorescent detection of nucleic acid hybrids was achieved in the 1950s (Rudkin and Stollar, 1977), but was soon replaced by nucleic acid probes. *In situ* hybridization initially involved probes labeled with radioisotopes, but their limitations lead to the development of fluorescent labeling. Radioactive probes did not have specific constant activity due to the decay of radioactivity, the resolution of radiography was also limited, long exposure time was required to cause impressions on the radiography film, and also radiolabeled probe was costly and hazardous and its use was governed by regulations. FISH, because of high sensitivity of detection, advanced resolution, speed and safety has taken over ISH as a tool in basic and applied research for simultaneous detection of multiple targets, quantitative analysis and live cell imaging.. The sensitivity of FISH depends on the size of probes, cytological target it is applied to, and on the state of chromatin condensation (Bowler *et al.*, 2004; Ohmido and Fukui, 2004; Kato *et al.*, 2005). FISH also allows direct observation of the behaviour of individual

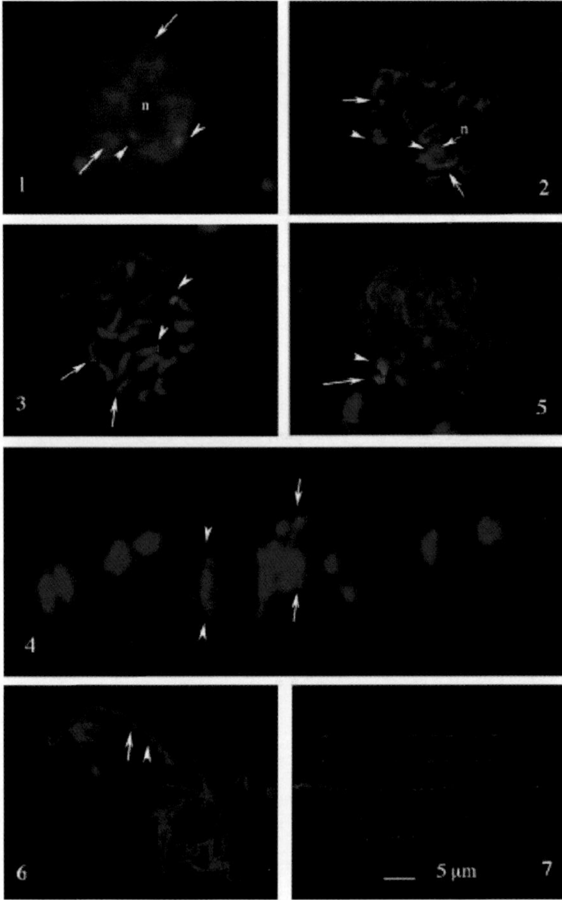

Fig. 2. The FISH results of 45S rDNA in rice The FISH signals of 45S rDNA as revealed on 1. interphase (n: nucleolus)? 2. mitotic prophase, 3. mitotic metaphase, 4. bivalent of meiotic metaphase I, 5. pre-pachytene chromosomes, 6. late pachytene chromosomes, 7. extended DNA fiber of *Indica* rice. (Arrowheads indicated the intense signals and arrows indicated the weak signals. (Source: Yun Li *et al.*, 2006)

chromosomes during interphase (Lichter *et al.*, 1988; Schwarzacher *et al.*, 1989, 1992). The technique has been applied to studies of early meiotic prophase from yeast to plants and mammals, thereby giving a comprehensive picture of chromosome behaviour immediately before and at meiotic prophase (Schwarzacher, 2003) (Fig. 2).

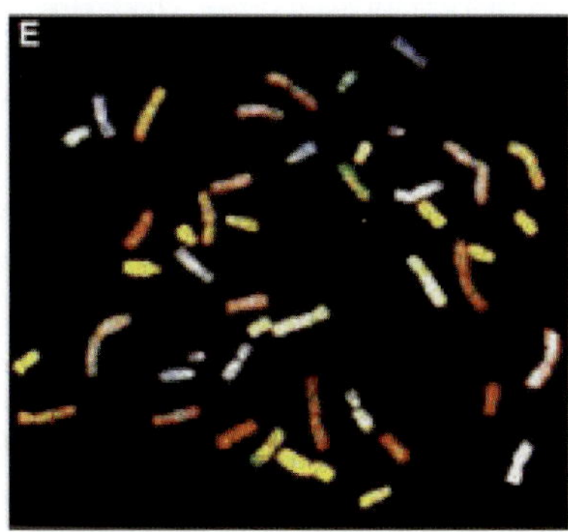

Fig. 3. Detection of 24 chromosomes using spectral imaging (Source: Macville *et al.*, 1997).

FISH has also been utilized successfully for chromosome mapping in many plant species. Efforts to integrate cytological and genetic maps have been made using FISH with repetitive and single-copy probes. Pedersen *et al.* (1995 to 1997) determined the cytological position of ten loci by FISH in barley. This allowed them to anchor parts of the genetic map onto the barley mitotic metaphase cytological map. Peterson *et al.* (1999) showed that FISH mapping can be used to determine unambiguous gene order.

The methodology involves the fixation and denaturation of the samples before hybridization to the DNA probe with a specific sequence. The cell wall and cytosol are removed, at least partially by mechanical or chemical (enzymatic) methods to permit the entry of the probe. The availability of many spectrally distinct chromophores, spanning the optical range of UV to near infrared and corresponding fluorescence microscopes that can precisely distinguish these dyes, help detect many DNA sequences in a single nucleus (Lam *et al.*, 2004) (Fig. 3). An extension of FISH technique, chromosome painting, helps decorate an entire chromosome. Chromosome specific probes of different colours are hybridized to target samples *i.e.* permeabilized fixed nuclei or chromosome spreads. Chromosome painting in mammals (Schrock, 1996), birds and insects has been used widely to characterize chromosome arrangements , but has been used in a limited manner in plants due to the abundance of dispersed repetitive sequences. It has been done in *Arabidopsis* (Lysak, 2003) taking advantage of the tremendous molecular, genetic and genomic information available for it;

unusual chromosomes like B chromosomes (Houen *et al.*, 2001) and in some hybrids, from a cross of two distinct parental species(Desel *et al.*, 2002) for characterizing the hybrid status of the progenies.

The problem of low resolution limit of 1Mb in metaphase chromosomes and 100 kb in interphase chromosomes (Mukai, 2004) has been circumvented by DNA fibre FISH method, which was first developed for mammalian system (Heng *et al.*, 1992)) and first applied to plant system by Fransz, *et al.* (1996). Isolated nuclei were treated with chemicals to break the membrane and release the fibres which spread on the slides and hybridized with suitable probes to generate the necessary information, the advantage being that the hybridization signals spread in a line on the DNA fibre giving one dimensional information which helps estimate the size of the target molecule (Yamamoto and Mukai, 2007). The resolution of FISH on extended DNA fibres has been reported to be 1 kb (Yamamoto and Mukai, 2005).

FISH is an invasive technique that cannot be applied to living cells (Lam *et al.*, 2004). Kanda (1998) first demonstrated in mammalian cells that stable labeling of chromatin and chromosomes can be done in live cells using Green Fluorescent Protein (GFP) as an *in vivo* tag by fusing it to H2B. In transgenic *Arabidopsis* plants, the expression of an H2B fusion protein with Yellow Fluorescent Protein , did not cause any developmental defects and such fusion could be used for noninvasive labeling of chromatin (Boisnard-Lorig *et al.*, 2001).

The popularity of assays based on complementarity and hybridization has increased significantly in the last decade and the new avenues of research made possible by these techniques demand continuous modification to make them widely applicable. In the present state also, they find use in very diverse areas of research, and some of their applications are cited below.

3.1.1 ISH for Transgene Localization

Genetic engineering has emerged as a vital component of several crop improvement programmes, and production of transgenic plants is now routine for many crop species. A number of techniques are now available to introduce foreign genes into the desired plant species. However, plant transgenic production is severely limited because of two factors: unpredictability of integration sites and lack of expression stability. Significant efforts have been made in order to understand the mechanisms of transgene integration in the host genome (Somers, 2004). A combination of sequencing, genetic approaches, and cell biology techniques has been used for characterization and visualization of transgene loci. A good amount of information has been gathered from the visualization of transgenes by

FISH, and the technique has made it possible to physically map transgene integration sites in the host genome. It is well-known that spatial organization of the gene in the nucleus has a strong impact on its expression; FISH analysis has proved to be significantly helpful in unraveling the relationship between higher order chromatin structure and transgene expression. It is a useful tool for analyzing the complex mechanisms of transgene integration in the host genome.

The first achievement of physically mapping transgenes using ISH was reported in *Crepis capillaris* (Ambros *et al.*, 1986*a*, 1986*b*) and in *Nicotiana tabacum* (Mouras *et al.*, 1987; Mouras and Negrutiu, 1989). Since then several workers have described the integration of transgene in the host genome using ISH. Moscone *et al.* (1996), used FISH and GISH in combination with DAPI counterstaining to determine the chromosome integration site and subgenomic allocation of a transgene insert in amphidiploid tobacco. The study proved to be highly useful in identifying the tobacco chromosome carrying the transgene, as well as in understanding how chromosomal location of the transgene influenced its expression. Leggett *et al.* (2000) used FISH to localize two transgenes (*gus* and *bar*) in the cultivated hexaploid oat *Avena sativa* L. cotransformed by microprojectile bombardment of embryogenic callus. The results of the study showed that in the two independent transformation events, the *gus* and *bar* genes had inserted in the same position relative to each other. The FISH analysis was very useful in discriminating between homozygous and heterozygous plants containing one or more inserted gene at the seedling stage. This analysis could be highly beneficial for breeding programmes that involve transgenics as time-consuming progeny testing would not be required in such cases. During the conventional ISH analysis either whole plasmid or linearized plasmid is used as probe. The size of the plasmid used as probe, and extent of labeling play important roles in transgene localization by ISH. Salvo-Garrido *et al.* (2001) designed an improved method for the detection of transgenes in barley. Instead of using the whole plasmid, 1 to 3 kb fragments of the plasmid were used as labeled probe for transgene localization. The fragmentation of the plasmid into smaller fragments led to its efficient labeling. The improved method gave optimal hybridization, minimal fluorescent background, and precise physical position of the transgene. Pedersen *et al.* (1997) used FISH to localize transgenes on metaphase chromosomes of barley, wheat, and triticale transformed by microprojectile bombardment. In all, a total of thirteen integration sites were detected in the nine lines analysed. The study was helpful in detecting even single-copy integrations. Dong *et al.* (2001) performed FISH analysis on nuclear chromosomal DNA at metaphase, interphase, meiotic prophase (pachytene) and on extended chromatin fibers (DNA fiber-FISH) and naked DNA molecules in rice to detect chromosomal location of the transgene. The study evaluated the relationship between local chromatin structure of the transgene integration site and its expression. Barro *et al.* (2003) analysed transgene integration sites and chromosome rearrangements in two transgenic lines of tritordeum (an amphiploid between *Triticum turgidum* cv. durum and *Hordeum chilense*). The study identified three integration sites and four translocations. The analysis of transgenic plants allowed the early detection of homozygous and heterozygous plants.

3.1.2 ISH for Localization of Low-copy or Single-Copy Genes

Genetic linkage maps are fundamental for the localization of genes involved in diverse functional aspects of the plant. Detection of single-copy genes on plant chromosomes has been difficult. However, ISH has been used for the localization of single-copy genes in many plant species. Bhatt *et al.* (1988) devised a rapid method for non isotopic *in situ* mapping of single copy genes directly on G-banded chromosomes. It is based on hybridizing biotinylated probes to metaphase chromosomes and involves regular light microscopy. The method is as sensitive as radioisotopic *in situ* hybridization, is faster and has much better resolution. It does not require sophisticated techniques such as reflection contrast, fluorescence or phase

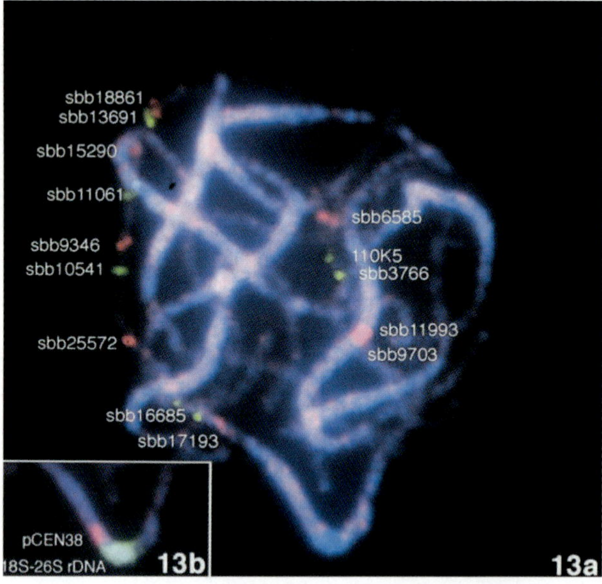

Fig. 4. FISH signals on sorghum chromosome *1* using a 14-BAC probe cocktail to pachytene chromosomes. The identity of each individual BAC used in the probe cocktail is given next to its representative FISH signal. (13b) The region around the centromere of 13a reprobed with CEN38 (centromere-associated probe) and 18S-28S ribosomal DNA. (Source: Islam Faridi, 2002)

microscopy as genes are visualized by regular light microscopy. Gustafson *et al.* (1990) conducted *in situ* hybridization to map low-copy DNA sequences in rye (*Secale cereale*). A 900-base-pair (bp) sequence from a cDNA clone of the rye endosperm-storage-protein gene *Sec-1* was hybridized to *Secale cereale* 'Blanco' chromosomes. Hybridization was seen on the short arm of chromosome 1R. The gene was genetically mapped on the same location previously. Though some cross-hybridization with *Sec-2* and *Sec-3* was also observed, but it was attributed to some homology with these genes and low-stringency conditions used during ISH. It was suggested that chromosome walking used in conjunction with ISH was an excellent tool for physical mapping of chromosomes. Fransz *et al.* (1996) used FISH to detect single-copy genes and chromosome rearrangements in *Petunia hybrida*. DNA sequences homologous to single-copy genes were labeled and used to detect DNA targets on metaphase chromosomes as small as 1.4 kb. Using a cDNA probe chalcone synthase gene (*chsA*) was mapped at the distal end of the short arm of chromosome 5 in the line V30. However, it was mapped to the long arm of chromosome 3 in the lines Mitchell and V26. The mapping of single-copy *chsA* gene on two different chromosomes in these lines suggested that a chromosomal rearrangement had taken place. Cabrera *et al.* (2002) used FISH to physically map the *Glu-1* loci controlling high-molecular weight (HMW) glutenin in common wheat and tritordeum (an amphiploid between *T. turgidum* cv. *durum* and *Hordeum chilense*). The probe used in the analysis was single-copy *Glu-D1-1d* gene coding the 1Dx5 HMW glutenin subunit. The application of FISH proved useful in studying homoeology among different genomes. Wang *et al.* (2006) undertook an ambitious study to construct a cytogenetic map of maize chromosome 9 using high resolution single-copy FISH. They developed a squash FISH procedure that successfully detected single-copy genes on maize (*Zea mays* L.) pachytene chromosomes. The method could easily resolve two sequences separated by approximately 100 kb; and a probe as small as 3.1 kb could be detected. The researchers localized nine genetically mapped single-copy genes on chromosome 9 in one FISH experiment. The study paved the way for integration of information from genetic maps and BAC contig-based physical map with cytological structure of chromosomes for the development of high-resolution cytogenetic maps (Fig. 4). Lamb *et al.* (2007) utilized single-copy gene detection by FISH for karyotyping in maize. They determined single-copy regions for each of the maize chromosome and used these regions to develop a small-target karyotyping system. A collection of probes that included one marker on each maize chromosome was assembled and used for chromosomal identification in maize and its relatives.

3.1.3 ISH for the Analysis of Satellite DNA

ISH has been successfully employed for the detection of repeat sequences or satellite DNA. It has been used to localize rDNA sequences, telomeres and other repeat sequences. Weiss and Scherthan (2002) used FISH to show that *Aloe* species contain vertebrate-like telomeric sequences rather than the canonical *Arabidopsis*-type $(TTTAGGG)_n$ telomere repeats. The study analysed the composition of *Aloe* telomeres by single-primer PCR and fluorescence *in situ* hybridization (FISH) with directly labeled *Arabidopsis*-type $(TTTAGGG)_{28-43}$ DNA probe, and with vertebrate-type $(TTAGGG)_{33-50}$ DNA and a $(C_3TA_2)_3$ peptide nucleic acid (PNA) probe. It was found that *Aloe* chromosomes did not give FISH signals with the *Arabidopsis*-type $(TTTAGGG)_{28-43}$ DNA probe, while *Nicotiana tabacum* contained *Arabidopsis*-type telomeric repeats. However, FISH with the highly specific vertebrate-type $(C_3TA_2)_3$ PNA (peptide nucleic acid) probe resulted in strong T_2AG_3-specific FISH signals at the ends of *Aloe* chromosomes. However, long $(TTAGGG)_{33-50}$ DNA probe highlighted *Aloe* chromosome ends, while this probe failed to reveal FISH signals on tobacco chromosomes. The results suggested that *Aloe* species have vertebrate-like telomere sequences; and plant telomeres, which were thought to lack GC-rich repeats, may in fact contain variant repeat types. Taking a cue from the work of Weiss and Scherthan (2002) where it was demonstrated that some plants of the order Asparagales did not have the 'typical' *Arabidopsis*-type $\{(TTTAGGG)_n\}$ telomere sequences, Sýkorová *et al.* (2003) determined which sequences were found in these species in place of the *Arabidopsis*-type telomeres. FISH analysis revealed that *Arabidopsis*-type telomeric sequence has been partly or fully replaced by the human-type telomeric sequence (TTAGGG)(n) in the members of Asparagales, and *Allium* species lack the human-type variant. The variants of minisatellite sequences of telomeric sequence of *Arabidopsis*, *Bombyx* (TTAGG)(n) and *Tetrahymena* (TTGGGG)(n) were also found to occur with a lower abundance in most cases. It was found that in these species the error rate of telomere synthesis was higher in comparison to the 'typical' plant telomere, and *Asparagales clade* is unified by a mutation that resulted in a switch from synthesis of *Arabidopsis*-like telomeres to human-like telomeres. Dechyeva *et al.* (2003) analysed several repetitive sequences of the *Beta procumbens* genome. Comparative FISH revealed remarkable differences in the chromosomal position of repeat sequences between *B. procumbens* and *B. vulgaris*, suggesting that these sequences were involved in the expansion or rearrangement of the intercalary *B. vulgaris* heterochromatin. It was also shown that these repetitive elements could be used as DNA probes to discriminate parental genomes in interspecific hybrids. Herrán *et al.* (2005) analysed the controversial molecular structure of telomeres of six species of Asparagales and two species of Liliales using Southern blot and FISH. The major findings of the study were: 1) It was found that two Liliales species displayed hybridization signals with oligonucleotides corresponding to telomere repeats of both plants and vertebrates; 2) Four Asparagales species lacked both the plant telomere repeats and the vertebrate telomere repeats; and 3) Two Asparagales species displayed positive hybridization with the vertebrate telomere repeats but not with the plant telomere repeats. Abd El-Twab and Kondo (2006) used FISH to map 5S and 45S rDNA, and *Arabidopsis*-type telomere sequence repeats in *Chrysanthemum zawadskii*. The analysis revealed 5-6 interstitial and 2 terminal signals at the 5S rDNA region, 10-14 terminal and 4-5 interstitial signals at the 45S rDNA region, and the telomeric signals at the chromosome ends. One important finding of the study was co-localization of the 5S and 45S rDNA regions on 2 terminal and 5 interstitial loci. This indicates natural hybridization with closely related genera and/or genome rearrangement. Dechyeva and Schmidt (2006) studied the molecular organization of terminal repeats of *Beta* species. The high-resolution FISH analysis revealed physical organization of *B. vulgaris* chromosome ends with telomeric DNA and subtelomeric satellites extending over 63 kb and 125 kb, respectively.

3.1.4 ISH for the Analysis of Chromosomal Aberrations

ISH has been extensively used for the detection and analysis of chromosomal aberrations both in plants and animals. Jovtchev *et al.* (2002) used telomere- and centromere-specific probes to analyse the cytogenetic effects of N-methyl-N-nitrosourea (MNU) on root tip meristem cells of barley. Chromatid aberrations (CA) and micronuclei (MN) were detected by FISH using the repetitive DNA probes. FISH also helped elucidate the origin of MN and discriminate between aneugenic and clastogenic effects. Maluszynska *et al.* (2003) used FISH to detect chromosomal aberrations in *Crepis capillaris*, a highly useful system for the assay of chromosome aberrations after mutagenic treatment. It has a simple karyotype with three pairs of distinct and large chromosomes. The genotoxicity of chemicals and environmental pollutants was assessed on the basis of frequency of structural chromosome aberrations and micronuclei in root meristem cells. FISH allowed the localization of chromosomal rearrangements not only in mitotic but also in interphase nuclei. The study used rDNA and telomeric sequences as probes to detect the aberrations in *C. capillaris* using FISH. Juchimiuk *et al.* (2007) used multicolour FISH to compare the cytogenetic effects of 2 chemical mutagens, N-nitroso-N-methylurea (MNU) and maleic hydrazide (MH), on root tip meristem cells of barley. The study involved the simultaneous use of 5S rDNA, 25S rDNA, telomeric and centromeric DNA sequences as probes to conduct FISH. The micronucleus (MN) test combined with FISH allowed the analysis of involvement of chromosome fragments in micronuclei formation. This facilitated the explanation of possible origin of mutagen-induced micronuclei. The frequency of occurrence of MN was different between the two mutagens as revealed by FISH. The study also made an attempt to identify the origin of chromosome fragments in mitotic anaphase.

3.1.5 Miscellaneous uses of ISH

Besides being used for transgene localization, low-copy or single-copy gene localization, satellite DNA analysis and detection of chromosomal aberrations, ISH has been used for diverse other purposes. Leitch *et al.* (1990) utilized GISH to distinguish the chromatin from the two parental genomes in the grass hybrid *Hordeum chilense* and *Secale africanum*. The analysis showed that chromatin of the two parental genomes did not intermix but occupied distinct domains. Fu *et al.* (2004) conducted GISH analysis of intergeneric somatic hybrids of *Citrus aurantium* L. and trifoliate orange *Poncirus trifoliata* (L.) Raf. Ploidy analysis verified that the hybrids were tetraploids with 2n=4 x=36. GISH analysis confirmed that 18 chromosomes came from trifoliate orange and the remaining 18 from *C. aurantium*. The study discussed the potential of GISH in *Citrus* somatic hybrid analysis. Tel-zur *et al.* (2004) used ISH for the study of genetic relationship among four diploid *Hylocereus* species. FISH was used to confirm the ploidy level of hybrids, and GISH analysis was helpful in the analysis of sequence composition of parents and the hybrids. Hua *et al.* (2006) utilized GISH to study parental genome separation in *Brassica carinata* and *Orychophragmus violaceus* hybrids. The key findings of the study were: 1) The morphological features of F_1 plants were similar to *B. carinata*. 2) GISH analysis revealed that *O. violaceus* genome complement was lost in the hybrid, and only intact *B. carinata* geneome was present in the hybrid between the two species. Cai *et al.* (2007) used GISH to compare the composition and methylation-variation of the nuclear and cytoplasmic genomes of somatic hybrids between *Festuca arundinacea* Schreb. and *Triticum aestivum* L. The results indicated that the hybrids contained introgressed nuclear and cytoplasmic DNA and methylation variations compared to both parents.

4.0 CONCLUSION

The two requirements for constructing a genetic map with good genome coverage and precise locus order are: large number of DNA markers, and analyses of large numbers of individuals. The genetic maps will provide DNA diagnostic tools that can be used to predict plant characteristics such as disease resistance, productivity, or quality, and this information can be integrated into classical plant breeding programmes aimed at development of new varieties. The recent years have seen several improvements in the methods of detection of markers, development of high-throughput techniques, and establishment of new kinds of markers. Multiplex PCR (which enables multiple marker loci to be tested simultaneously) and SNPs have significantly helped genetic mapping endeavours. Several factors need to be considered when choosing a molecular marker like availability of a marker system, time required to construct a map using a particular marker, level of polymorphism expected, reproducibility, transferability, size and structure of the population to be studied, marker inheritance, etc.

Genetic maps based on DNA markers are available for several economically important crops like rice, maize, wheat, barley, sunflower, potato, pea, bean, rye, millet, cotton, soybean, sorghum, cowpea, sunflower, alfalfa, coffee, sugarcane, etc. It is only with the availability of physical maps that the order of markers along chromosomes and their actual length in kilobase (kb) pairs could be precisely assigned. The ultimate physical map is the whole genome DNA sequence. Such a physical map is available for a few species such as *Arabidopsis thaliana* and rice. The availability of such physical maps covering the entire genome sequence would be very useful in better designing crop improvement programmes. The knowledge of gene sequences and how genes function will lead to development of stress-tolerant and high-yielding crops. In addition to providing lots of new genes to study, the plant genome could also provide clues towards the working of genetically modified crops and how they interact with other plants in the environment.

The advent of ISH for the construction of cytological maps has revolutionized the art and science of mapping. Both FISH and GISH have been used for diverse applications like gene mapping, transgene localization, satellite DNA localization, detection of chromosomal aberrations, etc. The technique has been refined since its first use in 1969 (Gall and Pardue, 1969). Several new practices have been incorporated in the traditional methodology of using a probe, labeling it, hybridizing it with the target DNA and then visualizing the signals. Player *et al.* (2001) developed a branched DNA *in situ* hybridization *(bDNA ISH)*. The technique is highly sensitive and can be used to detect low-copy or single-copy genes effectively. bDNA ISH, originally developed to detect human papillomavirus (HPV) DNA in whole cells, can be easily adopted to work with plant cells. It can be utilized to detect viral DNA in plant cells, and thereby serve as an efficient diagnostic method. Smolina *et al.* (2007) proposed the use of peptide nucleic acid-assisted rolling-circle amplification integrated with FISH to detect low-copy-number genomic DNA sequences in individual bacterial cells. A few alternatives to FISH have also been developed to overcome the problems associated with probe designing and time required to conduct FISH analyses. Kubaláková *et al.* (2001) developed primed *in situ* labeling (PRINS), which was found useful to detect high-copy tandem repeats on plant chromosomes. The study also developed a more sensitive variant of PRINS called cycling PRINS (C-PRINS) that has been used successfully to detect low copy repeats.

Cytomolecular mapping is done by relating molecular maps to chromosome structure as in tomato by Peterson *et al.* (1999). In species where physical maps are under elucidation, cytomolecular mapping can be used to position a complete chromosomal DNA molecule directly onto its corresponding SC and in such a case, each cytomolecular marker would serve as a point at which the DNA molecule would be "anchored" onto the framework of the chromosome (Draye, 2001)). Cytomolecular research can help identify genomic regions and features related to marked deviations from the mean relationship between recombinational and physical distance, compare the locations of kinetochores, knobs, nucleolus organizers, and other cytological features of different grasses such as maize and sorghum at unparalleled levels of resolution, and investigate the basis for unexpected levels of variation in diversity maps. A reference karyotype was generated for loblolly pine based on FISH using 18s-28s rDNA, 5s rDNA and *Arabidopsis*-type telomere repeat sequence (A-type TRS) by Islam-Faridi *et al.* (2007). The position and relative strength of the rDNA and telomeric sites made it possible to identify all chromosomes, providing a reference karyotype for use in comparative genome analysis and a cytomolecular map was also developed.

The knowledge gained from genetic, physical and cytogenetic maps would ultimately lead to the elucidation of plant biodiversity at the genic and genomic level, and would be useful in gaining a better understanding of the 200-million year history of flowering plant diversification from a common ancestor, and molecular-level events that contributed to the ability of flowering plants to colonize much of the earth.

5.0 REFERENCES

Abd El-Twab, M.H. and Kondo, K. 2006. FISH physical mapping of 5S, 45S and Arabidopsis-type telomere sequence repeats in *Chrysanthemum zawadskii* showing intra-chromosomal variation and complexity in nature. Chromosome Botany, 1(1):1-5.

Ali, H.B.M., Lysak, M.A. and Schubert, I. 2004. Genomic *in situ* hybridization in plants with small genomes is feasible and elucidates the chromosomal parentage in interspecific Arabidopsis hybrids. Genome, 47(5):954-960.

Ambros, P.F., Matzke, M.A. and Matzke, A.J.M. 1986a. Detection of a 17kb unique sequence (T-DNA) in plant chromosomes by *in situ* hybridization. Chromosoma, 94 (1):11–18.

Ambros, P.F., Matzke, M.A. and Matzke, A.J.M. 1986b. Localization of *Agrobacterium rhizogenes* T-DNA in plant chromosomes by *in situ* hybridization. EMBO Journal, 5 (9): 2073–2077.

Amiel, A., Elis, A., Manor, Y., Tangi, I., Fejgin, M. and Lishner. M. 1996. Fluorescence in situ hybridization (FISH) for the detection of trisomy 8 in acute myeloblastic leukemia. Leuk Lymphoma, 23:603-607.

Arrighi, F.E. and Hsu, T.C. 1971. Localization of heterochromatin in human chromosomes. Cytogenetics, 10(2): 81-86.

Barre, P., Layssac, M., D'Hont, A., Louarn, J., Charrier, A.,Hamon, S. and Noirot, M. 1998. Relationship between paternal chromosomic contribution and nuclear DNA content in the coffee interspecific hybrid *C. pseudozanguebariae* × *C. liberica* var *'dewevrei'*. Theoretical and Applied Genetics, 96(2):301–305.

Barro, F., Martín, A. and Cabrera, A. 2003. Transgene integration and chromosome alterations in two transgenic lines of tritordeum. Chromosome Research, 11(6):565-572.

Bennett, M.D. 1995. The development and use of genomic *in situ* hybridization (GISH) as a new tool in plant biosystematics. *In*: Brandham, P.E. and Bennett M.D. (*eds*) Kew Chromosome Conference IV. Royal Botanic Gardens, Kew, pp. 167–183.

Bhatt, B., Burns, J., Flannery, D. and McGee, J.O.D. 1988. Direct visualization of single copy genes on banded metaphase chromosomes by nonisotopic *in situ* hybridization. Nucleic Acids Research, 16 (9):3951-3961.

Boisnard-Lorig, C., Colon-Carmona, A., Bauch, M., Hodge, S., Doerner, P., *et al.* 2001. Dynamic analysis of the expression of HISTONE::YFP fusion protein in Arabidopsis show that syncytial endosperm is divided in mitotic domains. Plant Cell, 13:495-509.

Bowler, C., Benvenuto, G., Laflamme, P., Molino, D., Probst, A.V., Tariq, M. and Paszkowski, J. 2004. Chromatin techniques for plant cells. Plant Journal, 39 (5):776–789.

Bridges, C.B. 1916. Nondisjunction as proof of the chromosome theory of heredity. Genetics, 1(1–52):107–163.

Burr, B., Burr, F.A., Thompson, K.H., Albertson, M.C and Stuber, C.W. 1988. Gene mapping with recombinant inbreds in maize. Genetics, 118(3):519-526.

Cabrera, A., Martin, A. and Barro, F. 2002. *In-situ* comparative mapping (ISCM) of Glu-1 loci in Triticum and Hordeum. Chromosome Research, 10(1):49-54.

Cai, Y., Xiang, F., Zhi, D., Liu, H. and Xia, G. 2007. Genotyping of somatic hybrids between *Festuca arundinacea* Schreb. and *Triticum aestivum* L. Plant Cell Reports, 26(10):1809-1819.

Caspersson, T., Zech, L. and Johansson, C. 1970. Differential banding of alkylating fluorochromes in human chromosomes. Experimental Cell Research 60(3):315-319.

Collard, B.C.Y., Jahufer, M.Z.Z., Brouwer, J.B. and Pang, E.C.K. 2005. An introduction to markers, quantitative trait loci (QTL) mapping and marker-assisted selection for crop improvement: The basic concepts. Euphytica 142(1/2):169-196.

Dechyeva, D. and Schmidt, T. 2006. Molecular organization of terminal repetitive DNA in Beta species. Chromosome Research, 14(8):881-897.

Dechyeva, D., Gindullis, F. and Schmidt, T. 2003. Divergence of satellite DNA and interspersion of dispersed repeats in the genome of the wild beet *Beta procumbens*. Chromosome Research, 11(1):3-21.

Desel C, Jansen R, Dedong G and Schimdt T 2002. painting of parental chromatids in Beta hybrids by multicolor fluorescent *in situ* hybridization Annals of Botany. 89:171-181.

Doerge, R. 2002. Mapping and analysis of quantitative trait loci in experimental populations. Nature Reviews Genetics, 3(1): 43-52.

Dong, J., Kharb, P., Cervera, M., and Hall, T.C. 2001. The use of FISH in chromosomal localization of transgenes in rice. Methods in Cell Science, 23 (1-3):105-113.

Draye X, Lin Y-R, Qian X-y, Bowers, J.E., Burow, G.B., Morrell, P.L., Peterson, D.G., Presting, G.G., Ren S-x, Wing, R.A. and Paterson, A.H. (2001) Toward integration of comparative genetic, physical, diversity, and cytomolecular maps for grasses, using the Sorghum genome as a foundation. Plant Physiology 125:1325-1341.

Dutrillaux, B. and Lejeune, J. 1971. Sur une novelle technique d'analyse du caryotype human. C R Acad Sci Hebd Seances Acad Sci D, 272(20):2638-2640.

Echt, C., Knapp, S., and Liu, B.H. 1992. Genome mapping with non-inbred crosses using Gmendel 2.0. Maize Genetics Cooperation Newsletter, 66:27-29.

Fahleson, J., Lagercrantz, U., Mouras, A. and Glimelius, K. 1997. Characterization of somatic hybrids between *Brassica napus* and *Eruca sativa* using species-specific sequences and genomic *in situ* hybridization. Plant Science, 123(1-2): 133–142.

Fransz, P.F., Alonso-Blanco, C., Liharska, T.B., Peeters, A.J.M., Zabel, P. and Jong, J.H. de (1996) High-resolution physical mapping in *Arabidopsis thaliana* and tomato by fluorescence *in situ* hybridization to extended DNA fibres.The Plant Journal, 9:421-430.

Fransz, P.F., Stam, M., Montijn, B., Hoopen, R.T., Wiegant, J., Kooter, J.M., Oud, O. and Nanninga, N. 1996. Detection of single-copy genes and chromosome rearrangements in *Petunia hybrida* by fluorescence *in situ* hybridization. The Plant Journal, 9(5):767-774.

Fu, C.H., Chen, C.L., Guo, W.W. and Deng, X.X. 2004. GISH, AFLP and PCR-RFLP analysis of an intergeneric somatic hybrid combining Goutou sour orange and *Poncirus trifoliata*. Plant Cell Reports, 23(6):391-396.

Gall, J.G. and Pardue, M.L. 1969. Formation and detection of RNA-DNA hybrid molecules in cytological preparations. Proceedings of the National Academy of Sciences, USA, 63 (2):378-383.

Grodzicker, T., Williams, J., Sharp, P. and Sambrook , J. 1975. Physical mapping of temperature sensitive mutants of adenovirus. Cold Spring Harbor Symposia on Quantitative Biology, 39: 439-446.

Gustafson, J.P., Butler, E. and McIntyre, C.L. 1990. Physical mapping of a low-copy DNA sequence in rye (*Secale cereale* L.). Proceedings of the National Academy of Sciences, USA, 87(5):1899-1902.

Haldane, J.B.S. 1919. The combination of linkage values, and the calculation of distances between the loci of linked factors. Journal of Genetics, 8:299–309.

Harper, L.C. and Cande, W.Z. 2000. Mapping a new frontier; development of integrated cytogenetic maps in plants. Functional and Integrative Genomics, 1(2):89–98.

Hass-Jacobus, B.L., Futrell-Griggs, M., Abernathy, B., Westerman, R., Goicoechea, J.L., Stein, J., Klein, P., Hurwitz, B., Zhou, B., Rakhshan, F., Sanyal, A., Gill, N., Lin, J.Y., Walling, J.G., Luo, M.Z., Ammiraju, J.S., Kudrna, D., Kim, H.R., Ware, D., Wing, R.A., San Miguel, P. and Jackson, S.A. 2006. Integration of hybridization-based markers (overgos) into physical maps for comparative and evolutionary explorations in the genus Oryza and in Sorghum. BMC Genomics, 7:199.

He, P., Li, J.Z., Zheng, X.W., Shen, L.S., Lu, C.F., Chen, Y. and Zhu, L.H. 2001. Comparison of molecular linkage maps and agronomic trait loci between

DH and RIL populations derived from the same rice cross. Crop Science, 41(4):1240-1246.

Heng, H.H.Q., Squire, J., and Tsui, L.-C. (1992) High resolution mapping of mammalian genes by *in situ* hybridization to free chromatin. Proceedings of the National Academy of Sciences, USA, 89:9509-9513.

Herrán, DeLa R., Cuñado, N., Navajas-Pérez, R., Santos, J.L., Rejón, C.R., Garrido-Ramos, M.A. and Ruiz Rejón, M. 2005. The controversial telomeres of lily plants. Cytogenetics and Genome Research, 109(1-3):144-147.

Houen, A., Field, B.L. and Saunders, V.A. 2001 Microdissection and chromosome painting of plant B chromosomes Methods Cell science 23:115-124.

Hua, Y.W., Liu, M. and Li, Z.Y. 2006. Parental genome separation and elimination of cells and chromosomes revealed by AFLP and GISH analyses in a Brassica *carinata* x *Orychophragmus violaceus* cross. Annals of Botany, 97(6):993-998.

Islam-Faridi, M.N., Childs, K.L., Klein, P.E., Hodnett, G., Menz, M.A., Klein,R.R., Rooney, W.L., Mullet, J. E. , Stelly D. M., and Price H. J. 2002. A molecular cytogenetic map of sorghum chromosome 1:fluorescence *in situ* hybridization analysis with mapped bacterial artificial chromosomes. Genetics, 161:345–353.

Islam-Faridi, M. Nurul; Nelson, C. Dana and Kubisiak, Thomas L. 2007 Reference karyotype and cytomolecular map for loblolly pine (*Pinus taeda* L.) Genome. 50: 241-251

Jaccoud, D., Peng, K., Feinstein, D. and Kilian, A. 2001. Diversity arrays: a solid state technology for sequence information independent genotyping. Nucleic Acids Research, 29(4): e25.

Jordan, S.A. and Humphries, P. 1994. Single nucleotide polymorphism in exon 2 of the BCP gene on 7q31-q35. Human Molecular Genetics, 3(10):1915.

Jovtchev, G., Stergios, M. and Schubert, I. 2002. A comparison of N-methyl-N-nitrosourea-induced chromatid aberrations and micronuclei in barley meristems using FISH techniques. Mutation Research, 517(1-2):47-51.

Juchimiuk, J., Hering, B. and Maluszynska, J. 2007. Multicolour FISH in an analysis of chromosome aberrations induced by N-nitroso-N-methylurea and maleic hydrazide in barley cells. Journal of Applied Genetics, 48(2):99-106.

Kanda, T., Sullivan, K.F., Wahl, G.M. 1998. Histone-GFP fusion protein enables sensitive analysis of chromosome dynamics in living mammalian cells. Current Biology, 8:377-385.

Kato, A., Vega, J.M., Han, F., Lamb, J.C. and Birchler, J.A. 2005. Advances in plant chromosome identification and cytogenetics techniques. Current Opinion in Plant Biology, 8(2):148–154.

Knoll, J., Gunaratna, N. and Ejeta, G. 2008. QTL analysis of early-season cold tolerance in sorghum. Theoretical and Applied Genetics, 116(4): 577-587.

Konieczny, A. and Ausubel, F.M. 1993. A procedure for mapping Arabidopsis mutations using co-dominant ecotype-specific PCR-based markers. Plant Journal, 4(2):403-410.

Kubaláková, M., Vrána, J., Cíhalíková, J., Lysák, M.A. and Dolezel, J. 2001. Localisation of DNA sequences on plant chromosomes using PRINS and C-PRINS. Methods in Cell Science, 23(1-3):71-82.

Lam Eric , Kato, N. and Watanabe, K. 2004. Visualizing chromosome structure/ organization Annual Review of Plant Biology, 55:537-554

Lander, E.S., Green, P., Abrahamson, J., Barlow, A., Daly, M.J., Lincoln, S.E. and Newburg, L. 1987. MAPMAKER: an interactive computer package for constructing primary genetic linkage maps of experimental and natural populations. Genomics, 1(2):174–181.

Lamb, J.C., Danilova, T., Bauer, M.J., Meyer, J.M., Holland, J.J., Jensen, M.D. and Birchler, J.A. 2007. Single-gene detection and karyotyping using small-target fluorescence *in situ* hybridization on maize somatic chromosomes. Genetics, 175(3):1047–1058.

Lawrence, J.B., Villnave, C.A. and Singer, R.H. 1988. Sensitive, high resolution chromatin and chromosome mapping *in situ*: Presence and orientation of two closely integrated copies of EBU in a lymphoma line. Cell, 52(1):51-61.

Leggett, J.M., Sophie, J.P., Harper, J. and Morris, P. 2000. Chromosomal localization of cotransformed transgenes in the hexaploid cultivated oat *Avena sativa* L. using fluorescence in situ hybridization. Heredity, 84(1):46-53.

Leitch, A.R., Mosgöller, W., Schwarzacher, T., Bennett, M.D. and Heslop-Harrison, J.S. 1990. Genomic *in situ* hybridization to sectioned nuclei shows chromosome domains in grass hybrids. Journal of Cell Science, 95(3): 335-341.

Lichter, P., Boyle, A.L., Cremer, T. and Ward, D.C. 1991. Analysis of genes and chromosomes by non-isotopic *in situ* hybridization. Genetic Analysis of Technical Application, 8:24–35.

Lichter, P., Cremer, T., Borden, J., Manuelidis, L. and Ward, D.C. 1988. Delineation of individual human chromosomes in metaphase and interphase cells by in situ suppression hybridization using recombinant DNA libraries. Human Genetics, 80(3): 224–234.

Long, Y., Shi, J., Qiu, D., Li, R., Zhang, C., Wang, J., Hou, J., Zhao, J., Shi, L., Park, B.S., Choi, S.R., Lim, Y.P. and Meng, J. 2007. Flowering time quantitative trait Loci analysis of oilseed brassica in multiple environments and genomewide alignment with *Arabidopsis*. Genetics, 177(4):2433-2444.

Lysak, M.A. Pecinka, A. and Schubert, I. 2003. Recent progress in chromosome painting of Arabidopsis and related species Chromosome research 11:195-204.

Macas, J., Navratilova, A., Kubalakova, M. and Dolezel, J. 2006. PRINS on Plant Chromosomes. In: Pellestor, F. (ed.) - PRINS and In Situ PCR Protocols. Second Edition, Pp. 133-139. Humana Press, Totowa, New Jersey. Macville, M., Veldman, T., Padilla-Nash, H., Wangsa, D., O'Brien, P., Schrock, E. and Ried, T. 1997. Spectral karyotyping, a 24-colour FISH technique for the identification of chromosomal rearrangements. Histochem. Cell Biology, 108,299 -305.

Maluszynska, J., Juchimiuk, J. and Wolny, E. 2003. Chromosomal aberrations in *Crepis capillaris* cells detected by FISH. Folia Histochemica et Cytobiologica, 41(2):101-104.

Manly, K.F., Cudmore, Jr R.H. and Meer, J.M. 2001. Map Manager QTX, crossplatform software for genetic mapping. Mammalian Genome, 12(12):930–932.

Meagher, R.B., McLean, M.D. and Arnold, J. 1988. Recombination within a subclass of restriction fragment length polymorphisms may help link classical and molecular genetics. Genetics, 120(3):809-818.

Moscone, E.A., Matzke, M.A. and Matzke, A.J.M. 1996. The use of combined FISH/GISH in conjunction with DAPI counterstaining to identify chromosomes containing transgene inserts in amphidiploid tobacco. Chromosoma, 105(4): 231-236.

Mouras, A. and Negrutiu, I. 1989. Localization of the T-DNA on marker chromosomes in transformed tobacco cells by *in situ* hybridization. Theoretical and Applied Genetics, 78(5):715–720.

Mouras, A., Saul, M.W., Essad, S., and Potrykus, I. 1987. Localization by *in situ* hybridization of a low copy chimaeric resistance gene introduced into plants by direct gene transfer. Molecular and General Genetics, 207(2-3):204–209.

Mukai, Y. 2004. Fluorescence *in situ* hybridization *In*: Goodman RM (*ed*) Encyclopedia of Plant and Crop science pp. 467-471.

Ohmido, N. and Fukui, K. 2004. Recent advances in FISH analysis of plant chromosomes. Recent Research and Development in Biochemistry, 5(2):267–279.

Ohmido, N., Sato, S., Tabata, S. and Fukui, K. 2007. Chromosome maps of legumes. Chromosome Research, 15(1): 97-103.

Olson, M., Hood, L., Cantor, C. and Botstein, D. 1989. A common language for physical mapping of the human genome. Science, 245(4925):1434-1435.

Ortiz, J.P.A., Pessino, S.C., Bhat, V., Hayward, M.D. and Quarin, C.L. 2001. A genetic linkage map of diploid *Paspalum notatum*. Crop Science, 41(3):823-830.

Paran, I. and Michelmore, R.W. 1993. Development of reliable PCR-based markers linked to downy mildew resistance genes in lettuce. Theoretical and Applied Genetics, 85(8):985-993.

Paterson, A.H. 1996. Making genetic maps. *In*: Paterson, A.H. (*ed*) Genome mapping in plants. San Diego, California: Academic Press, Austin, Texas, pp. 23–39.

Paterson, A.H., Bowers, J.E., Burow, M.D., Draye, X., Elsik, C.G., Jiang, C.X., Katsar, C.S., Lan, T.H., Lin, Y.R., Ming, R. and Wright, R.J. 2000. Comparative genomics of plant chromosomes. Plant Cell, 12(9):1523-1539.

Pedersen, C., Zimny, J., Becker, D., Jähne-Gärtner, A. and Lörz, H. 1997. Localization of introduced genes on the chromosomes of transgenic barley, wheat and triticale by fluorescence *in situ* hybridization. Theoretical and Applied Genetics, 94(6-7):749-757.

Peterson, D.G., Lapitan, N.L.V., and Stack, S.M. 1999. Localization of single- and low-copy sequences on tomato synaptonemal complex spreads using fluorescence *in situ* hybridization (FISH). Genetics 152:427-439.

Player, A.N., Shen, L., Kenny, D., Antao, V.P. and Kolberg, J.A. 2001. Single-copy gene detection using branched DNA (bDNA) *in situ* hybridization. Journal of Histochemistry and Cytochemistry, 49(5):603-612.

Raina, S.N., Mukai, Y. and Yamamoto, M. 1998. *In situ* hybridization identifies the progenitor species of *Coffea arabica* (Rubiaceae). Theoretical and Applied Genetics, 97(8):1204-1209.

Raina, S.N. and Rani, V. 2001. GISH technology in plant genome research. Methods in Cell Science, 23(1-3):83-104.

Risch, N. 1992. Genetic linkage: Interpreting LOD scores. Science, 255(5046):803-804.

Rudkin, G.T. and Stollar, B.D. 1977. High resolution detection of DNA RNA hybrids *in situ* byb direct immunofluorescence Nature 265:472-473.

Salvo-Garrido, H., Travella, S., Schwarzacher, T., Harwood, W.A. and Snape, J.W. 2001. An efficient method for the physical mapping of transgenes in barley using *in situ* hybridization. Genome, 44(1):104-110.

Schrock E. du Manoir, S., Veldman, T., Schoell, B., Weinberg, J., *et al.* 1996. Multicolor spectral karyotyping of human chromosomes Methods in Molecular Biology, 28:167-176.

Schwarzacher, T. 2003. Meisosis, recombination and chromosomes: a review of gene isolation and fluorescent *in situ* hybridization data in plants. Journal of Experimental Botany, 54(380):11-23.

Schwarzacher, T., Anamthawat-Jensson, K., Harrison, G.E., Islam, A.K.M.R., Jia, J.Z., King, I.P., Leitch, A.R., Miller, T.E., Reader, S.M., Rogers, W.J., Shi, M. and Heslop-Harrison, J.S. 1992. Genomic *in situ* hybridization to identify alien chromosomes and chromosome segments in wheat. Theoretical and Applied Genetics, 84 (7-8):778–786.

Schwarzacher, T., Leitch, A.R., Bennett, M.D. and Heslop-Harrison, J.S. 1989. *In situ* localization of parental genomes in a wide hybrid. Annals of Botany, 64(3): 315–324.

Seabright, M. 1971. A rapid banding technique for human chromosomes. Lancet, 2(7731):971-972.

Semagn, K., Bjørnstad, Å. and Ndjiondjop, M.N. 2006a. Principles, requirements and prospects of genetic mapping in plants. African Journal of Biotechnology, 5(25):2569-2587.

Semagn, K., Bjørnstad, Å. and Ndjiondjop, M.N. 2006b. An overview of molecular marker methods for plant. African Journal of Biotechnology, 5(25):2540-2569.

Smolina, I., Lee, C. and Frank-Kamenetskii, M. 2007. Detection of low-copy-number genomic DNA sequences in individual bacterial cells by using peptide nucleic acid-assisted rolling-circle amplification and fluorescence *in situ* hybridization. Applied and Environmental Microbiology, 73(7):2324-2328.

Snowdon, R.J., Köhler, W., Friedt, W. and Köhler, A. 1997. Genomic *in situ* hybridization in Brassica amphidiploids and interspecific hybrids. Theoretical and Applied Genetics, 95(8):1320–1324.

Somers, D.A. and Makarevistch, I. 2004. Transgene integration in plants: poking or patching holes in promiscuous genomes? Current Opinion in Biotechnology, 15(2):126–131.

Spadoro, J.P., Payne, H., Lee, .Y and Resenstraus, J.J. 1990. Single copies of HIV proviral DNA detected by fluorescent *in situ* hybridization. Biotechniques, 9 (2):186-195.

Stam, P. 1993a. Construction of integrated genetic linkage maps by means of a new computer package: Join Map. Plant Journal, 3(5):739-744.

Stam, P. 1993b. Join Map Version 1.4: A computer program to generate genetic linkage maps. CPRO-DLO, Wageningen, The Netherlands.

Stam, P. and Van Ooijen, J.W. 1995. JoinMap ™ version 2.0: Software for the caluclation of genetic linkage maps. CPRO-DLO, Wageningen, The Netherlands.

Sturtevant, A. 1913. The linear arrangment of six sex-linked factors in Drosophila, as shown by their mode of association. Journal of Experimental Zoology, 14:43–59.

Sturtevant, A.H. 1965. A history of genetics. Harper and Row, New York.

Suiter, K.A., Wendel, J.F. and Case, J.S. 1983. Linkage-I: A Pascal computer program for the detection and analysis of genetic linkage. Journal of Heredity, 74(3): 203-204.

Sýkorová, E., Lim, K.Y. Kunická, Z., Chase, M.W., Bennett, M.D., Fajkus, J. and Leitch, A.R. 2003. Telomere variability in the monocotyledonous plant order Asparagales. Proceedings of Biological Sciences, 270(1527):1893–1904.

Tel-Zur, N., Abbo, S., Bar-Zvi, D. and Mizrahi, Y. 2004. Genetic relationships among Hylocereus and Selenicereus vine cacti (Cactaceae): Evidence from hybridization and cytological studies. Annals of Botany, 94(4):527-534.

Torada, A., Koike, M., Ikeguchi, S. and Tsutsui, I. 2008. Mapping of a major locus controlling seed dormancy using backcrossed progenies in wheat (*Triticum aestivum* L.). Genome, 51(6):426-32.

Trask, B.J. 1991. Fluorescence in situ hybridization: applications in cytogenetics and gene mapping. Trends in Genetics, 7:149-154

Vega JM, Abbo S, Feldman M, AND A 1994 Chromosome painting in plants: *In situ* hybridization with a DNA probe from a specific microdissected chromosome arm of common wheat Genetics, 91:12041-12045

Vos, P., Hogers, R., Bleeker, M., Reijans, M., Van De Lee, T., Hornes, M., Frijters, A., Pot, J., Peleman, J., Kuiper, M. and Zabeau M. 1995. AFLP: a new technique for DNA fingerprinting. Nucleic Acids Research, 23(21):4407-4414.

Vuysteke, M., Mank, R., Antonise, R., Bastiaans, E., Senior, M.L. and Stuber, C.W. 1999. Two high density AFLP linkage maps of *Zea mays* L.: analysis of distribution of AFLP markers. Theoretical and Applied Genetics, 99(6):921–935.

Wang, C.R., Harper, L. and Cande, W.Z. 2006. High-resolution single-copy gene fluorescence *in situ* hybridization and its use in the construction of a cytogenetic map of maize chromosome 9. The Plant Cell, 18(3):529-544.

Weiss, H. and Sherthan H. 2002. *Aloe* spp.—plants with vertebrate-like telomeric sequences. Chromosome Research, 10(2):155-164.

Welsh, J. and McClelland, M. 1990. Fingerprinting genomes using PCR with arbitrary primers. Nucleic Acids Research, 18(24):7213-7218.

Williams, J.G.K., Kublelik, A.R., Livak, K.J., Rafalski, J.A. and Tingey, S.V. 1990. DNA polymorphisms amplified by arbitrary primers are useful as genetic markers. Nucleic Acids Research, 18(22):6531-6535.

Xu, Z., Kohel, R.J., Song, G., Cho, J., Yu, J., Yu, S., Tomkins, J. and Yu, J.Z. 2008. An integrated genetic and physical map of homoeologous chromosomes 12 and 26 in Upland cotton (*G. hirsutum* L.). BMC Genomics, 9:108-115.

Yammamoto, M. and Mukai, Y. 2005. High resolution physical mapping of the secalin-1 locus of rye on extended DNA fibres Cytogenet Genome res 109:79-82.

Yammamoto, M. and Mukai, Y. 2007. Extended DNA fibre fish in plants : Visible messages from cell nuclei. The nucleus 50(3):439-452.

Yim, Y.S., Davis, G.L., Duru, N.A., Musket, T.A., Linton, E.W. and Messing J.W. 2002. Characterization of three maize bacterial artificial chromosome libraries toward anchoring of the physical map to the genetic map using high-density bacterial artificial chromosome filter hybriddization. Plant Physiolology, 130(4):1686-1696.

Yun Li Zong, Li Fu, Mei., Fang hu, Fang., Feng Huang, Shu., and Chun Song,Yun. 2006. Visualization of the ribosomal DNA (45S rDNA) of Indica rice with FISH on some phases of cell cycle and extended DNA fibers. BIOCELL 30(1):27-32.

Molecular Plant Breeding: Principle, Method and Application
Eds : R.K. Singh, Rajesh Singh, Guoyou Ye, A. Selvi and G.P. Rao
Studium Press LLC, Texas, USA, 2009, pp. 269-294

CHAPTER
10

Genomics and Gene Based Markers for Crop Improvement

A. SELVI[*1], *N.V.NAIR*[1] *and R. K. SINGH*[2]

ABSTRACT

Since the development of first molecular markers in 1980, a diverse array of molecular marker technologies have revolutionized conventional plant breeding efforts for crop improvement. Significant progress has been made in crop improvement through these classical markers or conventional random molecular markers (RDMs). Besides throwing light on organization, conservation and evolution of plant genomes, these markers have also aided geneticists and plant breeders to map QTLs for the traits of economic importance and to identify genes. Further advancements in genomics with high throughput sequencing methods and bioinformatics aided in the characterization of these genes and to date sequences of several genes are available in databases. The markers derived from the genes or ESTs are commonly called as functional markers or genic molecular markers (GMM). The availability of technologies for precise manipulation of these genes and their deployment is helping plant breeders in a way as never before for evolving better crop varieties. The following write-up focuses on the advancement of genomic tools for developing strategies for tagging genes and identifying candidate genes for traits of interest and their applications for improving crop plants.

Key Words: Random markers, Gene based markers, Functional markers

[1]*Sugarcane Breeding Institute, Coimbatore-641 007, Tamil Nadu, India*
[2]*Indian Institute of Sugarcane Research, Lucknow-226002, U.P., India*
[*]*Corresponding auther e-mail: selviathiappan@yahoo.co.in*

1.0 INTRODUCTION

Remarkable success has been achieved in the past fifty years in developing new improved crop varieties through plant breeding. Major emphasis was on sourcing genes contributing to better productivity and adaptability from related species and wild relatives through genetic manipulation at cultivar, interspecific or intergeneric level. Plant breeding has seen a major transition in the past decade as advances in biological sciences helped in evolving tools that can be applied to commonly accepted field techniques. Impressive advances in molecular genetics over the last two decades have provided a range of tools and techniques for analyzing and manipulating plant genomes. A large number of plant genes have been identified and their structural and functional aspects fully characterized. Integrating genomic tools have rendered plant breeding programs more focused, precise and less time consuming. The major developments in this area include the sequencing of the several plant genomes like *Arabidopsis*, rice, sorghum and poplar, the generation of expressed sequence tag (EST) databases, the development of microarray technologies, molecular markers, gene expression methodologies etc.

Genomic approaches are beginning to impact the conventional breeding processes. With the advent of genomic tools like DNA marker technology, several types of DNA markers both random and functional are now available to plant breeders and geneticists, helping them to overcome many of the limitations of conventional breeding. Advances in automated sequencing, methodologies for gene expression studies, development of computational abilities and algorithms have enabled the structural and functional characterization of several genes governing economically important traits and their further use in enhancing the breeding efficiency. It is now perceivable that this scientific development is heading towards sequence-based knowledge that would greatly improve the reliability and precision of breeding programs.

2.0 MOLECULAR MARKERS AS GENOMIC TOOLS

Of the several genomic tools that were developed in the past decade molecular markers deserve special mention. The DNA based markers were the first genomic tools that were developed and used for mapping several genes in a variety of crop species. Several molecular marker techniques were developed from the genome of the crop species as well as from random amplification of the genome. These techniques include the first generation restriction based markers like RFLP followed by the second generation amplification based markers like RAPD, AFLP, SSR, ISSR and the third generation sequence based markers like SNPs. The RFLP markers which are highly polymorphic, reproducible and are co-dominantly inherited but have been used less frequently owing to the laborious procedure involved. RAPDs and ISSRs had been the marker of choice in the nineties and soon interest had shifted to more reliable and reproducible marker systems like AFLPs and SSRs. An increasing number of agronomically important genes have been correlated using DNA based molecular markers. Molecular markers offered numerous advantages over conventional phenotype based alternatives, as they are stable and detectable in all tissues regardless of growth, differentiation or developmental status of the cell. They are not confounded by the environment, pleiotropic and epistatic effects. They have been used for several purposes including screening programs for selecting desirable individuals, introgression breeding, gene pyramiding etc. Marker assisted breeding has helped in the indirect selection of difficult traits at the seedling stage and has helped in the speeding up of conventional breeding in several crop plants.

Most of the above mentioned markers are developed from genomic DNA, and therefore they may arise from both the transcribed regions and the non-transcribed regions of the genome. These DNA-based markers derived from any region of the genome have also been described as RDMs. These markers when used for indirect selection are completely independent of any functional knowledge about the underlying DNA sequences. Thus even for tightly linked markers, the effectiveness of marker aided selection is greatly diminished by the occasional uncoupling of the marker from the trait during many cycles of meiosis in the breeding program. Also the application of such random markers for selection across populations has been limited.

Recently, interest have shifted towards the development of molecular markers from the genes that are responsible for the expression of phenotypic trait variations. These markers from transcribed region of the genome target the functional polymorphism in the gene sequences and allow selection in different genetic backgrounds which is not always possible with random markers.

3.0 GENOMIC RESOURCES

3.1 Genome and Gene Space Sequencing

The emphasis laid on genomics in the recent past had led to several gene discovery projects by way of genome sequencing, analysis of transcriptome, gene expression studies etc. This has resulted in the expanded databanks that are available for crop improvement. An increasingly large number of genes have been identified in both wet lab and by *in-silico* analysis of available databases. The identified gene sequences have been stored both in the form of genomic sequences and EST sequences, or as BAC clones as well as full length cDNA clones and genes.

The whole genome sequencing projects of *Arabidopsis*, rice and poplar has made available complete genome sequences of these species. Gene space sequencing that targets sequencing of long gene-rich regions containing many genes that are separated by long gene-poor regions has been carried out for plant species like maize (http://www.maizegenome.org/), sorghum, wheat (http://www.wheatgenome.org/), tomato (http://sgn.cornell.edu/help/about/tomato_sequencing.html), tobacco (http://www.intl-pag.org/13/abstracts/PAG13_P027.html), poplar (http://genome.jgi-psf.org/Poptr1/), Medicago (http://www.medicago.org/genome/) and lotus (http://www.kazusa.or.jp/lotus/). Several genomics based databases are now available to identify gene sequences that would help in molecular marker development (Table 1).

3.2 Expressed Sequence Tags

Apart from whole genome sequencing projects and gene space sequencing, strategies to study the transcriptome of several important crop species have led to the establishment of several expressed sequence tag (EST) sequencing projects for gene discovery. Generation of ESTs is a quick and simple strategy involving partial sequencing of 5′ or 3′ end of the cDNAs. A wealth of DNA sequence information has been generated from these projects and deposited in online databases like http://www.ncbi.nlm.nih.go (National Centre for Biotechnology Information), http://www.tigr.org (The Institute of Genome Research), http://www.ebi.ac.uk (European Bioinformatics Institute). The plant EST database at EMBL has exceeded over five million EST sequences. ESTs have been developed from more than 50 plant species and in each species more than 5000 ESTs are now available. These species represent important crops and their sequences are a potential source from which several valuable molecular markers that are of interest to plant breeders can be generated.

Several crop specific EST databases are available for crops like wheat (http://genome.arizona.edu, http://wheat.pw.usda.gov/genome, http: //wheat.pw.usda.gov/NSF/), barley (http://hordeum.ipk-gatersleben.de/est/est.html), sugarcane (http://sucest.lad.ic.unicamp.br/en/) etc. In species that lack EST resources, comparative mapping strategies have facilitated the isolation of genes from these species. GRAMENE (http://www.gramene.org) is a curated open source database that is available for comparative genome analysis for grasses. HarVEST (http://harvest.ucr.edu) is software developed to enable searching for differentially expressed ESTs among cDNA libraries and is oriented towards comparative genomics. However since the EST sequences are generated through partial sequencing of the 3′ or 5′ ends of cDNAs there is redundancy in the genes sequences that are obtained from the databases. The EST have to be clustered to identify unigenes from random EST sequences using bioinformatics tools. The NCBI UniGene Resources is an excellent system for automatically partitioning GenBank sequences into a non-redundant set of gene oriented clusters. UniGene Sets are available for wheat, barley and many other crop plants in the NCBI database. The TIGR Gene Indices includes ESTs clustered into Tentative Consensi (TC), with top 5 peptide hits, and alignment to rice BACs and *Arabidopsis* chromosomes. These unigenes are then utilized to develop molecular markers.

3.4 Bioinformatics

The advances made in bioinformatics have made it possible to acquire and organize large amounts of information and also allows the visualization of information from heterogeneous datasets. It facilitates both the analysis of genomic and post-genomic data, and the integration of data from the related fields of transcriptomics, proteomics, metabolomics and phenomics. Improved algorithms and increased computing power has helped to analyze DNA sequences, marker discovery and analyzing the information generated. It provides tools to integrate phenotypic and genotypic data and to derive meaningful conclusions for important agronomic traits.

Development of bioinformatic tools, databases and integration of information from different fields enable the identification of genes and gene products, and can elucidate the functional relationships between genotype and observed phenotype. Bioinformatic tools that are commonly used for datamining, gene identification, analysis and molecular marker development include MISA for identifying SSRs, AutoSNP used for SNP identification, SNP2CAPS-used for developing CAPs from SNP markers, TASSEL which is a tool for microarray data analysis, datamining and data visualization etc. Databases like NCBI, EMBL, Swissprot, AceDB, Plantmarkers, HarvEST, PEDANT, tools for datamining, Bioinformatic.net etc also provides multiple bioinformatics tools.

Table 1. Databases related to crop genomics

Crop related databases	Details	Websites
Barley Genomics	Databases on molecular markers, ESTs, maps, mutants barley BACs	http://barleygenomics.wsu.edu
Barley DB	Database with molecular markers, ESTs and QTL maps	http://uscrop.net/perl/ace/search/BarleyDB
MaizeDB	Comprehensive information source on the genetics and molecular biology of maize	http://www.agron.missouri.edu
ZmDB	Analysis tool for sequence, expression and phenotype data, online ordering for ESTs, and microarrays	http://www.zmdb.iastate.edu
MilletGenes	AceDB database with molecular markers, ESTs, QTL, maps	http://uscrop.net/perl/ace/search/MilletGenes
SorghumDB	AceDB database with molecular markers, ESTs, QTL, maps	http://algodon.tamu.edu/sorghumdb.html
Rye, *Triticum aestivum*	Gene Index	http://www.tigr.org
Wheat	compile and distribute SSR-containing ESTs	http://wheat.pw.usda.gov/ITMI/EST-SSR/
	SNP Development	http://wheat.pw.usda.gov/ITMI/WheatSNP

4.0 GENE BASED MARKERS

With the advent of high throughput sequencing and ever enlarging sequence databases gene based markers came into existence and more recently with the development of gene expression studies several functional markers are being identified and are proving to be efficient (Table 2). In

Table 2. Genic molecular markers in important plant species

General name	Species	Type of markers developed	References
Cereals and grasses			
Barley	*Hordeum vulgare*	EST-SSR, EST-SNP, EST-RFLP, cDNA-RFLP	Thiel *et al.* (2003), Rostocks *et al.* (2005), Varshney *et al.* (2006), Willsmore *et al.* (2006), Stein *et al.* (2007), Varshney *et al.* (2007)
Maize	*Zea mays*	cDNA-RFLP, EST-SNP	Gardiner *et al.* (1993), Chao *et al.* (1994), Picoult-Newberg *et al.* (1999), Falque *et al.* (2005)
Wheat	*Triticum aestivum*	EST-SSR, EST-SNP, cDNA-RFLP	Holton *et al.* (2002), Yu *et al.* (2004), Somers *et al.* (2003), Gao *et al.* (2004), Qi X. *et al.* (2004), Nicot *et al.* (2004)
Rice	*Oriza sativa*	EST-SSR, EST-SNP, cDNA-RFLP, Intron Length Polymorphism (ILP)	Causse *et al.* (1994), Harushima *et al.* (1998), Temnykh *et al.* (2001), Feltus *et al.* (2004), Wang *et al.* (2005)
Rye	*Secale cereale*	EST-SSR, EST-SNP	Hackauf and Wehling, (2002), Khlestkina *et al.* (2004), Varshney *et al.* (2007)
Sorghum	*Sorghum bicolor*	EST-SSR, cDNA-RFLP	Childs *et al.* (2001), Klein *et al.* (2003), Bowers *et al.* (2003), Ramu *et al.* (2006), Jayashree *et al.* (2006)
Lolium	*Lolium perenne*	EST-SSR	Faville *et al.* (2004)
Legumes			
White clover	*Trifolium repens*	EST-SSR	Barret *et al.* (2004)
Soybean	*Glycine max*	EST-SSR	Song *et al.* (2004), Zhang *et al.* (2004)
Fiber and oil seed crops			
Cotton	*Gossypium* sps.	EST-SSR	Zhang *et al.* (2005), Chee *et al.* (2004), Park *et al.* (2005)
Sunflower	*Helianthus* sps.	EST-SNP	Lai *et al.* (2005)
Fruit and vegetables			
Grape	*Vitis vinifera*	EST-SSR	Chen *et al.* (2006)
Kiwi fruit	*Actinidia chinensis*	EST-SSR	Fraser *et al.* (2004)
Raspberry	*Rubus* spp.	EST-SSR	Frary *et al.* (2005)
Tomato	*Lycopersicon esculentum*	EST-SSR	Frary *et al.* (2005)
Strawberry	*Fragaria* spp.	EST-SSR	Sargent *et al.* (2006)
Trees			
Pinus	*Pinus* ssp.	EST-SSR, ESTP	Cato *et al.* (2001)
Coffee	*Coffea* ssp.	EST-SSR	Bhat *et al.* (2005), Aggarwal *et al.* (2007)

contrast to random markers that are developed from any part of the genome, gene based markers are derived from polymorphic sites within gene sequences. These are also called 'Genic' markers. The gene based markers are further classed as Gene Targeted Markers (GTMs) and Functional Markers (FMs) depending on the functional characterization of the polymorphisms that are generated by these markers (Anderson and Luebberstedt, 2003).

4.1 Gene Targeted Markers (GTMs)

Gene target markers generate polymorphisms from gene sequences which are obtained from EST databases or genome sequences or cDNA sequences and the functions of these genes may be already known or unknown. Some common examples of GTMs are cDNA-RFLP, EST-simple sequence repeats (EST-SSRs) (Varshney *et al.*, 2005), EST-Single nucleotide polymorphisms (EST-SNPs) (Rafalski, 2002) and conserved orthologous sets of markers (COSs).

4.2 cDNA-Restriction Fragment Lymph Polymorphism (cDNA-RFLP)

The first genic markers that were developed were in the form of cDNA-RFLP (Graner *et al.*, 1991, Causse *et al.*, 1994). Any transcript derived polymorphic fragment can be cloned and used as a probe for developing gene based markers for the trait of interest. In the past several random markers have been developed from mRNA or cDNA from a number of plant species that are subjected to stress. The polymorphic fragments obtained from these experiments like differential display, cDNA-AFLP, or PCR products obtained by amplifying gene specific fragments from cDNA or genomic DNA can be cloned and used as probes. These probes have the potential of being used as heterologous probes across species and genus.

4.3 Expressed Sequence Tags-single Sequence Repeats (EST-SSRs)

The availability of ESTs sequences from many genomes has led to the mining of these sources using computational approaches and has permitted rapid and economical marker development programs. ESTs are ideal candidates for mining SSRs not only because of their availability in large numbers but also due to the fact that they represent expressed genes. Recent studies have observed that the frequency of microsatellites was significantly higher in ESTs than in genomic DNA in several plant species investigated (Morgante *et al.*, 2002; Toth *et al.*, 2000). The generation of SSRs from EST sequences allows the identification of polymorphic loci directly from sequence data, if the sequence information for the same gene is available from more than one genotype of the same species. Most of the efforts till date for finding SSRs in EST sequences use several bioinformatics tools that have been developed for mining SSRs from EST databases (Table-3). Some important ones are:

- ***Sputnik*** which is a simple program written in C programming language that searches DNA sequence files in FASTA format for microsatellite repeats (Abajian, 1994).
- ***FindPatterns*** is one of the programs available in the Genetics Computer Group (GCG), now Accelrys, package (www.accelrys.com). It looks through large data sets and identifies short nucleotide or amino acid patterns specified by the user.
- ***RepeatFinder*** is a web-based program specifically developed for the identification of SSRs (http://www.genet.sickkids.on.ca/~ali/repeatfinder.html). This program was originally developed for identifying repeats in a single input sequence, however, later upgraded to handle batch files containing multiple sequences.
- ***Simple Sequence Repeats Identification Tool*** (SSRIT) is a simple program available through Gramene/Genome databases portal at Cornell University (http://brie2.cshl.org:8082/gramene/searches/ssrtool). The program helps in the identification of "perfect" simple sequence repeats and can handle moderate-sized datasets. Recently websites have also been created for documentation, curation and transaction of EST-SSRs data.
- ***PlantSSR database*** is a major source of information on plant EST-SSRs, which was established at Clemson University Genomics Institute (CUGI).

Since EST-SSRs are derived from transcripts, they have been found useful for assaying the functional diversity in natural populations and germplasm collections. These markers are highly transferable to related species, and are thus useful for comparative mapping and evolutionary studies. When the EST-SSRs are generated from genes responsible for a phenotypic trait they are more effectively used for marker assisted selection.

4.4 Expressed Sequence Tags-single Nucleotide Polymorphism (EST-SNPs)

Recent developments in sequencing techniques had made the detection of single nucleotide polymorphisms (SNPs), which are the basis of most differences, between alleles more efficient. SNPs are generated by two methods. Direct sequencing of DNA segments that are amplified by PCR from several individuals is used for identification of SNP polymorphisms

(Gaut and Clegg 1993; Shattuck-Eidens *et al.,* 1990). PCR primers are designed from genes of interest, to amplify 400–700 bp segments of DNA from a diverse set of individuals that represent a population. The resulting sequences are aligned and polymorphisms are identified.

Table 3. Tools for database mining for SSRs

Script or program	References
MIcroSAtellite (MISA)	http://pgrc.ipk-gatersleben.de/misa/; Thiel *et al.* (2003)
SSRFinder	Gao *et al.* (2003)
BuildSSR	Rungis *et al.* (2004)
SSR Identification Tool (SSRIT)	Kantaty *et al.* (2002)
Tandem Repeat Finder (TRF)	Benson (1999)
Tandem Repeat Occurrence	Castelo (2002)
Locator (TROLL) CUGIssr	http://www.genome.clemson.edu/projects/ssr/
Sputnik C. Abajian;	http://abajian.net/sputnik/index.html
Modified Sputnik	Morgante *et al.* (2002)
Modified Sputnik II SSRSEARCH	http://wheat.pw.usda.gov/ITMI/EST-SSR/LaRota/

The second method involves *in-silico* methods to identify SNPs by aligning EST sequences derived from different genotypes and available in public databases. A large number of SNP mining tools to automate the process of SNP discovery are available (Table-4) and SNPs have been generated in a number of species. ESTs were generated by sequencing shoot apical meristem (SAM) cDNA from maize inbred lines. The computational tool PolyBayes was used to identify single-nucleotide polymorphisms in 454 EST sequences of maize (Barbazuk *et al.,* 2007). A comparative study of EST-SSR, EST-SNP and AFLP markers for evaluation of genetic diversity and conservation of genetic resources using wild, cultivated and elite barley cultivars revealed that EST-SNPs are the best class of markers for characterizing and conserving the gene bank materials (Varshney *et al.,* 2007).

4.5 Conserved Orthologous Set of markers (COS)

The increasing information on genomics and functional genomics from model plants like *Arabidopsis* and the evolution of new tools in functional genomics provides opportunities to develop gene based markers in crops where information is unavailable. COS markers represent orthologous genes from known sequences of a given species that can be used for related species through comparative genomics. It helps in identifying a set of genes conserved throughout evolution in both sequence and copy number in members of related species.

Table 4. Softwares used for *in-silico* mining of SNPs

Software tools for detecting SNPs automatically	Reference
PolyPhred	Nickerson *et al.*. (1997)
TRACE_DIFF	Bonfield *et al.* (1998)
PolyBayes	Marth *et al.* (1999)
AutoSNP	Barker *et al.* (2003)
SNP locator (SNPL)	In house VB program
PARSESNP	Taylor and Greene (2003)
SNiPpER	Kota *et al.* (2003)
Single Nucleotide Polymorphism Finder (SNPF)	http://jic-bioinfo.bbsrc.ac.uk/cereals/webstart/snpf/snpf.html
Quality SNP	Tang *et al.* (2006)
SEAN	Huntley *et al.* (2006)

SEAN: SNP prediction and display program utilizing **EST** sequence clusters

The EST database of tomato was computationally compared with the *Arabidopsis* genomic sequence and a set of conserved genes were identified as COS markers for tomato. These COS markers, 1,025 in numbers, represented functional genes and have shown to be conserved over a wide range of dicotyledonous plants. These genes were annotated, and most of them were identified to have putative functions that are associated with basic metabolic processes, such as energy-generating processes and the biosynthesis and degradation of cellular building blocks.

Similarly 1130 potential COS markers for Lettuce, 426 for Sunflower, 1860 for Tomato and 1413 for Corn were identified by screening 2185 sequences from *Arabidopsis* (Kozik and Michelmore 2002, http://cgpdb.ucdavis.edu/COS_Markers/COS_Markers.html). EST sequences of three drought tolerance related genes like chalcone synthase (CHS), dihydroflavanol-4-reductase (DHRF-1) and drought responsive element binding factor (DREB-1) from *Musa* were used to identify Cassava homologs that were screened against *Arabidopsis* genome database to identify COS markers (Castelblanco and Fregene, 2006). These markers have proved useful in comparative mapping among divergent genomes, and are useful for taxonomic studies and in deducing phylogenetic relationships between different genera and species.

4.6 Resistance Gene Analogues (RGAs):

Another important class of GTMs are resistance gene analogues. Several disease resistant genes from diverse sources are available now and are being used as markers thereby increasing gene based selections of superior genotypes. Genes conferring resistance to major classes of plant pathogens, including bacteria, virus, fungi and nematodes have been isolated from different plant species. Numerous genes involved in pathogen recognition, signal transduction and defense have been isolated. Nearly 40 resistance genes have been cloned during the past ten years. Many of these cloned genes are related in sequence and encode a limited number of functional classes.

Based on common molecular features, the R-genes are classified into several classes. The Class 1 encodes cytoplasmic receptor like proteins that contain a leucine rich repeat (LRR) domain and a nucleotide binding site (NBS). The genes that fall in this category are the *Arabidopsis RPS2, RPM1,* tomato *Prf* genes conferring resistance to *Pseudomonas syringae,* the *12-C* gene conferring resistance to fungus *Fusarium oxysporum,* the tobacco *N* gene conferring resistance to Tobacco mosaic virus, the rust resistance gene *L6* of flax and the *Arabidopsis RPP5* gene conferring resistance to downy mildew. Another class of resistance genes include the *Pto* which confers resistance to bacterial pathogen *Pseudomonas syringae* pv. *Tomato.* The *Pto* does not possess any LRR or NBS region and is dependent on the NBS-LRR containing protein *Prf* for its function. The third category includes the tomato *Cf-2* and *Cf-9* genes conferring resistance to the fungus *Cladosporium fulvum.* Yet another class of resistant genes consists of trans-membrane receptor with an extracellular LRR domain and an intracellular serine threonine kinase like the rice *Xa 21* gene.

4.7 Resistance Gene Homologue Polymorphism (RGHPs)

RGHPs target groups of resistance genes by PCR, using primers for conserved domains of resistance genes, such as the Leucine Rich Repeat (LRR) or the Nucleotide Binding Site (NBS), both involved in resistance mechanisms. These RGHPs are then used to identify linkage with known disease resistant loci for use in marker assisted selections as well as to clone the resistant genes. Many RHGPs have been located to chromosome regions containing major R genes as well as QTLs. The cosegregation of RHGPs with major disease resistant genes and quantitative trait loci (QTL) has been reported in several crops species (Table-5) (Pflieger *et al.,* 2001). The disease R-gene database (available on line from the National Center for Genome Resources web site: http://www.ncgr.org/research/rgenes) facilitates access to R-gene and R-gene-like sequence data collected from public sequence databases and protein databases. The second database contains information about genes for both pathogen recognition (resistance genes and homologs) and plant defense responses (defense genes) (Chittoor *et al.,* 1999).

Table 5. Resistance gene homologue polymorphisms (RGHPs) in plants and their co- segregations with major genes and/or QTLs involved in disease resistance

Plant species	Type of RGA	Resistance locus	Disease or pathogen	References
Arabidopsis thaliana	NBS	major R-genes and gene cluster	*Pseudomonas syringae* *Peronospora parasitica* Turnip crinckle virus *Albugo candida* *Erisyphe cichoracearum*	Aarts *et al.* (1998)
			Caulimovirus *Turnip crinckle virus* *Tobacco ring spot virus*	Speulman *et al.*, (1998)
Soybean	NBS	major R-genes	*Phytophthora sojae* *Microspora diffusa* *Bradyrhizobia japonicum* Potyvirus	Kanazin *et al.* (1996)
		QTL	Cyst nematode	Yu *et al*, (1996) Kanazin *et al.* (1996)
Common bean	NBS	R gene cluster and	*Colletotrichum lindemuthianum*	Geffroy *et al.* (1998)
	Kinase		*Xanthomonas*	Nodari *et al.* (1993)
		QTL Major R gene	*Uromyces appendiculatus*	Rivkin *et al.* (1999)
Sunflower	NBS Kinase	R gene cluster QTL	*Plasmopara halstedii* *Sclerotinia sclerotiorum*	Gentebittel *et al.* (1998)
Lettuce	NBS	Major R genes	*Bremia lactucae*	Woo *et al.* (1998)
Potato	NBS	Major R genes	*Globodera rostochiensis* *Phytophthora infestans* *Potato virus Y* *Potato virus A*	Leister *et al.* (1996)
			Potato virus A	Hamalainen *et al.* (1998)

contd....

Plant species	Type of RGA	Resistance locus	Disease or pathogen	References
Pepper	NBS	QTL	*Cucumber mosaic virus*	Pflieger *et al.* (1999)
	Kinase	QTL	*Potyvirus*	
Rapeseed	NBS	QTL	*Leptosphaeria maculans*	Pilet *et al.* (1999)
	Kinase	QTL	*Pyrenopeziza brassicae*	
		Major R genes	*Leptosphaeria maculans*	
Poncirus	NBS/LRR	Major R genes	*Citrus tristeza virus* *Citrus nematode*	Pilet (1999)
Sugar beet	NBS/LRR	Major R gene	*Cercospora rhizomania*	Weiland and Koch (2004)
Pepper	NBS/kinase	QTLs	*Phytopthora capsici*	Donnelly *et al.* (2005)
Maize	NBS/LRR	Major genes	*Sugarcane mosaic virus*	Quint *et al.* (2002)
Sugarcane	NBS/LRR	Major R genes and QTLs	*Puccinia melanocephala* *Ustilago scitaminia* *SCMV*	Rossi *et al.* (2003) McIntyre *et al.* (2005)
Wheat	RGAs	Major gene	*Puccinia striiformis*	Chen *et al.* (1998)
Rice	NBS	Major R gene	*Xanthomonas oryzae*	Leister *et al.* (1998)
		Gene cluster	*Pyricularia oryzae*	
		QTLs	*Magnaparthae grisea*	
Barley	NBS Kinase/LRR	Major R Genes Leaf rust R gene cluster	*Erysiphe graminis*	Leister *et al.* (1998) Chen *et al.* (1998)

5.0 FUNCTIONAL MARKERS (FM):

The development of functional markers requires functionally characterized genes, allele sequences from such genes, the identification of polymorphic, functional motifs that affect plant phenotype within these genes, and the validation of associations between DNA polymorphisms and trait variation. Functional markers are further classed as direct functional markers and indirect functional markers.

5.1 Direct Functional Markers

These markers are developed from gene sequences with known expression and hence it is of prime importance to establish proof of gene function affecting a particular phenotype. The most direct means of obtaining proof of sequence motif function is by comparing isogenic genotypes differing in single sequence motifs. At present, the most appropriate approach for generating isogenic lines in crops is by targeting induced local lesions in genomes (TILLING). TILLING provides point mutant alleles that are usually induced by chemical mutagens like ethyl methane sulphonate and these mutations are used in functional genomics and gene characterization (Fig. 1).

The ability to detect point-mutations in specific genes within a large population of mutagenized plants was first demonstrated by Claire McCallum from the Henikoff and Comai laboratories at the Fred Hutchinson Cancer Center and the University of Washington in Seattle and coined the word TILLING. Since then TILLING has been developed in several species including *Arabidopsis*, rice, maize, sorghum, wheat, barley, tomato, soybean, rapeseed etc. TILLING helps in generating a series of missense mutations in the alleles of a target gene. The polymorphisms that are generated by these mutations are compared with phenotypic variation to provide direct

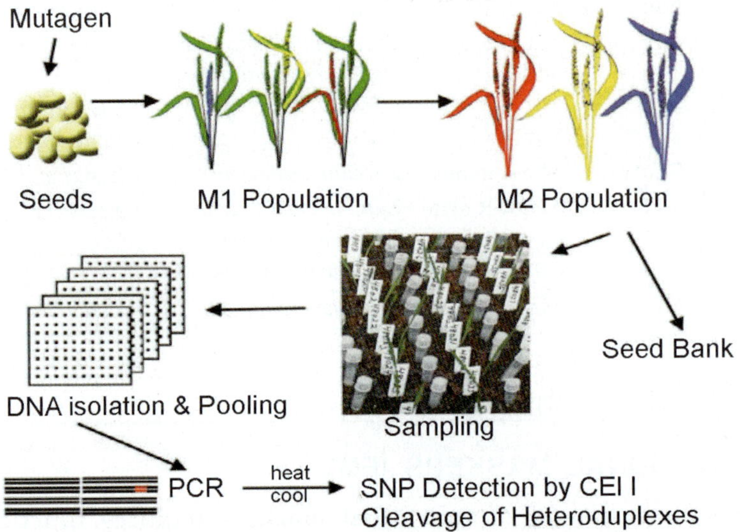

Fig. 1. The TILLING Method. Seeds are treated with a chemical mutagen to induce genetic variation, and then planted. The resulting M_1 population of plants is chimeric for mutations. Therefore, one seed from each M_1 is planted to create the M_2 population. M_2 DNA is extracted from leaf tissue and M_3 seeds from each plant are stored in a seed bank. DNA samples are pooled to increase throughput and PCR amplified with dye-labeled PCR primers specific to a target gene of interest. PCR products are denatured and allowed to reanneal to form heteroduplexes. Heteroduplex DNA is then cleaved by Cel I and analyzed. By this novel SNPs within a tilling population is discovered (source: Slade and Knauf, 2005)

functional markers. Thus TILLING represents a viable method by which spontaneous and induced mutants help in the direct identification of beneficial nucleotide and amino acid changes in genes with known functions that can further be used in the development of diagnostic functional markers for selection.

5.2 Indirect Functional Markers

Association studies have the potential to identify sequence motifs affecting trait expression. They provide indirect evidence for the function of a sequence motif. In association studies the polymorphisms that exist within the gene such as a few nucleotides differences or insertions/deletions (indels), are correlated to the phenotype of interest. With the advent sequencing methods it is possible to find the SNP variations that occur through out the gene and the linkage disequilibrium (LD) that exists between this SNPs. Based on the LD of the SNPs they can be grouped as haplotype SNPs and those haplotype SNPs that are involved in the functional variation of the gene can be identified. This strategy would considerably reduce the number of SNPs that has to be genotyped. The selected SNPs which are functional variants can be genotyped on a set of accessions to find associations with the phenotype of interest.

In one of the pioneering studies, nine sequence motifs in the *dwarf8* gene of maize were shown to be associated with variation for flowering time. A set of 92 inbred lines were genotyped and associations between the polymorphisms in the *dwarf8* gene and flowering time was established. These associations of the polymorphisms in the *dwarf8* gene aided in the selection of lines that exhibited early flowering by 7-11 days (Thornsberry *et al.*, 2001). Similarly associations of three haplotypes of the *StVe1* locus of potato that confers resistance to *V. alboatrum* were used to genotype 30 potato cultivars and one haplotype showed significant associations with *V. alboatrum* resistance (Simko *et al.*, 2004).

A recent study in grapes showed that allelic variation in the gene *VvmybA1* that is responsible for transcriptional regulation of anthocyanin biosynthesis, was associated with multiple classes of fruit color in the 200 accessions of cultivated grapevine that were studied. One SNP and three indels were found to be significantly associated with fruit skin color in the structured association analysis of pigmented accessions. All four polymorphisms were associated with genetic differences separating black or gray-skinned accessions from red and pink skinned accessions (This *et al.*, 2007). The use of functional motifs to correlate with phenotyopes requires comprehensive allele sequencing, a relatively low LD between haplotypes and a phenotypically well characterized population. Several studies have been carried out on LD decay and haplotype diversity (Dvornyk *et al.*, 2002;

van der Voort *et al.*, 2004; Oleson *et al.*, 2004; Neale and Savolainen 2004). In crops with low LD and high resolution of intragenic polymorphisms, association studies have the potential to identify sequence motifs that are correlated with trait variation.

6.0 APPLICATION OF GENIC MOLECULAR MARKERS

As we know, molecular markers have already shown their applications in a variety of ways in several plant species. Hence, now with the development of GMMS, it is possible to have a targeted approach for detection of nucleotide diversity in genes which control agronomic traits in plant populations. The GMMS implementation will prove quite useful and can be utilized in three main areas of plant breeding and genetics, which are outlined below:

6.1 Functional Genetic Diversity

Identification of diverse genotypes is the prerequisite for improvement of any trait in the crop plants. Further more, monitoring the genetic variability within gene pool of elite breeding material could make crop improvement more efficient by the direct accumulation of favored alleles. DNA markers are being increasingly utilized in cultivar development, quality control of seed production, measurement of genetic diversity for conservation and management, varietal identification and intellectual property protection (IPP). Recent studies have used molecular markers to help in identification of genetically diverse genotypes to use in crosses in cultivar improvement programme. These studies have more success than conventional selection programme in producing productive lines from plant introduction/exotic lines crosses with elite lines (Thompson *et al.*, 1998 a, b). Molecular markers have proven useful for assessment of genetic variation in germplasm collections (Hausmann *et al.*, 2004; Maccaferri *et al.*, 2006). Evaluation of germplasm with GMMs might enhance the role of genetic markers by assaying the variation in transcribed and known function gene, although there may be higher probability of bias owing to selection.

According to Ayers *et al.* (1997) the expansion and contraction of SSR repeats in genes of known function can be tested for association with phenotypic variation or, more desirably, biological function by using genic SSR markers for diversity studies. The SSRs may have role in gene expression or function as suggested by the presence of SSRs in transcripts of genes. However, it is yet to be determined whether any unusual phenotypic variation is associated with the length of SSRs in coding regions as was reported for several diseases in human (Cummings and Zoghbi, 2000). Similarly, the use of SNP markers for diversity studies may correlate the SNPs of coding *vs* non coding regions of the gene with trait variation. The variation associated with deleterious characters, however, is less likely to be represented in the germplasm collections of crop species than among natural populations because undesirable mutations are commonly culled from breeding populations (Cho *et al.*, 2000).

Several studies involving GMMs, especially genic SSRs, have been found useful for estimation of genetic relationship (Gupta and Rustgi, 2004; Varshney *et al.*, 2005a) and opportunities to examine functional diversity in relation to adaptive variation (Russel *et al.*, 2004) can be seen in several studies using GMMs. Very soon with the development of more GMMs in major crop species, genetic diversity studies will become more meaningful if functional genetic diversity were to be given more importance than the evaluation of unknown diversity. But use of the neutral traditional molecular markers will remain useful in situations where: (a) GMMS would not be available, and (b) to address some specific objectives *e.g.* neutral grouping of germplasm.

6.2 Cross Transferability Among the Species or Genera

The genic markers provide high degree of transferability among distantly related species which is one of the most important features of these markers while among the RDMs only RFLPs shows transferability. Transferability of GMM markers to related species or genera has now been demonstrated in several studies. A study based on analysis of ~ 1000 barely GMMS suggested a theoretical transferability of barley markers to wheat (95.2%), maize (69.3%), sorghum (65.9%), rye (38.1%) and even to dicot species (16.0%). In fact, *in-silico* analysis of GMMs of wheat, maize and sorghum with complete rice genome sequence data have provided a larger number of anchoring points among different cereal genomes as well as provided insight into cereal genome evolution (Salse *et al.*, 2004). Genic markers are now used to enrich the genetic maps of related crop species (Varshney *et al.*, 2004; 2005; 2007). Furthermore, genic markers from the related plant species offers the possibility to develop anchor or conserved orthologous sets (COS) for genetic analysis and breeding in different species. Based on these information workers identified a large repository of such COS markers and developed a database called "Plant Markers" (Rudd *et al.*, 2005).

6.3 Tagging and Mapping of Traits/QTLs

One of the most important applications of molecular markers in plant breeding is their use as diagnostic markers for the trait in the selection. However, if random markers (RDMs) are used there is a risk of losing the linkage through genetic recombination. Such type of situation is also happened in case of GMMs, when the polymorphism for the gene-targeted markers (GTMs) was discovered through one allele analysis without any further specifications of the polymorphic sequence motif are threatened by the same way (Rafalski and Tingey, 1993). While the functional markers (FMs) /DFMs or IFMs allow reliable application of markers in populations without prior mapping and the use of markers in mapped populations without risk of information loss owing to recombination, in comparison to random markers. GMM have been developed and mapped in several plant species. Since the development of FMs is expensive it can not be undertaken for all the traits and in all crop species.

The "transcript" or "gene" maps are the genetic maps, developed after mapping/integration of genic markers. Rostocks *et al.* (2005) have developed a "gene map" using a comprehensive set of >200 gene-based markers developed from candidate genes for drought tolerance in barely. Later, a "transcript map" of barley after integrating more than 1000 gene-based markers (GTMs) has been also developed (Stein *et al.*, 2007). Such molecular maps can not only be compared with those of other related plant species in an efficient manner but also provide gene based molecular markers associated with the trait of interest after the QTL analysis.

7.0 CONCLUSION

New genomic technologies are expensive to develop, and returns from the initial research can take time. However, once the knowledge reaches a critical level gains accelerate enormously. All the information and accumulation of knowledge gained since 1990 will allow breeders, in future to incorporate and stack useful genes into several crop species. Products of breeding supplemented with MAS using gene based markers are just now beginning to become available and work is continuing to maximize the utility of the sequence databases to integrate desirable genes in crop plants.

8.0 REFERENCES

Aarts, M., Hekkert, B., Holub, E., Beynon, J., Stiekema, W. and Pereira, A. 1998. Identification of R-gene homologous DNA fragments genetically linked to disease resistance loci in *Arabidopsis thaliana*. Molecular Plant-Microbe Interaction, 11:251–258.

Abajian, C. 1994. Sputnik. (http://espressosoftware.com/pages/sputnik.jsp)

Aggarwal, R.K., Hendre, P.S., Yarshney, R.K., Bhat, P.R., Krishna, K.Y. and Singh, L. 2007. Identification, characterization and utilization of EST-derived genic microsatellite markers for genome analyses of coffee and related species. Theoretical and Applied Genetics, 114:359-372.

Andersen, J.R. and Lubberstedt, T. 2003. Functional markers in plants. Trends in Plant Science, 8: 554–560.

Ann J. Slade and Vic C. Knauf. 2005. TILLING moves beyond functional genomics into crop improvement. Transgenic Research, 14:109–115

Ayers, N.M., McClung, A.M., Larkin, P.D., Bligh, H.F.J., Jones, C.A. and Park, W.D. 1997. Microsatellite and single nucleotide polymorphism differentiate apparent amylase classes in an extended pedigree of US rice germplasm. Theoretical and Applied Genetics, 94:773-781.

Barbazuk, W.B., Emrich, S. and Schnable, P.S. 2007. SNP Mining from Maize 454 EST Sequences Cold Spring Harb. Protoc., doi:10.1101/pdb.prot 4786.

Barker, G., Batley, J.O., Sullivan, H., Edwards, K.J. and Edwards, D. 2003. Redundancy based detection of sequence polymorphism in expressed sequence tag data using auto SNP. Bioinformatics, 19:421-422.

Barrett, B., Griffiths, A., Schreiber, M., Ellison, N., Mercer, C., Bouton, J., Ong, B., Forster, J., Sawbridge, T., Spangenberg, G., Bryan, G. and Woodfield, D. 2004. A microsatellite map of white clover. Theoretical and Applied Genetics, 109:596-608.

Benson, G. 1999. Tandem repeats finder: a program to analyze DNA sequences. Nucleic Acids Research, 27:573–580.

Bhat, P.R., Kumar, K.Y., Hendre, P.S., Kumar, R.P., Yarshney, R.K. and Aggarwal, R.K. 2005. Identification and characterization of expressed sequence tags-derived simple sequence repeats, markers from robusta coffee variety 'C x R' (an interspecific hybrid of *Coffea canephora* x *Coffea congensis).* Molecular Ecology Notes, 5:80-83.

Bonfield, J.K., Rada, C. and Staden, R. 1998. Automated detection of point mutations using florescent sequence trace subtraction. Nucleic Acids Research., 26:3404-3409.

Bowers, J.E., Abbey, C., Anderson, S., Chang, C., Draye, X., Hoppe, A.H., Jessup, R., Lemke, C., Lennington, J. and Li, Z. 2003. A high-density genetic recombination map of sequence-tagged sites for sorghum, as a framework for comparative structural and evolutionary genomics of tropical grains and grasses. Genetics, 165:367-386.

Castelblanco, W. and Fregene, M. 2006. SSCP-SNP based conserved ortholog set (COS) markers for comparative genomics in Cassava (*Manihot esculenta* Crantz). Plant Molecular Biology Reporter, 24:229-236.

Castelo, A.T., Wellington, M. and Gao, G.R. 2002. Troll-Tandem Repeat Occurrence Locator. Bioinformatics, 18:634-636.

Cato, S.A., Gardner, R.C., Kent, J. and Richardson, T.E. 2001. A rapid PCR based method for genetically mapping ESTs. Theoretical and Applied Genetics, 102:296-306.

Causse, M.A., Fulton, T.M., Cho, Y.G., Ahn, S.N., Chunwongse, J., Wu, K., Xiao, J., Yu, Z., Ronald, P.C., Harrington, S.E., Second, G., McCouch, S.R. and Tanksley, S. D. 1994. Saturated molecular map of the rice genome based on an inter-specific backcross population. Genetics, 138:1251-1274.

Chao, S., Baysdorfer, C., Heredia-Diaz, 0., Musket, T., Xu, G., Coe, Jr. E.H. 1994. RFLP mapping of partially sequenced leaf cDNA clones in maize. Theoretical and Applied Genetics, 88:717-721.

Chee, P.W., Rong, J.K., Williams-Coplin, D., Schulze, S.R. and Paterson, A.H. 2004. EST derived PCR-based markers homologues in cotton. Genome, 47:449-462.

Chen, C., Zhou, P.Y.A., Huang, S., Gmitter Jr. F.G. 2006. Mining and characterizing microsatellites from citrus ESTs. Theoretical and Applied Genetics, 112:1248-1257.

Chen, X., Line, R.F. and Leung, H. 1998. Resistance gene analogs associated with a barley locus for resistance to stripe rust. *In*: Heller SR (*ed.*), International Conference on the Status of Plant and Animal Genome Research VI. Abstracts. San Diego, CA.

Childs, K.L., Klein, R.R., Klein, P.E., Morishige, D.T. and Mullet, J.E. 2001. Mapping genes on an integrated sorghum genetic and physical map using cDNA selection technology. Plant Journal, 27:243-256.

Chittoor, J., Kukreja, K., Leung, H., Nelson, R., Hulbert, S. and Leach, J. 1999. A database of candidate genes for utilization QTL analyses of disease resistance in plants. *In*: Heller S.R. (*ed.*), International Conference on the Status of Plant and Animal Genome Research VII. Abstracts, San Diego, CA.

Cho, Y.G., Ishii,T., Temnykh, S., Chen, X., Lipovich, L., McCouch, S.R., Parl, W.D., Ayers, N. and Cartinhour, S. 2000. Diversity of microsatellites derived from genomic liberaries and GenBank sequences in rice (*Oryza sativa* L.). Theoretical and Applied Genetics, 100:713-722.

Cummings, C.J. and Zoghbi, H.Y. 2000. Fourteen and counting: unraveling trinucleotide repeat diseases. Human Molecular Genetics 9: 909-916.

Donnelly, L.M., Gomes, V.M., Ogundiwin, E.A., Glosier, B.R., Sidhu, G.S. and Prince, J.P. 2005. Pepper (*Capsicum sp.*) and *Phytophthora capsici*: Molecular genetic analysis of a host/pathogen system. Plant & Animal Genomes XIII Conference.

Dvornyk, V., Sirvio, A., Mikkonen, M. and Savolainen, O. 2002. Low nucleotide diversity at the pal1 locus in the widely distributed *Pinus sylvestris*. Molecular Biology Evolution, 19:179–188.

Falque, M., Decousset, L., Dervins, D., Jacob, A.M., Joets, J., Martinant, J.P., Raffoux, X., Ribiere, N., Ridel, C., Samson, D., Charcosset, A. and Murigneux, A. 2005. Linkage mapping of 1454 new maize candidate gene I sequences or sites with in genes loci. Genetics, 170:1957-1966.

Faville, M.J., Vecchies, A.C., Schreiber, M., Drayton, M.C., Hughes, L.J., Jones, E.S., Guthridge, K.M., Smith, K.F., Sawbridge, T., Spangenberg, G.C., Bryan, G.T. and Forster, W. 2004. Functionally associated molecular genetic marker map construction in perennial ryegrass (*Lahum perenne* L.). Theoretical and Applied Genetics, 110:12-32.

Feltus, F.A., Wan, J., Schulze, S.R., Estill, J.C., Jiang, N. and Paterson. 2004. An SNP resource for ice genetics and breeding based on subspecies *Indica* and *Japonica* genome alignments. Genetics, 148:479-494.

Frary, A., Xu, Y., Liu',1., Tedeschi, S.M.E. and Tanksley, S. 2005. Development of a set of PCR-based anchor markers encompassing the tomato genome and evaluation of their usefulness for genetics and breeding experiments. Theoretical and Applied Genetics,111:291-312.

Fraser, L.G., Harvey, C.F., Crowhurst, R.N. and De Silva, H.N. 2004. EST-derived microstellites from *Actinidia* species and their potential for mapping. Theoretical and Applied Genetics,108:1010-1016.

Gao, L., Tang, J., Li, H. and Jia, J. 2003. Analysis of microsatellites in major crops assessed by computational and experimental approaches. Molecular Breeding, 12: 245-261.

Gao, L.F., Jing, R.L., Huo, N.X., Li, Y., Li, X.P., Zhou, R.H., Chang, X.P., Tang, J.F., Ma, Z.Y. and Jia, J.Z. 2004. One hundred and one new microsatellite loci derived from ESTs (EST-SSRs) in bread wheat. Theoretical and Applied Genetics, 108:1392-1400.

Gardiner, J., Coe, M.E.H., Melia-Hancock, S., Hoisington, D.A. and Chao, S., 1993. Development of a core RFLP map in maize using an immortalized F2 population. Genetics, 134:917-930.

Gaut, B.S. and Clegg, M.T. 1993. Nucleotide polymorphism in the *Adh1* locus of pearl millet (*Pennisetum glaucum*) (Poaceae). Genetics, 135:1091-1097.

Geffroy, V., Sicard, D., de Oliveira, J., Sévignac, M., Cohen, S., Gepts, P., Neema, C., Langin, T. and Dron, M. 1999. Identification of an ancestral gene cluster involved in the coevolution process between *Phaseolus vulgaris* and its fungal pathogen *Colletotrichum lindemuthianum*. Molecular Plant-Microbe Interaction, 12:774–784.

Gentzbittel, L., Mouzeyar, S., Badaoui, S., Mestries, E., Vear, F., Tourvieille de Labrouhe, D. and Nicolas, P. 1998. Cloning of markers for disease resistance in sunflower, *Helianthus annuus* L. Theoretical and Applied Genetics, 96:519-525.

Graham, J., Smith, K., Mac Kenzie, K., Jorgenson, L., Hackett, C. and Powell, W. 2004. The construction of a genetic linkage map of red raspberry *(Rubus idaeus* subsp. *idaeus)* based on AFLPs, genomic-SSR and EST-SSR markers. Theoretical and Applied Genetics, 109:740-749.

Graner, A., Jahoor, A., Schondelmaier, J., Siedler, H., Pillen, K., Fischbeck, G., Wenzel, G. and Herrmann, R.G. 1991. Construction of an RFLP map of barley. Theoretical and Applied Genetics, 83:250-256.

Gupta, P.K. and Rustgi, S. 2004. Molecular markers from the transcribed/expressed region of the genome in higher plants. Functional and Integrative Genomics, 4: 139-162.

Hackauf, B. and Wehling, P. 2002. Identification of microsatellite polymorphisms in an expressed portion of the rye genome. Plant Breeding, 121:17-25.

Hämäläinen, J.H., Sorri, V.A., Watanabe, K.N., Gebhardt, C. and Valkonen, J.P.T. 1998. Molecular examination of a chromosome region that controls resistance to potato Y and A potyviruses in potato. Theoretical and Applied Genetics, 96: 1036–1043.

Harushima, Y., Yano, M., Shomura, A., Sato, M., Shimano, T., Kuboki, Y., Yamamoto, T., Lin, S.Y., Antonio, B.A. and Parco, A. 1998. A high-density rice genetic linkage map with 2275 Markers using a single F_2 population. Genetics, 148:479-494.

Hausmann, B.I., Hess, D.E., Omanya, G.O., Folkertsma, R.T., Reddy, B.V., Kayento, M., Welz, H.G. and Geiger, H.H. 2004. Genomic regions influencing resistance to the parasitic weed *Striga hermonthica* in two recombinant inbred populations of sorghum. Theoretical and Applied Genetics, 109: 1005-1016.

Holton, T.A., Christopher, J.T., McClure, L., Harker, N. and Henry, R.J. 2002. Identification and mapping of polymorphic SSR markers from expressed gene sequences of barley and wheat. Molecular Breeding, 9:63-71.

Huntley, D., Baldo, A., Johri, S. and Sergot, M. 2006. SEAN: SNP prediction and display program utilizing EST sequence clusters. Bioinformatics, 22:495-496.

Jayashree, B., Ramu, P., Prasad, P., Bantte, K., Hash, C.T., Chandra, S., Hoisington, D.A. and Varshney, R.K. 2006.A database of simple sequence repeats from cereal and legume expressed sequence tags mined in silico: survey and evaluation. In Silico Biology I., 6:0054.

Kanazin, V., Marek, L. and Shoemaker, R.C. 1996. Resistance gene analogs are conserved and clustered in soybean. Proceeding of National Academy of Sciences USA, 93:11746–11750.

Kantety, R.V., La Rota, M., Matthews, D.E. and Sorrells, M.E. 2002. Data mining for simple sequence repeats in expressed sequence tags from barley, maize, rice, sorghum and wheat. Plant Molecular Biology, 48:501-510.

Khlestkina, E.K., Than, M.H.M., Pestsova, E.G., Roder, M.S., Malyshev, S.V., Korzun, V. and Bomer, A. 2004. Mapping of 99 new microsatellite-derived loci in rye *(Secale cereale* L.) including 39 expressed sequence tags. Theoretical and Applied Genetics,109:725-732.

Klein, P.E., Klein, R.R., Vrebalov, J. and Mullet, I.E. 2003. Sequence-based alignment of sorghum chromosome 3 and rice chromosome 1 reveals extensive conservation of gene order and one major chromosomal rearrangement. Plant Journal, 34:605-622.

Kota, R., Rudd, S., Facius, A., Kolesov, G., Thiel, T., Zhang, H., Stein, N., Mayer, K. and Graner, A. 2003. Snipping polymorphism from large EST collections in barley (*Hordeum vulgare* L.). Molecular Genetics Genomics, 270:24-33.

Kozik, A. and Michelmore, R. 2002. Compositae Genome Project Database. http://cgpdb.ucdavis.edu/COS_Arabidopsis/

Lai, Z., Livingstone, K., Zou, Y., Church, S.A., Knapp, S.J., Andrews, J. and Rieseberg, L.H. 2005. Identification and mapping of SNPs from ESTs in sunflower. Theoretical and Applied Genetics,111:1532-1544.

Leister, D., Ballvora, A., Salamini, F. and Gebhardt, C. 1996. A PCR-based approach for isolating pathogen resistance genes from potato with potential for wide application in plants. Nature Genetics, 14:421–429.

Leister, D., Kurth, J., Laurie, D.A., Yano, M., Sasaki ,T., Devos, K., Graner, A. and Schulze-Lefert P. 1998. Rapid reorganization of resistance gene homologues in cereal genomes. Proceeding of National Academy of Sciences USA, 95:370-375.

Maccaferri, M., Sanguineti, M.C., Natoli, E., Arous-Ortega, J.L., Ben Salem, M., Bort, J., Chenenaoui, S., Deambrogio, E., Garcia, D.M.L.and De Montis, A. 2006. A panel of elite accessions of durum wheat (*Triticum durum* Defs) suitable for association mapping studies. Plant Genetic Resources, 4:79-85.

Marth, G.T., Korf, I., Yandell, M.D., Yeh, R.T., Zhijie, G., Zakeri, H., Stitziel, N.O., Hillier, L., Kwok, P.Y. and Gish, W.R. 1999. A general approach to single-nucleotide polymorphism discovery. Nature Genetics, 23:452-456.

McIntyre, C.L., Casu, R.E., Drenth, J., Knight, D., Whan,V.A., Croft, B.J., Jordan, D.R. and Manners J.M. 2005. Resistance gene analogues in sugarcane and sorghum and their association with quantitative trait loci for rust resistance. Genome, 48: 391-400.

Morgante, M., Hanafey, M. and Powell, W. 2002. Microsatellites are preferentially associated with nonrepetitive DNA in plant genomes. Nature Genetics, 30: 194–195.

Neale, D.B. and Savolainen, O. 2004. Association genetics ofcomplex traits in conifers. Trends Plant Science, 9:325–330.

Nickerson, D.A., Tobe, V.O. and Taylor, S.L. 1997. Polyphred: automating the detection and genotyping of single nucleotide substitutions using fluorescence-based resequencing. Nucleic Acids Research. 25:2745-2751.

Nicot, N., Chiquet, V., Gandon, B., Amilhat, L., Legeai, F., Leroy, P., Bernard, M. and Sourdille, P. 2004. Study of simple sequence repeat (SSR) markers from wheat expressed sequence tags (ESTs). Theoretical and Applied Genetics,109:800-805.

Nodari, R.O., Tsai, S.M., Guzman, P., Gilbertson, R.L. and Gepts, P. 1993. Toward an integrated linkage map of common bean. III. Mapping genetic factors controlling host-bacteria interactions. Genetics, 134:341–350.

Olsen, K.M., Halldorsdottir S.S., Stinchcombe, J.R., Weinig, C., Schmitt, J. and Purugganan, M.D. 2004. Linkage disequilibrium mapping of *Arabidopsis* CRY2 flowering time alleles. Genetics, 167:1361–1369.

Park, Y.H., Alabady, M.S., Ulloa, M., Sickler, B., Wilkins, T.A., Yu, J., Stelly, D.M., Kohel, R.J., El-Shihy, O.M. and Cantrell, R.G. 2005. Genetic mapping of new cotton fiber loci using EST-derived microsatellites in an interspecific recombinant inbred line cotton population. Molecular Genetic Genomics, 274:428-441.

Pflieger, S., Lefebvre, V. and Causse, M. 2001. The candidate gene approach in plant genetics: a review. Molecular Breeding, 7:275–291.

Pflieger, S., Lefebvre, V., Caranta, C., Blattes, A., Goffinet, B. and Palloix, A. 1999. Disease resistance gene analogs as candidates for QTLs involved in pepper/ pathogen interactions. Genome 42:1100–1110.

Picoult-Newberg, L., Ideker, T.E., Pow, M.G., Taylor, S.L., Donaldson, M.A., Nickerson, D.A. and Boyce-Jacino, M. 1999. Mining SNPs from EST databases. Genome Research, 9:167-174.

Pilet, M.L. 1999. Analyse génétique de la résistance du colza à la nécrose du collet et à la cylindrosporiose à l'aide des marqueurs moléculaires. Ph-D Thesis of Institut National Agronomique Paris-Grignon, France, 182 pp.

Qi, X., Pittaway, T.S., Lindup, T.S., Liu, S., Liu, H., Waterman, E., Padi, F.K., Hash, C.T., Zhu, J., Gale, M.D. and Devos, K.M. 2004. An integrated genetic map and a new set of simple sequence repeat markers for pearl millet, *Pennisetum glaucum.* Theoretical and Applied Genetics, 109:1485-1493.

Quint, M., Mihaljevic, R., Dussle, C., Xu, M., Melchinger, A., Lübberstedt, T. 2002. Development of RGA-CAPS markers and genetic mapping of candidate genes for sugarcane mosaic virus resistance in maize. Theoretical and Applied Genetics, 105:355-363.

Rafalski, J.A. and Tingey, S.V. 1993. Genetic diagnostics in plant breeding: RAPDs, microsatellites and machines. Trends in Genetics, 9:275-280.

Rafalski, A. 2002. Application of single nucleotide polymorphism in crop genetics. Current opinion in Plant Biology, 5:94-100.

Ramu, P., Bantte, K., Ashok, C.K., Jayashree, B., Rolf, T.F., Senthilvel, S., Mahalakshmi, V., Reddy, A.L., Fakrudin, B. and Hash, C.T. 2006. Development and mapping of EST-SSR markers in sorghum for comparative mapping with rice. *In*: Plant and animal genome XIV conference, San Diego, CA, USA, P 200 (http://www.intl-pag.org/14/abstractsIPAGI4]200.html).

Rivkin, M.I., Vallejos, C.E. and McClean, P.E. 1999. Disease resistance related sequences in common bean. Genome, 42:41–47.

Rossi, M., Araujo, P.G., Florence, P., Grasmeur, O., Dias, V.M., Hui, C., Sluys, M.A.V. and D'Hont, A. 2003. Genome distribution and characterization of EST derived sugarcane resistance gene analogs. Molecular Genetic Genomics, 269:406-419.

Rostoks, N., Mudie, S., Cardle, L., Russell, J., Ramsay, L., Booth, A., Svensson, J., Wanamaker, S., Walia, H., Rodriguez, E., Hedley, P., Liu, H., Morris, **J.,** Close, T., Marshall, D. and Waugh, R. 2005. Genome-wide SNP discovery and linkage analysis in barley based on genes responsive to abiotic stress. Molecular Genetics and Genomics, 274:515-527.

Rudd, S., Schoof, H. and Klaus, M. 2005. PlantMarkers- A database of predicted molecular markers from plants. Nucleic Acids Research, 33:D628-D632.

Rungis, D., Berube, Y., Zhang, J., Ralph, S., Ritland, C.E., Ellis, B.E., Douglas, C., Bohlmann. J. and Ritland, K. 2004. Robust simple sequence repeats markers for spruce *(Picea* spp.) from expressed sequence tags. Theoretical and Applied Genetics, 109:1283-1294.

Russel, J., Booth, A., Fuller, J., Harrower, B. and Hedley, P. 2004. A comparison sequence based polymorphism and haplotype content in transcrbed and anonymous regions of the barley genome. Genome, 47:389-398.

Salse, J., Piegu, B., Cooke, R. and Delseny, M. 2004. New *in silico* insight into the synteny between rice (*Oryza sativa* L.) and maize (*Zea mays* L.) highlights reshuffling and identifies new duplications in the rice genome. Plant Journal, 38:396-409.

Sargent, D.J., Clarke, J., Simpson, D.W., Tobutt, K.R., Aru's, P., Monfort, A., Vilanova, S., Denoyes-Rothan, B., Rousseau, M., Folta, K.M., Bassil, N.V. and Battey, N.H. 2006. An enhanced microsatellite map of diploid *Fraga ria.* Theoretical and Applied Genetics, 112:1349-1359.

Shattuck-Eidens, D.M., Bell, R.N., Neuhausen, S.L. and Helentjaris, T. 1990. DNA sequence variation within maize and melon: observations from polymerase chain reaction amplification and direct sequencing. Genetics, 126:207-217.

Simko, I., Haynes, K.G., Ewing, E.E., Costanzo, S., Christ, B.J. and Jones, R.W. 2004. Mapping genes for resistance to *Verticillium albo-atrum* in tetraploid and diploid potato populations using haplotype association tests and genetic analysis. Molecular Genetics and Genomics, 271:522–531.

Slade, A.J. and Knauf, S.C. 2005. TILLING moves beyond functional genomics and crop improvement. Transgenic Research, 14:109-115.

Somers, D.J., Robert, K., Moniwa, M. and Walsh, A. 2003. Mining single-nucleotide polymorphisms from hexaploid wheat ESTs. Genome, 46:431-437.

Song, Q.J., Marek, L.F., Shoemaker, R.C., Lark, K.G., Concibido, V.C., Delannay, X., Specht, I.E. and Cregan, P.B. 2004. A new integrated genetic linkage map of the soybean. Theoretical and Applied Genetics, 109:122-128.

Speulman, E., Bouchez, D., Holub, E.B. and Beynon, J.L. 1998. Disease resistance gene homologs correlate with disease resistance loci of *Arabidopsis thaliana*. Plant Journal, 14:467–474.

Stein, N., Prasad, M., Scholz, U., Thiel, T., Zhang, H., Wolf, M., Kota, R., Varshney, R., Perovic, D., Grosse, 1. and Graner, A. 2007. A I,OOO-ioci transcript map of the barley genome: new anchoring points for integrative grass genomics Theoretical and Applied Genetics, 114:823-839.

Tang J, Gao L, Cao Y and Jia J. 2006. Homologous analysis of SSR-ESTs and transferability of wheat SSR-EST markers across barley, rice and maize. Euphytica, 151:87-93.

Taylor, N.E. and Greene, E.A. 2003. PARSESNP: a tool for the analysis of nucleotide polymorphisms. Nucleic Acids Research, 31:3808-3811.

Temnykh, S., DeClerck, G., Lukashova, A., Lipovich, L., Cartinhour, S. and McCouch, S. 2001. Computational and experimental analysis of microsatellites in rice (*Oryza sativa* L.): frequency, length variation, transposon associations, and genetic marker potential. Genome, 14:1812-1819.

Thiel, T., Michalek, W., Varshney, R.K. and Graner, A. 2003. Exploiting EST database for the development and characterization of gene derived SSR-markers in barley (*Hordeum vulgare* L.). Theoretical and Applied Genetics, 106:411-422.

This, P., Lacombe, T., Cadle Davidson, M. and Owens, C.L. 2007. Wine grape (*Vitis vinifera* L.) color associates with allelic variation in the domestication gene Vvmyb A 1. Theoretical and Applied Genetics, 114:1432-2242.

Thompson, J.A., Nelson, R.L. and Vodkin, L.D. 1998a. Identification of diverse soybean germplasm using RAPD markers. Crop Science, 38:1348-1355.

Thompson, J.A. and Nelson, R.L. 1998b. Utilization of diverse germplasm for soybean yield improvement. Crop Science, 38:1362-1368.

Thornsberry, J.M., Goodman, M.M., Doebley, J., Kresovich, S., Nielsen, D. and Buckler, E.S. IV. 2001. Dwarf8 polymorphism associate with variation in flowering time. Nature Genetics, 28:286–289.

Toth, G., Gaspari, Z. and Zurka, J. 2000. Microsatellites in different eukaryotic genomes: Survey and analysis. Genome Research, 10:967-981.

Van der Voort, J.R., Sorensen, A., Lensink, D., Van der Meulen, M., Michelmore, R. and Peleman, J. 2004. Decay of linkage disequilibrium in the dm3 resistance-gene cluster of lettuce. *In*: Plant & Animal Genomes XII Conference, 10–14 January, Town & Country Convention Center, San Diego, CA, P748.

Varshney, R.K., Korzun, V.and Borner, A. 2004. Molecular maps in cereals: Methodology and preogress. *In*: Gupta PK, Varshney RK (*eds*.) Cereal genomics. Kluwer Academic Publishers, The Netherlands, pp 35-60.

Varshney, R.K., Mahendar, T., Aggarwal, R.K. and Borner, A. 2007. Genic molecular markers in plants: development and applications Varshney R.K and R.

Tuberosa (eds.), Genomics-Assisted Crop Improvement: Vol. 1: Genomics Approaches and Platforms, 13-29.

Varshney, R.K., Chabane, K., Hendre, P.S., Aggarwal, R.K. and Graner, A. 2007. Comparative assessment of EST-SSR, EST-SNP and AFLP markers for evaluation of genetic diversity and conservation of genetic resources using wild, cultivated and elite barleys. Plant Science, 173:638-649.

Varshney, R.K., Graner, A. and Sorrells, M.E. 2005. Genomic-assisted breeding for crop improvement. Trends in Plant Science, 10:621-630.

Varshney, R.K., Grosse, I., Hahnel, U., Siefken, R., Prasad, M., Stein, N., Langridge, P., Altschmied, L. and Graner, A. 2006. Genetic mapping and BAC assignment of EST-derived SSR markers shows non-uniform distribution of genes in the barley genome. Theoretical and Applied Genetics, 113:239-250.

Varshney, R.K., Marcel, T.C., Ramsay, L., Russell, J., Roder, M., Stein, N., Waugh, R., Langridge, P., Niks, R.E. and Graner, A. 2007. A high density barley micro satellite consensus map with 775 SSR loci. Theoretical and Applied Genetics, 114:1091-103.

Wang, H.Y., Liu, D.C., Yan, Z.H., Wei, Y.M. and Zheng, Y.L. 2005. Cytological characteristics of hybrid F_2 population between *Triticum aestivum* L. and *T. durum* with reference to wheat breeding. Journal of Applied Genetics, 46:365-369.

Weiland, J. and Koch, G. 2004. Sugarbeet leaf spot disease (*Cercospora beticola* Sacc.). Molecular Plant Pathology, 5:157-166.

Willsmore, K.L., Eckermann, P., Varshney, R.K., Graner, A., Langridge, P., Pallotta, M., heong, J. and Williams, K.J. 2006. New eSSR and gSSR markers added to Australian barley maps. Australian Journalof Agricultural Research, 57:953-959.

Woo, S.S., Sicard, D., Arroyo-Garcia, R., Ochoa, O., Nevo, E., Korol, A., Fahima, T. and Michelmore, R.W. 1998. Many diverged resistance genes of ancient origin exist in lettuce. *In*: Heller S.R. (*ed.*), International Conference on the Status of Plant and Animal Genome Research VI. Abstracts. San Diego, CA.

Yu, I.K., La Rota, M., Kantety, R.V. and Sorrells, M.E. 2004. EST derived SSR markers for comparative mapping in wheat and rice. Molecular Genetics and Genomics, 271:742-751.

Yu, Y.G., Buss, G.R. and Saghai Maroof, M.A. 1996. Isolation of a superfamily of candidate disease-resistance genes in soybean based on a conserved nucleotide-binding site. Proceedings of National Academy of Sciences USA, 93:11751–11756.

Zhang, L.Y., Bernard, M., Leroy, P., Feuillet, C. and Sourdille, P. 2005. High transferability of bread wheat EST-derived SSRs to other cereals. Theoretical and Applied Genetics, 111:677-687.

Zhang, W.K., Wang, Y.J., Luo, G.Z., Zhang, J.S., He, C.Y., Wu, X.L., Gai, J.Y. and Chen, S.Y. 2004. QTL mapping of ten agronomic traits on the soybean *(Glycine max* L. Merr.) genetic map and their association with EST markers. Theoretical and Applied Genetics, 108:1131-1139.

Molecular Plant Breeding: Principle, Method and Application
Eds : R.K. Singh, Rajesh Singh, Guoyou Ye, A. Selvi and G.P. Rao
Studium Press LLC, Texas, USA, 2009, pp. 295-319

Marker-Assisted Recurrent Backcrossing in Cultivar Development

GUOYOU YE[1*], *FRANCIS OGBONNAYA*[2] *and MAARTEN VAN GINKEL*[2]

ABSTRACT

Marker-assisted recurrent backcrossing (MARB), the combined use of marker-assisted selection (MAS) and recurrent backcrossing, is a versatile method for plant breeding and genetic studies. The benefits of MARB are well demonstrated and documented in theoretical and simulation studies, and confirmed by empirical applications. In MARB, markers are used during recurrent backcrossing to select for the presence of the target gene (foreground selection), to select against the donor genome contribution (background selection) to reduce the introgressed segment's size and thus potential linkage drag. MARB reduces the number of backcrossing needed to recover most of the recurrent parent genome in 3-4 generations, while one or two target genes are introgressed from the donor. For foreground selection markers are most useful for traits that are expensive and/or difficult to measure. Linkage drag associated with the target gene(s), when present, is difficult if not impossible to remove by phenotypic selection. In this chapter, we summarised some of the theoretical and simulation results of MARB, and provided comprehensive summary of the use of MARB in practical crop breeding.

Key Words: Background selection, Foreground selection, Linkage drag, Marker-assisted selection, Recurrent backcrossing.

[1] *Bundoora Centre, Biosciences Division, Department of Primary Industries Victoria, and Molecular Plant Breeding Cooperative Research Centre, 1 Park Drive, Bundoora Vic 3086, Australia*
[2] *The International Center for Agricultural Research in the Dry Areas (ICARDA), P.O. Box 5466, Aleppo, Syrian Arab Republic*
**Corresponding author e-mail: guoyou.ye@dpi.vic.gov.au*

1.0 INTRODUCTION

Recurrent backcrossing has long been used by breeders as an efficient method for improving a small number of unsatisfactory traits within an existing elite cultivar (Allard, 1999). In recurrent backcrossing, F_1 or BC individuals with the desired characteristics in heterozygous form are selected and then back-crossed to the recurrent parent for several generations (see Fig. 1 for an example scheme). The parental line carrying the target gene is usually called the "donor parent", while the one used as a parent in all the backcrossing generations is called as "recipient" or "recurrent parent". An important objective of recurrent backcrossing is to reduce the contribution of the donor genome, as the aim is to move just a few of its genes responsible for the target trait into the recurrent parents genetic background. Traditionally, recurrent backcrossing is mainly used to improve qualitatively inherited traits such as disease and insect resistance, since the presence of target trait genes must be confirmed by phenotyping in the resulting cross generation mostly at the individual level, and individual phenotypic

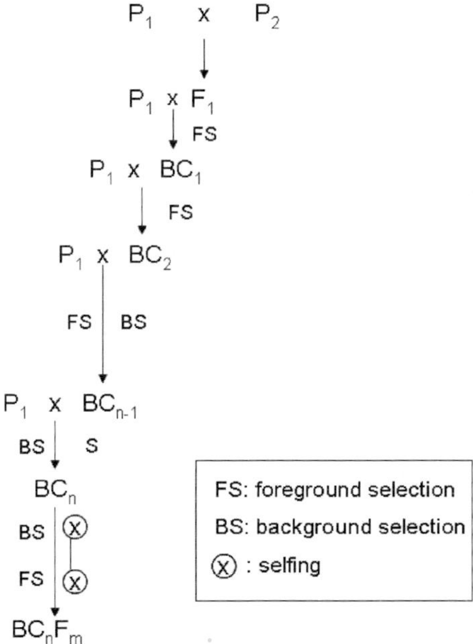

Fig. 1. An example recurrent backcrossing scheme with P_1 and P_2 as the recurrent and donor parent, respectively.

performance is a good indicator of the genotype only if genes have a major effect on phenotypic performance and the error of phenotyping is minimal (Ye and Smith, 2008).

The development of modern plant molecular and quantitative genetics in the last two decades has seen the rapid accumulation of markers tightly linked to or within the genes of agronomic interest. Marker-assisted selection (MAS), the incorporation of markers in the selection criterion for achieving breeding objectives, has been widely explored by plant breeders to further enhance the rate of genetic gain. In the context of recurrent backcrossing, MAS widened the applicability of recurrent backcrossing at least in the following aspects. Firstly, for traits that are simply inherited, but that are difficult or expensive to measure phenotypically, and/or that do not have a consistent phenotypic expression under certain specific selection conditions, the efficiency of phenotypic selection is low. The use of markers for foreground selection makes the transfer of target genes feasible and economic. Secondly, quantitative traits, which are generally not targeted by a recurrent backcrossing approach, can be improved using recurrent backcrossing, if major quantitative trait loci (QTL) affecting the trait have been identified. Thirdly, markers provide an effective option to control linkage drag and to speed up the recovery of recurrent genome and make the use of genes contained in unadapted resources easier (Tanksley, 1983; Ragot *et al.*, 1995; Visscher *et al.*, 1996; Hospital, 2001; Dekkers and Hospital, 2002; Frisch and Melchinger, 2001a; Ribaut *et al.*, 2002; Frisch and Melchinger, 2005; Kuchel *et al.*, 2005; Ribaut *et al.*, 2007).

MARB is routinely the method of choice for inbred line development targeted at improving traits controlled by major genes. In this chapter we provide comprehensive summary of the theory and use of MARB in practical crop breeding and provide some simulation results of MARB.

2.0 MARKER-ASSISTED FOREGROUND SELECTION (MAFS)

MAFS was proposed by Tanksley (1983). The presence of a target allele in an individual is diagnosed by monitoring the genotype with markers linked to the gene for alleles of the donor parent. This is a powerful tool for manipulation of oligogenic traits under numerous situations in plant breeding (Melchinger, 1990), but also for manipulation of QTL (Stuber, 1995). We know that the frequency of the desired allele is 0.25 in BC_1 population of two homozygous genotypes. The frequency of the desired allele can be increased by selection provided selection is effective. When linked marker(s) are used for selection, the frequency in the selected population can be substantially higher than 0.25 and approaches 1 (unity), which happens when the recombination rate is 0 (perfect or diagnostic marker).

If Q is the desired allele, the recombination frequencies between two flanking markers and the Q gene (r_1 and r_2) are known. Table 1 gives the expected genotypes and their frequencies in a BC_1 population created by

Table 1. Expected genotypes and their frequencies in a BC_1 population created by M1QM2/M1QM2 /m1qm2/m1qm2 // m1qm2/m1qm2

Genotype	Frequency equation	Genotype frequency: Recombination rate between target and flanking markers ($r_1 = r_2$)					
		0.010	0.050	0.100	0.200	0.300	0.400
M1QM2/m1qm2	$0.5(1-r_1)(1-r_2)$	0.490	0.451	0.405	0.320	0.245	0.180
M1Qm2/m1qm2	$0.5(1-r_1)r_2$	0.005	0.024	0.045	0.080	0.105	0.120
M1qM2/m1qm2	$0.5r_1r_2$	0.000	0.024	0.045	0.080	0.105	0.120
M1qm2/m1qm2	$0.5r_1(1-r_2)$	0.005	0.001	0.005	0.020	0.045	0.080
m1QM2/m1qm2	$0.5r_1(1-r_2)$	0.005	0.024	0.045	0.080	0.105	0.120
m1Qm2/m1qm2	$0.5r_1r_2$	0.000	0.001	0.005	0.020	0.045	0.080
m1qM2/m1qm2	$0.5(1-r_1)r_2$	0.005	0.024	0.045	0.080	0.105	0.120
m1qm2/m1qm2	$0.5(1-r_1)(1-r_2)$	0.490	0.451	0.405	0.320	0.245	0.180

r_1 and r_2 is the recombination rate between marker M_1 and M_2 and the target gene (Q)

crossing genotype M1QM2/M1QM2 and m1qm2/m1qm2 and backcrossing to m1qm2/m1qm2. The frequency of the Q allele in the BC_1 F_1 population after selection based on one marker and two equally distanced flanking markers for different recombination frequencies are computed and given in Fig. 2. It can be seen from Fig. 2 that (1) the tighter the linkage between the marker and the target gene, the higher the Q allele frequency in the selected

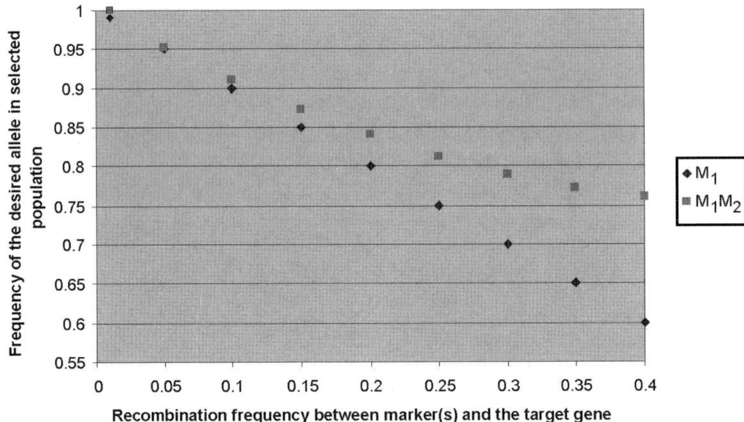

Fig. 2. Frequencies of the desired allele in the BC_1 population after selection using one linked marker or two equally distanced flanking markers.

population; (2) using flanking markers results in a higher Q allele frequency, particularly when markers are not tightly linked.

Knowing the frequency of the desirable genotype in the population, the minimum population size required to ensure with a probability (*i.e.* 99%) that at least m individuals with the desired genotype are obtained can easily be computed using well established method (Melchinger, 1990). For foreground selection, all genotypes with the desired allele are satisfactory in the sense that the desired genotype of the target gene (target genotype) can be recovered in later generations. However, the probabilities of recovering the target genotype in later generations from these satisfactory genotypes are different. Moreover, the genotypes that are finally derived from the different satisfactory genotypes will not be the same for genes other than the targeted one. Frisch *et al.* (1999a, b) detailed how to effectively explore these differences.

MAFS is mainly used when the effects of target alleles are difficult or impossible to measure phenotypically. For instance, when the target is inherited recessively, the presence or absence of the target gene in a backcross individual cannot by definition be known by observing its phenotypic performance. Traditionally, a phenotypic assay of progeny generated either by selfing of a cross between donor and recurrent parent or their F_1 crossed to the donor parent is used to determine whether an individual is kept or discarded (Allard, 1999). This is not only costly but time-consuming as well.

Similarly, when breeders identify a new resistant gene and would like to transfer it into an existing resistant cultivar containing other resistance genes to increase the durability, MAFS can effectively be used, since phenotyping may, depending on the circumstances, not be sufficiently accurate to identify the presence of the additional resistance gene. Individuals with the resistant phenotype may not indicate the presence of the added target resistant gene, since the effect of the target gene may be totally or partially masked by the other resistance genes. This is one of the reasons why the application of pyramiding multiple resistance genes for the same disease is not widely successfully deployed, although its potential as a strategy for the development of cultivars with durable resistance has long been recognised. This is very important if a breeder wants to transfer a gene conferring resistance to a devastating disease, which is not yet present in his/her testing environment, in order to prepare for the future (when virulences within the pathogens may be combined) or increase the adoption of his/her cultivars by growers in areas with the disease. The breeder will not be able to observe the resistance phenotypically due to obvious reasons mentioned above, and phenotyping has to be done by a collaborator in an environment, where the additional virulence is present.

MAFS will also be advantageous if phenotypic assays are more expensive than marker assays. For instance, the bio-assays for measuring resistance to nematodes in soybean (Young, 1999), tomato (Tanksley, 1983) and wheat (Eagles *et al.*, 2001) are expensive and unreliable, and breeding for resistance has been very slow. However, cultivars with resistance have been developed by using those markers that are tightly linked to the resistance genes in all three crops. These may be among the best examples of the practical use of markers in crop breeding.

3.0 MARKER-ASSISTED BACKGROUND SELECTION

It is well-known that if selection is applied for the desired characteristics only (foreground selection), the proportion of donor genome for all chromosomes except the one carrying the target gene is expected to be reduced by on average one-half at each backcross generation. On the chromosome carrying the target gene, the reduction of the donor genome is slower due to selection for the presence of the target gene, resulting in linked sequences also being selected. Marker-assisted background selection (MABS) was proposed by Young and Tanksley (1989) to accelerate recovery of the recurrent parent genome (RPG). In MABS, individuals are selected which are homozygous for the alleles of the recurrent parent at as many marker loci as are available in the hope of covering the entire genome. A key parameter in quantifying the effectiveness of MABS is the donor genome content (DGC). To estimate the genomic composition of individuals using markers, the most basic estimate of donor DGC could be to score the genotype at the markers, and then estimate DGC from the ratio of markers heterozygous for the donor allele over the total number of marker scores. This is a simple estimate that has the major drawback of being highly dependent upon the placement of markers along the genome. If markers are evenly spread and not too far apart from each other, the estimate is acceptable. However, if markers are not evenly distributed, weighting them equally is clearly not the best solution. Young and Tanksley (1989b) computed DGC by taking into account distances between markers. A chromosomal segment flanked by two markers of the donor type (DD) is considered as 100% donor type, a chromosomal segment flanked by two markers of recipient type (RR) is considered as 0% donor type, and a chromosomal segment flanked by one marker of donor type and one marker of recipient type (DR) is considered as 50% donor type. Servin *et al.* (2002) proposed an efficient algorithm for the computation of multilocus genotypes in complex pedigrees. This allows the calculation of the probability of the presence of the donor type, given the genotypes at the markers and their locations at any point of a segment flanked by two markers. Averaging over all possible positions between the two markers provides an estimate of DGC. They demonstrated that the estimate of Young and Tanksley (1989b) is not always correct: In DD segments, DGC is below 100% due to possible double crossovers between the markers. This error is minimal in BC_1 and increases in more advanced BC generations. Likewise in RR segments, DGC can be more than 0% due to possible double recombination between the markers. This error is maximal in BC_1 and decreases in more advanced BC generations. In a DR segment, DGC is exactly 50% in BC_1, but decreases to below 50% in advanced BC generations.

Since in recurrent backcrossing the number of markers of the recipient type increases with BC generations, even with no selection for the presence of the markers, it is expected that many segments are of the DR type, and hence the overall error might be important. When estimating the DGC in a chromosomal segment flanked by two markers at a given generation, not only the genotypes of the two markers at that generation are informative. In fact, the genotypes of the two markers in previous generations are also important, and so are the genotypes of non-flanking markers. Taking this additional information into account allows in some cases for some gain in precision of the estimate of the most probable genotype at any point in the segment. A computer program (Grafgen) was developed to estimate the genomic composition of individuals issued from complex inbred pedigrees and graphically presents the genotype (precise graphic genotype) (Servin and Hosptial, 2002).

MABS has been investigated by various authors using simulation (Hospital *et al.*, 1992; Openshaw *et al.*, 1994; Visscher *et al.*, 1996; Frisch *et al.*, 1999a, b). Tanksley *et al.* (1989) stated that a sufficiently high proportion of the RPG is recovered after three generations of MABS. Hospital *et al.* (1992) expected a saving of two backcross generations because of MABS. Frisch *et al.* (1999b) demonstrated that the number of backcross generations required for the introgression of one target gene was reduced by two to four backcross generations. Frisch and Melchinger (2001a) showed that a saving of three backcross generations due to MABS is a realistic goal for simultaneous introgression of two genes.

Several useful points for practical breeders deriving from these theoretical and simulation studies are (1) four independent markers per chromosome not carrying the target gene are enough for background selection; (2) the use of equally spaced markers reduces the population size required; (3) background selection is more efficient if it is applied in an advanced generation. The latter is the case because the selected individual contributes only half of its genome to the progeny in the next backcross generation, and as a result the reduced donor genome from background selection has a carry-over rate of one-half to the next backcross generation; (4) using a larger population size in advanced generation is advantageous, if genotyping cost is high. With the advance of backcross generations many marker loci become fixed for the recurrent allele and do not need to be genotyped; and (5) multiple –stage selection at each generation by exploring different types of marker genotypes can be used to reduce the number of marker genotyping required. However, considering that the cost of genotyping is largely the cost of DNA isolation as opposed to the cost of additional marker assays, this point might be less important.

4.0 REDUCTION OF LINKAGE DRAG

As discussed above, on the chromosome carrying the target gene the reduction of the donor genome is slower due to the selection for the presence of the target gene. This is particularly true for the chromosome region immediately surrounding the target gene. This may result in the phenomenon known as linkage drag, when genes for a negative trait(s) are closely associated with the introgression of the target gene. Generally linkage drag is assumed to be negative, but of course it need not be. Linkage drag is identified as the main cause for the differences between the recipient line and the converted line (Zeven *et al.*, 1983). Obviously, minimizing the size of the introgressed segment from the donor parent can be an effective way to eliminate/reduce linkage drag. Theoretical results (Stam and Zeven, 1981) show that the donor segment attached to the target allele remains surprisingly large even after many generations of conventional backcrossing. Young and Tanksley (1989) found that up to 51 cM of the chromosome segment attached to a resistance gene after six backcross generations in tomato. Tightly linked markers flanking the target gene can be used to reduce the length of donor chromosome segment attached to the target gene and thereby potentially linkage drag (Frisch and Melchinger, 2001b; Hospital, 2001). The reduction of chromosome segment depends on the flanking marker distance, population size and the number of breeding generations (recombination duration). Frisch *et al.* (1999a) developed equations for calculating the minimal population size needed to obtain an individual with at least one carrier of the target allele and being homozygous for the recurrent parent allele at one or both flanking markers. Hospital (2001) and Frisch and Melchinger (2001b) derived the probability distribution of the size of donor chromosome segments around the introgressed gene. The probability distribution function was then used to derive the expected segment length and variance, which can be used to investigate the effect of marker distance. The general conclusions in the context of practical breeding are (1) a small flanking marker distance is advantageous. Heterozygosity at tightly linked foreground selection markers results in a high probability that an individual carries the target gene. Moreover, homozygosity at tightly linked background selection markers results in a short donor chromosome segment around the target gene. The basic conclusion is that selecting for distant markers over several successive backcross generations cannot provide a better reduction of linkage drag than using close markers. Unsurprisingly using very close markers appears the best way to reduce linkage drag substantially. However, the population size needed to obtain recombinant genotypes increases rapidly with the reduction of marker distance and thus the genotyping cost is increased by using closer markers. (2) When flanking makers are used, symmetric marker brackets (*i.e.* the flanking markers that are equally distant from the target gene) are preferable.

This not only reduces the required population size, but also reduces the probability that a selected recombinant has a relatively large intact donor chromosome segment. (3) When the distance between the flanking markers is short (<20 cM), the number of backcross generations performed has little impact on the reduction of donor segment length. (4) The probability of having a smaller intact segment is greater with selection in an early generation than with selection in an advanced generation, because crossover events in subsequent generations after selection may result in the reduction of the intact chromosome segment. However, the required population size to obtain the desired recombinant genotype may be prohibitive. (5) To reduce the population size required it is generally more profitable to allow three or more successive backcrosses. For close markers, the probability of double recombination is much lower than the probabilities of single recombination, and thus the population size needed to obtain a double recombinant in a single backcross generation is much lower than twice the population size needed to obtain a single recombinant. Therefore, total population size can be considerably reduced if selection is conducted in two generations, selecting in the first generation a single recombinant on one side of the target gene and then selecting in the second generation for a single recombinant one the other side. Allowing more than two generations permits an even further reduction of the total number of individuals needed.

Foreground selection for the target genes and background selection for the reduction of the contribution of donor genome and linkage drag must be combined to obtain an acceptable cultivar using recurrent backcrossing. Using the results of the theoretical studies mentioned above as guidelines, the logical steps of marker assisted selection in a backcross generation are (1) Select individuals carrying the target allele. Perfect markers or the closest flanking markers are most useful. (2) Select individuals homozygous for recurrent parent genotype at loci close to the target gene or markers linked to it, (3) Select individuals homozygous for recurrent parent genotype at few (*i.e.* 2) marker loci on the chromosome carrying the target allele, and (4) Select individuals that are homozygous for recurrent parent genotype at other marker loci of the other chromosomes (3- 4 independent markers per chromosome). It is clear from these studies that it is easier to reduce the contribution of donor genome for chromosomes not involving the target gene. Moreover, although it makes sense to control linkage drag by reducing the introgressed segment size, segment size does not necessarily correspond to the presence or absence of linkage drag. In other words, a shorter segment may cause serious linkage in some cases, while no obvious linkage drag is caused by a fairly long segment in other cases. Therefore, it might be easier if we separate the background control and the reduction of segment size. This implies that lines with the target gene are developed first by foreground selection and background selection for chromosomes not carrying the target gene and then phenotypically tested for key agronomic traits to see whether there is significant undesirable linkage drag. Nevertheless, it is always beneficial to identify the recombinants between the target gene and its close flanking markers if they arise. Therefore, it is advisable that markers flanking to the target gene should always be screened first, although the presence of recombination is not used as a condition to select against donor genome. The best line can be used as donor to reduce the segment size around the target gene by recurrent backcrossing if the undesirable linkage drag is proven significant. This strategy will avoid spending time and resources in reducing/eliminating non-existing or unimportant linkage drag. It also has the potential to utilise the possible beneficial alleles linked to the target allele. When the program is set up to reduce linkage drag through reducing the size of the segment containing the target allele, the closest available flanking markers should be used. Several generations of backcrossing are required to reduce the total number of individuals to be screened. Background selection for reducing the contribution of donor genome from regions other than that surrounding the target allele is not required. Continuous selection on the flanking markers should be applied to identify the recombinant individuals as soon as possible, since the exact generation at which the recombination took place is difficult if not impossible to predict. Hospital and Decoux (2002) developed a computer program for optimising a multi-generation backcross program aimed at reducing segment size attached to a target gene by finding the minimal total population size (across generations) given the specified overall probability of success.

5.0 SUCCESSFUL APPLICATIONS OF MARB IN BREEDING PRACTICE

5.1 Introgression of a Single Major Gene

Ragot *et al.* (1995) provided a good example of gene introgression using MAS for foreground and background selection. By using the by-defintion diagnostic or perfect marker to monitor the Bt gene in the transgenic parent, and random markers for selecting for the recurrent parent genome, they not only transferred the Bt gene into elite maize lines, but also confirmed the theoretical prediction that the use of markers to speed up the recovery of the recipient genome provides a gain in time equivalent to two generations of backcrossing. In this study the target gene is inserted into the genome of the transgenic line by genetic transformation and linkage drag is not of concern and the marker used for foreground selection is in the transgenic construct (no recombination).

Pelemand and van der Voort (2003) provided a comprehensive example of the removal of undesirable linkage drag by MAS in the development of a novel lettuce variety resistant to the aphid *Nasonovia ribisnigri*. This aphid is a major problem in field-grown lettuce areas in Europe and California causing reduced and abnormal growth in addition to spread of viral diseases. Resistance could be introgressed from a wild relative, *Lactuca virosa*, by recurrent backcrossing. However, the new germplasm even after many rounds of backcrossing was of very poor quality, bearing chlorotic leaves and greatly reduced yield. Markers flanking the introgression segment were used to select individuals that are recombinant in the vicinity of the gene among more than 2000 F_2 plants. By testing the selected individuals for resistance and the absence of the negative characteristics at F_3, an individual bearing recombination events very close to each side of the gene was identified, which did not have the linkage drag. The linkage drag was caused by recessive genes at both sides of the resistance gene, which resulted in the failure of classical phenotype-based selection methods.

The development of a converted version of rice restorer line 'Minghui 63' is an example of marker-assisted gene introgression using both foreground selection for the target gene and background selection for reducing chromosome segment length attached to the target gene (linkage drag) and the recovery of the recurrent genome (Chen *et al.*, 2000). 'Minghui 63' is a very popular rice restorer line for hybrid production in China. However, it is susceptible to rice bacterial blight (BB) caused by *Xanthomonas oryzae* pv. *oryzae* (Xoo), one of the most devastating rice diseases. Using the isogenic line 'IRBB21' as donor Chen *et al.* (2000) successfully introgressed *Xa21*, a gene, which confers wide spectrum resistance to BB into 'Minghui 63'. The MAS deployed consisted of a marker located at 0.8 centimorgans (cM) from the *Xa21* locus on one side, and another marker at 3.0 cM from the gene on the other side. A total of 128 restriction fragment length polymorphism (RFLP) markers, evenly distributed on the 12 chromosomes, were used to recover the genetic background of 'Minghui 63'. The entire scheme took three generations of backcrosses and one generation of selfing to complete. In this scheme, the progeny of each backcross was first selected for the presence of the *Xa21* gene (foreground selection) by means of both PCR and disease inoculation. The *Xa21*-containing individuals in the BC_1F_1 were selected for recombination between *Xa21* and either of the flanking marker loci (background selection for reducing attached segment length). In BC_2F_1, the *Xa21*-containing individuals were selected for recombination between *Xa21* and the other marker locus. The *Xa21*-containing plants in the BC_3F_1 were assayed with a large number of molecular markers covering the entire rice genome to identify individuals that were homozygous for the Minghui 63 genotypes at all marker loci, except the *Xa21* locus. The selected individuals were then self-fertilized to produce individuals that were homozygous for the *Xa21* gene at this locus, thus completing the breeding procedure. The resulting improved version of Minghui 63 was at molecular level exactly the same as the original, except for a fragment of less than 3.8

cM in length surrounding the *Xa21* locus. Both the new version Minghui 63(*Xa21*) and its hybrid with 'Zhenshan 97A' showed the same spectrum of BB resistance as the donor parent. Field examination of a number of agronomic traits showed that the improved version was identical to Minghui 63, when there was no disease stress. Under heavily diseased conditions, the improved version showed significantly higher grain weight and spikelet fertility than Minghui 63.

Another good example is the incorporation of the *opaque2* gene along with phenotypic selection for kernel modification in the background of an early maturing normal maize inbred line, V25 (Babu *et al.*, 2005). The normal maize protein is of poor nutritional quality due to a deficiency in two essential amino acids (lysine and tryptophan) and a high leucine–isoleucine ratio. The maize mutant *opaque2* has enhanced nutritional quality, but very poor for yield and overall agronomic performance. Great efforts have been made by breeders world-wide to combine the yield and other agronomic traits of the elite inbreds with the good quality of *opaque2*. Babu *et al.* (2005) achieved this goal by a two generation marker-based backcross breeding program. Foreground selection was conducted using the *opaque2* specific SSR marker, *umc1066*. Flanking markers *bnlg2160* and *bnlg1200* (4.2 and 3.8 cM from the *opaque2* locus) were used to identify recombinants with reduced attached chromosome segment length. Whole genome background selection was conducted using 77 SSR markers spanning all the bin locations in a maize SSR consensus map. The tryptophan concentration in endosperm protein was significantly enhanced in all the three classes of kernel modification *i.e.*, less than 25%, 25–50% and more than 50% opaqueness. BC_2F_3 lines developed from the hard endosperm kernels were evaluated for desirable agronomic and biochemical traits in replicated trials and the best line was chosen to represent the quality protein maize (QPM) version of V25, with a tryptophan concentration of 0.85% in protein. In addition to obtaining the converted line, this study also demonstrated the use of the prediction equations proposed by Frisch *et al.* (1999a, b) and Hospital *et al.* (1992) for the determination of optimum population size of backcross generations. It also highlighted the importance of phenotypic selection among the lines with target genes for other agronomic traits.

5.2 Introgression of Two or More Major Genes for Disease/ Insect Resistance

The development of rice lines with bacterial blight (BB) resistance is a good example of marker-assisted introgression of multiple genes. So far, roughly 29 BB resistance genes have been identified. None are effective individually against all the pathotypes, though some of the genes such as *Xa4* confer resistance to many pathotypes. Some of these genes have been incorporated into modern rice varieties and used for development of near-isogenic lines. Cultivars with one or more BB resistance genes have been developed by conventional backcrossing methods and used in different rice growing regions. MAS lines with multiple resistance genes have also been successfully developed.

A three-gene line, IRBB59 (with *xa5, xa13,* and *Xa21*), was used as donor to transfer the three BB resistance genes into three new plant type lines with high yield potential, IR65598-112 and the two sister lines IR65600-42 and IR65600-96 (Sanchez *et al.,* 2000). Sequence tagged site (STS) markers for all the three resistance genes from the previously identified RFLP and RAPD markers developed by Huang *et al.,* (1997) and Sanchez *et al.,* (2000) were used for foreground selection. F_1 plants were obtained between the donor and the three recurrent parents and were advanced up to the BC_3 generation by MAS. Starting from the BC_1F_1, and in each of the following BCF_1 generations, approximately 50 plants were genotyped. From these, plants carrying resistant alleles of the three target resistance genes (based on their marker genotypes) and that were phenotypically similar to the recurrent parents were selected as the parents for the next backcross until BC_3F_1. The selected BC_3F_1 plants for each of the recurrent parents were selfed to produce BC_3F_2 seed. Based on phenotypic similarity to their recurrent parents, BC_3F_2 plants were selected for homozygosity at the STS marker genotypes and phenotyped for their reactions to the relevant Xoo races. The BC_3F_3 NILs having more than one BB resistance gene showed a wider resistance spectrum and manifested increased levels of resistance to the Xoo races, as compared with those having a single BB resistance gene. The resultant plants had a very high degree of similarity to their respective recurrent parents, suggesting that the phenotypic selection in every backcross generation was effective. The results of these studies clearly demonstrated the usefulness of MARB in pyramiding genes for BB resistance, particularly for recessive genes, such as *xa-5* and *xa13* that are difficult to select through conventional breeding in the presence of a dominant gene such as *Xa-21*.

Similarly, Singh *et al.* (2001) transferred the same three BB resistance genes, *xa5, xa13* and *Xa21*, into the elite cultivar 'PR106', which is widely grown in the State of Punjab, India. IRBB22, carrying all three genes in IR24 background, was used as donor parent. Lines of PR106 with introgressed genes were evaluated after inoculation with 17 isolates of the pathogen from the Punjab and six races of Xoo from the Philippines. Genes in combination were found to provide high levels of resistance to the predominant Xoo isolates from the Punjab and six races from the Philippines. Lines of PR106 with two and three BB resistance genes were also evaluated under natural conditions at 31 sites in commercial fields. The combination of genes provided a wider spectrum of resistance to the pathogen population prevalent in the region. Only one of the BB isolates, PX04, was virulent on the line carrying *Xa21*, but it was avirulent on the lines carrying the *a5* and *xa13* genes in combination with *Xa21*. However, the performance of the pyramided lines on other agronomic traits, particularly in comparison with the recurrent parent, was not reported. One of the pyramided lines containing all the three genes, named as SS113, was used by Sundaram *et al.* (2007) as a donor to introgress the resistance genes into the cultivar 'Samba Mahsuri' (BPT5204), which is a medium slender grain *indica* rice variety and is very popular with farmers and consumers across India, because of its high yield and excellent cooking quality. At each backcross generation, markers closely linked to the three genes were used to select plants possessing these resistance genes (foreground selection) and microsatellite markers polymorphic between donor and recurrent parent were used to select plants that have maximum contribution from the recurrent parent genome (background selection). A selected BC_4F_1 plant was selfed to generate homozygous BC_4F_2 plants with different combinations of BB resistance genes. The three-gene pyramided and two-gene pyramided lines exhibited high levels of resistance against the BB pathogen. Under conditions of BB infection, the three-gene pyramid lines exhibited a significant yield advantage over Samba Mahsuri, the recurrent parent. Multi-location testing demonstrated that these lines retain the excellent grain and cooking qualities of Samba Mahsuri without compromising the yield. One of these lines has been recommended for release as a commercial variety by the Variety Identification Committee of the Indian Council of Agricultural Research (ICAR). This study demonstrated that background selection with a limited number of polymorphic microsatellite markers (50), in conjunction with four backcrosses, is sufficient to recover the yield and quality characteristics of the recurrent parent, which was consistent with the theoretical and simulation results. The markers used for background selection did not include those that are tightly linked to the target genes, since the donor line already had good agronomic performance.

Joseph *et al.* (2004) screened 13 NILs of rice with different BB resistance genes and gene combinations against four isolates of the pathogen from the Basmati regions of India and identified *Xa4, xa8, xa13* and *Xa21* as effective against all the isolates tested. Two or more of these genes in combination imparted enhanced resistance, as expressed by reduced average lesion length, in comparison to individual genes. The two-gene pyramid line IRBB55 carrying *xa13* and *Xa21* was found to be equally effective as three/four gene pyramid lines. IRBB55 was then used as donor parent to transfer *xa13* and *Xa21* into Pusa Basmati-1, the most popular high yielding rice variety. Recombinants having enhanced resistance to BB, basmati quality and desirable agronomic traits were identified. Unlike studies outlined above, this study used only a single backcross generation and extensive selection for agronomic traits were conducted in three selfing generations to recover the genome of the elite parent.

5.3 Introgression of QTL

Most of the agronomic traits such as yield are quantitatively inherited. Manipulating these traits is difficult because of their intrinsic complexities: polygenic control, pleiotropy, epistasis, and gene-by-environment interaction (G x E). Since QTL with major effects are easily manipulated by empirical breeding practices and may already be fixed in many breeding lines, it would be more productive to use marker technology as a means for placing greater emphasis on those QTL that show only relatively minor but additive or multiplicative effects (Stuber *et al.*, 1999). When the effects of QTL are of small effect several QTL have to be manipulated simultaneously to achieve significant improvement. Reported applications of QTL introgression for the development of cultivars with improved performance of quantitative traits are few, although introgression of multiple QTL has been extensively used for the validation of previous mapped QTL.

Toojinda *et al.* (1998) successfully introgressed two QTL for stripe rust resistance in barley into a genetic background different from the one used to map the QTL. The effects of both QTL were confirmed and additional QTL were detected in the new background, including some resistance alleles brought in by the susceptible parent.

Four target chromosomal regions containing five QTL for pest resistance (ascysugar accumulation) were successfully introgressed from wild tomato into cultivated tomato (Lawson *et al.*, 1997). However, the level of ascysugar accumulation resistance in the progeny introgressed for the five QTL was lower than expected and was also lower than the interspecific F_1 hybrid.

Sebolt *et al.* (2000) performed marker-assisted introgression of two QTL for seed protein concentration identified in *Glycine soja* accessions in cultivated soybean. Only one QTL was confirmed and the other QTL may have been lost during backcrossing. When the confirmed QTL was transferred in three different backgrounds it had no effect in one background.

Chee *et al.* (2001) transferred a QTL for grain protein concentration (GPC) in emmer wheat into an adapted durum wheat background. An inbred line with high GPC and other desirable agronomic characters selected from the recombinant inbred lines derived from LDN (DIC-6B)/VIC population was crossed to 'Renville', a good quality, high yielding durum cultivar. Recombinant inbred lines were developed by single seed descent and used to confirm the presence and location of this QTL.

Ahmadi *et al.* (2001) introgressed two QTL for resistance to rice yellow mottle virus identified in highland cultivar into a lowland rice cultivar. In total three backcrossing and three selfing generations were used. The donor was a double haploid (DH) line selected from the original mapping population. One marker per QTL was used for selecting the QTL. Background selection was conducted using markers on the chromosomes without the target QTL for recovery of the recurrent genome in BC_1 and BC_3 generations. Phenotypic screening for resistance segregation in the selfing generation after selection in backcrossing generation was used to ensure the QTL were not lost.

Shen *et al.* (2001) developed near-isogenic lines (NILs) containing QTL associated with rice root traits on rice chromosomes 1, 2, 7, and 9 (designated as targets 1, 2, 7, and 9) identified in previous mapping studies. The donor parents were four doubled haploid lines that had the desirable alleles at the target QTL and > 50% of the recipient (IR64) genome. Several BC_3F_3 lines with one or two QTL were obtained by MAS. Among the four QTL, one exhibited the expected effect in the progeny, one was finally revealed as a false positive, one segment was shown to contain two QTL in repulsion phase that reduced its effect and one segment did not exhibit the expected effect. They also found the association of the NILs with some non-target traits. For instance, increased height and reduced tiller number per plant were detected for two of the three target-1 NILs. Three of the five target-2 NILs had increased height and reduced tiller number. Most target-7 NILs had significantly increased height; some of them had either more or less tillers. All target-9 NILs had significantly reduced tiller number. Three of the four NILs with introgressed targets 1 and 7 QTL were significantly taller than IR64.

Bouchez *et al.* (2002) reported the introgression of favourable alleles at three quantitative trait loci (QTL) for earliness and grain yield among maize elite lines. Introgression started from a selected RIL, which was crossed three times to one of the original parents and then self-fertilized, leading to BC_3S_1 progenies. Markers were used to assist both foreground and background selection at each generation. The marker-assisted introgression proved successful at the genotypic level in the sense that lines with all targeted markers were obtained. In addition, QTL positions were generally sustained in the introgression background. For earliness, the magnitude and sign of the QTL effects were in good agreement with those expected from initial RIL analyses. Conversely, for yield, important discrepancies were observed in the magnitude and sign of the QTL effects observed after introgression.

Yousef and Juvik (2002) successfully selected on three markers linked to QTL that enhanced seedling emergence in sweet corn. Three RFLP marker alleles linked to QTL that enhanced seedling emergence identified in an $F_{2:3}$ sweet corn mapping population were used to transfer these QTL into three elite commercial sweet corn inbreds. A recombinant inbred line derived from the original mapping population was used as a donor parent. The introgressed QTL alleles were observed to enhance seedling emergence in the BC_2F_1 generation as was observed in the original $F_{2:3}$ mapping population. In this study, a combination of QTL linked to *umc139* and *php200689* markers resulted in the highest seedling emergence compared with other combinations including all three of the beneficial marker-QTL alleles together.

Lecomte *et al.* (2004) introgressed five chromosome regions strongly involved in organoleptic quality attributes of tomato into three different recipient lines through marker-assisted selection. All the favourable alleles for the quality traits were provided by the same parental tomato line. Three improved lines were obtained after three backcrossing and two selfing generations. Breeding efficiency strongly varied according to the recipient parent, and significant interactions between QTL and genetic backgrounds were shown for all of the traits studied. About 50 % of the QTL were confirmed in each background and new QTL were detected. The QTL with largest effect were the most stable.

Thabuis *et al.* (2004) successfully transferred the favourable alleles of four QTL for resistance to *Phytophthora capsici*, which was identified in a small-fruited pepper line by three cycles of marker-assisted backcrossing using a bell pepper line as recipient. A DH line selected from the original mapping population was used as donor. Two populations, derived by selfing the plants selected after the first selection cycle, were genotyped and evaluated phenotypically for their resistance level. The additive and epistatic effects of the four resistance factors were re-detected and validated in these populations. A decrease of the positive effect for the moderate-effect QTL and of the epistatic interaction was observed.

Steele *et al.* (2006) conducted a MARB breeding programme to improve the root morphological traits, and thereby drought tolerance, of the Indian upland rice variety, 'Kalinga III', which had not previously been used for QTL mapping. The donor parent was Azucena, an upland *japonica* variety from Philippines. Five segments on different chromosomes were targeted for introgression; four segments carried QTL for improved root morphological traits (root length and thickness) and the fifth carried a recessive QTL for aroma. Two crosses between BC_3 lines were used to stack the five targets. The target segment on chromosome 9 significantly increased root length under both irrigated and drought stress treatments, confirming that this root length QTL from Azucena functions in a novel genetic background. No significant effects on root length were found at the other four targets. Azucena alleles at the locus RM248 delayed flowering. Selection for the recurrent parent allele at this locus produced early-flowering NILs that were suited for upland environments in eastern India.

Ragot *et al.* (2006) reported the results of a MARB experiment aimed at improving grain yield under drought conditions in tropical maize. The introgression increased grain yield and reduced the asynchrony between male and female flowering under water-limited conditions. Eighty-five per cent of the recurrent parent's genotype at non-target loci was recovered in only four generations of backcrossing by screening large segregating populations (2200 individuals) for three of the four generations. Selected MABC-derived BC_2F_3 families were crossed with two testers and evaluated under different water regimes. Mean grain yield of MARB-derived hybrids was consistently higher than that of control hybrids (crosses from the recurrent parent to the same two testers as the MARB-derived families) under severe water stress conditions. Under those conditions, the best five MARB-derived hybrids yielded, on average, at least 50% more than control hybrids. Under mild water stress (defined as resulting in <50% yield reduction), no difference was observed between MARB-derived hybrids and the control plants.

The combined use of high-throughput genotyping and marker-assisted backcross was adopted by Bai *et al.* (2007) to transfer the major Fusarium head blight (FHB) QTL from Sumai 3 and its derivatives into locally adapted hard winter wheat with minor FHB-resistance QTL. Three crosses were made between Sumai 3 derived soft red wheat lines and three locally adapted hard winter wheat cultivars (Harding, Wesley and Trego). About 80 BC_2F_2 plants homozygous for the 3BS QTL were selected from each backcross population based on closely linked markers. BC_2F_3 lines were evaluated in greenhouse for Type II resistance and 135 highly resistant and 87 moderate resistant lines were identified. These materials have the potential to develop marketable FHB resistant HWW cultivars and useful germplasm lines.

Neeraja *et al.* (2007) demonstrated the use of marker assisted backcross (MAB) to develop germplasm with enhanced submergence tolerance in rice. Submergence stress regularly affects 15 million hectares or more of rainfed lowland rice areas in South and Southeast Asia. A major QTL on chromosome 9, *Sub1*, has provided the opportunity to apply MAB to develop submergence tolerant versions of rice cultivars that are widely grown in the region. In this study, molecular markers that were tightly linked with *Sub1*, flanking *Sub1*, and unlinked to *Sub1* were used to apply foreground, recombinant, and background selection, respectively, in backcrosses between a submergence-tolerant donor and the widely grown recurrent parent Swarna. By the BC_2F_2 generation a submergence tolerant plant was identified that possessed Swarna type simple sequence repeat (SSR) alleles on all fragments analyzed except the tip segment of rice chromosome 9 that possessed the *Sub1* locus. A BC_3F_2 double recombinant plant was identified that was homozygous for all Swarna type alleles except for an approximately 2.3–3.4 Mb region surrounding the *Sub1* locus. The results showed that the mega variety Swarna could be efficiently converted to a submergence tolerant variety in three backcross generations, in just two to three years. Polymorphic markers for foreground and recombinant selection were identified for four other important varieties to develop a wider range of submergence tolerant varieties to meet the needs of farmers in the flood-prone regions.

6.0 FACTORS NEED TO BE CONSIDERED WHEN DESIGNING A MABR PROGRAM

In the general context of marker-assisted selection, the incorporation of MAS in recurrent backcrossing brings both opportunities and challenges. The better control of the intended selection objective offered by markers comes at a cost of increased complexity of decision-making. Several factors must be considered when deciding whether a MARB or a conventional recurrent backcrossing (CRB) strategy should be taken. To be useful to plant breeders, gains made from MARB must be more cost-effective than gains through traditional breeding or MARB must generate significant time savings, which would justify the additional cost involved. Assuming that markers for foreground and background selection are available and reliable genotyping systems are well established, the following four factors need to be considered, as discussed by Morris *et al.* (2003). (1) The relative cost of phenotypic versus marker screening. (2) The time saved by MARB. (3) The size and temporal distribution of benefits associated with accelerated release of improved germplasm and (4) The availability to the breeding program of operating capital. When MAS is more cost-effective, MARB is clearly superior to conventional recurrent backcrossing (CRB). When MAS is more costly than phenotypic selection, the trade-off between time and cost of MARB makes it necessary to evaluate MARB in economic terms. Morris *et al.* (2003) explored the relationship between time and costs by developing a simple model of recurrent backcrossing. Two measures of the project's worth were used: the net present value of the discounted streams of costs and benefits, and the internal rate of return to the investment. Using the maize line conversion program of CIMMYT as an example, they demonstrated that the ranking of the MARB and conventional recurrent backcrossing (CRB) differed depending on the measures of the project's worth. MARB generated the highest net present value, whereas CRB generated the highest internal rate of return. Therefore, the attractiveness of MARB depends on the availability of investment capital. If investment capital is available and the increased cost of MARB does not affect other ongoing breeding projects, MARB is preferred. Otherwise, CRB is the better choice.

Many factors affect the success of a QTL introgression program. The most important factors are QTL-by-environment interaction (QEI) and QTL-by-genetic background interaction (QGI). For gene introgression to be effective it should be aimed at QTL with good stability across the target population of environments. QTL mapping studies have shown that QTL with large effect also tend to be more stable across environments. This suggests that

MARB should target major QTL. The interactions between QTL and between QTL and genetic backgrounds are more difficult to qualify, let alone quantify. Given that the effects of the genetic background on the trait of interest could vary independently of the introgressed regions, it is necessary to introgress QTL in several recipient lines to increase the chance of breeding success. The precision of QTL mapping is also an important factor. Since introgression is generally targeted QTL with relatively large effects, it is unlikely that the target QTL is a false positive. However, the effect of a QTL can be different from its estimate in the original mapping population even if QEI and QGI are absent. If QTL were not precisely mapped, large regions of donor chromosomes were transferred. This has at least two possible consequences. First, the chromosomal segments transferred hold not just one but several genes. Recombination between those genes can modify the effect of the introduced segments (Hospital, 2005). There are many examples where fine-mapping of the detected QTL results in the identification of several genes (Eshed and Zamir, 1995; Monna *et al.*, 2002; Steinmetz *et al.*, 2002; Christian and Keightley, 2004). Second, unfavourable linkage drag may be caused by the unintentional introduction of undesirable alleles. Therefore, QTL should to be precisely mapped before starting the introgression process. The a-priori use of NILs for QTL confirmation is therefore recommended.

The combined effects of these factors may lead to complete failure of an MAS-based introgression program. For instance, Reyna and Sneller (2001) tested the effects of three beneficial yield QTL identified from the northern (USA) soybean cultivar 'Archer' in southern background and testing environments. Four sets of NILs for each QTL were derived from northern heterozygous F_6 plants identified from the crosses of Archer x Asgrow A5403 and Archer x Pioneer 9641. None of the marker effects were significant for any of the three QTL for yield, height, and maturity, when averaged over all sets or for southern individual sets. Similarly, in barley Kandemir *et al.* (2000) evaluated the effects of three previously identified grain yield QTL on chromosomes 2S (2HS), 3C (3HC) and 5L (1HL) for their potential to increase yields of high-quality malting barley without disturbing it's favourable malting quality profile. NILs were developed by introgressing QTL from the high-yielding cultivar 'Steptoe' to the superior malting quality, moderate-yielding cv.'Morex'. None of the three QTL studied altered the measured yield of the recipient genotype *per se*, although QTL 2S and QTL-3 affected yield-related traits. The study of Bouchez *et al.* (2002) mentions that one high-yielding allele putatively detected in a low-yielding parent finally exhibited an effect opposite to expectations in the breeding backcross population.

7.0 CONCLUSION

MARB, the combined use of MAS and recurrent backcrossing, greatly widens the applicability of recurrent backcrossing for cultivar improvement. It also serves as an important tool for the genetic study of complex traits. Breeding involving a much wider range of traits can be improved using MARB. Similar to conventional recurrent backcrossing, traits controlled by single or few genes are the most rewarding area of MARB application.

When QTL are to be introgressed, several difficulties arise. It is more difficult to select for the presence of QTL since QTL location can only be estimated with a given (im)precision (Visscher *et al.*, 1996). To improve this situation requires using more markers and optimizing the positions of these markers with respect to the uncertainty of the true QTL location (Hospital and Charcosset, 1997). Once the introgression is achieved, it must be checked that the effect of the QTL in the new genetic background is the same as the effect estimated originally or at least of a magnitude to make it competitive to other methods. It is unlikely that one single QTL for a quantitative trait could explain enough genetic variation ("percentage of trait variance explained") to justify the economic effort associated with a marker assisted introgression program. It is most efficient if several QTL are introgressed simultaneously. This necessitates larger population sizes for foreground selection and reduces the possibility of background selection. Both theoretical and experimental studies showed that the introgression of up to five desired donor chromosome regions using linked markers was very much feasible (Hospital and Charcosset, 1997; Koudande *et al.*, 2000).

8.0 REFERENCES

Ahmadi, N., Albar, L., Pressoir, G., Pinel, A., Fargette, D. and Ghesquière, A. 2001. Genetic basis and mapping of the resistance to Rice yellow mottle virus. III. Analysis of QTL efficiency in introgressed progenies confirmed the hypothesis of complementary epistasis between two resistance QTLs. Theoretical and Applied Genetics, 103:1084–1092.

Allard, R. 1999. 'Principles of Plant Breeding'. 2nd Edition, Wiley and Sons: New York.

Babu, R., Nair, S.K., Kumar, A., Venkatesh, S., Sekhar, J.C., Singh, N.N., Srinivasan, G. and Gupta, H.S. 2005. Two-generation marker-aided backcrossing for rapid conversion of normal maize lines to quality protein maize (QPM). Theoretical and Applied Genetics, 111:888-897.

Bai, G., St. Amand, P., Zhang, D., Ibrahim, A., Baenziger, P., Bockus, B. and Fritz, A. 2007. Improvement of FHB Resistance of Hard Winter Wheat through Marker-assisted Backcross. ASA-CSSA-SSSA, New Orleans, Louisiana.

Bernardo, R. 2004. What proportion of declared QTL in plants are false? Theoretical and Applied Genetics, 109:419–424.

Bouchez, A., Hospital, F., Causse, M., Gallais, A. and Charcosset, A. 2002. Marker-assisted introgression of favourable alleles at quantitative trait loci between maize elite lines. Genetics, 162:1945–1959.

Chee, P.W., Elias, E.M., Anderson, J.A. and Kianian, S.F. 2001. Evaluation of a high grain protein QTL from *Triticum turgidum* L. var. dicoccoides in an adapted durum wheat background. Crop Science, 41:295-301.

Chen, S., Lin, X.H., Xu., C.G. and Zhang, Q. 2000. Improvement of bacterial blight resistance of 'Minghui 63', an elite restorer line of hybrid rice, by molecular marker-assisted selection. Crop Science, 40:239-244.

Christians, J.K. and Keightley, P.D. 2004. Fine mapping of a murine growth locus to a 1.4-cM region and resolution of linked QTL. Mammalian Genome, 15:482–491.

Hospital, F. and Decoux, G. 2002. Popmin : a program for the numerical optimization of population sizes in market-assisted back-cross programs. Journal of Heredity, 93:383-384.

Dekkers, J. and Hospital, F. 2002. The use of molecular genetics in the improvement of agricultural populations. Nature Review Genetics, 3:22–32.

Eagles, H.A., Bariana, H.S., Ogbonnaya, F.C., Rebetzke, G.J., Hollamby, G.J., Henry, R.J., Henschke, P.H. and Carter, M. 2001. Implementation of markers in Australian wheat breeding. Austrian Journal of Agricultural Research, 52: 1349–1356.

Eshed, Y. and Zamir, D. 1995. An introgression line population of *Lycopersicon pennellii* in the cultivated tomato enables the identification and fine mapping of yield-associated QTL. Genetics, 141:1147–1162.

Frisch, M. and Melchinger, A.E. 2005. Selection theory for marker-assisted backcrossing. Genetics, 170:909-917.

Frisch, M., Bohn, M. and Melchinger, A.E. 1999a. Minimum sample size and optimal positioning of flanking markers in marker-assisted backcrossing for transfer of a target gene. Crop Science, 39:967-975.

Frisch, M., Bohn, M. and Melchinger, A.E. 1999b. Comparison of selection strategies for marker-assisted backcrossing of a gene. Crop Science, 39:1295-1301

Frisch, M. and Melchinger, A.E. 2001a. The length of the intact donor chromosome segment around a target gene in marker -assisted backcrossing. Genetics, 157: 1343–1356.

Frisch, M., Melchinger, A.E. 2001b. Marker-assisted backcrossing for simultaneous introgression of two genes. Crop Science, 41:1716–1725.

Hospital, F. 2001. Size of donor chromosome segments around introgressed loci and reduction of linkage drag in marker-assisted backcross programs. Genetics, 158:1363-1379.

Hospital, F. 2005. Selection in backcross programmes. Philosphical Transactions of The Royal Society (Biological Sciences), 360:1503-1512.

Hospital, F. and Charcosset, A. 1997. Marker-assisted introgression of quantitative trait loci. Genetics, 147:1469-1485.

Hospital, F., Chevalet, C. and Mulsant, P. 1992. Using markers in gene introgression breeding programs. Genetics, 132:1199-1210.

Huang, N., Angeles, E.R., Domingo, J., Magpantay, G., Singh, S., Zhang, G., Kumaravadivel, N., Bennet, J. and Khush, G.S. 1997. Pyramiding of bacterial blight resistance genes in rice: marker-assisted selection using RFLP and PCR. Theoretical and Applied Genetics, 95:313-320.

Joseph, M., Gopalakrishnan, S., Sharma, R.K., Singh, V.P., Singh, A.K., Singh, N.K. and Mohapatra, T. 2004. Combining bacterial blight resistance and Basmati quality characteristics by phenotypic and molecular marker-assisted selection in rice. Molecular Breeding, 100:1-11.

Kandemir, N., Jones, B.L., Wesenberg, D.M., Ullrich, S.E. and Kleinhofs, A. 2000. Marker-assisted analysis of three grain yield QTL in barley (*Hordeum vulgare* L.) using near isogenic lines. Molecular Breeding, 6:157-167.

Koudande, O.D., Iraqi, F., Thomson, P.C., Teale, A.J. and Van Arendonk, J.A. 2000. Strategies to optimize marker-assisted introgression of multiple unlinked QTL. Mamm. Genome, 11:145-150.

Kuchel, H., Ye, G., Fox, R. and Jefferies, S. 2005. Genetic and genomic analysis of a targeted marker-assisted wheat breeding strategy. Molecular Breeding, 16: 67-78.

Lawson, D.M., Lunde, C.F. and Mutschler, M.A. 1997. Marker-assisted transfer of acylsugar-mediated pest resistance from the wild tomato, *Lycopersicon pennellii*, to the cultivated tomato, *Lycopersicon esculentum*. Molecular Breeding, 3:307-317.

Lecomte, L., Duffé, P., Buret, M., Hospital, F. and Causse, M. 2004. Marker-assisted introgression of five QTLs controlling fruit quality traits into three tomato lines revealed interactions between QTLs and genetic backgrounds. Theoretical and Applied Genetics, 109:658–668.

Monna, L., Lin, H.X., Kojima, S. and Sasaki, T. 2002. Genetic dissection of a genomic region for a quantitative trait locus, Hd3, into two loci, Hd3a and Hd3b, controlling heading date in rice. Theoretical and Applied Genetics, 104:772–778.

Morris, M., Dreher, K., Ribaut, J.M., and Khairallah, M. 2003. Money matters (II): cost of maize inbred line conversion schemes at CIMMYT using conventional and marker-assisted selection. Molecular Breeding, 11:235-247.

Neeraja, C.N., Maghirang-Rodriguez, R., Pamplona, A., Heuer, S., Collard, B.C.Y., Septiningsih, E.M., Vergara, G., Sanchez, D., Xu, K., Ismail, A.M. and Mackill, D.J. 2007. A marker-assisted backcross approach for developing submergence-tolerant rice cultivars. Theoretical and Applied Genetics, 115:767-776.

Openshaw, S.J., Jarboe, S.G. and Beavis, W.D. 1994. Marker-assisted selection in backcross breeding. *In* Proceedings of the Symposium "Analysis of Molecular Marker Data", Corvallis, OR. 5–6 Aug. 1994. Am. Soc. Hortic. Sci. and Crop Science society of America. American Society of Horticultural Science.

Peleman, J.D. and van Der Voort, J.R. 2003. The challenges in marker assisted breeding. *In*: Eucarpia leafy vegetables (van Hintum ThJL, Lebeda A, Pink D, Schut JW *eds*). CGN pp 125-130.

Popmin : a programme for the numerical optimization of population sizes is marker-assited backcross programs. The Journal of Heridity, 93(5):383-384.

Ragot, M., Biosiolli, M., Delbut, M.F., DellOrco, A., Malgarini, L., Thevenin, P., Vernoy, J., Vivant, J., Zimmermann, R. and Gray, G. 2006. Marker-assisted backcrossing: A practical example. *In*: Berville A and Tersac M (*eds*) Techniques et Utilizations des Marqueurs Moleculaires. INRA Editions, Versailles, pp 29-31.

Ribaut, J.M. and Michel Ragot, M. 2007. Marker-assisted selection to improve drought adaptation in maize: the backcross approach, perspectives, limitations, and alternatives. Journal of Experimental Botany, 58:351-360.

Reyna, N. and Sneller, C.H. 2001. Evaluation of marker-assisted introgression of yield QTL alleles into adapted soybean. Crop Science, 41:1317–1321.

Ribaut, J.M., Jiang, C. and Hoisington, D. 2002. Simulation experiments on efficiencies of gene introgression by backcrossing. Crop Science, 42:557-565.

Melchinger, A.E. 1990. Use of molecular markers in breeding for oligogenic disease resistance. Plant Breeding, 104:1–19.

Sebolt, A.M., Shoemaker, R.C. and Diers, B.W. 2000. Analysis of a quantitative trait locus allele from wild soybean that increases seed protein concentration in soybean. Crop Science, 40:1438–1444.

Servin, B., Dillmann, C., Decony, G. and Hospital, F. 2002. MDM : A program to compute fully informative genotype frequency in complex breeding schemes. Journal of Heredity, 93(3):227-228.

Servin, B. and Hospital, F. 2002. Optimal positioning of markers to control genetic background in marker-assisted backcrossing. Journal of Heredity, 93:214-217.

Sanchez, A.C., Brar, D.S., Huang, N., Li, Z. and Khush, G.S. 2000. Sequence Tagged Site marker-assisted selection for three bacterial blight resistance genes in rice. Crop Science 40:792–797.

Shen, L., Courtois, B., McNally, K.L., Robin, S. and Li, Z. 2001. Evaluation of near-isogenic lines of rice introgressed with QTLs for root depth through marker-aided selection. Theoretical and Applied Genetics, 103:75-83.

Singh, S., Sidhu, J.S., Huang, N., Vikal, Y., Li, Z., Brar, D.S., Dhaliwal, H.S. and Khush, G.S. 2001. Pyramiding three bacterial blight resistance genes (xa5, xa13, Xa21) using marker-assisted selection into indica rice cultivar PR106. Theoretical Applied Genetics, 102:1011–1015.

Steinmetz, L.M., Sinha, H., Richards, D.R., Spiegelman, J.I., Oefner, P.J., McCusker, J.H. and Davis, R.W. 2002. Dissecting the architecture of a quantitative trait locus in yeast. Nature, 416:326–330.

Stuber, C.W., Polacco, M. and Senior, M.L. 1999. Synergy of empirical breeding, marker-assisted selection, and genomics to increase crop yield potential. Crop Science, 39:1571-1583.

Sundaram, R.M., Vishnupriya, M.R., Biradar, S.K., Laha, G.S., Reddy, G.A., Rani, N.S., Sarma, N.P. and Sonti, R.V. 2007. Marker assisted introgression of bacterial blight resistance in Samba Mahsuri, an elite indica rice variety. Euphytica, 10.1007/s10681-007-9564-6.

Stam, P. and Zeven, A.C. 1981. The theoretical proportion of the donor genome in near-isogenic lines of self-fertilizers bred by backcrossing. Euphytica, 30: 227-238.

Steele, K.A., Price, A.H., Shashidhar, H.E. and Witcombe, J.R. 2006. Marker-assisted selection to introgress rice QTLs controlling root traits into an Indian upland rice variety. Theoretical and Applied Genetics, 112:208–221.

Tanksley, S.D. 1983. Molecular markers in plant breeding. Plant Molecular Biology Reports, 1: 3-8.

Tanksley, S.D. and Nelson, J.C. 1996. Advanced backcross QTL analysis: A method for simultaneous discovery and transfer of valuable QTL from unadapted germplasm into elite breeding. Theoretical Applied Genetics, 92: 191-203.

Tanksley, S.D., Young, N.D., Patterson, A.H. and Bonierbale, M.W. 1989. RFLP mapping in plant breeding: New tools for an old science. Biotechnology, 7: 257-264.

Thabuis, A., Palloix, A., Servin, B., Daube'ze, A.M., Signoret, P., Hospital, F. and Lefebvre, V. 2004. Marker assisted introgression of 4 *Phytophthora capsici* resistance QTL alleles into a bell pepper line: validation of additive and epistatic effects. Molecular Breeding, 14:9-20.

Toojinda, T., Baird, E., Booth, A., Broers, L., Hayes, P., Powell, W., Thomas, W., Vivar, H. and Young, G. 1998. Introgression of quantitative trait loci (QTLs) determining stripe rust resistance in barley: an example of marker-assisted line development. Theoretical and Applied Genetics, 96:123–131.

Toojinda, T., Tragoonrung, S., Vanavichit, A., Siangliw, J.L., Pa-In, N., Jantaboon, J., Meechai Siangliw, M. and Fukai, S. 2004. Molecular breeding for rainfed lowland rice in the Mekong Region. 4th International Crop Science Congress.

Visscher, P.M., Haley, C.S. and Thompson. R. 1996. Marker-aided introgression in backcross breeding programs. Genetics, 144:1923-1932.

Ye, G. and Smith, K.F. 2009. Marker-assisted gene pyramiding for cultivar development. Plant Breeding Review (In press).

Young, N.D. 1999. A cautiously optimistic vision for marker-assisted breeding. Molecular Breeding, 5: 505-510.

Young, N.D. and Tanksley, S.D. 1989. RFLP analysis of the size of chromosomal segments retained around the Tm-2 locus of tomato during backcross breeding. Theoretical and Applied Genetics, 77: 353-359.

Yousef, G.G. and Juvik, J.A. 2002. Enhancement of seedling emergence in sweet corn by marker-assisted backcrossing of beneficial QTL. Crop Science, 42:96–104.

Zeven, A.C., Knott, D.R. and Johnson, R. 1983. Investigation of linkage drag in near isogenic lines of wheat by testing for seedling reaction to races of stem rust, leaf rust and yellow rust. Euphytica, 32:319–327.

Molecular Plant Breeding: Principle, Method and Application
Eds : R.K. Singh, Rajesh Singh, Guoyou Ye, A. Selvi and G.P. Rao
Studium Press LLC, Texas, USA, 2009, pp. 321-342

CHAPTER 12

Marker-Assisted Gene Pyramiding in Cultivar Development

GUOYOU YE[1]* *and KEVIN F. SMITH*[2]

ABSTRACT

Gene pyramiding, which aims to assemble multiple desirable genes into a single genotype, is a commonly used method in breeding for self-pollinated crops. Traditionally, the main use of gene pyramiding is to improve an existing elite cultivar through introgression of a few genes of large effects from other sources, since the presence of the target genes has to be monitored by phenotyping, which is only effective for major genes. Depending on the trait and inheritance of the targeted genes, gene pyramiding may require much labour, time and material resources. The development of modern plant molecular techniques and quantitative genetics in the last two decades has dramatically widened the applicability of gene pyramiding. It provides enhanced knowledge of the genetics of the breeding traits and of the relative genomic location of functionally related as well as neutral markers associated with the genes responsible for the traits. It facilitates the identification of genes with large effect for traits which are traditionally regarded as quantitative and not targeted by gene pyramiding program. Marker-assisted selection reduces/eliminates extensive phenotyping, makes the pyramiding of genes with very similar phenotypic effects possible, and reduces the breeding duration. Marker-assisted gene pyramiding is now the method of choice for cultivar development targeted at improving traits controlled by major genes.

Key Words: Gene pyramiding, Introgression lines, Marker-assisted selection, Molecular markers, Quantitative trait loci

[1] *Bundoora Centre, Biosciences Division, Department of Primary Industries Victoria, 1 Park Drive, Bundoora Vic 3086, Australia.*
[2] *Hamilton Centre, Biosciences Division, the Department of Primary Industries, Mount Napier Road, Hamilton, Vic 3300, Australia*
* *Corresponding author e-mail: guoyou.ye@dpi.vic.gov.au*

1.0 INTRODUCTION

Gene pyramiding is a breeding method aimed at assembling multiple desirable genes from multiple parents into a single genotype (Allard, 1999). The end product of a gene pyramiding program is a genotype with all of the target genes. Generally speaking, the objectives of gene pyramiding include: (1) enhancing trait performance by combining two or more complementary genes, (2) remedying deficits by introgressing genes from other sources, (3) increasing the durability of disease and/or disease resistance, and (4) broadening the genetic basis of released cultivars.

Traditionally, gene pyramiding is mainly used to improve qualitative traits such as disease and insect resistance. This is associated with the fact that the presence of target trait genes must be confirmed by phenotyping mostly at the individual level and that individual phenotypic performance is a good indicator of the genotype only if genes have a major effect on phenotypic performance and the error of phenotyping is minimal. In addition to the reliability of phenotyping at individual level other factors influencing the success of gene pyramiding are the inheritance model of the genes for the target traits, linkage and/or pleiotropism between the target trait and other traits. For instance, allelic genes cannot be combined in the same genotype. The effect conferred by a recessive gene cannot be evaluated on heterozygous individuals and progeny testing is required. Therefore, any improvement in the knowledge of the trait genetics (inheritance, genetic relationship etc) and techniques for inferring genotype-phenotype relationship will be useful.

The development of modern plant molecular and quantitative genetics in the last two decades has the potential to revolutionise what has mostly been experienced-based empirical plant breeding. This development has enhanced our knowledge of the genetics of the breeding traits and the relative genomic location of functionally related and neutral markers associated with the genes responsible for the traits. It has also widened several aspects of the practical application of gene pyramiding. Firstly, for traits that are simply inherited, but that are difficult or expensive to measure phenotypically, and/or that do not have a consistent phenotypic expression under certain specific selection conditions, marker-based selection is more effective and /or economic than phenotypic selection (Paterson *et al.*, 1991; Stuber *et al.*, 1999; Eagles *et al.*, 2001; Dekkers and Hospital, 2002; Langridge and Barr, 2003; Dubcovsky, 2004). Secondly, traits which are traditionally regarded as quantitative and not targeted by gene pyramiding program can be improved using gene pyramiding if major genes affecting the trait are identified (Ashikari and Matsuoka, 2006). Thirdly, genes with very similar phenotypic effects, which are impossible or difficult to combine in single genotype using phenotypic selection, can be pyramided through marker-assisted selection (MAS). Marker-assisted gene pyramiding is currently the method of choice for cultivar development targeted at improving traits controlled by major genes. In this chapter, some of the successful stories of gene pyramiding in cultivar development are reviewed. The basic principle and guidelines for designing an efficient marker-based gene pyramiding strategy are also summarised.

2.0 APPLICATIONS OF GENE PYRAMIDING IN BREEDING PRACTICE

2.1 Pyramiding Major Genes for Disease/Insect Resistance

Gene pyramiding has long been used by breeders to develop cultivars with multiple resistance genes for an insect or pathogen. It is hoped that a variety with multiple resistance genes is more durable since the insect or pathogen has to overcome all of the resistance genes simultaneously in order to survive on the new variety. Pyramiding multiple qualitative alleles in single genotypes may also be an approach to increasing the level of resistance relative to that conferred by a single qualitative resistance locus and multigenic qualitative resistance may also lead to greater durability. Experimental results are inconclusive with regard to the effectiveness of pyramiding major resistance genes. For instance, Klopper and Pretorius (1997) investigated the effects of wheat leaf rust gene combinations in lines *Lr13* + *Lr34* (T34-13), *Lr13* + *Lr37* (T13-37) and *Lr34* + *Lr37* (T34-37). They found higher levels of resistance in the combination lines T13-37 and T34-37 than in the lines with the individual genes. In the T34-13 line, no increased resistance to pathotype UVPrt13 was apparent from assessment of the infection types in the glasshouse. Precise measurements of its resistance components showed, however, that it had a longer latent period and smaller uredinia and its resistance was highly effective in the field. Significant restriction of fungal growth during early post-infection stages occurred in the gene combination lines T34–13, T13–37 and T34–37. Colony size in these lines was also significantly reduced compared with that in the single gene lines and the leaf rust-susceptible line, when either or both pathotypes possessed avirulence for one of the *Lr* genes. However, Porter *et al.* (2000) tested the effectiveness of several greenbug resistance genes and their combinations against biotypes E, F, G, H, and I and found that pyramiding provided no additional protection over that conferred by the single resistance genes. Nevertheless, the most successful application of gene pyramiding is found in breeding for pest and disease resistance.

A good example of pyramiding major resistance genes with the aid of molecular markers is the development of rice lines with bacterial blight (BB) resistance. So far, roughly 29 resistance genes have been identified. None are effective individually against all the pathotypes, though some of the genes such as *Xa4* confer resistance to many pathotypes. Some of these genes have been incorporated into modern rice varieties and used for development of near-isogenic lines. Cultivars with one or more BB resistance genes have been developed by conventional backcrossing methods and used in different rice growing regions. Lines through MAS with multiple resistance genes have been successfully developed.

Yoshimura *et al.* (1995) developed markers for four BB resistance genes. Using these linked markers they selected lines homozygous for pairs of resistance genes, *Xa4 +xa5* and *Xa4 +Xa10.* Lines carrying *Xa4 +xa5* and *Xa4 +Xa10* were evaluated for reaction to eight strains of the BB pathogen, representing eight pathotypes and three genetic lineages. It was found that the lines carrying pairs of genes were resistant to more of the isolates than their single-gene parental lines. Lines carrying *Xa4 +xa5* were more resistant to isolates of race 4 than either of the parental lines, while no such effects were seen for*Xa4 +Xa10.* Thus, combinations of resistance genes may provide broader spectra of resistance. Huang *et al.* (1997) developed lines containing up to four BB resistance genes using markers. Four isogenic lines and their recurrent parent IR24, and a line containing two *Xa4/xa5* developed by Yoshimura *et al.* (1995) were used as parents. Intermediate lines with two resistance genes were also used in late crossing to accumulate more genes. Lines with three or four genes were developed by crossing between two-gene lines and one-gene or two- gene lines. The pyramided lines having three or four genes in combination also showed an increased and wider spectrum of resistance to bacterial blight than those having a single resistance gene. A three-gene line, IRBB59 (with *xa5, xa13,* and *Xa21*), was used as donor to transfer the three BB resistance genes into three new plant type lines with high yield potential, IR65598-112 and the two sister lines IR65600-42 and IR65600-96 (Sanchez *et al.,* 2000). Sequence tagged site (STS) markers for all the three resistance genes from the previous identified RFLP and RAPD markers developed by Huang *et al.* (1997) and Sanchez *et al.* (2000) were used for foreground selection. F_1 plants were obtained between the donor and the three recurrent parents and were advanced up to the BC_3 generation by MAS. Starting from the BC_1F_1, and in each of the following BCF_1 generations, approximately 50 plants were genotyped. From these, plants carrying resistant alleles of the three target resistance genes (based on their marker genotypes) and those phenotypically similar to the recurrent parents were selected as the parents for the next backcross until BC_3F_1. The selected BC_3F_1 plants for each of the recurrent parents were selfed to produce BC_3F_2 seed. Based on phenotypic similarity to their recurrent parents, BC_3F_2 plants were selected for homozygosity at the STS marker genotypes and phenotyped for their reactions to the six Xoo races. The BC_3F_3 NILs having more than one BB resistance gene showed a wider resistance spectrum and manifested increased levels of resistance to the Xoo races, as compared with those having a single BB resistance gene. The resultant plants had very high degree of similarity to their respective recurrent parents, suggesting that the phenotypic selection in every backcross generation was effective. The results of these studies clearly demonstrated the usefulness of MAS in gene pyramiding for BB resistance, particularly for recessive genes, such as *xa-5* and *xa13* that are difficult to select through conventional breeding in the presence of a dominant gene such as *Xa-21*.

Similarly, Singh *et al.* (2001) transferred the same three BB resistance genes, *xa5*, *xa13* and *Xa21*, into the elite cultivar 'PR106', which is widely grown in Punjab, India. IRBB22 with all three genes in IR24 background was used as donor parent. Lines of PR106 with pyramided genes were evaluated after inoculation with 17 isolates of the pathogen from Punjab and six races of Xoo from Philippines. Genes in combination were found to provide high levels of resistance to the predominant Xoo isolates from Punjab and six races from Philippines. Lines of PR106 with two and three BB resistance genes were also evaluated under natural conditions at 31 sites in commercial fields. The combination of genes provided a wider spectrum of resistance to the pathogen population prevalent in the region. Only 1 of the BB isolates, PX04, was virulent on the line carrying *Xa21* but avirulent on the lines having *xa5* and *xa13* genes in combination with *Xa21*. However, the performance of the pyramided lines on other agronomic traits, particularly in comparison with the recurrent parent was not reported. One of the pyramid lines contained all the three genes, named as SS113, was used by Sundaram *et al.* (2008) as donor to introgress the resistance genes into cultivar 'Samba Mahsuri' (BPT5204), which is a medium slender grain indica rice variety and very popular with farmers and consumers across India because of its high yield and excellent cooking quality. At each backcross generation, markers closely linked to the three genes were used to select plants possessing these resistance genes (foreground selection) and microsatellite markers polymorphic between donor and recurrent parent were used to select plants that have maximum contribution from the recurrent parent genome (background selection). A selected BC_4F_1 plant was selfed to generate homozygous BC_4F_2 plants with different combinations of BB resistance genes. The three-gene pyramid and two-gene pyramid lines exhibited high levels of resistance against the BB pathogen. Under conditions of BB infection, the three-gene pyramid lines exhibited a significant yield advantage over Samba Mahsuri. Multi-location testing demonstrated that these lines retained the excellent grain and cooking qualities of Samba Mahsuri without compromising the yield. One of these lines has been recommended for release as a commercial variety by the variety identification committee of the Indian Council of Agricultural Research.

Joseph *et al.* (2004) screened 13 NILs of rice with different BB resistance genes and gene combinations against four isolates of the pathogen from the

Basmati regions of India and identified *Xa4, xa8, xa13* and *Xa21* as effective against all the isolates tested. Two or more of these genes in combination imparted enhanced resistance as expressed by reduced average lesion length in comparison to individual genes. The two-gene pyramid line IRBB55 carrying *xa13* and *Xa21* was found equally effective as three/four gene pyramid lines. IRBB55 was then used as donor parent to transfer *xa13* and *Xa21* into Pusa Basmati-1, the most popular high yielding rice variety. Recombinants having enhanced resistance to BB, Basmati quality and desirable agronomic traits were identified. Unlike studies outlined above, this study used only a single backcross generation and extensive selection for agronomic traits were conducted in three selfing generations to recover the genome of the elite parent.

Cereal cyst nematode (CCN), *Heterodera avenae*, is a significant pathogen of wheat. Nine genes, designated *Cre1* to *Cre8* and *CreR,* in both hexaploid wheat and its relatives have been identified as sources of resistance to CCN (Mc Intosh *et al.,* 2003). Diagnostic markers for *Cre1* and *Cre 8* have been developed and are being employed successfully to pyramid the two resistance genes in Australia (Eagles *et al.,* 2001; Ogbonnaya *et al.*, 2001a, b). Barloy *et al.* (2006) pyramided the two resistance genes *CreX* and *CreY* identified in *Ae. variabilis* Accession No. 1 into a wheat background through MAS. CCN bioassays with the Ha12 pathotype showed that the level of resistance of the pyramided line was significantly higher than that of *CreX* and *CreY* single introgression lines, but lower than that of *Ae. variabilis*. The *CreY* gene, carried by line X8, seemed to confer a higher level of resistance to the Ha12 pathotype than the *CreY* gene carried by line D. The differences in the number of cysts between the pyramided lines and *Ae. variabilis* may be due to the lower expression of *CreX* and *CreY* genes in the wheat background or the possibility that more than two genes are involved in CCN resistance in *Ae. variabilis*. Markers linked with wheat rust genes *Sr38/ Lr37/Yr17* and *Sr39* have been implemented by the National Cereal Rust Control Program at the University of Sydney, Australia, to accumulate several resistance genes (Eagles *et al.,* 2001).

Liu *et al.* (2000) developed three lines with two of the three powdery mildew resistance genes in the background of elite wheat cultivar 'Yang158'. In this study, near-isogenic lines (NILs) were first developed using 'Yang158' as recurrent parent and lines with different resistance genes as donor parents. They were then used as parents for pyramiding genes. Pyramided lines with two of the three powdery mildew resistance gene combinations, *Pm2 + Pm4a, Pm2 + Pm21, Pm4a + Pm21* were obtained by MAS in the F_2 generation. The pyramided lines showed good uniformity in morphological and other non-resistance agronomic traits. Since more than four backcross generations and intensive selection for recurrent characteristics were conducted in the development of the NILs used as parents in pyramiding, the process of developing pyramided lines is simplified.

Hittalmani *et al.* (2000) developed rice line with two or three major blight resistance genes (*Pi*1, *Piz*-5 and *Pi*ta) using three NILs and RFLP markers. Each of the NILs carried the major genes *Pi*1, *Piz*-5 and *Pi*ta, respectively, in the background of the susceptible recurrent parent CO39. Three single-pair crosses made between the NILs and F_2 plants homozygous for resistance genes were identified based on the parental banding pattern of RFLP probes. Individual plants identified as carrying homozygous resistance genes *Pi*1+*Piz*-5 and *Pi*1+*Pi*ta were further crossed with each other and F_2 plants with all the three genes were identified. The plants carrying the two- and three-gene combinations that were tested for resistance to leaf blast in the Philippines and India indicated that combinations including *Piz*-5 have enhanced resistance than when it is present alone.

Sharma *et al.* (2004) constructed a gene-pyramided *japonica* line, in which two Brown planthopper (BPH) (*Nilaparvata lugens* Stål) resistance genes *Bph1* and *Bph2* on the long arm of chromosome 12 independently derived from two *indica* resistance lines were combined. The parent line containing *Bph1*and the one containing *Bph2* were both homozygous elite introgression lines developed by transferring the resistance genes from *indica* donor cultivars. A single F_1 plant heterozygous for the linked markers was selected. MAS was continued in F_2, F_3 and F_4 populations and a homozygous recombinant line was obtained. BPH bioassay showed that the resistance level of the pyramided line was equivalent to that of the *Bph1*-single introgression line, which showed a higher level of resistance than the *Bph2*-single introgression line.

He *et al.* (2004) reported the development of the improved versions of the two parental lines of Shanyou 63, an elite hybrid rice cultivar widely grown in China (Zhenshan 97 and Minghui 63) using MAS and genetic transformation. (1) two wide-spectrum BB resistance genes, *Xa21* and *Xa7*, were incorporated into the restorer line Minghui 63. (2) Minghui 63 was transformed with a *Bt* ä-endotoxin gene to improve the stem borer resistance. (3) *Xa21* and *Bt* genes were combined into a line with the Minghui 63 background. (4) Two genes, *Pi1* and *Pi2*, showing broad-spectrum resistance to fungi blast, were introgressed into Zhenshan 97. (5) Two genes, for brown planthopper resistance, *Qbph1* and *Qbph2*, were introgressed into Zhenshan 97. These versions of improved lines are being combined in various ways to make new hybrids to meet the needs of rice production.

Barone *et al.* (2005) pyramided several resistance genes in the same tomato variety. Two NILs, each possessing at least 3 resistant genes (Momor, which was resistant to tobacco mosaic virus, *Verticillium dahliae* and *Fusarium oxysporum* f.sp. *radicis-lycopersici*, and Motelle, which was resistant to *V. dahliae*, F. *oxysporum* f.sp. *lycopersici*, *Stemphylium* sp. and *Meloidogyne incognita*), were intercrossed and selfed for many generations. Selection for resistance was performed using molecular markers linked to the genes to fix them at the homozygous level in the same genotype. After the F_4 generation, various genotypes which carried all resistance genes at homozygous condition were obtained.

Shi *et al.* (2006) pyramided *Rsv*1, *Rsv*3, and *Rsv*4 for Soybean Mosaic Virus (SMV) resistance. A population of 84 lines derived from J05 (*Rsv*1, *Rsv*3) x V94-5152 (*Rsv*4) were developed, and six specific SSR markers were identified for SMV resistance genes. Two SSR markers Sat_154 and Satt510 were used for selecting lines having the *Rsv*1 gene, Satt560 and Satt726 for *Rsv*3, and Sat_254 and Satt542 for *Rsv*4. These SSR markers allowed for identification and selection of specific lines and individual plants containing different genes and for distinction of the homozygous and heterozygous lines or individual plants for all three resistance loci. Individual plants with homozygous alleles at three genetic loci (*Rsv*1*Rsv*1, *Rsv*3*Rsv*3 and *Rsv*4*Rsv*4) have been identified and new soybean germplasm is expected to be released with three genes combined for SMV resistance.

Wei *et al.* (2007) described the breeding of transgenic rice restorer line for multiple resistance against bacterial blight, striped stem borer (SSB) (*Chilo suppressalis*) and herbicide. Two stable transgenic rice lines, Zhongguo91 containing *cry1Ab* gene (for insect resistance) and *bar* gene (for tolerance of herbicide), and Yujing6 containing *Xa21* gene for BB resistance, were developed by PIG gene gun bombardment and *Agrobacterium*-mediated transformation, respectively. Two elite restorers, T773-1 expressing *cry1Ab* and *bar* genes and T773-2 expressing *Xa21* gene, were obtained by five successive backcrossing to the elite restorer line Hui773. The transgenic restorer line T773 with good agronomic traits and obvious multiple resistance to BB, SSB and herbicide were then developed by selection in four selfing generations of the cross between T773-1 and T773-2. Bioassay, selectable marker and reporter genes were used in selection process. The hybrid (F_1 generation) produced from the cross between T773 and a corresponding male sterile line Zaohua2A maintained resistance to BB, leaffolder and SSB, and showed significant heterosis.

Gal *et al.* (2007) reported the use of markers to incorporate the major leaf rust (*Puccinia triticina*) resistance genes (Lr genes) currently effective in Hungary (Lr9, Lr24, Lr25, Lr29) into winter wheat cultivars. The Lr genes were identified using STS, SCAR and RAPD markers closely linked to them. Investigations were made on how these markers could be utilised in plant breeding and near-isogenic lines resembling the recurrent cultivar but each containing a different Lr gene were developed to form the initial stock for the pyramiding of resistance genes. The results indicated that the MAS technique elaborated for resistance genes Lr24, Lr25 and Lr29 can be applied simply and effectively in wheat breeding, while the detection of the Lr9 marker was uncertain.

Saghai Maroof *et al.* (2008) developed pyramided soybean lines containing two or three soybean mosaic virus (SMV) resistance genes (Rsv1, Rsv3, and Rsv4) through MAS. The two-gene and three-gene isolines of Rsv1Rsv3, Rsv1Rsv4 and Rsv1Rsv3Rsv4, acted in a complementary manner, conferring resistance against all strains of SMV, whereas isolines of Rsv3Rsv4 displayed a late susceptible reaction to selected SMV strains.

2.2 Pyramiding Major Genes for Other Traits

Using simple sequence repeat (SSR) markers linked to the thermo-sensitive genetic male sterility (tms) genes in rice, Nas *et al.* (2005) developed two-gene and three-gene pyramided lines using lines containing different tms genes (IR80775-46 with *tms2* and *tms5*, and IR80775-21 with *tms2*, *tgms* and *tms5*).

Sattari *et al.* (2007) used two sequence-tagged site (STS) markers (RG140/*Pvu*II and S10019/*Bst*UI) to select for two major *Rf* genes (*Rf3* and *Rf4*) governing fertility restoration of cytoplasmic male sterility in rice. The combined use of markers associated with these two loci improved the efficiency of screening for putative restorer lines from a set of elite lines.

3.0 PYRAMIDING QTL

Stuber (1992, 1999) reported increased grain yield in maize by introgression using six parents containing favourable chromosome segments. Three backcross generations (two marker-facilitated) were used for the transfer of subsets of the identified chromosomal segments into the target lines, B73 and Mo17. This was followed by two generations of marker-facilitated selfing to fix the introgressed segments. However, all six target segments were obtained in any given line. The "enhanced" lines were then crossed in appropriate combinations and the "enhanced" single crosses were evaluated in replicated yield tests. On the basis of 4 yr of testing, yields of the best "enhanced" B73 x "enhanced" Mo17 hybrids exceeded the original B73 x Mo17 hybrid and high yielding commercial hybrids by 8 to 10% (628–1004 kg ha-1). They found that there appears to be some indication that there may no advantage in transferring more than two to four segments. In fact, there is some indication that there could be a disadvantage. They offered several explanations for this observation: first increasing the number of transferred segments may have replaced the recipient genome with an excessive amount of linked donor chromosomal segments that may cause a deleterious effect. Second, epistatic interactions between a larger number of introgressed segments may result in a negative effect. Third, favourable epistatic complexes in coupling phase (*e.g.*, between recurrent parent alleles) could be disrupted.

Castro *et al.* (2003a, b) developed double haploid (DH) lines combining two barely strip rust (BSR) resistance QTL alleles from the accession "Calicuchima-sib" on chromosomes 4 (4H) and 7 (5H) (QTL4 and QTL7), and a BSR resistance QTL allele from the cultivar Shyri on chromosome 5 (1H) (QTL5). Results on seedling resistance validated the effects and locations of QTL4 and QTL5, but the QTL7 did not have a significant effect on disease symptom expression. The presence of resistance alleles at both loci substantially increased the probability of recovering the resistant phenotype. In mapping populations, a resistance allele at a QTL on chromosome 6(6H) was necessary for resistance, in conjunction with a resistance allele at either QTL4 or QTL5. In the DH lines studied in this experiment, resistance alleles were also necessary at two QTL, but the two QTL are on chromosomes 4(4H) and 5(1H). In other words, an allele at QTL4 or QTL5 can substitute for an allele on chromosome 6(6H). Results on adult resistance validated the effects of resistance alleles at all the three QTL regions on disease severity and area under disease progress curve and these QTL explain 94% of the genetic variation of the trait expression. A comparison of QTL effects as estimated in the source mapping populations and in the derived lines reveals changes in magnitude of effect.

Using two doubled-haploid (DH) experimental lines (BCD47 contains resistance alleles at the QTL on chromosomes 4H and 5H and BCD12 on 1H) developed by (Castro *et al.*, 2003a) as donors of the resistance alleles, Richardson *et al.* (2006) developed a set of QTL introgression lines in a susceptible background, which represent disease resistance QTL combined in one-, two-, and three-way combinations in a susceptible background. These lines were used to measure four components of disease resistance: latent period, infection efficiency, lesion size, and pustule density. Pyramiding multiple QTL alleles led to higher levels of resistance in terms of all components of QR except latent period. There were linear reductions in infection efficiency, lesion size, and pustule density as more resistance alleles were added to individual genotypes, but resistance pyramiding did not increase latent period. The introgression of the same resistance alleles at the same QTL into different lines did not always lead to the same resistance. There was more variance among lines within the quantitative resistance allele introgression classes and no variance among lines within the qualitative resistance gene introgression group. Pyramids of multiple resistance QTL alleles where the 4H QTL was present led to lower infection efficiency, lesion size, and pustule density.

Gur and Zamir (2004) demonstrated that tomato yield can be increased dramatically by pyramiding three independent yield-promoting genomic regions from the drought-tolerant *S. pennellii*. The yields of hybrids that were parented by the pyramided genotypes were more than 50% higher than that of a control market-leader variety under both wet and dry field conditions.

Ashikari *et al.* (2005) identified major QTL for grain number (*Gn1*) and QTL for plant height (*Ph1*) using the progeny from the cross between the *japonica* rice 'Koshihikari' and the *indica* rice 'Habataki'. NILs in the Koshihikari' genetic background were developed and used to combine both beneficial traits. Two lines were crossed and a pyramiding line carrying *Gn1* and *Ph1(sd1)* was selected from the progenies using MAS. The pyramided line showed increased grain production (23%) and reduced plant height (20%) compared with 'Koshihikai'.

Guo *et al.* (2005) pyramided two major QTL for high fibre strength in cotton identified in an elite fiber germ-plasm line 7235 (*Gossypium hiusutum* L.), the pyramiding lines have significantly higher fibre strength.

Recently Miedaner *et al.* (2006) reported the pyramiding of three donor QTL for FHB resistance originating from two non-adapted sources into elite European spring wheat lines. Lines carrying the donor QTL alleles individually and in combinations were developed and the eVects of the introduced QTL on expression of resistance were evaluated. All three individual QTL alleles significantly reduced the DON content and FHB severity compared to the contrasting susceptible controls. Combination of the 3B- and 5A-QTL alleles produced the highest effect for both traits (Miedaner *et al.*, 2006).

A large scale marker-assisted gene pyramiding program in rice is undergoing in the International Rice Research Institute (IRRI) and the Chinese Academy of Agricultural Sciences (CAAS) to improve drought tolearance. Introgression lines containing unrelated QTLs for drought tolerance are crossed to produce segregating F_2 populations, which are then screened under both stress and non-stress conditions to identify superior individuals that have desirable QTL or QTL combinations and good yield potential. At IRRI, 10 F_2 populations developed this way were screened under very severe lowland drought, resulting in 560 drought tolerance pyramiding lines. In CAAS, a total of 56 F_2 populations were developed, from which 1207 drought tolerance pyramiding lines were selected (Li *et al.*, 2007).

Ando *et al.* (2008) developed a rice line containing two major QTL, qSBN1 (for secondary branch number on chromosome 1) and qPBN6 (for primary branch number on chromosome 6). The pyramided line produced more spikelets than those containing only one of the QTL.

4.0 MAIN FACTORS AFFECTING GENE PYRAMIDING

4.1 Characteristics of the Target Traits/Genes

When the genes to be pyramided are functionally well characterised, gene pyramiding will be more successful. For qualitative traits controlled by one or a few genes, the identification of the genes and tightly linked markers is easier, provided phenotyping is carefully conducted. When the target genes are QTL with moderate or small effects, pyramiding may be less successful due to the following reasons. Firstly, the identified QTL may be more likely to be a false positive. Secondly, inaccurate QTL localizations result in the need to select for more marker loci covering large genomic segments to be certain that target QTL alleles are retained in selected progeny (Hospital and Charcosset, 1997). Thirdly, QTL effects may be specific to a particular genetic background. Moreover, markers identified for a QTL can be ineffective in monitoring the QTL since the marker-QTL association might be different from population to population. Fourthly, more QTL need to be pyramided to achieve a significant improvement.

4.2 Reproductive Characteristics

The propagation capability of a crop is determined by the number of seeds produced by a single plant. This capacity determines the population size applicable if seed has to be collected from only a single plant. In a gene pyramiding program, in most generations this is the case, since the chance of selecting two or more individuals of exactly the same genotype in previous generation is very low. For example, although a fairly large F_2 population can be obtained by collecting seed from many F_1 plants of the cross between two homozygous parents, F_3 generation seed can only be collected from a single plant. The fact that F_1 plants of the cross between two homozygous parents are genetically the same can also be used to increase the size of a progeny population of the F_1 plants of two crosses (double cross) or of the F_1 plants of one cross and an inbred line (Three-way cross or testcross).

The efficiency of hybridization may be an important constraint for some crop species. When wild relatives are used as the donor of desirable genes, many more reproduction related constraints may exist including cross incompatibility between the wild species and cultivated crop. F_1-hybrid sterility, infertility of the segregating generations, reduced recombination between the chromosomes of the two species. Appropriate techniques that may include chemical treatment and immature embryo culture for overcoming these problems must be established.

4.3 A Breeder's Capability to Identify the 'Desired' Genotypes

It is obvious that the desirable genes must be present in all generations leading to the target genotype. To ensure the presence of the target genes individuals of desired genotype (which may change with generation advance) must be identified among all individuals in each generation. Breeder's capability to identify the desired genotypes has been greatly enhanced by the use of tightly linked or diagnostic markers. It might be appropriate to consider the importance of marker and trait gene linkage here.

The efficiency of marker-based gene pyramiding will decrease substantially if the markers are not perfectly or tightly linked with useful trait genes. Association of a marker with a trait allele and consequently the reliability of the marker-based selection decreases with increasing cycles of meiosis. With a recombination value of r between a marker and trait allele, the probability of this linkage being maintained across m cycles of meiosis is equal to $(1 - r)^m$. To keep this probability higher than a certain critical value, say P, m must not exceed lnP/ ln(1 - r), suggesting that a phenotypic test should be performed every m generation of selection to confirm the persistence of the initial linkage. Put in another way, the probability of losing the target allele by recombination is $1 - (1- r)^t$. For example, if the marker locus exhibits 10% recombination with the target gene, there is a 10% chance of losing the target allele each generation, and a 27% chance of losing the target allele after three generations of meiosis. However, if the recombination frequency is 1%, there is only a 3% chance of losing the target allele after three generations of meiosis.

When tightly linked markers are not available, selection on a pair of markers flanking the target locus can be very effective. If two marker loci M_1 and M_2 flank the target locus, one would select progeny that have both M_1 and M_2 alleles. The probability of losing the target allele with flanking marker selection is equal to the probability of selecting a double recombinant progeny from among the doubly heterozygous backcross progeny. If the flanking loci have recombination frequencies r_1 and r_2, respectively, with the target locus, the probability of losing the target allele due to double crossovers within the selected region is $\frac{r_1 r_2}{1 - r_1 - r_2 + 2r_1 r_2}$. This probability can be much lower than the probability of losing the target allele based on selection for a single marker. For example, if the flanking markers each have 10% recombination frequency with the target locus, there is only a 1.2% chance of losing the target allele after a single generation. In any case, with tighter linkage, the chance of losing the target allele is reduced. However, this requires more plants to be tested and higher cost per plant. It is imperative to use markers that are tightly linked with trait genes.

Multiple marker loci closely linked to the target gene, permits discrimination on the basis of the haplotype of several markers rather than just the genotype at one marker. For example, Cregan *et al.* (1999) developed two SSR markers tightly linked to the *rhg1* gene. Neither marker could alone distinguish all resistant from all susceptible genotypes, because of identity in state alleles shared by some resistant and susceptible lines, but the two markers together could discriminate almost all resistant and susceptible lines. One resistant cultivar carried the susceptible allele at both loci, presumably due to recombination between marker and resistance loci during line development. Thus, recombination can change the linkage phase between markers, but if MAS is used first to select putatively resistant lines, followed by phenotypic evaluation of resistance, the linkage phase will remain intact in all selected progeny. Therefore, MAS can be self-reinforcing, ensuring that the same set of markers will be effective in future crosses.

4.4 Operating Capital

All breeding programmes are operated within the limits of available operating capital. Therefore, reducing the overall cost is always an important consideration when choosing a strategy. In addition to the use of the most economic mating and testing approaches, other factors affecting the cost also need to be considered. In the context of gene pyramiding, cost effects both what can be achieved and the means of achieving it. Increasing the number of generations (duration) will reduce the pressure on population size required in each generation and may result in the reduction of the total cost. However, increasing the duration delays the release of the new cultivar and consequently reduced market share. The well-known trade-off between duration and cost in breeding has no exception in gene pyramiding. To find an optimum balance between duration and cost is desired but very difficult to achieve. In practice, the best strategy may be defined as the one that enabled the breeder to achieve the objectives with the shortest duration and within a fixed expected investment.

5.0 DESIGNING AN EFFICIENT GENE PYRAMIDING PROGRAM

To have a successful gene pyramiding requires (1) that the locations of a series of genes of interest (target genes) and thus the linkage relationship between them is known; (2) that a target genotype for these genes is defined prior to selection as the genotype with favourable alleles at all loci of interest; (3) that the genotype of an individual can be identified by these genes or markers linked to them; (4) that a collection of lines containing all the target genes are available. Marker discovery through QTL mapping using classical segregating populations, introgression line libraries and association genetics are discussed in Chapters 11 and 16. The development of parental lines for gene pyramiding using recurrent backcross is discussed in Chapter 11.

An efficient gene pyramiding program will lead to the development of near-homozygous breeding lines that are fully homozygous for the desirable alleles of the target genes using the minimum number of generations of selection and the lowest genotyping and phenotyping costs. The total cost and duration are the two principal criteria when pyramiding strategies are designed and compared. Both the phenotyping and genotyping costs can be roughly quantified using the total number of plants to be screened for phenotypic performance and genotypic status, respectively. Therefore, the cost of a strategy can be roughly estimated with the sum of minimal population size at each of the generations. When the number of parental lines containing the desirable genes (founding parents) is more than three, more than one crossing scheme can result in the generation of the target genotype. Therefore, the gene pyramiding scheme can be divided into two parts. The first part is aimed at cumulating one copy of all target genes in a single genotype (called root genotype). The second part is aimed at fixing the target genes into a homozygous state, that is, to derive the target genotype from the root genotype. Sevrin *et al.* (2004) called these two parts pedigree and fixation, respectively. Fig. 1 is an example of a gene-pyramiding scheme cumulating six target genes from six parental lines. Ye and Smith (2008a, b) discussed in detail the design of an efficient gene pyramiding program and suggested some general guidelines for increasing the efficiency. The key points are summarised below.

The fixation step should be considered first to ensure that the number of genes to be pyramided is realistic. For the fixation step one- step selection conducted only after homozygous individuals are obtained is considered first. The most commonly used methods for the production of homozygous individuals are the development of recombinant inbred lines (RIL), and doubled-haploid (DH) population. Therefore, it is advisable to investigate the feasibility of achieving the objective using RIL or DH. With the recombination frequencies between the target genes, the proportion of the desired genotype in the RIL or DH population can be worked out and the minimum population size required can be determined using equation 4. Either RIL or DH can be adopted if this size of population is practically achievable and the cost of genotyping it is affordable, although it may not be the optimal scheme.

If neither RIL nor DH options are feasible, repeated selection in more than one subsequent segregating generation is required. Selection in sequential generations of individuals that have an increasing number of the desired alleles fixed at the desired loci while heterozygous at the remaining desired loci increases the frequency of the targeted recombinant through accumulated recombination. The objective is to identify a selection scheme that leads to the production of the target genotype using the minimum number of generations and the practically allowable population sizes in each of the generations. This can be achieved by maximising the number of genes to be fixed at each generation within the limits of applicable population size. Ye *et al.* (2007) showed how to define such a selection strategy in steps. When crossing to another genotype can be easily and cheaply conducted, crossing to other genotype instead of self-pollination might be used to reduce population size. At least three options can be explored. (1) Cross to a founding parent. The advantage of crossing to a

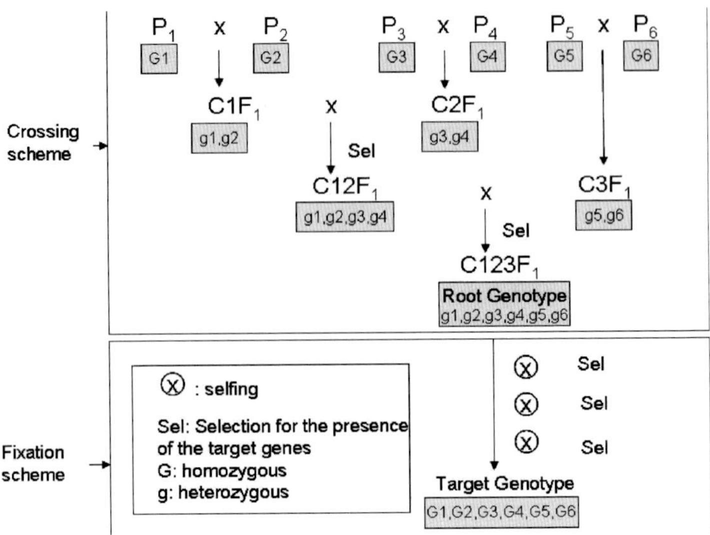

Fig. 1. Example of a gene-pyramiding scheme cumulating six target genes

founding parent is that the probability of obtaining a genotype that is homozygous for the target genes brought by the founding parent but heterozygous for the other targets is high. Hence, that target gene need not be fixed subsequently, increasing the probability of getting the target genotype. (2) Cross to a blank line containing none of the favourable alleles. The use of blank line increases the chance of obtaining a genotype carrying all favourable alleles in coupling and thereby increasing the frequency of target genotype in subsequent generations. (3) Cross between selected individuals. If a satisfactory genotype with at least one copy of the desirable alleles in all target loci is missing at any generation, crossing two plants with the best complementary genotypes can be used to secure the program. Crossing between complementary genotypes may also speed up the breeding process even if a satisfactory genotype is present.

Since the number of possible satisfactory crossing schemes increases fast with the number of genes, even with the computer program, it is impossible to evaluate all the satisfactory schemes when the number of loci are more than a dozen (Servin *et al.* 2004). The guidelines developed by Ye and Smith (2008a, b) based on previous studies (*i.e* Ishii and Yonezawa, 2007a,b) are as follows.

(1) Founding parents with fewer target markers enter the schedule at earlier stages.

(2) A cross that invokes a strong repulsion linkage should be performed as early as possible

(3) More crosses should be conducted at each generation if genotyping cost is low and the practically applicable population size is large

(4) One cross per generation is required if the practically applicable population size is small or genotyping cost is high

(5) Using backcrossing or testcross before assembling more genes

6.0 CONCLUSION

Traits controlled by major genes should be the primary target of gene pyramiding. QTL mapping studies clearly demonstrated that only QTL with large effects can be estimated and positioned accurately using an affordable size of mapping population. Moreover, it is also true that major genes are usually more stable in different genetic backgrounds and environments. Traits for which there is convincing and unequivocal evidence for the presence of two or more additive or complementary genes, pyramiding these genes into a common genotype would help maximise the character expression or gains from selection. Pyramiding such genes could also broaden the genetic basis of cultivars.

Pest resistance conferred by single or few genes are the most rewarding area of gene pyramiding application. To pyramid multiple resistance genes into a single cultivar, breeders must be able to monitor the effects of these genes, which is not always possible through phenotypic measurements since the effect of one gene may be affected by the present of other genes due to epistasis and/or the masking effect. The use of MAS makes it possible to identify plants with various numbers of resistance genes with very similar resistance performance. It also makes it possible to select for recessive genes without progeny testing. Therefore, marker-assisted pyramiding is being actively used in breeding for qualitative resistance and is proven fruitful.

Major genes for agronomically important traits are more likely to be identified in wild relatives rather than the elite gene pool used by breeders. This is because the repeated exploration of the elite gene pool by active breeding should have already fixed most of the major genes. Therefore, it seems that gene pyramiding strategy may suit the exploration of wild germplasm.

The pyramiding of QTL has been less successful in terms of achieving expected improvement. Hospital (2005) listed the following reasons for the unexpected results when MAS is applied to QTL introgression. (1) the putative QTL is false positive; (2) QTL expression is testing environment specific (QTL-by-environment interaction (QEI); (3) QTL are interacting among themselves or with genetic background effect (epistasis), and (4) The chromosomal segments detected as QTL hold not just one but several genes.

These can also be used to explain the unexpected results in gene pyramiding. QTL with small effects are more likely to be false positive. Therefore, pyramiding should target QTL with relatively large effects. Genotype-by-environment interaction (GEI) is a well known phenomenon of quantitative traits and thus it is not surprised that QTL underlying these traits are also sensitive to testing environments. For gene pyramiding to be effective it should aimed at QTL with good stability across the target population of environments. QTL mapping experiments with multiple testing environments suggested that major QTL also tend to be stable across environment. This again favours the use of major QTL in pyramiding. The interactions between QTL and between QTL and genetic background are more difficult to handle in practice. This is particularly true when QTL are from multiple sources. It might be unrealistic to expect that all QTL from different sources will act additively and the chance to observe QTL-by-QTL interaction is higher. Since the direction of QTL-QTL interaction is unpredictable, step-wise pyramiding of the target QTL by using NILs containing the target QTL in the same target background is recommended. Overall, the main reason for the less success of QTL pyramiding is that the target QTL are not well characterised. We recommend the use of NILs for QTL confirmation and step-wise pyramiding.

The genotyping/phenotyping cost, practically applicable segregating population size, and the length of breeding duration need to be considered when the crossing and selection strategy is designed. The most important factor is the number of genes to be pyramided because the population size necessary to fix the target genes increases exponentially with the number of target loci. In practice, the number of target genes should be kept low so that the overall cost and the length of breeding duration are acceptable. When several favourable genes are originally hosted by only two or three parents, the crossing scheme can be easily determined. The target genotype is obtained by selecting the most promising genotypes (genotype with the highest number of genes homozygous for the desirable alleles and heterozygous for all other target loci) in each of the selfing generations. Therefore, pyramiding will be more straightforward if parental lines with complementary sets of homozygous loci are available. When genes are dispersed between many parents, crossing scheme needs to be chosen so that the root genotype can be obtained quickly at low cost.

As for all applications of MAS, the consistence of the linkage phase between the target gene and its linked markers across multiple populations presents a serious challenge for selecting for the target gene using markers. Markers linked to the QTL identified by linkage mapping using one or a few populations may or may not be useful in gene pyramiding because different subsets of QTL will be polymorphic in each population, and the linkage phases between the marker and QTL alleles can differ even between closely related genotypes. The linkage phrase tends to be more consistent if the source of QTL is from a gene pool which is distinct from the one used by the breeders. Thus, markers linked to novel alleles from exotic germplasm or wild relatives are more likely to be successfully implemented (Tanksley and McCouch, 1997). It is expected that the use of IL for the identification and pyramiding of favourable genes contained in wild relatives into an elite parent will be prove to be an efficient method for improving quantitative traits. It is usually true that the tighter the linkage the more consistent the linkage phrase across populations. With the rapid development of plant functional genomics functional markers will become available for more traits and crops (Andersen and Lübberstedt, 2003) and greatly facilitate the application of marker-based gene pyramiding in cultivar development.

When gene pyramiding is essential to be used as a strategy to utilise novel genes /QTL identified, it makes sense to think of gene identification, validation and pyramiding as the components of an integrated process. Ideally, the gene identification and validation process should also lead to the creation of good parental lines for later pyramiding. This can be achieved by using introgression lines or the advanced backcrossing QTL mapping methods. Ye and Smith (2008a, b) discussed the intergration of gene identification, validation and pyramiding through advanced backcrossing QTL mapping and the use of introgression lines.

7.0 REFERENCES

Allard, R.W. 1999. Principle of Plant Breeding (2nd Edn). John Wiley and Sons, New York, 272 pp.

Andersen, J.R. and Lübberstedt, T. 2003. Functional markers in plant. Trends in Plant Science, 8:554-560.

Ando, T., Yamamoto, T., Shimizu, T., Ma, X.F., Shomura, A., Takeuchi, Y., Lin S.Y. and Yano M. 2008. Genetic dissection and pyramiding of quantitative traits for panicle architecture by using chromosomal segment substitution lines in rice. Theoretical and applied genetics, 116:881-890.

Ashikari, M. and Matsuoka, M. 2006. Identification, isolation and pyramiding of quantitative trait loci for rice breeding. Trends in Plant Science, 11:344-350.

Ashikari, M., Sakakibara, H., Lin, S.Y., Yamamoto, T., Takashi, T., Nishimura, A., Angeles, E.R., Quian, Q., Kitano, H. and Matsuoka, M. 2005. Cytokinin oxidase regulates rice grain production. Science, 309:741-745.

Barone, A., Ercolano, M.R., Langella, R., Monti, L. and Frusciante, L. 2005. Molecular marker-assisted selection for pyramiding resistance genes in tomato. Advances in Horticultural Science, 19:147-152.

Barloy, D., Lemoine, J., Abelard, P., Tanguy, A.M., Rivoal, R. and Jahier, J. 2006. Marker-assisted pyramiding of two cereal cyst nematode resistance genes from *Aegilops variabilis* in wheat. Molecular Breeding, 20:31-40.

Castro, A.J., Chen, X.M., Hayes. P.M., and Johnston, M. 2003a. Pyramiding quantitative trait locus (QTL) alleles determining resistance to barley stripe rust: effects on resistance at the seedling stage. Crop Science, 43:651-659.

Castro, A.J., Chen, X., Corey, A., Filichkina, T., Hayes, P.M., Mundt, C., Richardson, K., Sandoval-Islas, S. and Vivar, H. 2003b. Pyramiding and validation of quantitative trait locus (QTL) alĺeles determining resistance to barley stripe rust: Effects on adult resistance. Crop Science, 43:2234-2239.

Cregan, P.B., Mudge, J., Fickus, E.W., Danesh, D., Denny, R. and Young N.D. 1999. Two simple sequence repeat markers to select for soybean cyst nematode resistance conditioned by the *rhg1* locus. Theoretical and Applied Genetics, 99:811-818.

Dekkers, J.C.M. and Hospital, F. 2002. The use of molecular genetics in the improvement of agricultural populations. Nature Review Genetics, 3:22-32.

Dubcovsky, J. 2004 Marker assisted selection in public breeding programs: The wheat experience. Crop Science, 44:1895-1898.

Eagles, H.A., Bariana, H.S., Ogbonnaya, F.C., Rebetzke, G.J., Hollamby, G.J., Henry, R.J., and Henschke, P.R. 2001. Implementation of markers in Australian wheat breeding. Australian Journal of Agricultural Research, 52:1349-1356.

Gal, M., Vidad, G., Uhrin, A., Bedo, Z. and Veisz, O. 2007. Incorporation of leaf rust resistance genes into winter wheat genotypes using marker-assisted selection. Acta Agronomica Hungarica, 55:1491-156.

Guo, W.Z., Zhang, T.Z., Ding, Y.Z., Zhu, Y.C., Shen, X.L. and Zhu, X.F. 2005. Molecular Marker Assisted Selection and Pyramiding of Two QTLs for Fiber Strength in Upland Cotton. Yi Chuan Xue Bao, 12:1275-85.

Gur, A. and Zamir, D. 2004. Unused natural variation can lift yield barriers in Plant breeding. PLOS Biology, 2:1610-1615.

He, Y., Li, X., Zhang, J., Jiang, G., Liu, S., Chen, S., Tu, J., Xu, C., and Zhang, Q. 2004. Gene pyramiding to improve hybrid rice by molecular marker techniques. *4th International Crop Science Congress*, Brisbane, September 28-2 October, Australia.

Hittalmani, S., Parco, A., Mew, T.V., Zeigler, R.S. and Huang, N. 2000. Fine mapping and DNA marker-assisted pyramiding of the three majors genes for blast resistance in rice. Theoretical and Applied Genetics, 100:1121-1128.

Hospital, F. 2005. Selection in backcross programmes. Philosophical Transactions of the Royal Society B 360:1503-1512.

Huang, N., Angeles, E.R., Domingo, J., Magpantay, G., Singh, S., Zhang, G., Kumaravadivel, N., Bennet, J. and Khush G.S. 1997. Pyramiding of bacterial blight resistance genes in rice: Marker-assisted selection using RFLP and PCR. Theoretical and Applied Genetics, 95:313-320.

Ishii, T. and Yonezawa, K. 2007a. Optimization of the marker-based procedures for pyramiding genes from multiple donor lines: I. Schedule of crossing between the donor lines. Crop Science, 47:537-546.

Ishii, T. and Yonezawa, K. 2007b. Optimization of the marker-based procedures for pyramiding genes from multiple donor lines: II. Strategies for selecting the objective homozygous plant. Crop Science, 47:1878-1886.

Joseph, M., Gopalakrishnan, S., Sharma, R.K., Singh, V.P., Singh, A.K., Singh, N.K. and Mohapatra, T. 2004. Combining bacterial blight resistance and Basmati quality characteristics by phenotypic and molecular marker-assisted selection in rice. Molecular Breeding, 100:1-11.

Kloppers, F.J. and Pretorius, Z.A. 1997. Effects of combinations amongst genes *Lr13, Lr34* and *Lr37* on components of resistance in wheat to leaf rust. Plant Pathology, 46:737-750.

Langridge, P. and Barr, A. 2003. Preface to 'Better Barley Faster: the Role of Marker Assisted Selection'. Australian Journal of Agricultural Research, 54:1-4.

Li, Z.K., Gao, Y.M., Zheng, T.Q., Xu, J.L., Fu, B.Y., Sun. Y. and Zhu, L.H. 2007. Improving Drought tolerance of rice by designed QTL pyramiding. Molecular Plant Breeding, 5:205-206.

Liu, J., Liu, D., Tao, W., Li, W., Wang, S., Chen, P., Cheng, S. and Gao, D. 2000. Molecular marker-facilitated pyramiding of different genes for powdery mildew resistance in wheat. Plant Breeding, 119:21-24.

Meintosh, Rot., Yamozabi, Devos, K.M., Dubeovsky, J.W., Rogers W.J. and Appels, R. 2003. Catalogue of gene symbols for wheat. In proceedings of the Tenth International wheat Genetics Symposium, Volume of Paestum, Italy.

Miender, T., Wildes, F., Steiner, B., Buerstmayr, H., Korzun, N and Elomeyer, E. 2006. Stacking quantitative trait loci (QTL) for Fusarium head blight resistance from non-adopted sources in an European elite spring wheat background and asscesing than effects on deoxynivallnol (DON) content and disease severity. Theoretical and Applied Genetics, 12:562-569.

Nas, T.M.S., Sanchez, D.L., Diaz, G., Mendioro, M.S. and Virmani, S.S. 2005. Pyramiding of thermosensitive genetic male sterility (TGMS) genes and identification of a candidate *ims5* gene in rice. Euphytica, 145:67-75.

Ogbonnaya, F.C., Seah, S., López-Brana, I., Jahier, J., Delibes, A., and Lagudah, E.S. 2001a. Molecular-genetic characterisation of a new nematode resistance gene in wheat. Theoretical and Applied Genetics, 102:623-629

Ogbonnaya, F.C., Subrahmanyam, N.C., Moullet, O., Majnik, J. de, Eagles, H.A., Brown J.S., Eastwood, R.F., Kollmorgen, J., Appels, R., and Lagudah E.S. 2001b. Diagnostic markers for cereal cyst nematode resistance in bread wheat. Australian Journal of Agricultural Research, 52:1367-1374.

Paterson, A.H., Tanksley, S.D. and Sorrells, M.E. 1991. DNA markers in plant improvement. Advances in Agronomy, 46:39-90.

Poter, D.R., Burd, J.D., shufran, K.A., and Webster, J.A. 2000. Efficacy of pyramiding greenbug (Homoptera: Aphididae) resistance genes in wheat. Journal of Economic Entomology, 93:1315-1318.

Richardson, K.L., Vales, M.I., Kling, J.G., Mundt, C.C. and Hayes, P.M. 2006. Pyramiding and dissecting disease resistance QTL to barley stripe rust. Theoretical and Applied Genetics, 113:485-495.

Saghai Maroof, M.A., Jeong, S.C., Gunduz, I., Tucker, D.M., Buss, G.R. and Tolin, S.A. 2008. Pyramiding of Soybean Mosaic Virus Resistance Genes by Marker-Assisted Selection. Crop science, 48:517-526.

Sanchez, A.C., Brar, D.S., Huang, N., Li, Z. and Khush, G.S. 2000. Sequence Tagged Site marker-assisted selection for three bacterial blight resistance genes in rice. Crop Science, 40:792-797.

Servin, B., Martin, O.C., Mezard, M. and Hospital, F. 2004. Toward a theory of marker-assisted gene pyramiding. Genetics, 168:513-523.

Sharma, P., Torii, A., Takumi, S., Mori, N. and Nakamura, C. 2004. Marker-assisted pyramiding of brown planthopper (*Nilaparvata lugens* Stål) resistance genes *Bph1* and *Bph2* on rice chromosome 12. Hereditas, 140:61-69.

Shi, A., Chen, P., Zheng, C., Hou, A. and Zhu, S. 2006. Gene pyramiding for soybean mosaic virus resistance using microsatellite markers. 2006 International meeting of ASA-CSSA-SSSA, 13 November, 2006.

Singh, S., Sidhu, J.S., Huang, N., Vikal, Y., Li, Z., Brar, D.S., Dhaliwal, H.S. and Khush, G.S. 2001. Pyramiding three bacterial blight resistance genes (*xa5, xa13, Xa21*) using marker-assisted selection into indica rice cultivar PR106. Theoretical Applied Genetics, 102:1011-1015.

Stuber. C.W., Polacco, M. and Senior, M.L. 1999. Synergy of empirical breeding, marker-assisted selection, and genomics to increase crop yield potential. Crop Science, 39:1571-1583.

Stuber, C.W. and Sisco, P.H. 1992. Marker-facilitated transfer of QTL alleles between elite inbred lines and responses in hybrids pp 104-113. *In*: 46th Annual Corn and Sorghum Research Conference, American Seed Trade Assoc Ed.

Sundaram, M., Vishnupriya, R., Biradar, K.S., Lala, S., Reddy, G.A. Rani, N.S., Sarma, P. and Sonti, R.V. 2008. Marker assisted introgression of bacterial blight resistance in Samba Mahsuri, an elite indica rice variety. Euphytica, 160:411-422.

Tanksley, S.D. and McCouch, S.R. 1997. Seed banks and molecular maps: unlocking genetic potential from the wild. Science, 277:1063-1068.

Wei, Y., Yao, F., Zhu, C., Jiang, M., Li, G., Song, Y., and Wen, F. 2007. Breeding of transgenic rice restorer line for multiple resistance against bacterial blight, striped stem borer and herbicide. Euphytica in press.

Yoshimura, S., Yoshimura, A., Iwata, N., McCouch, S.R., Abenes, M.L., Baroidan MR, Mew T.W. and Nelson R.J. 1995. Tagging and combining bacterial blight resistance genes in rice using RAPD and RFLP markers. Molecular Breeding, 1:375-387.

Ye, G., Moody, D., Livinus, L. and van Ginkel, M. 2007. Designing an optimal marker-based pedigree selection strategy for parent building in barley in the presence of repulsion linkage, using computer simulation. Australian Journal of Agricultural Research, 58:243-251.

Ye, G. and Smith, K.F. 2008a. Marker-assisted gene pyramiding for inbred line development: Basic principles and practical guidelines. International Journal of Plant Breeding, 2:1-10.

Ye, G. and Smith, K.F. 2008b. Marker-assisted gene pyramiding for inbred line development: Practical applications. International Journal of Plant Breeding, 2: 1-22.

Molecular Plant Breeding: Principle, Method and Application
Eds : R.K. Singh, Rajesh Singh, Guoyou Ye, A. Selvi and G.P. Rao
Studium Press LLC, Texas, USA, 2009, pp. 343-365

CHAPTER

13

Marker Assisted Selection for Improvement of Quality Traits in Crop Plants

GYAN P. MISHRA[1*], *SHAILESH K. TIWARI*[2], *RAGHWENDRA SINGH*[1] *and SHASHI BALA SINGH*[1]

ABSTRACT

Like other agronomic traits, improvement of quality traits depends on the availability of sufficient variability for the targeted traits in the germplasm. The rapid development of molecular biology and functional genomics provides breeders with tools that have potential to further improve breeding efficiency for quality traits through marker-assisted selection (MAS). It can be used for both the selection of individual quality traits and the pyramiding of important quality characters simultaneously into the same improved genotype. Recent advances in rice genomics research have made it possible to identify and map a number of genes through linkage to DNA markers. Improving grain and end-product quality such as bread- making quality in wheat, vitamin 'A' enrichment as beta carotene in rice ("Golden rice"), selection for malting quality in barley etc. are the few examples. The quality protein maize (QPM) holds superior nutritional and biological value and many studies had demonstrated the effectiveness of SSR markers in QPM genotype discrimination and analysis of genetic relationships. For fiber quality linkage disequilibrium (LD) based association mapping has been explored as an alternative tool to dissect and exploit the natural genetic diversity from cotton germplasm. The evaluation of organoleptic quality of tomato is difficult because it requires physical, chemical and sensory analyses. Several QTLs controlling the variation of tomato quality traits have been detected and used in MAS. In the present scenario genomics and proteomics approaches will be instrumental in developing crops with better nutritional quality. In the present chapter, application of molecular marker technologies has been discussed in the improvement of quality traits in crop plant giving some classical examples.

Key Words: MAS, Quality traits, QTL, Genomics, Proteomics

[1]*Defence Institute of High Altitude Research, DRDO, C/o 56 APO, Leh, 901205, India*
[2]*National Bureau of Plant Genetic Resources, New Delhi, India*
**Corresponding author e-mail: gyan.gene@gmail.com*

1.0 INTRODUCTION

In most field crops, end-use quality has played a secondary role in breeding programs; to be employed as a selection criterion only after improvement in yield per hectare has been achieved. Nevertheless, the very complexity of crop quality has earned it the attention of plant breeders, agronomists, forage scientists, plant pathologists, entomologists, and grain, fiber, and food scientists. There are many aspects of crop quality, which can be defined differently according to, for example, crop species, geographical region, and intended use of crop or crop product. Some attributes that define crop quality include: nutritional value (amino acid composition, protein content, micronutrients, vitamins, secondary metabolites, nutraceuticals, etc.), consumer preference (flavour, texture, colour, grain size/shape), pre- and post-harvest and industrial/technological characteristics (fibre traits, sucrose content, storage quality, sprouting, oil content, starches, processing, bread-making).

The market value of most crops is determined by various factors. Quality is a subjective, complex attribute, with many different components determined by the interactions between many genetic factors and growing environments. The consumer preference of products with particular quality characteristics varies from region to region. Although breeding for high yielding varieties in many crops have contributed greatly to world agriculture in the past decades, improvement of grain quality has been slow. The complex genetics of quality traits has led to their being difficult to improve through conventional plant breeding. Furthermore, the complexity in assessing some quality characters aggravates the difficulties in improving crop quality.

As for other agronomic traits, improvement of quality traits depends on the availability of sufficient variability for the targeted traits. Naturally existed genetic variations in the gene primary and secondary pools of a crop are the main sources of genetic variations breeders have been relying on for conventional breeding. The level of variability in some quality traits has been increased using mutagenesis. However, the full potential of induced mutations has not been realized in plant breeding for quality traits because of the difficulty in screening large mutagenised populations for infrequent mutations generating desirable crop quality alleles. The exploitation of mutated genes has been restricted to those that have easily identifiable phenotypes (*e.g.* waxy mutation in rice, barley and other crops). Recent development of screening procedures for quality traits in mutagenised populations will allow the more efficient identification of novel and useful variants (Shu, 2004). Such screening procedures and the development of markers linked to the induced mutant alleles will facilitate their incorporation into breeding programmes

The rapid development of molecular biology and functional genomics provides breeders with tools that have potential to further improve breeding efficiency through marker-assisted selection (MAS). As discussed in many chapters of this book, breeders are actively exploring these new technologies to achieve their breeding objectives quickly and economically. Grain and end-product quality traits may be suitable candidates for the implementation of MAS. Quality assessments are time consuming and costly, and they may require larger samples than are available in early generations of a breeding program. These factors limit the effectiveness of phenotypic selection to maintain or improve quality. Researchers at different labs now use different aspects of biotechnology (MAS, genomics, proteomics etc.) to introduce various desired quality traits in different crops. In this chapter, a few good examples of quality improvement using MAS are summarized. The potential of matabolomics and proteomics are discussed.

2.0 MARKER-ASSISTED BREEDING

The advent of molecular marker techniques now makes it possible to tag alleles conferring desirable quality traits. MAS offers great promise for both the selection of individual quality traits and the pyramiding of a number of important quality characters simultaneously into the same improved genotype. Marker-assisted (or molecular-assisted) breeding provides a dramatic improvement in the efficiency with which breeders can select plants with desirable combinations of genes. DNA markers have been used for transferring quality genes to cultivated varieties, assisting selection of complex multi-gene traits (such as flavor), aiding evaluation of regionally and seasonally optimized varieties (Suslow *et al.*, 2001).

The availability of markers tightly linked to or resided in the genes affecting target traits is the prerequisite for MAS. In the past few years there has been much progress in the development of strategies to discover new plant genes. In large part, these developments derive from four experimental approaches:

1. Genetic and physical mapping in plants and the associated ability to use map-based gene isolation strategies (Coe *et al.*, 2002).
2. Transposon tagging which allows the direct isolation of a gene *via* forward and reserve genetic strategies (Walbot, 2000).
3. Protein-protein interaction cloning that permits the isolation of multiple genes contributing to a single pathway or metabolic process (Pelletier and Sidhu, 2001) and
4. Through bioinformatics/genomics, the development and use of large expressed sequence tags (ESTs) databases, which are easier to generate than long tracts of genomic sequences and provide large-scale information on the gene complement of maize (Fernandes *et al.*, 2002), rice and Medicago.

3.0 APPLICATIONS OF MAS FOR SOME OF THE IMPORTANT QUALITY TRAITS IN CROP PLANTS

3.1 Amylose Content in Rice

Recent advances in rice genomics research and completion of the rice genome sequence have made it possible to identify and map precisely a number of genes through linkage to DNA markers. Noteworthy examples of some of the genes tightly linked to markers are resistance to or tolerance of blast, bacterial blight, improved agronomic and grain quality traits etc. MAS can be used for monitoring the presence or absence of these genes in breeding populations and can be combined with conventional breeding approaches. The use of cost-effective DNA markers derived from the fine mapped position of the genes for important agronomic traits and MAS strategies will provide opportunities for breeders to develop high-yielding, stress-resistant, and better-quality rice cultivars (Jena and Mackill, 2008).

Phillips (2001) found the first marker (after rice genome sequencing) which regulates amylose, a component of starch. In rice, high amylose means that the grains are firm and separate, and low starch means we can eat it with chopsticks because it sticks together. A problem in breeding new varieties is that the air temperature while rice is growing influences the amount of amylose that the plant produces. By diagnosing rice in breeding programs with DNA markers, however, scientists can accurately decide whether to keep working with progeny from a cross, or to cease selection. Breeders don't have to worry about unusual weather giving false reads on a potentially good variety.

Long-te-fu (LTF) and Zhan-shan 97 (ZS) are two key female parents for the generation of *Indica* hybrid rice, which have greatly contributed to the achievement of rice production in China. However, the high amylose content (AC) in the endosperm, controlled by the *Waxy* (*Wx*) gene encoded granule-bound starch synthase I, of both lines results in poor cooking and eating quality of the milled rice. Previous studies have shown that AC was correlated with the ability to excise intron 1 from the leader sequence of the *Wx* transcript, and which is responsible by a single nucleotide polymorphism (G or T) located at the first nucleotide of the splice donor site of *Wx* intron 1. Thus, a CAPS marker was subsequently developed, and with this MAS, Qiao-Quan *et al.* (2006) successfully introgressed the *Wx*-TT locus of rice cultivars with good quality intermediate AC into LTF-B and ZS-B. These were subsequently introduced into their relevant male-sterile lines (LTF-A and ZS-A) to generate improved *Indica* hybrids. In the selected lines LTF(tt)-B and ZS(tt)-B, the AC was reduced to a relatively low level (15%). Consequently, the hybrids crossed from the selected lines had dramatically reduced amylose levels.

3.2 Beta Carotene Content: Golden Rice

The ability to transfer cloned genes allows plant breeders to use genes from essentially any source as tools for crop improvement. For example, to enable rice grains to accumulate beta-carotene (which is converted into vitamin A when consumed by animals) and create the so-called "Golden rice". Golden rice was created by Ingo Potrykus of the Institute of Plant Sciences at the Swiss Federal Institute of Technology, working with Peter Beyer of the University of Freiburg. The project started in 1992 and at the time of publication in 2000, golden rice was considered a significant breakthrough in biotechnology as the researchers had engineered an entire biosynthetic pathway.

Golden rice was created by transforming rice with two beta-carotene biosynthesis genes:

1. *psy* (phytoene synthase) from daffodil (*Narcissus pseudonarcissus*)
2. *crt1* from the soil bacterium *Erwinia uredovora*

(The insertion of a *lyc* (lycopene cyclase) gene was thought to be needed but further research showed that it is already being produced in wild-type rice endosperm.)

The *psy* and *crt1* genes were transformed into the rice nuclear genome and placed under the control of an endosperm specific promoter, so that they are only expressed in the endosperm. The exogenous *lyc* gene has a transit peptide sequence attached so that it is targeted to the plastid, where geranylgeranyl diphosphate formation occurs. The bacterial *crt1* gene was an important inclusion to complete the pathway, since it can catalyze multiple steps in the synthesis of carotenoids, while these steps require more than one enzyme in plants (Hirschberg, 2001). The end product of the engineered pathway is lycopene, but if the plant accumulated lycopene the rice would be red. Recent analysis has shown that the plant's endogenous enzymes process the lycopene to beta-carotene in the endosperm, giving the rice the distinctive yellow colour for which it is named (Schaub *et al.*, 2005). The original Golden rice was called SGR1, and under greenhouse conditions it produced 1.6 µg/g of carotenoids. Golden rice was developed as a fortified food to be used in areas where there is a shortage of dietary vitamin A. In 2005 a new variety called *Golden Rice 2* was announced which produces up to 23 times more beta-carotene than the original variety of golden rice (Paine *et al.*, 2005), but none of the variety is currently available for human consumption, till date.

3.3 Grain and End Product Quality in Wheat

Improving grain and end-product quality such as bread- making quality is a major target in wheat breeding programs. Genetic progress from conventional breeding has proven to be low for most of the quality-related characteristics. The requirements of large sample and the cost involved in assessing quality-related characteristics by conventional standard test such as full-scale mill and bakery testing makes it impossible to measure them routinely and results in considerably reduced selection pressure applicable. DNA marker screening of these quality traits may be performed on leaf materials at early stage and before grain setting (De Bustos *et al.*, 2000).

3.3.1. Glutenin

Wheat gluten contains both high molecular weight (HMW-GS) and low molecular weight (LMW-GS) glutenin subunits. The HMW-GS are key factors in bread making quality (BMQ) since they are major contributors of glutenin elasticity and polymer formation of wheat dough. The effects of HMW-GS on dough properties (strength and elasticity) may be additive or synergistic with significant interactions with LMW-GS subunits (Beasley *et al.*, 2002). The HMW-GS are encoded by polymorphic genes at the *Glu 1* loci that are present on the long arm of group 1 chromosome (Payne and Lawrence, 1983). At each locus (*Glu-A1, GluB1, GluD1*) there are two tightly linked HMW-GS genes; one of them is x-type with higher molecular weight and the other is y-type. The presence of different allelic composition of the HMW-GS in one specific wheat variety is one of the most important genetic factors in determining the bread making quality (Payne *et al.*, 1987). For example, wheat varieties containing allelic compositions of (Dx5 paired with Dy10) at the *Glu-D1* locus will form stronger dough than those containing (Dx2 paired with Dy12). Due to the large contribution of allelic interaction in bread making quality, Békés *et al.* (2006) suggested targeting different allelic combinations rather than individual glutenin allele in developing new lines with certain quality attributes.

DNA markers for HMW-GS allelic composition have proven to be superior to electrophoresis of SDS-PAGE in some instances (D'Ovidio and Anderson, 1994). Moreover, this assay can be used to select individual plants within populations and to overcome the environmental variation originating from both field and laboratory. Data presented by Uthayakumaran *et al.* (2006) showed that the MAS for the HMW-GS are especially valuable tool for breeding programs since the information about the BMQ can be obtained at an earliest stages of breeding program, thus poor-quality lines are not propagated. Moreover, MAS for the HMW and LMW glutenin alleles is widely performed by breeding programs throughout the world to select for improved dough characters. Blechl *et al.* (2007) suggested that contribution of allele–allele interactions, and different allelic combinations should be targeted rather than the individual glutenin alleles in breeding program to develop new lines with certain quality attributes, especially to improve extensibility. The main reason given was the significant differences found among the values, which described the contribution of HMW- GS×LMW-GS interactions on extensibility.

The association between molecular markers and BMQ was investigated by Manifesto *et al.* (1998) in a cross between two wheat cultivars with the same high M_r-glutenin subunits but significantly different BMQ. A segregant F_2 population was generated after crossing Klein 32 and Chinese Spring, and the BMQ of each $F_{2:3}$ families were estimated using sodium dodecyl sulfate (SDS) sedimentation and mixograms. The same families were characterized for 11 polymorphic loci using restriction fragment length polymorphisms (RFLP) and simple sequence repeats (SSR). These loci were specifically selected for their complete or close linkage to storage protein gene families. No significant differences in BMQ were detected at *XGlu-B1* and *XGlu-A1* loci using RFLP markers. Highly significant ($P<0.01$) differences in all BMQ parameters were detected for *XGli-B1* and *XGlu-B3* loci on chromosome arm 1BS. The increase in the number of Klein 32 alleles at these loci determined a linear increase in sedimentation and mixogram values. It was not possible to differentiate the effect of *XGli1* from that of *XGlu3* because of the close linkage between these two loci. These two loci, considered together, explained from 11 to 15% of the variation in BMQ observed in this cross.

3.3.2 QTLs for various quality traits in wheat

To genetically dissect QTLs for quality traits such as grain and flour protein content and gluten strength (evaluated by mixograph and SDS sedimentation volume), an F_1-derived doubled haploid (DH) population of 185 individuals was developed. A genetic map was constructed based on 167 marker loci, consisting of 160 microsatellite loci, three HMW glutenin subunit loci: *Glu-Al, Glu-B1* and *Glu-D1,* and four STS markers. A total of 26 QTLs for quality-related traits were identified. The largest QTL clusters, consisting of up to nine QTLs, were found on chromosomes 1D and 4D. HMW glutenin subunits at *Glu-1* loci had the largest effect on BMQ. However, other genomic regions also contributed genetically to BMQ (Huang *et al.*, 2006).

Breseghello *et al.* (2005) conducted a study to identify genomic regions related to differences in milling and baking quality between a soft and a hard cultivar of hexaploid wheat. A population of 101 double-haploid lines was used to generate a genetic map which included 320 markers in 43 linkage groups and spanned 3555 cM. The effect of qualitative variation for kernel texture, caused by the segregation of the hardness gene, was controlled by regression on texture class. The residual variance was used for composite interval mapping, and QTLs on 1A, 1B, 1A/D, 2A, 2B, 2D, 3A/B, 4B, 5B and 6B were detected. The effect of some QTLs was opposite to the direction expected on the basis of parental phenotypes. The hard wheat parent contributed alleles favorable for soft wheat varieties at QTLs on 1AS,L, 1BL-2, and 6B, whereas the soft parent contributed alleles for higher protein content at QTLs on 2BL-1, 4B-1, and 6B and higher flour yield on 2BL-2 and 4B-2. Their results indicated that hard x soft wheat crosses have considerable potential for improving milling and baking quality of either class.

A set of 187 doubled haploid lines was explored for QTLs for three BMQ tests: hardness, protein content and strength of the dough (W of alveograph) by Perretant *et al.* (2000). About 350 molecular and biochemical markers were used to establish the genetic map. For hardness, they confirmed a previously tagged major QTL on chromosome 5DS, and two additional minor QTLs were found on chromosome 1A and 6D, respectively. For protein content two main QTLs were identified on chromosomes 1B and 6A, respectively. For W, three consistent QTLs were detected: two at the same location as those for hardness, on chromosomes 1A and 5D; the third one on chromosome 3B.

3.4. MAS for Malting Quality in Barley (*Hordeum vulgare*)

Brewers are reluctant to change malting barley cultivars due to concerns of altered flavor and brewing procedures. Selection for malting quality is a major breeding objective for many breeding programs. Malting quality is measured by micro-malting and micro-mashing, which is time consuming, and resource-intensive. Characters that affect malting quality (*i.e.* malt extract content, α- and β-amylase activity, diastatic power, malt β-glucan content, malt β-glucanase activity, grain protein content, kernel plumpness, and dormancy) are quantitatively inherited and variously influenced by the environment (Zale *et al.*, 2000).

Considerable QTL analyses have been performed in recent years on a number of crosses. A minimum of 168 malting quality QTLs representing 19 malting quality traits have been mapped in nine mapping populations. QTL regions are spread across each of the 7 barley chromosomes with concentrations especially within chromosomes 1, 2, 4, 5 and 7. Whereas, there is remarkable QTL conservation in some chromosome regions among crosses, some regions hold unique QTLs as well. It is also noteworthy that there are many overlapping QTLs, especially but not surprisingly, of related traits. Malt extract QTLs are almost always coincident with component traits such as carbohydrate hydrolytic enzyme activities. In some cases widely conserved QTL chromosome regions may be targets for selection to maintain malting quality, but selection for unique regions may lead to new improvements (Zale *et al.*, 2000).

The flanking markers, Brz and Amy2 (for the QTL on chromosome 1), and WG622 and BCD402B for (the QTL on chromosome 4) identified using doubled haploid (DH) lines derived from the cross Steptoe x Morex were used in selection by Han *et al.* (1997). They compared four alternative selection strategies; phenotypic selection, genotypic selection, tandem genotypic and phenotypic selection, and combined phenotypic and genotypic selection applied to a population consisting of 92 DH lines derived from the same cross Steptoe x Morex crosses. MAS for QTL1 including both tandem genotypic and phenotypic selection, and the combined phenotypic and genotypic selection was more effective than phenotypic selection. However, MAS for QTL2 was not as effective as phenotypic selection, since the effects of QTL2 in the population for selection were very small.

Igartua *et al.* (2000) conducted marker-based selection (MBS) for four regions of the genome found to affect several grain and malt quality traits in the two-row barley cross 'Harrington'/'TR306'. Molecular marker were used to select 47 lines from 410 Harrington/TR306 lines that had not been used in the original mapping experiment. Four groups of lines were selected on the basis of their genotype at marker loci in two regions on chromosome 7 (5H) with QTL affecting kernel weight and plumpness, grain protein, extract ß-glucan content, the difference between fine-grind and coarse-grind extract, soluble protein, diastatic power, α-amylase activity and fine-grind extract, and in regions on chromosomes 3 (3H) and 6 (6H) with QTL affecting extract ß-glucan content and fine-coarse difference. Grain and malt quality traits of these lines were determined from grain grown in five field environments in western Canada. The results confirmed the presence of QTL on chromosome 7 affecting all traits previously reported. MBS for two regions on chromosome 7 was effective in identifying phenotypically superior lines, and the magnitudes of the combined effects for these regions were close to the estimates calculated in the mapping experiment. Combined selection for extract ß-glucan content and fine-coarse difference at all four QTL regions was more effective than selection for only the two regions on chromosome 7, although the presence of QTL on chromosomes 3 and 6 could not be firmly confirmed.

Ayoub *et al.* (2003) used MBS to manipulate a malt quality characteristic, α-amylase activity, in a barley breeding population. MBS was applied among $F_{2:3}$ lines from the cross 'Morex'/'Labelle', targeting Morex alleles in a region of chromosome 7(5H) that had previously been found to affect α-amylase activity in the cross 'Steptoe'/Morex. The target region was represented by two polymerase chain reaction (PCR) based markers. Selected lines were grown in field plots in two years. Selection for the Morex allele at two PCR based markers on chromosome 7(5H) was effective in increasing α-amylase activity. It was concluded that marker-based selection for a quantitative trait locus could be effective even when applied in a population other than the mapping population.

Coventry *et al.* (2003) investigated the usefulness of MBS to improve diastatic power. Desirable alleles for quantitative trait loci and structural genes involved in enhanced diastatic power and activity of its component hydrolytic enzymes identified in Alexis, Amagi Nijo, Harrington, Haruna Nijo, and Sloop were targeted. Six unmapped breeders' populations involving these donor sources of malting quality were used for MAS. For each population, individual lines were pooled into classes separated on the basis of either the presence or absence of malting quality parent marker alleles at each of 9 identified loci (QTL or structural genes). Diastatic power, α-amylase, and α-amylase activities were determined for each line, and used to compare alternative marker allele class means. Lines carrying malting parent marker alleles at a chromosome 5H locus *abg463* were associated with 21–44% higher α-amylase activity levels, depending on the cross. The malting parent alleles at the chromosome 4H *Bmy1* locus were associated with increased diastatic power and α-amylase activity. A simple PCR based marker detecting the *Bmy1* locus was found to be effective in screening for improved diastatic power, α-amylase activity, and thermostability. Lines carrying malting parent alleles at the chromosome 2H *Bmy2* locus produced differences in diastatic power and â-amylase activity that, after adjusting for the correlated effect of malt protein, became non-significant. The Alexis allele of the chromosome 1H *EBmac501* locus was associated with significant differences in all traits for a population carrying this source.

Schmierer *et al.* (2005) wanted to develop high yielding near isogenic lines that maintain traditional malting quality characteristics by transferring QTL associated with yield, *via* molecular marker-assisted backcrossing, from the high yielding cv. Baronesse to the North American two-row malting barley industry standard cv. Harrington. For transfer, they targeted Baronesse chromosome 2HL and 3HL fragments presumed to contain QTL that affect yield. Analysis of genotype and yield data suggest that QTL reside at two regions, one on 2HL (ABG461C-MWG699) and one on 3HL (MWG571A-MWG961). Based on yield trials conducted over 22 environments and malting analyses from 6 environments, they selected one isogenic line (00-170) that has consistently produced yields equal to Baronesse while maintaining a Harrington-like malting quality profile.

3.5 Quality Protein Maize (QPM) in maize (*Zea mays*)

Maize endosperm consisting of approximately 9-12 per cent protein is, however, deficient in two essential amino acids *viz.*, lysine and tryptophan, which leads to poor net protein utilization and low biological value of traditional maize varieties. Breeding for improved protein quality in maize began in the mid-1960s with the discovery of mutants, such as *opaque-2*, that produce enhanced levels of lysine and tryptophan, the two amino acids deficient in maize endosperm proteins. However, adverse pleiotropic effects imposed severe constraints on successful exploitation of these mutants. Interdisciplinary and concerted research efforts led to amelioration of the negative features of the opaque phenotype, and the rebirth of 'Quality Protein Maize' (QPM). QPM holds superior nutritional and biological value and is essentially interchangeable with normal maize in cultivation and kernel phenotype (Prasanna *et al.*, 2001).

A set of 23 Quality Protein Maize (QPM) lines, including 13 lines developed in India and 10 lines at CIMMYT (International Maize and Wheat Improvement Center), Mexico, was analyzed by Kassahun and Prasanna (2003) using microsatellite or simple sequence repeat (SSR) markers. Polymorphic profiles for 36 SSR loci have aided in differentiating the QPM inbred lines. The study resulted in identification of SSR markers, such as bnlg439, phi037, bnlg125, dupssr34 and bnlg105, with high polymorphism information content in the selected QPM genotypes. Detection of 30 unique/rare SSR alleles could contribute to effective differentiation of 14 of the 23 QPM inbreds. An opaque-2 specific microsatellite marker, phi057, also facilitated differentiation of opaque-2 carrying QPM inbreds from the non-opaque genotypes. Analysis using SSR markers indicated high levels of heterozygosity in majority of the Indian QPM lines and in one CIMMYT QPM inbred, CML188. Cluster analysis using SSR data, followed by canonical discriminant analysis, clearly distinguished the Indian QPM inbreds from those developed at CIMMYT. The cluster patterns were largely in congruence with the available pedigree information of the QPM inbreds studied. The study demonstrates the effectiveness of SSR markers in QPM genotype discrimination and analysis of genetic relationships (Kassahun and Prasanna, 2003) .

Babu *et al.* (2004) reported the conversion of four normal inbreds *viz.*, CM212, CM141, CM145 and V25, which are parents of three single cross hybrids into high quality protein versions using marker-assisted backcrossing. Three SSR markers located within *opaque2* were used for foreground selection. Two SSR markers *viz.*, *phi057* and *umc1066* were co-dominant while *phi112* was dominant in nature. Upon screening of appropriate back cross populations (BC) with *phi057* and *umc1066*, around 50 per cent plants in each BC population were found to be heterozygous for the *opaque-2* (Qq) gene. The selected BC_2 progenies were advanced to selfing generation and similar molecular marker analysis was performed to identify the desirable homozygous recessive individuals (qq) in the selfed populations. The converted maize lines had twice the amount of lysine and tryptophan than the native lines and hence could be used as potential food and feed alternatives in resource-poor and marginal areas in general and hill ecosystems in particular. They also demonstrated that foreground selection in an early (BC_1) generation combined with background selection at a later generation (BC_2) along with the phenotypic selections for quantitative/continuously distributed traits would result in rapid genetic gain in a cost effective manner.

Babu *et al.* (2004) described in detail the marker-assisted backcrossing scheme for the conversion of an early maturing normal maize inbred line, V25. Foreground selection was conducted using the *opaque2* specific SSR marker, *umc1066*. Flanking markers *bnlg2160* and *bnlg1200* (4.2 and 3.8 cM from the *opaque2* locus) were used to identify recombinants with reduced attached chromosome segment length. Whole genome background selection was conducted using 77 SSR markers spanning all the bin locations in a maize SSR consensus map. The tryptophan concentration in endosperm protein was significantly enhanced in all the three classes of kernel modification *i.e.*, less than 25%, 25–50% and more than 50% opaqueness. BC_2F_3 lines developed from the hard endosperm kernels were evaluated for desirable agronomic and biochemical traits in replicated trials and the best line was chosen to represent the quality protein maize (QPM) version of V25, with a tryptophan concentration of 0.85% in protein.

3.6 Fiber Quality Improvement in Cotton (*Gossypium* sp.)

Fiber quality is the key breeding objective of breeding programs worldwide. The initial focus is on improving fiber characteristics such as fiber length, uniformity and strength that are important in spinning. Approaches are also being explored to improve properties of cotton fiber that would add value to the overall fiber processing industry and be of real benefit to consumers (Dever and Hamill, 2005). Conventional breeding method approaches include germplasm access with a focus on fiber quality, intense selection pressure throughout the breeding process, modifications to pedigree breeding methodology and the development of a fiber selection index that includes more measurable fiber properties. Fiber quality improvements through molecular approaches have gained much attention recently.

Until very recently, little was known about the molecular aspects leading to specific cotton fiber properties and few research tools were available to probe cotton fiber quality.Although potential genetic diversity exists in *Gossypium* genus, it is largely 'underutilized' due to photoperiodism and the lack of innovative tools to overcome such challenges.

Linkage disequilibrium (LD) based association mapping has been explored as an alternative tool to dissect and exploit the natural genetic diversity conserved within cotton germplasm collections. Abdurakhmonov *et al.* (2008) reported the extent of genome-wide LD and association mapping of fiber quality traits by using microsatellite markers in a total of 285 exotic *Gossypium hirsutum* accessions, comprising of 208 landrace stocks and 77 photoperiodic variety accessions. They demonstrated the existence of useful genetic diversity within exotic cotton germplasm. In their germplasm set, 11–12% of SSR loci pairs revealed a significant LD. At the significance threshold ($r^2 = 0.1$) a genome-wide average of LD declines within the genetic distance at < 10 cM in the landrace stocks germplasm and > 30 cM in variety germplasm. Genome wide LD at $r^2 = 0.2$ was reduced on average to ~ 1–2 cM in the landrace stock germplasm and 6–8 cM in variety germplasm, providing evidence of the potential for association mapping of agronomically important traits in cotton. Between 6% and 13% of SSR markers were found to be associated with the main fiber quality traits. The SSR markers associated with fiber quality traits in diverse cotton germplasm, which broadly covered many historical meiotic events, should be useful to effectively exploit potentially new genetic variation by using MAS programs.

3.7 Organoleptic Quality in Tomato (*Lycopersicon esculentum*)

Improving organoleptic quality of fresh market tomato fruit has become an important objective for tomato breeders. The evaluation of organoleptic quality of tomato fruit requires physical, chemical and sensory analyses, which are expensive and difficult to assess. Therefore, their practical use in phenotypic selection is difficult. Five chromosome regions strongly involved in organoleptic quality attributes were then chosen to be introgressed into three different recipient lines through MAS. A marker-assisted backcross (MABC) strategy was performed by Lecomte *et al.* (2004), as all the favorable alleles for quality traits were provided by the same parental tomato line, whose fruit weight and firmness were much lower than those of the lines commonly used to develop fresh market varieties. Three improved lines were obtained after three backcrossing and two selfing generations. Breeding efficiency strongly varied according to the recipient parent, and significant interactions between QTLs and genetic backgrounds were shown for all of the traits studied.

Several QTLs controlling the variation of tomato quality traits have been detected using a recombinant inbred line population derived from a cross between a cherry tomato chosen for its good flavor and a line with bigger fruits but poor taste. A MAS scheme was then set in order to transfer the five most important QTLs involved in fruit quality into three recurrent lines. The backcross optimisation (population size, number and position of markers) is used, taking into account both theoretical and practical aspects (Causse *et al.*, 2002; 2004).

Color is among the most important attributes of tomato for processing into whole and diced products. Both color and color uniformity are greatly affected by Yellow Shoulder Disorder (YSD), a ripening disorder that results in discoloration of the proximal end tissues of the fruit. Darrigues *et al.* (2008) show that lycopene and beta-carotene concentrations are reduced by

18% and 22%, respectively, in fruits affected by YSD. In order to elucidate the genetic basis of YSD, they developed single nucleotide polymorphisms (SNPs) as molecular markers for application in three inbred backcross populations derived from either *Solanum lycopersicum* × *S. lycopersicum* or *S. lycopersicum* × *S. pimpinellifolium* crosses. SNP discovery for application in these populations is based on both analyses of large public EST databases and on hybridization to a custom oligonucleotide array. The array was hybridized with target cDNA from *S. lycopersicum* (Ohio 7814) and *S. pimpinellifolium* (LA1589). They developed algorithms to detect outliers and identified 1,296 potential SNPs. These putative SNPs are being verified by sequencing, screened for utility as markers on a collection of 99 *S. lycopersicum* lines and wild relatives and applied to the genetic dissection of YSD. Implementing SNP-based marker technology has the potential to dramatically alter the approach of genetic characterization.

3.8 Oil Quality in Soybean

Finding QTLs, which influence seed protein and oil pool determinant, is an interesting preposition because soybean is cultivated for its high protein and oil content. Soybean oil can be idealized for consumption as food oil by reducing the linolenic (18:3) content. Modification of the fatty acid composition of soybean seeds to lower linolenic acid levels can improve oil stability and flavour and eliminate the need for hydrogenation (Dutton *et al.*, 1951; Lui and White, 1992). The production of oxidatively stable soybean oil without hydrogenation is of increasing importance because recent studies have linked the consumption of trans fatty acids found in hydrogenated oils with an increased risk of coronary heart disease (Hu *et al.*, 1997). Linolenic acid levels in soybean are governed by Fan locus (Spencer *et al.*, 2003). Bilyeu (2003) have identified 3 omega-3-fatty acid desaturase genes that contribute to soybean linolenic acid levels. Molecular markers for defects in the three genes will enhance soybean-breeding program for low linolenic acid.

4.0 FUNCTIONAL GENOMICS IN QUALITY IMPROVEMENT

The "Genomic Era" was ushered in by rapid advances in nucleic acid sequencing, making it possible to sequence whole genomes (all the genetic material in the chromosomes of a particular organism) relatively quickly (Lee and Lee, 2000). The Genome Projects accelerated development of the various techniques and technologies involved in genomics, and opened up new "-omics" fields, notably Proteomics (the study of the complement of proteins expressed in a given cell, tissue or organism under particular conditions at a particular time).

While extensive collections of maize ESTs have been assembled by the commercial sector, approximately 154,500 ESTs derived from 20 cDNA libraries have been deposited in the ZMDB database (http://www.zmdb.iastate.edu/zmdb/EST/) of the NSF Maize Gene Discovery Project. DNA microarray technology (Brown and Botstein, 1999) represents a collection of promising tools for the discovery of mRNA level controls of complex pathways and may shed light on pathway interactions, the understanding of which is essential for successful metabolic engineering of crop plants.

The project "A Functional Blueprint of the *Zea mays* Endosperm Cell Factory" funded by the EU examines in considerable detail the transcriptome and proteome of the developing maize endosperm. This information will be used to target distinctive, previously uncharacterized endosperm specific genes which will be knocked out *via Mutator* transposon tagging. Characterization of the normal and modified endosperm will provide further research material for the academic laboratories involved, as well as material for the plant breeders and food processors to include in their respective research or product development pipelines.

Proteomes and proteomics is a new frontier in biosciences. The "proteome" is the collection of thousands of proteins that make up an organism or the individual cells of an organism. The proteome is even more complex than the transcriptome which is all the RNAs encoded by the genome. A given protein may have more than one function and frequently operates as part of any complex system. The need to go beyond nucleic acid analysis, and reach an understanding of total protein expression and regulation in biological systems is motivating the field of 'Proteomics' (Dutt and Lee, 2000). Proteomics will be instrumental in developing crops with improved agronomic traits, improving the nutritional quality, and identifying high valued proteins in complex mixtures of proteins (Han and Wang, 2008).

Proteomic approach is being used by US government scientists to improve the quality of wheat flour ingredients. The focus is to characterise the hundreds of different proteins in the wheat kernels. The role of the protein ranges from storing carbohydrates to protecting kernels against insects. Scientists believe that discovering more about the work of these proteins and how they are affected by the heat, soil nutrients, and other environmental conditions in which the plant is grown could lead to even better flours for tomorrow. Gluten proteins are the most abundant and most studied. Researchers already know that these proteins have a premier role in influencing flour's quality. But still scientists know very little about wheat kernels' so-called metabolic proteins, which occur in much smaller amounts. It is known, however, that these mostly mysterious proteins are essential to a kernel's growth. For example, wheat plants need metabolic proteins to form the gluten proteins and to make starch. (http://www.bakeryandsnacks.com).

Eivazi *et al.* (2008) analyzed the genetic diversity among 10 Iranian bread wheat (*Triticum aestivum*) genotypes using 12 quality traits and 294 proteome markers. Based on comparative analysis of 2-D maps, 241 out of 294 protein spots were detected in all the genotypes. Among these protein spots, the expression level of 177 showed significant quantitative changes ($P<0.01$) in at least one genotype compared to the others. They identified 13 common and 8 genotype specific protein spots using MALDI TOF mass spectrometry (MS). The majority of the identified proteins were involved in metabolism, energy, protein destination and storage, and plant defence mechanisms. Their results revealed that the genotypes differed for quality traits and proteome markers. The average genetic diversity based on quality traits (0.684 with a range of 0.266–0.997) was higher than the proteome (0.464 with a range of 0.264–0.870) markers. In addition to the genetic diversity assessment, specific proteins with known function were detected uniquely for the studied genotypes. The results suggest that the classification based on quality traits and proteome markers of these wheat genotypes will be useful for wheat breeders to plan crosses for positive traits.

Functional genomics and related technologies are now being used to revisit the difficult questions of cell wall polysaccharide biosynthesis in barley to gain better understanding of factors affecting malting quality. Polysaccharide biosynthesis involves synthase enzymes that had proved impossible to purify and characterize by classical biochemical methods. In addition to the resolution of their structure, the genes encoding for the more extensively studied hydrolase enzymes, responsible for cell wall breakdown and starch degradation, are being identified. These studies will generate knowledge of how characteristics such as thermo-stability might be enhanced to improve malting and brewing performance (Kristensen *et al.*, 1999; Ziegler, 1999). The structural and functional studies can be linked back to protein and transcript profiling expression patterns, and ultimately to the families of genes that a breeder might wish to conserve or alter in a breeding programs.

A proteomic approach was also used by Mihr *et al.* (2005) to investigate characteristics involved in tomato fruit quality at the protein level. Total protein extracts of the tomato pericarp were separated by 2D gel electrophoresis. Two NILs differing in the concerned segment of chromosome 2 as well as the two parental lines of the RIL population were compared during several fruit developmental stages. There are more differences observed between the parental lines than between the NILs. Several proteins of interest were analyzed by MALDI-TOF or ESI-MS/MS and candidate genes for the above mentioned characteristics were obtained. Some of these genes are well known to be linked to fruit development processes but many of them are related to stress and defence response (Mihr *et al.*, 2000).

5.0 PATENTS FOR FEW IMPORTANT QUALITY TRAITS

5.1 4-Ketocarotenoids in Flower Petals (Patent # WO03080849)

The formation of a carotenoid compound containing a 4-keto-beta-ionene ring such as astaxanthin or canthaxanthin in flowers, and particularly in the corolla and reproductive parts of a flower of a higher plant whose flowers produce a carotenoid compound containing a beta-ionene ring such as beta-carotene or zeaxanthin, but otherwise do not produce astaxanthin or canthaxanthin is disclosed. One or more genes controlled by a promoter are inserted (transformed) into a higher plant. The inserted gene encodes a chimeric enzyme including (a) a carotenoid-forming enzyme that is at least a ketolase. That gene is operatively linked to (b) a plastid-directed transit peptide. Some higher plants to be transformed produce at least zeaxanthin or beta-carotene in their flowers prior to transformation, whereas other plants produce little if any coloured carotenoid pigments prior to transformation and are transformed with a cassette of carotenoids-forming genes (Hauptmann *et al.*, 2003).

5.2 Method for Promoting Fatty Acid Synthesis in Plant (Patent # JP2002335786)

The present invention provides a new method for promoting the synthesis of fatty acid in a plant. The amount of protein of carboxyl transferase and beta subunit encoded by accD gene is increased by introducing a promoter sequence of a gene highly expressed in chloroplast at the upstream of an *E.coli* type acetyl CoA carboxylase accD gene by chloroplast transformation technique. The amount of protein of other subunit constituting acetyl CoA carboxylase is also increased by this process. Since acetyl CoA carboxylase is the key enzyme of the first stage of fatty acid synthesis, the synthesis of fatty acid can be promoted by the method of the present invention. The transformed vegetable produced by the method exhibits remarkable promotion of fatty acid synthesis, prolonged life of the leaf, increased yield of seeds and improved productivity of the plant body (Sasaki *et al.*, 2002).

5.3 Preparing Transgenic Leguminous Plants with Increased Protein Content (Patent # WO0175128)

The invention relates to a method for the production of leguminous plants with increased protein content in the seeds and longer seed filling duration, by means of introduction of recombinant DNA molecules. Said recombinant DNA molecules are introduced into the plant, by means of a transformation system and comprise a DNA sequence from the plant, expressed in plants, the genetic product of which inhibits a protein in the seed with the enzymatic activity of an ADP glucose pyrophosphorylase (AGP) and/or a plastid phosphoglucomutase (pPGM) and, optionally, the regulatory sequence of a seed-specific promoter in leguminous plants. Furthermore, at least one selection marker gene is separately transferred, which is subsequently removed again. The plants which display increased protein content and a lengthier seed-filling duration are chosen (Weber *et al.*, 2001).

6.0 CONCLUSION

Crop quality improvement is gaining unprecedented importance in both developed and developing countries. Products with improved quality give the farmer added value and a competitive market advantage, which in turn will result in improved human welfare and increased farm income. Thus, the improvement of quality characters in crop plants has great potential to alleviate problems caused by poverty and malnutrition through both direct (food quality and quantity) and indirect effects (income stability, etc) that affect farmer's social and economic status.

By utilizing genetic variation, the composition of the grain/fruit can be altered for both the quantity and quality (structure and chemical diversity) of starch, protein and oil throughout grain/fruit development. The ability of plant breeders/plant scientists to use existing genetic variation and to identify and manipulate commercially important genes is greatly enhanced by the development in molecular biology and functional genomics.

As demonstrated by the examples reviewed in this chapter and other chapters in this book, the use of MAS have the potential to overcome some of the rate-limiting factors associated with phenotypic selection such as requirement of large samples, expensive and time- consuming testing. MAS technology provides substantially increased opportunities to alter the composition of grains/fruits to better match current and emerging food and industrial uses. The first such modifications to enter production and trade are likely to be major changes in the oil and protein composition of oilseeds, followed by protein and starch modifications in cereals and pulses, to better fit these products to specific end uses.

Employing high-throughput genomic, proteomic and metabolomic approaches to the examination of plant genomics research will develop improved understanding of systems that govern plant growth, development and performance. It has the potential to link genes to traits and thus transform plant breeding from an art to a science. The ability of the geneticists to discover new genes and to manipulate genetic variation at the level of specific genes offers the potential to tailor genetic variation for the production of precisely designed specialty variety in the future. Ultimately, the extent to which advances in functional genomics will be embraced and adopted by breeders depends heavily on their ability to "keep up" with the rapidly moving field, and on the economics of doing so.

It is clear that molecular marker technology provides tremendous capability for improvement of plant performance and diversification of product quality characteristics beyond what can be achieved by conventional plant breeding. Introduced and managed correctly, MAS based crops should provide one of most effective means of achieving further improvements in crop quality, and better fitting crop products to market requirements.

7.0 REFERENCES

Abdurakhmonov, I.Y., Kohel, R.J., Yu, J.Z., Pepper, A.E., Abdullaev, A.A., Kushanov, F.N., Salakhutdinov, I.B., Buriev, Z.T., Saha, S., Scheffler, B.E., Jenkins, J.N. and Abdukarimov, A. 2008. Molecular diversity and association mapping of fiber quality traits in exotic *G. hirsutum* L. germplasm. Genomics, 92(6):478-487.

Ayoub, M., Armstrong, E., Bridger, G., Fortin, M.G. and Mather, E. 2003. Marker based selection in barley for a QTL region affecting α-amylase activity of malt. Crop science, 43:556-561.

Babu, E.R., Mani, V.P. and Gupta, H.S. 2004. Combining high protein quality and hard endosperm traits through phenotypic and marker assisted selection in maize. 4th International Crop Science Congress 25th Sept – Oct 1st, Brisbane, Australia.

Babu, R., Nair, S. K., Prasanna, B.M. and Gupta, H.S. 2004. Integrating marker assisted selection in crop breeding: Prospects and Challenges. Current Science, 87(5): 607-619.

Beasley, H.L., Uthayakumaran, S., Stoddard, F.L., Patridge, S.J., Daqiq, L., Chong, P. and Békés, F. 2002. Synergistic and additive effects of three high molecular weight glutein subunit loci. II. Effects on wheat dough functionality and end-use quality. Cereal Chemistry, 79:301-307.

Békés, F., Kemény, S. and Morell, M. 2006. An integrated approach to predicting end-product quality of wheat. European Journal of Agronomy, 25:155-162.

Bilyeu, K.D., Palavalli. L., Sleper, D.A. and Beuselinck, P.R. 2003. Three microsomal omega-3-fatty acid desaturase genes contribute to soybean linolenic acid levels. Crop Science. 43:1833-1838.

Blechl, A., Lin, J., Nguyen, S., Chan, R., Anderson, O. and Dupont, F. 2007. Transgenic wheats with elevated levels of Dx5 and/or Dy 10 high molecular weight glutein subunits yield doughs with increase mixing strength and tolerance. Journal of Cereal Science, 45:172-183.

Breseghello, F., Finney, P.L., Gaines, C., Andrews, L., Tanaka, J., Penner, G. and Sorrells, M.E. 2005. Genetic loci related to kernel quality differences between a soft and a hard wheat cultivar. Crop Science, 45:1685-1695.

Brown, P.O. and Botstein, D. 1999. Exploring the new world of the genome with DNA microarray. Nature Genetics (Supplementary), 21:33-37.

Causse, M., Saliba-Columbani, V., Lecomte, L., Duffé P., Rousselle P. and Buret, M. 2002. QTL analysis of fruit quality attributes in fresh marked tomato: a few chromosome regions control the variation of sensory and instrumental traits. Journal of Experimental Botany, 53:2090-2098.

Causse, M., Lecomte, L., Baffert, N., Duffe, P. and Hospital, F. 2004. Marker-assisted selection for the transfer of QTLs controlling fruit quality traits into tomato elite lines. Journal of Environmental Quality, 33:1576-1577.

Coe, E., Cone, K., McMullen, M., Chen, S.S., Davis, G., Gardiner, J., Liscum, E., Polacco, M., Paterson, A., Sanchez-Villeda, H., Soderlund, C. and Wing, R. 2002. Access to the maize genome: an integrated physical and genetic map. Plant Physiology, 128:9-12.

Corentry, S.J., Collins, H.M., Barr, A.R., Jefferies, S.P., Chalmers, K.J., Logue, S.J. and Langridge, P. 2003. Use of putative QTLs and structural genes in marker assisted selection for diastatic power in malting barley (*Hordeum vulgare* L.). Australian Journal of Agricultural Research, 54(12):1241-1250.

D'Ovidio, R. and Anderson, O.D. 1994. PCR analysis to distinguish between alleles of a member of a multigene family correlated with wheat bread-making quality. Theoretical and Applied Genetics, 88:759-763.

Darrigues, A., Yang, W. and Francis, D.M. 2008. DNA-microarray detection of molecular markers for improving color and nutritional quality in tomato. ISHS Acta Horticulturae 789: XV Meeting of the EUCARPIA Tomato Working Group.

De Bustos, A., Rubio, P. and Jouve, N. 2000. Molecular characterisation of the inactive allele of the gene *Glu-A1* and the development of a set of AS-PCR markers for HMW glutenins of wheat. Theoretical and Applied Genetics, 100:1085-1094.

Dever, J.K. and Hamill, E.M. 2005. Breeding: Approaches to Fiber Quality Improvement. Systems Conference Presentations. http://www.cottoninc.com/2005ConferencePresentations/BreedingFiberQualityImprovement/ .

Dutt, M.J. and Lee, K.H. 2000. Proteomic analysis. Current Opinion in Biotechnology, 11:176-179.

Dutton, H.J., Lancaster, C.J., Evans, C.D. and Crown, J.C. 1951. The flavour problem of soybean oil VIII linolenic acid. Journal of American Society of Oil Chemists, 28:115-118.

Eivazi, A.R., Naghavi, M.R., Hajheidari, M., Pirseyedi, S.M,, Ghaffari, M.R., Mohammadi, S.A,, Majidi, L., Salekdeh, G.H. and Mardi, M. 2008. Assessing wheat genetic diversity using quality traits, amplified fragment length polymorphisms, simple sequence repeats and proteome analysis.Annals of Applied Biology, 152:81-91.

Fernandes, J., Brendel, V., Gai, X., Lal, S., Chandler, V.L., Elumalai, R.P., Galbraith, D.W., Pierson, E.A. and Walbot, V. 2002. Comparison of RNA expression profiles based on maize expressed sequence tag frequency analysis and micro-array hybridization. Plant Physiolology., 128:896-910.

Han, F., Romagosa, I., Ullrich, S.E., Jones, B.L., Hayes P.M. and Wesenberg, D.M. 1997. Molecular marker-assisted selection for malting quality traits in barley. Molecular Breeding, 3(6):427-437.

Han, J.Z. and Wang, Y.B. 2008. Proteomics: present and future in food science and technology. Trends in Food Science and Technology, 19(1):26-30.

Hauptmann, R., Eisenreich, R., Eschenfeldt, W. and Khambatta, Z. 2003. 4-Ketocarotenoids In Flower Petals. Ball Horticultural Company, USA. Patent # WO03080849.

Hirschberg, J. 2001. Carotenoid biosynthesis in flowering plants. Current Opinion in Plant Biology, 4: 210-218.

http://www.bakeryandsnacks.com;http://www.bakeryandsnacks.com/Formulation/Proteomic-research-to-improve-flour-quality

http://www.zmdb.iastate.edu/zmdb/EST

Huang, X.Q., Cloutier, S., Lycar, L., Radovanovic, N., Humphreys, D.G., Noll, J.S., Somers, D.J. and Brown P.D. 2006. Molecular detection of QTLs for agronomic and quality traits in a doubled haploid population derived from two Canadian wheat (*Triticum aestivum* L.). Theoretical and Applied Genetics, 113(4):753-766.

Hu, F.B., Stampfer, M .J., Manson,C.H. and Willett,W.C. 1997. Dietary fat intake and the risk of coronary heart disease in woman. The New England Journal of Medicine, 337:1491-1499

Igartua, E., Edney, M., Rossnagel, B.G., Spaner, D., Legge, W.G., Scoles, G.L., Echstein, P.E., Penier, G.A., Tinker, N.A., Briggs, K.G., Falk, D.E. and Mather, D.E. 200. Marker-based selection of QTL affecting grain and malt quality in two-row barely. Crop Science, 40:1426-1433.

Jena, K.K. and Mackill, D.J. 2008. Molecular markers and their use in marker-assisted selection in rice. Crop Science, 48(4):1266-1276.

Kassahun, B. and Prasanna B.M. 2003. Simple sequence repeat polymorphism in Quality Protein Maize (QPM) lines. Euphytica, 129(3):337-344

Kristensen, M., Lok, F., Planchot, V., Svendsen, I., Leah, R. and Svensson, B. 1999. Isolation and characterization of the gene encoding the starch debranching enzyme limit dextrinase from germinating barley. Biochimica et Biophysica Acta, 1431:538-546.

Lecomte, L., Duffé, P., Buret, M., Servin, B., Hospital, F. and Causse, M. 2004. Marker-assisted introgression of five QTLs controlling fruit quality traits into three tomato lines revealed interactions between QTLs and genetic backgrounds. Theoretical and Applied Genetics, 109(3):658-68.

Lee, P.S. and Lee, K.H. 2000. Genomic analysis. Current Opinion in Biotechnology, 11:171-175.

Lui, H.K., and White.P.J. 1992. Oxidative stability of soybean oils with altered fatty acid compositions. Journal of American Society of Oil Chemists, 69:528-532

Manifesto, M.M., Feingold, S., Hopp, H.E., Schlatter, A.R. and Dubcovsky, J. 1998. Molecular markers associated with differences in bread-making quality in a cross between bread wheat cultivars with the same high M_r glutenins. Journal of Cereal Science, 27(3):217-227.

Mihr, C., Faurobert, M., Bouchet, J.P. and Causse, M., Paw³owski, T., Negroni, L., Sommerer, N. and Rossignol, M. 2005. Proteome analysis of organoleptic quality in tomato.http://www1.montpellier.inra.fr/proteome/pdf/Mihr%20et%20al%20Acta%20Hortic%202005.pdf.

Paine, J.A., Shipton, C.A., Chaggar, S., Howells,R.M., Kennedy, M.J., Vernon, G., Wright, S.Y., Hinchliffe, E., Adams,J.L., Silverstone, A.L. and Drake, R. 2005.

Improving the nutritional value of Golden Rice through increased pro-vitamin A content. Nature Biotechnology, doi:10.1038/nbt1082.

Payne, P.I. and Lawrence, G.J. 1983. Catalogue of alleles for the complex gene loci *Glu-A1*, *Glu-B1* and *Glu-D1* which code for highmolecular weight subunits of glutenin in hexaploid wheat. Cereal Research Communication, 11:29-35.

Payne, P.I., Nightingale, M.A., Krattinger, A.F. and Holt, L.M. 1987. The relationship between HMW glutenin subunit composition and bread-making quality of British grown wheat varieties. Journal of the Science of Food and Agriculture, 40:51-65.

Pelletier, J. and Sidhu, S. 2001. Mapping protein-protein interactions with combinatorial biology methods. Current Opinion in Biotechnology, 12:340-347.

Perretant, M.R., Cadalen, T., Charmet, G., Sourdille, P., Nicolas, P., Boeuf, C., Tixier, M.H., Branlard, G. and Bernard, S. 2000. QTL analysis of bread-making quality in wheat using a doubled haploid population. Theoretical and Applied Genetics, 100(8):1167-1175.

Phillips, K. 2001. «Scientists use DNA markers for transfer of desired traits into improved varieties». Planet Rice.net; http://www.biotech-info.net/DNA_markers.html

Prasanna, B.M., Vasal, S.K., Kassahun, B. and Singh. N.N. 2001. Quality Protein Maize. Current Science, 81(10):1308-1318.

Qiao-Quan, L., Qian-Feng, L., Xiu-Ling, C., Hong-Mei, W., Shu-Zhu, T., Heng-Xiu, Y., Zong-Yang, W. and Ming-Hong, G. 2006. Molecular marker-assisted selection for improved cooking and eating quality of two elite parents of hybrid rice. Crop Science, 46:2354-2360.

Sasaki, S. Yokota, A. and Tsubura Yuka, Y.T. 2002. Method for Promoting Fatty Acid Synthesis in Plant, NARA Institute of Science and Technology, Japan. Patent # JP2002335786.

Schaub, P., Al-Babili, S., Drake, R., and Beyer, P. 2005. Why Is Golden Rice Golden (Yellow) Instead of Red?. Plant Physiology, 138:441–450.

Schmierer, D.A., Kandemir, N., Kudrna, D.A., Jones, B.L., Ullrich, S.E. and Kleinhofs. A. 2005. Molecular marker-assisted selection for enhanced yield in malting barley. Molecular Breeding, 14(4):463-473.

Shu, Q. 2004. Plant Breeding and Genetics. FAO/IAEA Co-ordinated research project. http://www.iaea.org/nafa/d2/crp/d2-crop-quality.html.

Spencer, M.M., Pantalone, V.R., Meyer, E.J., Landau-Ellis, D. and Hyten Jr D.L. 2003. Mapping the /Fas/locus controlling stearic acid content in soybean. Theoretical and Applied Genetics, 106:615-619.

Suslow, T.V., Thomas, B.R. and Bradford, K.J. 2001. Special Topic Focus: Plant Research Excerpts from "Biotechnology Provides New Tools for Plant Breeding". http://www.scienceboard.net/community/perspectives.41.html.

Uthayakumaran, S., Listiohadi, Y., Baratta, M., Batey, I.L. and Wrigley, C.W. 2006. Rapid identification and quantitation of high molecular weight glutenin subunits. Journal of Cereal Science, 44:34-39.

Walbot, V. 2000. Saturation mutagenesis using maize transposons. Current Opinion in Plant Biology, 3:103-107.

Weber, H., Saalbach, I., Giersberg, M. and Hoffmeister, P. 2001. Preparing transgenic leguminous plants with increased protein content, useful *e.g.* as fodder, comprises transformation with specific genes, do not contain marker genes, Institut für Pflanzengenetik und Kulturpflanzenschung, Germany. Patent # WO0175128.

Zale, J.M., Clancy, J.A., Ullrich, S.E., Jones, B.L., Hayes, P.M. and the North American Barley Genome Mapping Project. 2000. Summary of barley malting quality QTLs mapped in various populations. Barley Genetics Newsletter, 30:44.

Ziegler, P. 1999. Cereal*Beta*-Amylases. Journal of Cereal Science, 29:195-204.

Molecular Plant Breeding: Principle, Method and Application
Eds : R.K. Singh, Rajesh Singh, Guoyou Ye, A. Selvi and G.P. Rao
Studium Press LLC, Texas, USA, 2009, pp. 367-408

CHAPTER

14

Molecular and Physiological Approaches for Drought Management in Crops

GYAN P. MISHRA[1], R.K. SINGH[2], RAGHWENDRA SINGH[1] and SHASHI BALA SINGH[1]

ABSTRACT

Plant adaptation to environmental stresses is dependent upon the activation of cascades of molecular networks involved in stress perception, signal transduction, and the expression of specific stress- related genes and metabolites. Consequently, engineering genes that protect and maintain the function and structure of cellular components can enhance tolerance to stress. Success in improving drought tolerance will largely be dependent on the accuracy of plant phenotyping and the capacity to determine the GxE interactions. These are the two major components that affect the prediction of the allelic value on the plant phenotype in new populations. For complex traits like drought, selections based on markers alone is not highly effective. There is need to implement strategies based on integration of phenotypic selection along with marker assisted selection. No quick-fix is available, but pathways do exists. Combining genetic mapping with association studies has great potential in molecular breeding for drought tolerance in crop plants. Our limited knowledge of stress-associated metabolism remains a major gap in our understanding; therefore, comprehensive profiling of stress-associated metabolites is most relevant to the successful molecular breeding of stress-tolerant crop plants. Unraveling additional stress-associated gene resources, from both crop plants and highly tolerant model plants, will enable future molecular dissection of stress tolerance mechanisms in important crop plants.

Key Words: Abiotic stresses, Drought, Molecular marker, Marker assisted selection, HSP.

[1]*Defence Institute of High Altitude Research, DRDO, C/o 56 APO, Leh, India*
[2]*Indian Institute of Sugarcane Research, Lucknow-226002, UP, India*
Corresponding author e-mail: gyan.gene@gmail.com

1.0 INTRODUCTION

Plants differ from animals in many aspects, but the important may be that plants are more easily influenced by environment than animals. Plants have a series of fine mechanisms for responding to environmental changes, which has been established during their long period evolution and artificial domestication. These mechanisms are involved in many aspects of anatomy, physiology, biochemistry, genetics, development, evolution and molecular biology, in which the adaptive machinery related to molecular biology is the most important. The elucidation of it will extremely and purposefully promote the sustainable utilization of plant resources and make the best use of its current potential under different scales. This molecular mechanism at least include environmental signal recognition (input), signal transduction (cascades of biochemical reactions are involved in this process), signal output, signal responses and phenotype realization, which is a multi-dimensional network system and contain many levels of gene expression and regulation. We will focus on the molecular adaptive machinery of plants under abiotic stresses and draw a possible blueprint for it. The issues and perspectives are also discussed.

Stress is a major component of natural selection in soil ecosystems. The most prominent abiotic stress factors in the field are temperature extremes (heat, cold), dehydration (drought), high salinity and specific toxic compounds such as heavy metals. Organisms are able to deal with these stresses to a certain extent, which determines the limits of their ecological amplitudes (Fig. 1). Functional genomic tools are now becoming available

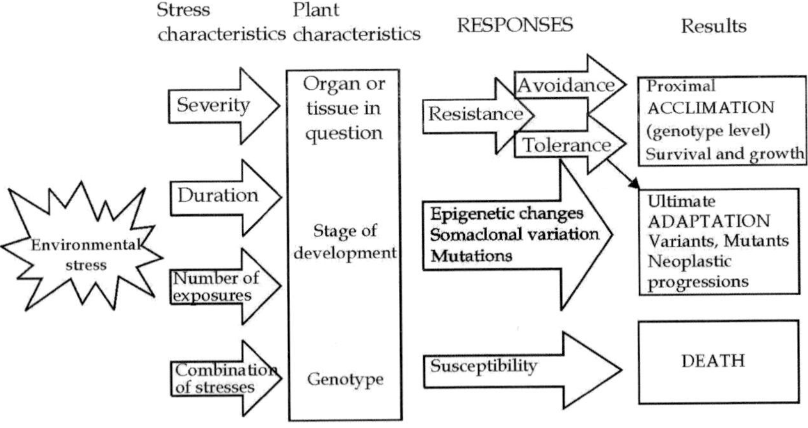

Fig. 1. Stress its response and ultimate result on plant (Source : Prasad, 2008)

to study stress in ecologically relevant soil organisms. The picture emerging from signalling pathways and transcription factors identified in transcription profiling studies suggests that there is a large overlap of genomic responses to drought, salinity and cold; however, heat and heavy metals trigger different stress response pathways. The heat shock response and the oxidative stress response seem to represent universal components of the environmental stress response (ESR). Furthermore, the commonality across plants and animals seems to be higher in effector genes than in transcriptional regulators. Finally, adaptation to stress factors in soil seems to evolve through enhanced constitutive transcription of otherwise stress responsive genes both in plants and animals. Cold, drought and salinity are those environmental stressors which affect plants in many respects and which, due to their wide-spread occurrence cause the most fatal economic losses in agriculture (Fig. 2). No wonder that the effects of these stressors have been addressed in a countless number of studies ranging from the molecular to the whole plant level, from a description of the damages to mechanisms of tolerance and hardening. All three forms of abiotic stress affect the water relations of a plant on the cellular as well as whole plant level causing specific as well as unspecific reactions, damages and adaptation reactions.

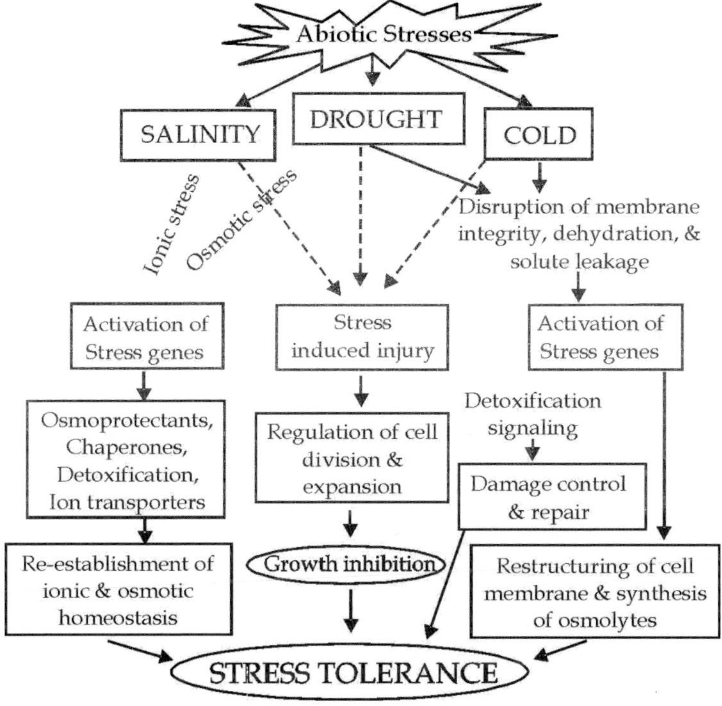

Fig. 2. Abiotic stresses and tolerance mechanisms in plants (Source : Tuteja, 2008)

The remarkable ability of plants to adapt under many different adverse environments is a fascinating process. Research into the physiology and metabolism of these plants under different stress not only fosters better understanding of the evolutionary processes that have created the diversity of life as it exists on earth, but also has economic implications for agricultural biotechnology and the development of novel products. In the present era, genomics has emerged and has developed a wide range of tools to identify genes that play roles in specific pathways. However, relating individual genes and alleles to agronomic traits is still quite challenging. The capacity to sequence genomes and the availability of novel molecular tools have now catapulted biological research into eras of genomics and post-genomics, creating an opportunity to apply genomic techniques for crop improvement. This has led plant scientists to search for such models among the relatives of Arabidopsis (*Arabidopsis thaliana*), the most universally used species in molecular plant research owing to its many technical advantages and the wealth of available biological information. Further, plants may experience different types of stress at different developmental stages and their mechanisms of response to stress may vary in different tissues (Queitsch *et at*, 2000). The initial stress signals (*e.g.*, osmotic or ionic effects, or changes in temperature or membrane fluidity) would trigger downstream signaling processes and transcription controls, which activate stress-responsive mechanisms to reestablish homeostasis and protect and repair damaged proteins and membranes. Inadequate responses at one or more steps in the signaling and gene activation processes might ultimately result in irreversible damages in cellular homeostasis and destruction of functional and structural proteins and membranes, leading to cell death (Vinocur and Altman, 2005). Transgenic technology has become an essential tool for studying plant biology and for the development of novel plant varieties that have been cultivated extensively in some regions of the world. It has had a profound impact on the rapid development of plant biology by providing means of producing gene-tagged populations, cell markers to study plant development and the technology to study gene function.

2.0 STRESS IMPOSITION AND CHARACTERIZATION

Many groups of drought stress responsive genes have been identified and characterized, including chaperones, reactive oxygen species (ROS) scavenging enzymes, proteins involved in osmolyte metabolism and genes involved in drought specific signaling cascades. Identification and characterization abiotic stress responsive genes and their validation, has become a critical requirement for enhancing agricultural production.

For identification of stress responsive genes and their characterization, the following scheme of events is generally followed:

1. Stress imposition at whole plant level, generation of stress plant material and characterization
2. Isolation of RNA for analyzing stress effect at molecular level and stress gene discovery
3. Expression analysis of stress genes
4. Generation of stress specific cDNA libraries and screening
5. Annotation and analysis of gene sequences

Accurate methods of stress imposition are crucial for the optimal expression of stress responsive genes. Conditions that mimic the gradual development of stress in the field level are best obtained by gravimetric approaches. Plants samples are collected at the stress level where optimum transcript expression is observed and are used for RNA isolation and further expression analyses. RNA isolated is also used for creation of stress specific cDNA libraries. Both stress cDNA and the stress specific libraries are used as base material for isolation of stress responsive genes. The isolated stress genes need to be sequenced, annotated and analysed using bioinformatics tools available.

2.1 Isolation of RNA for Stress Gene Discovery

Accuracy of transcriptome expression and analysis depends on the purity of the RNA. Cellular RNases should be inactivated as quickly as possible at the very first stage in the extraction process.

2.2 Studies on the Expression of Stress Responsive Genes

For expression analysis, the RNA needs to be converted into cDNA by reverse transcription. The cDNA is then used as template to study the expression of specific genes by semi-quantitative RT PCR or quantitative real time PCR. Alternatively, the RNA can directly be blotted for northern hybridization.

2.3 cDNA Synthesis or Reverse Transcription

The enzymatic conversion of poly (A+) mRNA to complementary DNA (cDNA) is known as reverse transcription.

2.4 Semi Quantitative Reverse Transcriptase Polymerase Chain Reaction (RT-PCR)

Total RNA is reverse transcribed to cDNA in a reaction mix containing reverse transcriptase with random hexamer and dNTP mix. The reverse transcription is performed using the oligodT primers. The enzyme is inactivated quick chilled. PCR is performed using cDNA mix. The reaction mixture is divided into four parts and each aliquot is amplified for different cycles (20, 25, 30, 35 cycles). The tubes are taken out at different cycles-20, 25, 30, 35 cycles. The samples are kept at RT and final extension is performed at a time. All the products are run on a single gel.

2.5 Northern hybridization

Northern hybridization is used to measure the amount and size of RNAs transcribed from eukaryotic genes and to estimate their abundance.

2.6 Real-time PCR (qRT-PCR) to Analyze Stress-specific Genes Expression

The expression pattern of mRNA or transcript level of gene can be quantified by several techniques like, northern hybridization, ribonuclease protection assays (RPA), *in-situ* hybridization and other PCR based methods such as RT-PCR and qRT-PCR. Real-time polymerase chain reaction, also called quantitative real time polymerase chain reaction (qRT-PCR) or kinetic polymerase chain reaction, is a laboratory technique used to quantify and amplify a specific part of a given DNA molecule concurrently. It is used to determine whether or not a specific sequence is present in the sample and if it is present, the number of copies in the sample. It is the real-time version of quantitative polymerase chain reaction (q-PCR), itself a modification of polymerase chain reaction. The procedure follows the general pattern of PCR, but the DNA is quantified after each round of amplification; this is the "real-time" aspect of it. Two common methods of quantification are the use of fluorescent dyes (such as ethidium bromide, and SYBR Green) that intercalate the double-strand DNA, and modified DNA oligo-nucleotide probes (Taqman probe, molecular beacons) that fluoresce when hybridized with a complementary DNA. Frequently, real-time polymerase chain reaction is combined with reverse transcription polymerase chain reaction to quantify low abundance messenger RNA (mRNA), enabling a researcher to quantify relative gene expression at a particular time, or in a particular cell or tissue type.

2.7 Quantification of Transcripts/template DNA

Two strategies are commonly employed to quantify the results obtained by real-time PCR, the standard curve method and the comparative threshold (Ct) method. These are discussed briefly below.

2.7.1 Standard curve method

In this method, a standard curve is first constructed from RNA of known concentration. This curve is then used as a reference standard for extrapolating quantitative information for mRNA targets of unknown concentrations.

2.7.2 Comparative Ct Method

This involves comparing the Ct values of the samples of interest with a control or calibrator such as a non-treated sample or RNA from normal tissue. The Ct values of both the calibrator and the samples of interest are normalized to an appropriate endogenous housekeeping gene. If the plot of cDNA dilution versus delta Ct is close to zero, it implies that the efficiencies of the target and housekeeping genes are very similar. If a housekeeping gene cannot be found whose amplification efficiency is similar to the target, then the standard curve method is preferred.

2.8 Generation of Stress Specific cDNA Libraries and Screening

Induction of drought resistance in plants involves a complex network of signal perception, amplification and transduction of the signal, involving a set of differentially expressed genes. To understand the molecular regulation of these processes, the relevant subsets of differentially expressed genes of interest must be identified, cloned and characterized.

2.8.1 Subtractive hybridization and plaque lift

Subtractive cDNA hybridization is a novel approach to identify and isolate cDNAs of differentially expressed genes (Hara *et al.*, 1991). Numerous cDNA subtraction methods have been reported but the basic principle involves hybridization of cDNA from one population (tester) to excess of mRNA from another population (driver) and then separation of the unhybridized fraction (target) from hybridized common sequences. The separation of the target fraction is achieved by hydroxylapatite chromatography avidin-biotin binding or Oligo (dT) 30-latex beads (Hara *et. al.*, 1991). Traditional procedures have been successful in some cases but often are technically demanding, labour-intensive, requires large amounts of mRNA and are not suited for the identification of rare messages (Hara *et al.*, 1991). A specialized form of subtractive hybridization, the "Gene Expression Screen" can detect both upregulated and downregulated transcripts. A number of protocols have been devised in recent years to simplify and expedite the process of transcript identification by subtractive hybridization. Clonetech PCR-Select cDNA subtraction (Clonetech, USA), subtractor kit (Invitrogen, USA), the SMART cDNA subtraction kit (Stratagene, USA) are some of the new cDNA subtraction procedures with relevant modifications. They can increase the detection sensitivity by 10 to 100-fold and make the identification of quite rare genes possible.

One of the cDNA subtraction protocols called Suppression Subtractive Hybridization (SSH) has been widely used to identify differentially expressed genes (Diatchenko *et al.*, 1996). A key feature of this method is simultaneous subtraction and normalization that makes it possible to equalize abundance of target cDNAs in the subtracted population. As a result, rare, differentially expressed transcripts can be enriched by -1000 fold. The subtractive hybridization protocol is discussed in this section which involves hybridization of drought mRNA (tester) with biotinylated single stranded cDNA. The unhybridized stress specific fraction is converted to double stranded cDNA and cloned into a lambda UNI-ZAP XR vector by following the ZAP-cDNA synthesis kit (Stratagene, USA).

2.8.2 Screening of stress cDNA library for identification of stress genes

By subtractive hybridization approach the pool of stress specific double stranded cDNA molecules are identified and cloned into UNI ZAPXR vector. The ligated mixture is later packaged into GigaGoldIII packaging extract. Each vector molecule will have cDNA of one of the genes expressed under stress conditions. These genes have to be screened to identify the stress specific gene expressed during the drought stress. This section explains the protocol to be followed to identify the cloned stress genes.

3.0 FUNCTIONAL GENOMICS

Functional genomics is a recent topic in the field of molecular biology that attempts to describe gene (and protein) functions and interactions. Unlike genomics and proteomics, functional genomics focuses on the dynamic aspects such as gene transcription, translation, and protein-protein interactions, as opposed to the static aspects of the genomic information such as DNA sequence or structures. There are various approaches which are now being used to study gene functions in plants such as:

a. *In-silico* prediction of plant gene function,

b. Forward and Reverse genetic studies:

   i. Upregulation studies include Activation tagging, Over expression, Gain in function studies etc.

   ii. Down regulation studies include Transposon tagging, T-DNA Tagging, Physical and Chemical mutagenesis, Tilling, Site directed mutagenesis, Post-Transcriptional gene silencing (PTGS) etc.

c. Other approaches such as Microarray, Proteomics etc.

Many different terminologies are used in functional genomics. The term gene silencing is a general term describing epigenetic processes of gene regulation. Gene silencing term is generally used to describe the "switching off' of a gene by a mechanism other than genetic modification. That is, a gene which would be expressed (turned on) under normal circumstances is switched off by machinery in the cell. As we know gene(s) is regulated at either the transcriptional or post-transcriptional level. In general, Transcriptional gene silencing is the result of histone modifications, creating an environment of heterochromatin around a gene that makes it inaccessible to transcriptional machinery (including TFs and RNA polymerases). Post-transcriptional gene silencing is the result of mRNA of a particular gene being destroyed, which prevents translation process to form active gene product.

3.1 Post Transcriptional Gene Silencing (PTGS) Approaches to Study Gene Function

A common mechanism of post-transcriptional gene silencing is RNAi. Both transcriptional and post-transcriptional gene silencing are used to regulate endogenous genes. Mechanisms of gene silencing also protect the organism's genome from transposons and viruses. Gene silencing thus may be part of an ancient immune system protecting from such infectious DNA elements. Although most the above approaches are being followed for functional genomics in plants, the techniques involving PTGS is being widely been used to study gene function. PTGS is a process during which accumulation of an RNA species in the cytoplasm is suppressed based on its homology to an introduced gene (Klahre *et al.*, 2002). This is known as RNA silencing and was first discovered in plants (Lindbo and Dougherty, 1992). This phenomenon has been observed in different organisms, including fungi, insects, nematodes, fish and mice. It is believed that such a phenomenon normally function in viral defense and regulation of gene expression in plants (Burch-Smith *et al.*, 2004).

Using the principle of PTGS, there are two main approaches to study the gene function: RNA interference (RNAi) and Virus induced gene silencing (VIGS).

3.2 RNA Interference (RNAi)

As mentioned above, RNAi is a mechanism for RNA-guided regulation of gene expression in which double-stranded RNA ribonucleic acid inhibits the expression of genes with complementary nucleotide sequences. Conserved in most eukaryotic organisms, the RNAi pathway is thought to have evolved as a form of innate immunity against viruses and also plays a major role in regulating development and genome maintenance. The RNAi pathway has been particularly well-studied in certain model organisms such as the nematode worm *Caenorhabditis elegans*, the fruit fly *Drosophila melanogaster*, and the flowering plant *Arabidopsis thaliana*.

3.3 Virus Induced Gene Silencing (VIGS)

In plants VIGS systems have developed to suppress gene transcripts in plants. The VIGS occurs when plants are infected with a virus carrying target sequences with homology to a host nuclear gene. The virus infection triggers the cytoplasmic degradation of any RNA with sufficient homology to the target sequence. The phenotype of the plant silenced by VIGS for a particular gene mimics the phenotype of loss of function mutant. Many studies have demonstrated that VIGS is a powerful tool to determine gene function (Liu *et al.,* 2002; Burch-Smith *et al.,* 2004).

3.3.1 Mechanism

It has been shown that when a plant virus infects a host cell it activates an RNA-based defense that is targeted against the viral genome. The sequence of RNA degradation is triggered by double stranded RNA (dsRNA) synthesized during viral replication, and occurs in a two-step process. In the first step, double stranded RNA (dsRNA) is processed into shorter, 21-25 nucleotide long sense and antisense units (called short interfering (si) RNAs, siRNA) with the help of an enzyme called Dicer. These small RNAs, called short interfering RNA or siRNA, in the second step forms a complex with RISC protein (RNA inducing silencing complex), act as guide sequences to identify homologous transcripts and target them for destruction. An RNase complex is generally guided by base pairing of the siRNAs so that it specifically targets single-stranded (ss) target RNA that is similar to the dsRNAs. The dsRNA causes the siRNA/RNase complex to target the viral single-stranded RNA. In nut shell the enzyme deer frims double stranded RNA, to form small interfering RNA or microRNA. These processed RNAs are incorporated into the RNA induced silencing complex (RISC), which targets messenger RNA to prevent translation (Hammond *et al.,* 2000).

It has been reported that as the rate of viral RNA replication increases, the viral dsRNA and siRNA would become more abundant and the viral ssRNA would be targeted intensively. This results in reduction in virus multiplication. Based on the above principles, VIGS vectors have been developed. In simple, VIGS is a virus vector technology that exploits this RNA defense by host plants. The viral cDNA is modified to include non viral host gene insert. When the insert is derived from a host gene, the siRNAs would target the RNase complex to the corresponding host mRNA and the symptoms in the infected plant would exhibit the loss *of* the function mutant (Liu *et al.,* 2002) Fig. 3.

Majority of the cases RNA viruses have been modified and used as VIGS vectors. It has been shown that amongst the different vectors, Tobacco Rattle Virus (TRV a bipartite positive sense RNA virus) based VIGS vectors are most suitable for plant functional genomics (Liu *et al., 2002)* in model systems like *Nicotiana benthamiana.* A good VIGS vector should be able to infect the plant rapidly and uniformly including meristamatic region of the plant. In addition, it should not produce any strong suppressor of silencing.

The TRV based vectors have several advantages over other virus vectors. This vector has been successfully used to silence genes in *Nicotiana benthamiana* and tomato (Liu *et al.*, 2002). For constructing TRV-VIGS, the cDNA clones of RNA1 and RNA2 have been inserted in two different T-DNA expression cassettes. The cassette containing RNA2 have been constructed with multiple cloning sites (MCS) to allow the cloning *of* target gene sequences for VIGS (Liu *et al.*, 2002).

3.3.2 Applications of VIGS in stress tolerance studies

Virus-induced gene silencing (VIGS) is a rapid and robust method for determining and studying the function of plant genes or expressed sequence tags (ESTs). However, only a few plant species are amenable to VIGS. There is a need for a systematic study to identify VIGS-efficient plant species and to determine the extent of homology required between the heterologous genes and their endogenous orthologs for silencing. Recently, the application of this technique has been extended to characterize the genes and cellular processes involved in abiotic-stress tolerance, and in particular drought and oxidative stress. Because abiotic-stress tolerance is multigenic, identification and characterization of genes involved in this process is challenging. VIGS could become one among the several potential tools in understanding the relevance of these stress-responsive genes. Development of VIGS protocols for the use of heterologous gene sequences as VIGS-inducers has extended its applicability to analyze genes of VIGS recalcitrant plant species (Senthil *et al.*, 2008).

Senthil and Udayakumar (2006) developed a low moisture-stress protocols in which the plants experienced the desired stress level when silencing the target stress gene using VIGS was at a maximum. The functional relevance of a groundnut (*Arachis hypogea*) subtracted-stress cDNA clone putative leaf was examined by VIGS in tomato. A 400 bp fragment of *lea4* was cloned into tobacco rattle virus-based VIGS vector to trigger post-transcriptional gene silencing by Agrobacterium-mediated inoculation in tomato plants. In silenced plants only *lea4* transcripts showed a substantial decline, whereas the expression of other known stress-responsive genes such as apx (ascorbate peroxidase) and elip (early light-induced protein) were unaltered. Under moderate moisture stress, the silenced plants showed enhanced susceptibility as measured by cell viability, superoxide radical activity, and cell osmotic adjustment. This approach illustrates the potential benefits of VIGS in identifying functional relevance of low moisture stress-responsive genes. It is also demonstrated that heterologous probes with a fairly high degree of homology to the native genes can be used to study the functional relevance of stress-responsive genes using VIGS.

3.4 Usefulness of Drought Responsive Genes of *Chlamydomonas reinhardtii*

Mechanisms controlling plant abiotic stress tolerance are multigenic. With advances in genomics, several stress responsive genes have been identified and many stress databases provide clues about the nature of stress genes. About 100,000 expressed sequence tags responding to abiotic stresses were isolated from model plants (Wang *et al.*, 2004). Annotated information from diverse stress libraries provide leads about the nature of expressed genes and their putative function, but their relevance and functionality under stress is yet to understood. Although the functional significance of a gene in stress tolerance can be elucidated using mutants and other loss of function studies, overexpression studies can provide additional information about whether ox not the gene expressed is relevant for imparting higher level of tolerance.

Many model systems are used to study the function of plant stress genes like Arabidopsis, tobacco, yeast etc. In addition *Chlamydomonas reinhardtii* could be other potential system to study the function of plant genes. It is being used as an experimental organism for studying the genetics and molecular biology of a number of cellular processes, like photosynthesis and chloroplast development, flagellar assembly and motility. *Chlamydomonas* grows rapidly both photoautotrophically or heterotrophically and is amenable to classical genetic analysis. It is haploid during vegetative growth and allows any mutation to be immediately detected. *Chlamydomonas* can be stably transformed by particle bombardment, electroporation, agitation with glass beads and *Agrobacterium*. Apart from this, *Chlamydomonas* is also a suitable organism to efficiently study stress effects. Differential response of *Chlamydomonas* to abiotic stresses has been reported. Stress effect in *Chlamydomonas* can be measured by cell growth, chlorophyll content, estimation of malondialdehyde (MDA), osmolytes (Siripomadulsil *et al.*, 2002) and also by phenotypic observation.

3.5 Improvement of Water Use Efficiency in Rice by Expression of HARDY, an *Arabidopsis* Drought and Salt Tolerance Gene

Since fresh water is limited and a dwindling global resource; efficient water use is the need for food crops having high water demands, such as rice, or for the production of sustainable energy biomass. It has been established that expression of the *Arabidopsis* HARDY (HRD) gene in rice improves water use efficiency, the ratio of biomass produced to the water used, by enhancing photosynthetic assimilation and reducing transpiration. Increased shoot biomass under well irrigated conditions and an adaptive increase in root biomass under drought stress were observed in drought tolerant, low water consuming rice plant. The HRD gene, an AP2/ERF-like transcription factor, identified by a gain-of-function Arabidopsis mutant hrd-D having roots with enhanced strength, branching, and cortical cells, exhibits drought resistance and salt tolerance, accompanied by an enhancement in the expression of abiotic stress associated genes. Thicker leaves with more chloroplast-bearing mesophyll cells were produced in *Arabidopsis* due to HRD over expression and in rice, there is an increase in leaf biomass and bundle sheath cells that probably contributes to the enhanced photosynthesis assimilation and efficiency. The results exemplify application of a gene identified from the model plant Arabidopsis for the improvement of water use efficiency coincident with drought resistance in the crop plant rice (Karaba *et al.*, 2007)

4.0 PHENOTYPING OF COMPLEX PHYSIOLOGICAL TRAITS

4.1 Heat Tolerance

Crop species and cultivars differ in their sensitivity to high temperature and other stresses. High temperature affects membrane linked processes and the enzyme functions leading to the creation of heat induced oxidative stress. At whole plant level, these damages lead to reduced photosynthesis and other metabolic events leading to altered growth and reproduction. Plants overcome high temperature stress by adapting several physiological and biochemical mechanisms. Plants have both inherent ability to survive at high temperature and ability to acquire tolerance to otherwise lethal temperatures. This thermo-tolerance can also be induced by exposing the organisms to a gradual increase in temperature during which they acquire tolerance. The acquired thermo-tolerance to tolerate the lethal temperature is through the production and accumulation of stress shock proteins (Vierling, 1991) which most often act as molecular chaperones to protect the cells against stress effects.

4.1.1 Mechanism of heat tolerance

Plants manifest different mechanisms for surviving under elevated temperatures, including long-term evolutionary phenological and morphological adaptations and short-term avoidance or acclimation mechanisms such as changing leaf orientation, transpirational cooling, or alteration of membrane lipid compositions. In many crop plants, early maturation is closely correlated with smaller yield losses under high temperatures, which may be attributed to the engagement of an escape mechanism. Even plants growing in their natural distribution range may experience high temperatures that would be lethal in the absence of this rapid acclimation response. Furthermore, because plants can experience major diurnal temperature fluctuations, the acquisition of thermo-tolerance may reflect a more general mechanism that contributes to homeostasis of metabolism on a daily basis.

Elucidating the various mechanisms of plant response to stress and their roles in acquired stress tolerance is of great practical and basic importance. Some major tolerance mechanisms, including ion transporters, osmoprotectants, free-radical scavengers, late embryogenesis abundant proteins and factors involved in signaling cascades and transcriptional control are essentially significant to counteract the stress effects (Wang *et al.*, 2004). Series of changes and mechanisms, beginning with the perception of heat and signaling and production of metabolites that enable plants to cope with adversaries of heat stress, have been proposed (Fig. 3). Heat stress effects are notable at various levels, including plasma membrane and biochemical pathways operative in the cytosol or cytoplasmic organelles. Initial effects of heat stress, however, are on plasmalemma, which shows more fluidity of lipid bilayer under stress. This leads to the induction of Ca^{2+} influx and cytoskeletal reorganization, resulting in the up regulation of mitogen activated protein kinases (MAPK) and calcium dependent protein kinase (CDPK). Signaling of these cascades at nuclear level leads to the

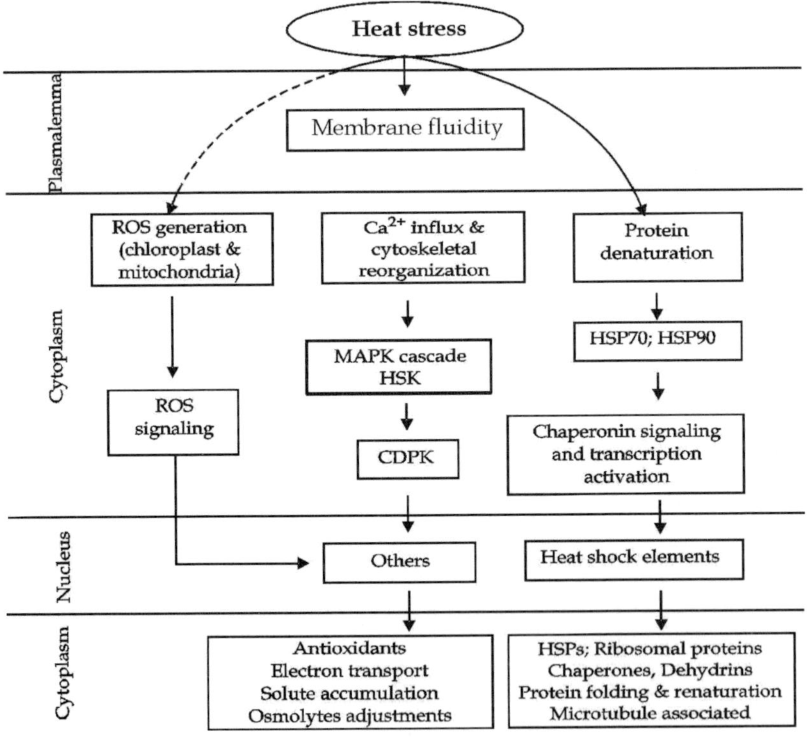

Fig. 3. Heat stress tolerance mechanism at cellular level (Source : Raju, 2008)

production of antioxidants and compatible osmolytes for cell water balance and osmotic adjustment. Production of ROS in the organelles (*e.g.*, chloroplast and mitochondria) is of great significance for signaling as well as production of antioxidants (Bohnert *et al.*, 2006). The antioxidant defense mechanism is a part of heat-stress adaptation, and its strength is correlated with acquisition of thermo-tolerance. Accordingly, in a set of wheat genotypes, the capacity to acquire thermo-tolerance was correlated with activities of CAT and SOD, higher ascorbic acid content, and less oxidative damage (Sairam and Tyagi, 2004). One of the most closely studied mechanisms of thermo-tolerance is the induction of HSPs, which comprise several evolutionarily conserved protein families. However, each major HSP family has a unique mechanism of action with chaperonic activity. The protective effects of HSPs can be attributed to the network of the chaperone machinery, in which many chaperones act in concert. An increasing number of studies suggest that the HSPs/ chaperones interact with other stress-response mechanisms (Wang *et al.*, 2004). The HSPs/ chaperones can play a role in stress signal transduction and gene activation as well as in regulating cellular redox state. They also interact with other stress-response mechanisms such as production of osmolytes and antioxidants (Panchuk *et al.*, 2002).

Membrane lipid saturation is considered as an important element in high temperature tolerance. Currently, it is unknown whether a higher or a lower degree of membrane lipid saturation is beneficial for high-temperature tolerance (Klueva *et al.*, 2001). The contribution of lipid and protein components to membrane function under heat stress needs further investigation. Localization of LMW-HSPs with chloroplastic membrances upon heat stress suggests that they play a role in protecting photosynthetic electron transport (Heckathorn *et al.*, 1998). An important component of thermo-tolerance is changes in gene expression. Heat stress is known to swiftly alter patterns of gene expression, inducing expression of the HSP complements and inhibiting expression of many other genes. The mRNAs encoding non heat- stress-induced proteins are destabilized during heat stress. Heat stress may also inhibit splicing of some mRNAs. Earlier it was hypothesized that HSP-encoding mRNAs could not be processed properly due to the absence of introns in the corresponding genes. Subsequently it was shown that some HSP-encoding genes have introns and, under heatstress conditions, their mRNAs were correctly spliced (Visioli *et al.*, 1997). However, the mechanism of preferential post-transcription modification and translation of HSP encoding mRNA under heat stress remains yet to be elucidated.

4.1.2 Synthesis and expression of heat stress proteins

Expression of stress proteins is an important adaptation to cope with environmental stresses. Most of the stress proteins are soluble in water and therefore contribute to stress tolerance presumably *via* hydration of cellular structures (Wahid and Close, 2007). Although heat shock proteins (HSPs) are exclusively implicated in heat-stress response, certain other proteins are also involved.

4.1.2.1 Heat shock proteins (HSPs)

Synthesis and accumulation of specific proteins are ascertained during a rapid heat stress (Schoffl *et al.*, 1999). Induction of HSPs seems to be a universal response to temperature stress, being observed in all organisms ranging from bacteria to human (Vierling, 1991). Plants of arid and semi arid regions may synthesize and accumulate substantial amounts of HSPs. Certain HSPs are also expressed in some cells under cyclic or developmental control. In this case, the expression of HSPs is restricted to certain stages of development, such as embryogenesis, germination, pollen development and fruit maturation (Prasinos *et al.*, 2005). In higher plants, HSPs are usually induced under heat shock at any stage of development and major HSPs are highly homologous among distinct organisms (Vierling, 1991). HSP-triggered thermo-tolerance is attributed to the observations that (a) their induction coincides with the organism under stress, (b) their biosynthesis is extremely fast and intensive, and (c) they are induced in a wide variety of cells and organisms. Three classes of proteins, as distinguished by molecular weight, account for most HSPs, *viz.*, HSP90, HSP70 and low molecular weight proteins of 15-30 kDa. The special importance of small HSPs in plants is suggested by their unusual abundance and diversity. The proportions of these three classes differ among plant species. HSP70 and HSP90 mRNAs can increase ten-fold, while low molecular weight (LMW) HSPs can increase as much as 200-fold.

Immuno-localization studies have determined that HSPs normally associate with particular cellular structures, such as cell wall, chloroplasts, ribosomes and mitochondria. For instance, a heat-tolerant maize line (ZPBL-304) exhibited increased amounts of chloroplast protein synthesis elongation factor under heat stress, which was related to the development of heat tolerance (Moriarty *et al.*, 2002). Similarly, in tomato plants under heat stress, HSPs aggregate into a granular structure in the cytoplasm, possibly protecting the protein biosynthesis machinery (Miroshnichenko *et al.*, 2005). Presence of HSPs can prevent denaturation of other proteins caused by high temperature. The conformational dynamism and aggregate state of small HSPs may be crucial for their functions in thermo-protection of plant cells from detrimental effects of heat stress (Schöffl *et al.*, 1999; Iba, 2002). The ability of small HSPs to assemble into heat shock granules (HSGs) and their disintegration is a prerequisite for survival of plant cells under continuous stress conditions at sub lethal temperatures (Miroshnichenko *et al.*, 2005). In response to high temperatures, specific HSPs have been identified in different plant species. For example, HSP68, which is localized in mitochondria and normally constitutively expressed, was determined to have increased expression under heat stress in cells of potato, maize, tomato, soybean and barley. Another HSP identified in maize is a nucleus-localized protein, HSP101, which belongs to the campylobacter invasion antigen B (CiaB) protein sub-family, whose members promote the renaturation of protein aggregates, and are essential for the induction of thermo tolerance. Levels of HSPl0l increased in response to heat shock, more abundantly in developing tassel, ear, silks, endosperm and embryo and less abundantly in vegetative and floral meristematic regions, mature pollen, roots and leaves (Young *et al.*, 2001). In addition, heat treatment increases the level of other maize HSPs, which are associated with plant ability to withstand heat stress. For example, 45-kDa HSP was found to play a major role in recovery from heat stress (Ristic and Cass, 1992).

The mechanism by which HSPs contribute to heat tolerance is still enigmatic though several roles have been ascribed to them. Many studies assert that HSPs are molecular chaperones insuring the native configuration and functionality of cell proteins under heat stress. There is considerable evidence that acquisition of thermo-tolerance is directly related to the synthesis and accumulation of HSPs. For instance, HSPs provide for new or distorted proteins to fold into shapes essential for their normal functions. They also help shuttling proteins from one compartment to another and transporting old proteins to "garbage disposals" inside the cell. Among others, HSP70 has been extensively studied and is proposed to have a variety of functions such as protein translation and translocation, proteolysis, protein folding or chaperoning, suppressing aggregation, and reactivating denatured proteins (Zhang *et al.*, 2005). Recently, dual role of LMW HSP21 in tomato has been described as protecting PSII from oxidative damage and involvement in fruit color change during storage at low temperatures (Neta-Sharir *et al.*, 2005). In many plant species, thermo-tolerance of cells and tissues after a heat stress is pretty much dependent upon induction of HSP70, though HSP101 has also been shown to be essential (Schöffl *et al.*, 1999). One hypothesis is that HSP70 participates in ATP dependent protein unfolding or assembly/ disassembly reactions and it prevents protein denaturation during heat stress (Iba, 2002). Evidence for the general protective roles of HSPs comes from the fact that mutants unable to synthesize them or the cells in which HSP70 synthesis is blocked or inactivated are more susceptible to heat injury. Further, LMW-HSPs may play structural roles in maintaining cell membrane integrity in plants.

4.1.2.2 Other heat induced proteins

Besides HSPs, there are a number of other plant proteins, including ubiquitin, cytosolic Cu/Zn-SOD and Mn-POD (Brown *et al.*, 1993), whose expressions are stimulated upon heat stress. For example, in *Prosopis chilensis* and soybean under heat stress, ubiquitin and conjugated-ubiquitin synthesis during the first 30 min of exposure emerged as an important mechanism of heat tolerance. Some studies have shown that heat shock induces Mn peroxidase, which plays a vital role in minimizing oxidative damages (Iba, 2002). In a study on *Chenopodium murale,* when leaf proteins extracts from thylakoid and stromal fractions were subjected to heat stress it was determined that Cu/Zn-SOD from stromal fraction was more heat tolerant than Cu/Zn-SOD from thylakoid, and this was responsible for chloroplastic stability under heat stress (Khanna-Chopra and Sabarinath, 2004). In another study, a number of osmotin like proteins induced by heat and nitrogen stresses, collectively called Pir proteins, were found to be over expressed in the yeast cells under heat stress conferring them resistance to tobacco osmotin (an antifungal) (Yun *et al.*, 1997). Late embryogenesis abundant (LEA) proteins can prevent aggregation and protect the citrate synthase from desiccating conditions like heat- and drought-stress (Goyal *et al.*, 2005). Using proteomics tool, Majoul *et a1.* (2003) determined enhanced expressions of 25 LEA proteins in hexaploid wheat during grain filling. Geranium leaves exposed to drought and heat stress revealed expression of dehydrin proteins (25-60 kDa), which indicated a possible linkage between drought and heat-stress tolerance. Recently, three low-molecular-weight dehydrins have been identified in sugarcane leaves with increased expression in response to heat stress (Wahid and Close, 2007). Function of these proteins is apparently related to protein degradation pathway, minimizing the adverse effects of dehydration and oxidative stress during heat stress (Schöffl *et al.*, 1999).

5.0 ACQUIRED TOLERANCE

Unlike animals, plants are unable to escape from adverse environmental conditions. However, plants have evolved to survive under such conditions by adapting to the environments in which they originate and grow. Plants exposed to some periods of moderate stress often induce tolerance which results in the ability to survive more extreme conditions. However, there is a limit beyond which the acclimatization cannot permit survival and this can be termed as the 'adaptive range'. One approach to understand the ability of plants to acclimatize and tolerate the environmental stresses is to identify stress-induced changes in gene expression. It has been known since few decades that particular proteins, whose synthesis is induced by stress conditions, are critical in the survival of plants. Gene activation, transcription and translation that often occur during the acclimatization process are thought to be involved in the induction of stress tolerance (Vierling, 1991).

In several studies, it has been found that upon acclimation, a significant increase in HSP's (HSP 70, HSP 104, HSP 90 and HSP 18.1) occurred which confer thermo-tolerance to seedlings upon their exposure to relatively at high temperatures. Even in some of the studies, it has been shown that the small HSP's like HSP 18.1 and HSP 17.9 also were up regulated. Synthesis and localization of these HSP's and other stress genes has been shown to trigger several physiological and biochemical processes including membrane stability (Grover *et al.*, 2000). Decreased electrolyte leakages, increased chlorophyll stability, reduced free radicals generation, maintenance of protein synthesis and greater incorporation of amino acids to proteins have also been shown in a number of systems upon induction. This clearly indicates that the observed thermo-tolerance in the tolerance genotypes/lines is through the production HSP's which act as molecular chaperones protecting the functional proteins. In one of the studies, the contrasting sunflower showed differential accumulation of HSP's and recovery growth indicating the role of HSP' s in imparting thermo tolerance.

6.0 INDUCTION RESPONSE

Several studies on genetic variability have provided compelling evidences that genetic variability is mainly due to differential expression of stress responsive genes. The gene expression occurs only during early stages of stress, referred to as induction stress (Vierling, 1991). Under natural conditions, plants are exposed to a sub-lethal stress (referred to as induction stress) before being subjected to severe stress. Several earlier studies have indeed shown that the plants exposed to induction stress develop the ability to withstand the otherwise lethal temperatures and this phenomenon of withstanding severe temperature is often referred to as 'acquired thermo tolerance'.

During induction stress, the stress-signaling pathway is triggered on, which brings in the expression of an array of stress responsive genes. Thus, relevance of a physiological or biochemical trait for thermo tolerance can best be studied by pre-exposure of seedlings/ plants to a sub-lethal acclimation temperature. This has been shown in a number of systems like perl millet and sorghum, pea, sunflower and others. In all these systems, it has been clearly shown that the stress responsive genes would express only during the sub-lethal induction and the differential expression of these genes is the reason for the observed genetic variability for stress tolerance (Senthil Kumar *et al.*, 2007). Therefore, to screen a genotype for thermo tolerance, it is necessary to expose it to the induction stress before its exposure to the lethal stress.

6.1 Temperature Induction Response (TIR) Technique

According to this technique, the seedlings are exposed to induction temperature before their exposure to severe temperature. The tolerant genotypes showed higher survival rate and more recovery growth as compared to the susceptible genotypes/lines (Senthil Kumar *et al.*, 2007). The observed tolerance upon induction is not only seen at the seedling level, but also at the plant level. This is evident from the fact that, the plants when they are exposed to sub lethal temperature before their exposure to relatively a high temperature often exhibited normal growth compared to the plants exposed directly to high temperature.

Principle: The principle in this technique is, initially the seedlings are exposed to mild temperature (Induction temperature) following which the seedlings are exposed to a severe challenging temperature. Recovery growth is determined as a measure of tolerance. Only the tolerant lines/genotypes could survive and put on recovery growth while the susceptible do not. Thus, this approach forms one of the potential screening techniques to evaluate genetic variability for intrinsic stress tolerance. Groundnut, sunflower, pigeonpea and such other systems clearly indicated that the susceptible and tolerant genotypes could be identified only when challenged with high temperature stress following acclimation temperature. Similar response was also demonstrated in pearl millet and sorghum by Howarth *et al.* (1997). This clearly demonstrates the validity of this technique in screening and selecting intrinsically tolerant genotypes/lines from a segregating population.

6.2 Cross Talk among the Stresses

In spite of diverse signal perception and transduction mechanisms that may operate during stresses like temperature, drought and salinity, the gene products involved in cell protection against extreme environmental conditions can be similar. Further there is certain degree of cross talk between stresses signifying that there is interaction in signaling events for acquired tolerance. Chilling injury has been shown to be substantially reduced by heat shock treatment before exposure to chilling temperature. This phenomenon is termed as 'heat shock-induced chilling tolerance' (Li *et al.*, 2003). The correlation of heat shock-induced chilling tolerance and accumulation of HSPs and APX in plants has been characterized. Sunflower hybrids selected as thermo tolerant based on the TIR have shown tolerance to salinity stress and PEG-induced osmotic stress. Even abscissic acid signaling mutants (abil and abi2) and the UV sensitive mutant uvh 6, showed strongest defects in acquired thermo tolerance in terms of root growth and seedling survival. Ethylene signaling mutants (ein2 and etrl) and reactive oxygen metabolism mutants (vtcl, vtc2, npq1 and cad2) were more defective in basal tolerance than acquired thermo tolerance, especially under high light (Larkindale *et al.*, 2005). This suggests that there are common sets of mechanisms operating under different types of stresses, which may be driven by different sets of genes. Larkindale *et al.* (2005), identified the sunflower hybrids as thermo-tolerant through TIR which showed high leaf area under moisture stress and also upon recovery. A strong positive relationship seen between relative thermo tolerance based on TIR and the yield performance under rainfed conditions (Al-Ouda, 1999) signify that common mechanisms of adaptation exists across stresses. From these studies it can be concluded that TIR could be an initial screening method to identify tolerant types for other abiotic stresses.

7.0 ISOTOPE RATIO MASS SPECTROMETER (IRMS) AND ITS APPLICATION

Stable isotopes as surrogates for physiological traits. The past decade has seen a rapid expansion in the use of stable isotopes at natural abundance in ecological research. Geochemists, Oceanographers and Climatologists have developed a rigorous theoretical and empirical basis for the integration of stable isotopes into studies of global element cycles, past climatic conditions, hydrothermal vent systems and tracking rock sources. Though relatively quite recent, plant biologists, ecologists and environmental chemists are currently developing the theoretical frameworks and the empirical database for the use of stable isotopes to study plants and animals. Consequently, isotope analysis is rapidly becoming a standard tool for physiologists, ecologists and others studying element or material cycles in environment.

The abundance of stable isotopes of various elements are such that the heavier isotope is present to a very small extent ranging from 0.015% for Deuterium to 1.1% for ^{13}C. The isotopic compositions of naturally occurring compounds differ significantly. Such variations in the isotopic compositions of compounds arise due to the transformation processes. For instance, the heavy and the lighter isotopes of carbon significantly differ in the extent of solubility, diffusion and also in reactivity in enzyme catalyzed reactions. This differential participation of the isotopic forms of an element leads to a significant alteration in the isotopic composition of the end product compared with the starting compound. Since chemical and physical processes lead to this alteration in isotopic composition, determination of the isotopic composition should indeed be a reflection of the transformation process. The isotopic composition is normally expressed as a ratio of the heavy to the lighter isotope of the elements. Since the natural abundance of the heavier isotope are extremely small, this ratio is a very small value. Thus, determination of elements in such small quantities requires extremely precise measurement systems and protocols.

Mass Spectrometry is one such tool for the determination of the stable isotopic composition in various materials. Since the measurement of the ratio of the heavy to the lighter isotopes of an element is the prerequisite in understanding various processes, an isotope ratio mass spectrometer (IRMS) is used for this purpose. A group of elements in any complex molecule are separated based on their atomic mass by a mass spectrometer. This separation is achieved through ionization of the elements followed by the acceleration through a strong magnetic field. When charged elements are accelerated through a magnetic field, the straight line paths of the molecules now assume a trajectory whose radius is directly proportional to the mass of the molecule. Thus, for the determination of isotopic mass ratios, the samples need to be ionized, accelerated, separated and ultimately determined. Several different types of mass spectrometers are available based on each of these. The most common ones are based on the method of ionization through electron impact and separation through a magnetic sector. For organic sample mass spectrometry, the sample is suitably converted into its constituent gases and introduced into the ion source. However, this procedure requires large amounts of the sample and is laborious and time consuming. This method of isotopic ratio determination is often referred to as Off-line method.

Though the off-line method of sample preparation and introduction into the ion source can yield highly precise results, this method is cumbersome and large number of samples cannot be determined. With the advancement of technology of combustion and purification of gases, the analysis of stable isotope ratios can now be performed on-line. Here, the combustion, purification and the introduction of the relevant gas is performed as a chain of processes. This procedure is often referred to as continuous Flow mass spectrometry (CF-IRMS). The continuous flow mass detection using the IRMS requires accurate Sample preparation, subsequent elemental analysis using an appropriate elemental analyzer interfaced with an IRMS for the determination of mass to charge ratio.

Applications: Experimental evidence is presented to show that the ^{18}O enrichment in the leaf biomass and the mean (time-averaged) transpiration rate are positively correlated in groundnut and rice genotypes. The relationship between oxygen isotope enrichment and stomatal conductance (gs) was determined by altering gs through ABA and subsequently using contrasting genotypes of cowpea and groundnut. The Peclet model for the ^{18}O enrichment of leaf water relative to the source water is able to predict the mean observed values well, while it cannot reproduce the full range of measured isotopic values. Further, it fails to explain the observed positive correlation between transpiration rate and ^{18}O enrichment in leaf biomass. Transpiration rate is influenced by the prevailing environmental conditions besides the intrinsic genetic variability. As all the genotypes of both species experienced similar environmental conditions, the differences in transpiration rate could mostly be dependent on intrinsic gas. Therefore, it appears that the Delta ^{18}O of leaf biomass can be used as an effective surrogate for mean transpiration rate. Further, at a given vapour pressure difference, Delta ^{18}O can serve as a measure of stomatal conductance as well (Sheshshayee *et al.*, 2005).

Isotope signatures: A potential approach to identify desirable C_3 lines expressing C_4 genes. C_3 plants expressing C_4 genes will have two sources of CO_2 *i.e.* atmospheric and from C_4 fixation. The delta ^{13}C of the biomass will be close to that of C_4 species. So, delta ^{13}C a potential indicator of functioning of C_4 mechanism in C_3 plants. Carbon isotope ratios can be used as a robust approach to identify putative transformants.

8.0 EVALUATION OF TRANSGENICS WITH STRESS TOLERANCE GENES

Plant transformation technology has become a versatile platform for cultivar improvement as well as for studying gene function in plants. This success represents the culmination of many years of effort in tissue culture improvement, in transformation techniques and in genetic engineering. Although gene transfer technology has become routine in working with several plant species, in others the limiting step is not the transformation itself but rather the lack of efficient regeneration protocols. To tackle the problems pertaining to regeneration in recalcitrant crops, alternate methods to minimize or eliminate the steps of regeneration are being standardized. These are called the *in planta* transformation protocols. Research with *Arabidopsis* has benefited from the development of high throughput transformation methods that avoid plant tissue culture. In particular, the development of the *Agrobacterium mediated* vacuum infiltration method has had a major impact on *Arabidopsis* research. *In planta* transformation methods have also been standardized for rice (Supartana *et al*, 2005), buckwheat (Kojima *et al.*, 2000), kenaf (Kojima *et al.*, 2004) and mulberry (Ping *et al.*, 2003). In all the crops, *Agrobacterium* is directed towards either the apical meristem or the meristems of axillary buds. One such viable *in planta* transformation protocol has also been standardized for other crops (Rohini and Sankara Rao, 2002). The strategy essentially involves *in planta* inoculation of embryo axes of germinating seeds and allowing them to grow into seedlings *ex vitro.* These *in planta* transformation protocols are advantageous over other methods because they do not involve regeneration procedures and therefore the tissue culture-induced somaclonal variations are avoided. A list of genes, their mechanisms and genetically modified plant species implicated in response to stress is given in the Table 1.

8.1 Analysis of the Transgenics

After the transformation by using any of the methods, the plants have to be analysed to determine the transgenic nature. Once the plants grow large enough, they may be assayed for 'trans-gene' integration, expression and efficacy. The assay for gene expression is conducted using methods consistent with the transgene coding sequence and desired results.

8.1.1 Analysis of marker genes

Plants are generally transformed with the gene of interest and marker genes that are called as, selectable and screenable markers for easy analysis of the transgenic plants. The selectable markers are generally the antibiotic resistance genes while the screenable marker genes include β-glucuronidase, luciferase, and genes involved in anthocyanin biosynthesis and green fluorescent protein.

8.1.2 Polymerase chain reaction

PCR is performed as one of the first proofs for the presence of the gene in the transformants. Primers are designed either for the gene or the vector separately or in combination or for the marker genes. In case of trans genes that are present in the plants also, primers are synthesized such that one

Table 1. Mechanism, genes and genetically modified plant species implicated in plant response to abiotic stress

Mechanism	Genes	Species
Transcription control	*CBF1*	*Arabidopsis thaliana*
	DREB1A	*Arabidopsis thaliana*
	CBF3	*Arabidopsis thaliana*
	CBF1	*Lycopersicon esculentum*
Compatible solute proline	*P5CS*	*Nicotiana tabacum*
	ProDH	*Arabidopsis thaliana*
Myoinositol	*IMT1*	*Nicotiana tabacum*
sorbitol	*stpd1*	*Nicotiana tabacum*
Antioxidants and detoxification	*CuZn-SOD*	*Nicotiana tabacum*
	Mn-SOD or Fe-SOD	*Medicago sativa*
Ion transport	*SOS1*	*Arabidopsis thaliana*
HSPs and molecular chaperones	*Hsp17.7*	*Dacus carota*
	Hsp 21	*Arabidopsis thaliana*
LEA type proteins	*COR15a*	*Arabidopsis thaliana*
	WCS19	*Arabidopsis thaliana*

of the primer set will belong to the vector, for *e.g.*, promoter or terminator so that, the amplified product is sure to be from the T-DNA and not an artifact.

8.1.3 Southern blot analysis

Southern analysis (Southern, 1975) is performed to check: the integration and copy number of the trans gene in the transformants.

8.1.4 Analysis for the expression of the trans genes

The next step, after confining the presence of the genes of interest by PCR and the copy number by southern analysis is to analyse the expression and efficacy of the transgenic plants for the new conferred trait. This is generally done by

1. ELISA/western blot for the expression of the trans gene
2. Bioassays for the efficacy of the trans gene

9.0 PHYSIOLOGICAL CHARACTERIZATION

In the case of transgenics developed for abiotic stress tolerance, physiological analysis of the transformants is carried out to choose the best performing plants against any stress. In transgenic plants developed for herbicide tolerance, preliminary analysis for herbicide tolerance is carried out at two levels, *viz.*, seedling and whole plant level.

9.1 Seedling Level

Two days old seedlings are incubated with different concentrations of glyphosate starting from 500, 1000, 2000, 3000 to 5000 ppm for 24 hours and put for recovery in water. The concentration of glyphosate which is not tolerable by the wild type plants is used as the screening concentration of the transgenics. Those seedlings that survive upon subjecting to the higher concentration of glyphosate is grown in small bowls containing soil rite, covered with polythene cover following the traditional hardening protocol of keeping the plants inside the polythene cover for a few days (15-20 days) and then allowing them to grow by transferring them to pots and raising them in the pots.

9.2 Whole Plant Level

After the establishment of the primarily screened seedlings in the pots, the next level screening is on the plants. Two types of screening can be performed at plant level.

1. Swabbing the leaves with glyphosate
2. Spraying a known concentration of glyphosate

In the leaf swabbing technique, the wild type plants are first swabbed with different concentrations of glyphosate and observed for chlorosis, the concentration of glyphosate that produces chlorosis on wild type plants is used to screen the transgenic plants. Those plants that don't show the chlorosis and are tolerant to the known level of glyphosate, are taken for further analysis.

Selection of the best performing plants can also be done by spraying a known concentration of glyphosate and checking for tolerance symptoms. The concentration of the glyphosate for spraying is generally the same as that used for swabbing.

In the treated plants, physiological differences between the tolerant, the susceptible and the wild type plants is characterized by looking for

(a) chlorophyll content (b) membrane integrity, and (c) relative water content

10.0 MOLECULAR BREEDING

Genetic studies of natural populations are dependent on the availability of polymorphic markers. Information about gene flow, allele frequencies and mating systems that are crucial in population biology were obtained by combining information from several loci.

10.1 Association Mapping

An approach to use Natural Allelic Diversity to evaluate Gene Function

Association analysis can be used to relate natural variations at candidate gene level with agronomic phenotypes. Association approaches in plants can provide very high resolution and can evaluate a wide range of alleles rapidly. The issues related to experimental design, germplasm sample, population structure, and statistical analysis necessary for association analysis in plants is discussed. In a high diversity species such as rice, association analysis has the potential to map quantitative trait loci (QTL) with up to 5000 times better resolution than mapping with standard F_2 populations. In addition, association approaches may survey tens of alleles whereas standard mapping approaches survey a maximum of two alleles. Association approaches do not require special mapping populations, but rely on the extensive history of mutation and recombination to dissect a trait. The structure of linkage disequilibrium (LD) (the correlation between polymorphisms) and evaluation of selection is a key to utilizing association analysis (Nordborg *et al.*, 2002) (Fig. 4).

The use of extant natural diversity provides advantages in resolution and breadth of survey, but can also add difficulties in accurately assessing the true cause of an association. The most serious false positives can result when unlinked markers produce a positive association because of underlying population structure. The complex breeding history of most crops and limited gene flow in most wild plants create population stratification within the germplasm. In recent years, a few statistical methods have been developed that use independent marker loci as a means of detecting and correcting for population structure (Reich and Goldstein, 2001). These methods work on the assumption that population structure should affect all loci in a similar manner.

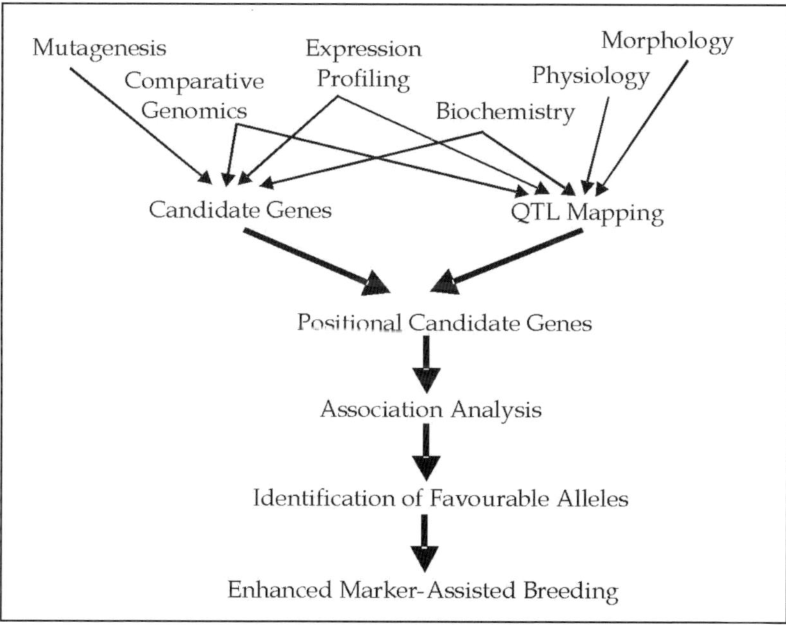

Fig. 4. Association approach for enhanced marker assisted breeding (Source: Prasanna, 2008).

A standard procedure for carrying out association analysis is as follows:

1. Select markers, so that they are equally distributed through the genome.
2. Choose germplasm that will capture the bulk of diversity present. If possible, inbred lines should be used.
3. Score phenotypic traits in replicated trials.
4. Amplify the fragments, identify and score the polymorphism
5. Obtain diversity estimates and evaluate patterns of selection
6. Statistically evaluate associations between genotypes and phenotypes taking population structure into account.

10.2 Interpretation of Genotype-Phenotype Associations

Once an association is empirically determined, the validity of the association must be ascertained.

1. Which associating polymorphisms most likely control the trait? First, it is critical that genotypes be rechecked and results should be examined to determine if phenotypic outliers are driving associations. Association studies will often find multiple polymorphisms that significantly associate. Carefully examining the LD structure surrounding the association can help identify this suite of polymorphisms, and where more sampling may be needed. Although the most significant site is the most likely case for the association, many of the slightly less significant sites could actually be the functional cause of the phenotypic variation. Breaking the polymorphisms into likely functional (biologically significant) versus likely silent is useful in developing lists of sites for future evaluation.

2. The most straightforward way to prove an association is to evaluate the candidate polymorphisms in an entirely different population sample. Only polymorphisms that are closely linked to the cause of a phenotype should be significant in a second study. It is important that the population structure of the second sample is truly independent of the first sample. Only the candidate polymorphisms need to be retested in the new sample.

3. In some cases associations will suggest a molecular or biochemical analysis with molecular biology and biochemistry can be very productive, but it should be warned that association studies could be picking up on effects that only explain a few percent of the variation. Many biochemical and molecular approaches may not be quantitatively sensitive enough to detect such small changes at a molecular level.

4. Final proof of the association can be obtained through marker assisted selection and production of near isogenic lines (NILS).

Application: In order to find candidates genes for low temperature and drought stresses, 13 genes encoding for transcription factors and upstream regulators were screened by amplification and SSCP on six parental genotypes of three barley mapping populations ('Nure' × 'Tremois', 'Proctor' × 'Nudinka', and 'Steptoe' × 'Morex'), and mapped as newly developed STS, SNP, and SSCP markers. A new consensus function map was then drawn using the three maps above, including 16 regulatory candidate genes (CGs). The positions of barley cold and drought tolerance quantitative trait loci (QTLs) presently described in the literature were added to the consensus map to find positional candidates from among the mapped genes. A cluster of six Hv*CBF* genes co-mapped with the *Fr-H2* cold tolerance QTL, while no QTLs for the same trait were positioned on chromosome 7H, where two putative barley regulators of *CBF* expression, *ICE1* and *FRY1*, found by homology search, were mapped. Four out of 12 drought tolerance QTLs of the consensus map are associated with regulatory CGs, on chromosomes 2H, 5H, and 7H, and two QTLs with effector genes, on chromosomes 5H and 6H. The results obtained could be used to guide MAS applications, allowing introduction into an ideal genotype of favourable alleles of tolerance QTLs (Tondelli *et al.*, 2006).

11.0 MAINTENANCE OF POSITIVE CARBON BALANCE

Under stress situation for better performance of any C_3 plant it should maintain its positive carbon balance. This can be maintained through either reduction in its photorespiration (PR) or increasing the carboxylation reaction. Further this can be achieved through either improved kinetics of rubisco or increasing the substrate CO_2 through CCM. Thus option to increase the carbon gain under stress in C_3 plant is through a better rubisco or super charging of photosynthesis (CCM).

11.1 Expressing C_4 Genes in C_3 Species may be a Reality

Since long many scientists were trying to express the C_4 genes in C_3 species so as to make it more efficient. Coordinated expression of relevant C_4 pathway genes is now possible in the C_3 species. Recent progress in cloning and expression provides leads for the coordinated expression of relevant genes in target organelle of some C_3 species. But the next step is how to identify the transformants expressing C_4 genes in C_3 plants that fixes CO_2 by C_4 pathway? For that one approach could be the use of isotope signature for C_4 transformant identification.

11.2 Single Cell C_4 Mechanism Suggests Possibilities to Express C_4 in C_3

Reasonable progress has been made in recent times likely to get some leads to address oxidative stress by CCM option (Fig. 5).

11.3 Lessons from Micro-Algae

CA-HCO_3 mediated CCM functions without Kranz anatomy. This unique mechanism was observed in two single cell C_4 species of microalgae *i.e. Borszczowia & Bienertia* (Fig. 6).

12.0 AQUAPORINS

Dehydration and salt stress require changes in water flow to allow cells and tissue to adapt to the stress situation. The rate of water flux into

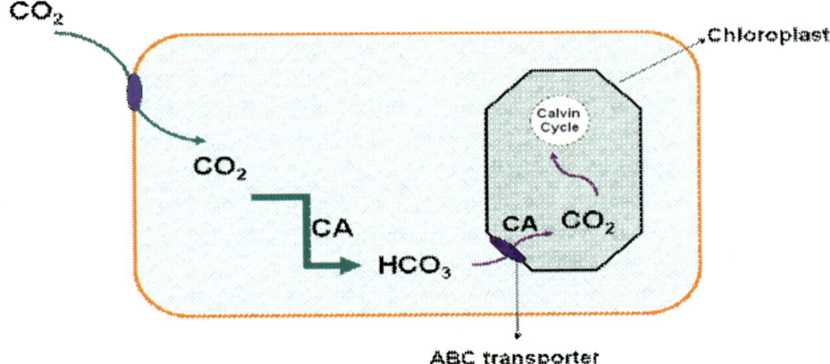

Fig. 5. Single cell C_4 mechanism (Source : Udaya Kumar, 2008)

or out of cells is determined by the water potential gradient that acts as the driving force for transport and by the water permeability of the membrane. Aquaporin proteins facilitate osmosis by forming water-specific pores as an alternative to water diffusion through the lipid bilayer, thus increasing the water permeability of the membrane. Aquaporins are members of a large super-family of membrane spanning proteins, the major intrinsic proteins

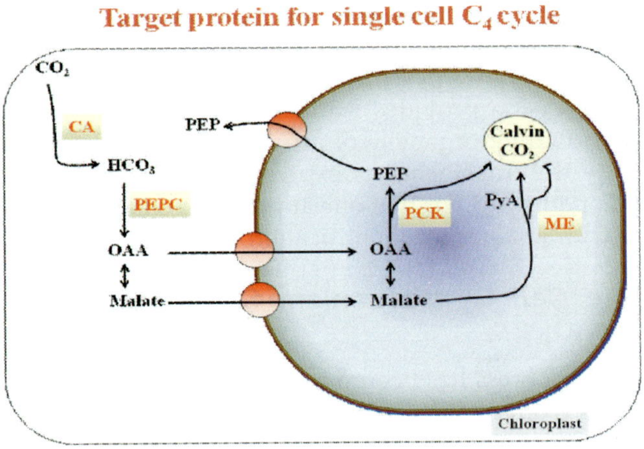

Target protein for single cell C_4 cycle

Fig. 6. Single cell C_4 mechanism in microalgae (Source : Udaya Kumar, 2008)

(MIPs). In plants, aquaporins localized in the tonoplast are called tonoplast intrinsic proteins (TIPs), while those in the plasma membrane are PIPs. Across different plant species aquaporins have very conserved structures and hence are encoded by similar DNA sequences. These functions were inferred from the expression patterns of specific aquaporin isoforms or from the blocking effects of mercury on water transport through plant tissues. There are several reports that aquaporin genes are induced by dehydration and salt stress; this may trigger greater osmotic water permeability and facilitate water flux (Yamaguchi-Shinozaki *et al,* 1992).

13.0 DEHYDRINS

Conditions that result in cellular dehydration such as drought and frost induce the expression of dehydrins, which also accumulate in seeds during maturation where they are known as late embryogenesis abundant (LEA) proteins. Dehydrins vary over a wide range of molecular masses from 9 to 200 kD, they are thermostable and contain a high proportion of glycine and lysine residues. Typical structural features of dehydrins are highly conserved sequences, termed K-, S-, and Y-segments (Allagulova *et al.*, 2003). Although dehydrins are widely distributed and have been found in vascular plants, mosses, ferns, lichens and algae, their molecular functions are not well understood as they do not catalyze any metabolic reaction. A high portion of random coil structures affects their exceptional water binding capacity and the conserved segments give rise to amphipathic á-helices which form lipid binding domains and thus can associate with, and protect lipid aggregates and hydrophobic domains of proteins. The ubiquitous 'K'-segment of dehydrins shows similarity to a class A2 amphipathic á-helix, resembling the lipid-binding domain found in apolipoproteins (Campbell and Close, 1997). Because of these physicochemical properties and a clear correlation of the accumulation of dehydrins with dehydration of a plant, a protective function is obvious. The classical view therefore considers dehydrins as membrane stabilizers but other targets, such as proteins and nucleic acids likewise associate with dehydrins.

14.0 STRESS AND HELICASES

Helicases are ubiquitous enzymes that catalyze the unwinding of energetically stable duplex DNA (DNA helicases) or duplex RNA secondary structures (RNA helicases). Most helicases are members of DEAD-box protein superfamily and play essential roles in basic cellular processes such as DNA replication, repair, recombination, transcription, ribosome biogenesis and translation initiation. Therefore, helicases might be playing an important role in regulating plant growth and development under stress conditions by regulating some stress-induced pathways. There are now few reports on the up-regulation of DEAD-box helicases in response to abiotic stresses. Recently, salinity-stress tolerant tobacco plants have already been raised by overexpressing a helicase gene, which suggests a new pathway to engineer plant stress tolerance (Sanan-Mishra *et al.*, 2005). Presently the exact mechanism of helicase-mediated stress tolerance is not understood. Furthermore, it is becoming increasingly evident that an RNA helicase can also perform more than a single physiological function. Although RNA helicases are associated with a diverse range of biotic cellular functions, there have been relatively few reports of RNA helicase involvement in cellular response to abiotic stress. An updated list of plant stress-responsive helicases is given in the Table 2.

Table 2. Plant stress-responsive helicases: An Update

Plant	Gene	Induction/ response	Reference	Remarks
Arabidopsis	*FL25A4*	Cold-induced; Not Characterized	Seiki *et al.*, 2001	Putative helicase
Arabidopsis	*CBF3;*	Transcriptional activator and chilling responsive	Gong *et al.*, 2002	Putative helicase
Arabidopsis	*los-4-2;*	Cold-induced; mRNA export	Gong *et al.*, 2005	Putative helicase
Apocynum venetum	*AvDH1;*	Salt and cold	Liu *et al.*, 2008	DNA helicase
Rice	*OsBIRH1*	Defence responses against pathogen infection and oxidative stresses.	Li *et al.*, 2008	RNA helicase

15.0 OSMOREGULATION AND STRESS TOLERANCE

Osmoregulation is considered as one of the best mechanisms of abiotic stress tolerance for crops exposed to dehydration stress specifically at early stages of plant development. The osmolytes including polyols, sugars, quaternary ammonium compounds and proline accumulate in significant amounts and help in maintaining osmotic potential, ionic balance, membrane integrity and protection of chromatin and labile proteins under stress conditions (Kavikishore *et al.*, 2005). In addition to acting as osmoprotectant, proline also serves as a sink for energy to regulate redox potentials, as a hydroxy radical scavenger (Smirnoff and Cumbes, 1989), as a solute that protects macromolecules against denaturation, as means of reducing the acidity in the cell and acts as a storage compound and nitrogen source for rapid growth after stress.

15.1 Water Mining-Molecular Approach to Improve Root Traits

Molecular controls of constitutive and adaptive pathways of root morphogenesis were poorly understood. To elucidate the genetic control of root system architecture several attempts are made in rice and maize (Hochholdinger *et al.*, 2006) and major emphasis is on plasticity of the plant root development.

15.1.1 QTL mapping of constitutive root developmental traits

There are different types of rice *e.g* Indica, Japonica, basmati deep water and floating types which can be classified into different groups based on the variability in not traits. Indica (Group 1) had thin, highly branched roots with narrow vessels and a low root to shoot ratio. Japonica type (Group 6) had thick, less branched long roots and larger root to shoot ratio. Aus group (Group 2) was intermediate, root profile similar to Japonica, but thin roots. Deep water and floating rice (Group 3 & 4) and basmati rice (group 5) root thickness and root distribution closer to Indica type. Variability within group was limited, suggesting the noticed genetic variability is not sufficient (esp group 2 *i.e.* Aus & 6 *i.e.* Japonica). For most traditional upland genotypes already deep rooted, it may be difficult to find donors in the *O. sativa* sps.

Inference: Wild species did not appear to be a better source of alleles for improved adoptive root distribution than the upland adapted Japonica cultivar (*O. sativa*). In Japonica cutivars IR 60080, IR 65907, IRAT 104 had increased root mass at depth and IRAT 212, WAB 56-50 had increased total root mass. Under stress and these associated with increased stomatal conductance and leaf extension. Increased root growth at depth under stress in some entries may relate to ABA production or ABA sensitivity. For example in maize also ABA produced in drying roots promoted root growth but inhibited shoot growth (Sharp *et al.*, 2004). Direct root measurements are difficult, leaf extension (after 12 to 16 days of drying), stomatal conductance (after 20 days of drying) can be used as measurable traits. In addition osmotic adjustments (OA) and increased leaf membrane stability can be used to improve drought tolerance in rice. In this case, wild species are a valuable source of improved alleles for this trait.

15.1.2 Mapping QTL's for root morphology

QTL's for deep root morphology were found only in the 3 Japonica/ Indica populations but not in Indica/Indica population. For thick roots QTL's were identified in 3 Japonica/Indica but not identified in Indica/ Indica population. Data suggest that some of the QTL's for root traits identified in Japonica/Indica populations, are not found at the same intervals in Indica/Indica cross.

Research is now in progress to quantify the relationships between improved constitutive and adaptive root traits, water extraction, nutrient uptake, grain yield using progenies from QTL mapping. QTL's for deep root morphology and root thickness traits over different mapping populations indicate a potential for marker aided selection (MAS) for these traits. The QTL's identified so far are constitutive. It's important to tackle the QTL's for adaptive response induced by water stress too. For both constitutive and inducible QTL's there is a need to check whether they can influence yield under stress. For this, NIL (Near Isogenic Lines) would represent a useful tool simplifying the genetic analysis.

15.2 Putative Physiological Traits Applied in Breeding for Drought Tolerance

There are only certain putative physiological traits which can be used in the breeding for drought tolerance. These are emergence characteristics/ vigor, nutrient acquisition/uptake efficiency, leaf development & photosynthesis, water use efficiency (WUE) component traits, deep root development, canopy temperature/stomata regulation, osmotic adjustment/ relative water content, hormonal control (ABA, GA, ethylene), stay green/ senescence, grain number maintenance, grain fill duration and rate, harvest index under drought, yield and its components under drought etc.

15.3 Meta-Analysis of QTL Data

QTL mapping experiments yield heterogeneous results due to the use of different genotypes, environments and sampling variation. Meta-analysis is the combining data from several sources into a single study. Compilation of QTL mapping results yields a more complete picture of the genetic control of a trait and reveals patterns in organization of trait variation. Meta-analysis was first suggested for analysis of statistical evidence in QTL mapping by Lander and Kruglyak (1995). It is first applied in human and animal QTL analysis. Adaptation to both arid and favorable conditions can be combined in the same genotype. QTL mapping can be used to test the association between productivity and quality under water deficit with a suite of traits often found to differ between genotypes adapted to arid versus well-watered conditions. Genomic tools and approaches may expedite breeding of genotypes that respond favorably to specific environments.

16.0 MAS FOR DROUGHT TOLERANCE IN MAIZE

A number of different MAS approaches do exist for the improvement of polygenic traits in maize. In an experiment, Riboot and Ragot (2007) compared the results of a marker-assisted back cross (MABC) selection with alternative MAS strategies for improving grain yield under drought canditions in tropical maize. The introgression favorable alleles at five target regions inverted in the expression of yield components and flowering traits increased grain yield and reduced the asynchrony between male and female femering under limited water conditions. Eighty-two percent of the recurrent percents genotype at non-target loci was received in only four generations of MABC by screening large segregating population (2200 individuals) for three of the four generations. Selected QTLs explained

~38% of the total phenotypic variance for anthesis to silking interval (ASI) in maize. A cluster of QTLs for flower traits and yield components was identified on chromosome 3, where CML247 displayed the favourable alleles. After evaluating 80 hybrids (70 MABC-derived and 10 controls) during 1998-2001, 30 were selected based on their agronomic traits and yield performance under drought stress. (Ribaut and Ragot, 2007).

16.1 Results of MAS for Drought Tolerance in Maize

Mean grain yield of MABC-derived hybrids was consistently higher than that of control hybrids (crosses from the recurrent parent to the same two testers as the MABC derived families) under severe water stress conditions. Under sever stress, the best five MABC derived hybrids yielded, on average, at least 50% more than control hybrids. Under mild water stress, defined as resulting in <50% yield reduction, no difference was observed between MABC-derived hybrids and the control plants, thus confirming that the genetic regulation for drought tolerance is dependent on stress intensity. MABC conversions involving several target regions are likely to result in partial rather than complete line conversion.

16.2 MAS can only Complement Phenotypic Selection

MAS experiment based only on the QTLs involved in the expression of yield components would not be the most efficient because only a few of the QTLs are stable across environments. A MAS experiment should consider the QTLs involved in the expression of secondary traits of interest correlated with yield under drought. An efficient MAS strategy should take into account the most suitable QTLs from different traits as an 'index' (Ribaut *et al.*, 1998, 2004).

Even with all the accumulated wisdom, why is the overall impact of MAS in developing drought stress tolerant cultivars has so far been disappointing?

Probable Reasons: Choice of parental lines used for QTL discovery are mostly based on differences for morpho-physiological traits of interest rather than agronomic value, limitations of experimental mapping populations, reliability of phenotypic assays, inconsistency in QTL detection across environments/genetic backgrounds, ineffective detection of epistatic QTL interactions and ineffective MAS schemes for dealing with multiple QTLs.

***Recent Emphasis*:** Improving the efficiency and reliability of phenotyping, verification of the beneficial QTL alleles, multiparental crosses, association mapping, enriching linkage maps with function-specific genes (functional maps) and identification of the 'eQTLs'.

16.3 What Traits to Measure?

Leaf growth, ABA, root growth, water use efficiency, specific leaf area, radiation use efficiency, ASI, flower/ear number, flower fertility, yield, senescence/stay green, enzyme activities etc. Some of the traits analyzed in some recent QTL mapping studies were given in the Table 3.

Table 3. Traits analyzed in some recent QTL mapping studies (Tuberosa & Salvi, 2006)

Species	Cross	Environment	Main trait
Arabidopsis	Landsberg x Cape Verde	Greenhouse	$^{12}C/^{13}C$, flowering time, stomatal conductance, transpiration efficiency
Cotton	*G. hirsutum x G. barbadense*	Field	$^{12}C/^{13}C$, osmotic potential, canopy temperature, dry matter, seed yield
Rice	CT9993 x IR62266	Field	Root morphology, plant height, grain yield
Maize	F2 x F252	Field	Silking date, grain yield, yield stability
Barley	Tadmor x Er/Apm	Field	$^{12}C/^{13}C$, osmotic adjustment, leaf relative water content, grain yield
Wheat	SQ1 x Chienese Spring	Field	Water use efficiency, grain yield

16.4 Transcriptome Analysis

Transcriptome analysis for identifying 'eQTLs' influencing the level of expression of a particular gene, allelic differences at both cis-regulatory regions and trans-acting regulatory factors. Caution: High costs, better suited to studies involving limited number of samples; *e.g.*, NILs, BDLs etc. and/or bulked samples from the tails of a mapping population

16.5 Searching for QTL Interactions through GMM

Genotype matrix mapping (GMM) is better than multiple regression or multifactor dimensionality reduction (MDR) approaches because it is capable of detecting epistatic QTLs in the absence of main effect QTLs or interactions among more than three loci. Comparison of many QTL interactions at once is vital for a comprehensive understanding of QTL expression affecting a complex trait (Isobe *et al.*, 2007)

17.0 CONCLUSION

The aim of generating stress tolerant crop plants has been tackled since many years using classical breeding and targeted gene transfer. However, the achieved progress is still small. This is due to the fact that the damage and tolerance or hardiness are multifactorial syndromes rather than result of a single reaction or gene. Nevertheless, tackling of the primary stress reactions by gene transfer can also alleviate the secondary strains and therefore generate plants with a higher stress tolerance than those which have been genetically tailored against a specific secondary strain, *e.g.* by the enhanced production of radical scavengers. Over-expression of a transcription factor or transcriptional activator of a multigene-sequence whose products directly or indirectly confer improved stress tolerance turned out as the most promising strategy to cope with the fact of a strain syndrome.

Drought, high temperature and salinity are some of the common adverse environmental factors that limit the productivity in crop plants. Hence, development of intrinsically tolerant crops is essential to combat the situations. In nature, genetic variability exists in cellular tolerance to different abiotic stresses, which can be exploited for crop improvement programs. In view of the shrinking land and water resources, production of higher crop yields per units of land, water, energy and time has become a key area of research. Biotechnology will play an increasingly important role in strengthening food, water and health security systems. Generation of crops that yield high even under abiotic stress conditions, especially drought stress will have positive implications on improving agricultural production. Increasing knowledge about the molecular and genetic basis of plant growth and development and the ability to use this understanding to develop new processes and products for agriculture, has resulted in the "Gene Revolution".

In essence, expression of stress proteins is an important adaptation toward heat-stress tolerance by plants. Of these, expression of low and high molecular weight HSPs, widely reported in a number of plant species, is the most important one. These proteins show organelle- and tissue-specific expression with deduced function like chaperones, folding and unfolding of cellular proteins and protection of functional sites from the adverse effects of high temperature. Among other stress proteins, expression of ubiquitin, Pir proteins, LEA and dehydrins has also been established under heat stress. A main function of these proteins appears to be protection of cellular and sub-cellular structures against oxidative damage and dehydrative forces. Research efforts in the following areas should be intensified: (1) transporter genes and signaling pathways affecting ion homeostasis under salt stress; (2) the biochemistry and regulation of metabolic pathways leading to the synthesis of compatible solutes and maintenance of metabolic activity under stress; and (3) the genes involved in protein protection and injury repair during extreme salt shock, drought, and cold stress.

18.0 REFERENCES

Allagulova, C.R., Gimalov, F.R., Shakirova, F.M. and Vakhitov, V.A. 2003. The plant dehydrins: Structure and putative functions. Biochemistry (Moscow), 68:945–951.

Bohnert, H.J., Gong, Q., Li, P. and Ma, S. 2006.Unraveling abiotic stress tolerance mechanisms—getting genomics going. Current Opinion in Plant Biology, 9 (2):180-8.

Burch-Smith, T.M., Anderson, J.C., Martin, G.B. and Dinesh-Kumar, S.P. 2004. Applications and advantages of virus-induced gene silencing for gene function studies in plants. Plant Journal, 39:734–746.

Campbell, S.A. and Close, T.J. 1997. Dehydrins: genes, proteins, and association with phenotypic traits. New Phytologist, 137:61–74.

Diatchenko, L., Lau, Y.F.C., Campbell, AP., Chenchik, A., Moqadam, F., Huang, B., Lukyanov, S., Lukyanov, K., Gurskaya, N., Sverdlov, E.D. and Siebert P.D. 1996. Suppression Subtractive Hybridization. Proceedings of the National Academy of Sciences, USA, 93:6025-6030.

Feussner, I. and Wasternack, C. 2002. The lipoxygenase pathway. Annual Review of Plant Biology, 53:275–297.

Gong, Z., Dong, C.H., Lee, H., Zhu, J., Xiong, L., Gong, D., Stevenson, B. and Zhu, J.K. 2005. A DEAD box RNA helicase is essential for mRNA export and important for development and stress responses in Arabidopsis. The Plant Cell, USA, 17:256–267.

Gong, Z., Lee, H., Xiong, L., Jagendorf, A., Stevenson, B. and Zhu, J.K. 2002. RNA helicase-like protein as an early regulator of transcription factors for plant chilling and freezing tolerance. Proceedings of the National Academy of Sciences, USA, 99:11507–11512.

Goyal, K., Walton, L.J. and Tunnacliffe, A. 2005. LEA proteins prevent protein aggregation due to water stress. Biochemistry Journal, 388:151–157.

Grover, A., Agarwal, M., Agarwal, S.K., Sahi, C. and Agarwal, S. 2000. Production of high temperature tolerant transgenic plants through manipulation of membrane lipids. Current Science, 79(5):557-558.

Hammond, S., Bernstein, E., Beach, D. and Hannon, G. 2000. An RNA-directed nuclease mediates post-transcriptional gene silencing in Drosophila cells. Nature, 404(6775):293–296.

Hara, E., Kato, T., Nakada, S., Sekiya, S. and Oda, K. 1991. Subtractive cDNA cloning using oligo(dT)-30-latex and PCR: Isolation of cDNA clones specific to undifferentiated human embryonal carcinoma cells, Nucleic Acids Research, 19(25):7097-7104.

Heckathorn, S.A., Downs, C.A., Sharkey, T.D. and Coleman, J.S. 1998. The small, methionine-rich chloroplast heat-shock protein protects photosystem II electron transport during heat stress. Plant Physiology, 116:439–444.

Hochholdinger, F., Sauer, M., Dembinsky, D., Hoecker, N., Muthreich, N., Saleem, M. and Liu, Y. 2006. Proteomic dissection of plant development. Proteomics, 6:4076–4083.

Howarth, C.J., Pollock, C.J and Peacock, J.M. 1997. Development of laboratory-based methods for assessing seedling thermotolerance in pearl millet. New Phytology, 137:129-139.

Iba, K. 2002. Acclimative response to temperature stress in higher plants: approaches of gene engineering for temperature tolerance. Annual Review of Plant Biology, 53:225–245.

Isobe, S., Nakaya, A. and Tabata, S. 2007. Genotype matrix mapping: searching for quantitative trait loci interactions in genetic variation in complex traits. DNA Research, 14(5):217-225.

Karaba, A., Dixit, S., Greco, R., Aharoni, A., Trijatmiko, K.R., Marsch-Martinez, N., Krishnan, A., Nataraja, K.N., Udayakumar, M. and Pereira, A. 2007. Improvement of water use efficiency in rice by expression of HARDY, an Arabidopsis drought and salt tolerance gene. Proceedings of the National Academy of Sciences, USA, 104(39):15270–15275.

Kavikishore, P.B., Sangam, S., Amrutha, R. N., Srilakshmi, P., Naidu, K. R., Rao, K.R.S.S., Rao, S., Reddy, K.J., Theriappan, P. and Sreenivasalu, N. 2005. Regulation of proline biosynthesis, degradation, uptake and transport in higher plants. Its implications in plant growth and abiotic stress tolerance. Current Science, 88(3):424-436.

Khanna-Chopra, R. and Sabarinath, S. 2004. Heat-stable chloroplastic Cu/Zn superoxide dismutase in *Chenopodium murale*. Biochem. Biophysics Research Communication, 320:1187–1192.

Klahre, U., Crété, P., Leuenberger, S.A., Iglesias, V.A. and Meins, F. Jr. 2002. High molecular weight RNAs and small interfering RNAs induce systemic posttranscriptional gene silencing in plants. Proceedings of the National Academy of Sciences, USA, 99:11981–11986.

Klueva, N.Y., Maestri, E., Marmiroli, N. and Nguyen, H.T. 2001. Mechanisms of thermotolerance in crops. *In*: A.S. Basra (*Ed.*) Crop Responses and Adaptations to Temperature Stress, Food Products Press, Binghamton, NY, pp. 177–217.

Lander, E. and Kruglyak, L. (1995). Genetic dissection of complex traits: guidelines for interpreting and reporting linkage results. Nature Genetics, 11:241–247.

Larkindale, J., Hall, J.D., Knight, M.R. and Verling, E. 2005. Heat stress phenotypes of Arabidopsis mutants implicate multiple signaling pathways in the acquisition of thermotolerance. Plant Physiology, 138:882–897.

Li, D., Liu, H., Zhang, H., Wang, X. and Song, F. 2008. OsBIRH1, a DEAD-box RNA helicase with functions in modulating defence responses against pathogen infection and oxidative stress Journal of Experimental Botany, 59(8):2133-2146.

Li, H.Y., Chang, C.S., Lu, L. Su., Liu, C.A., Chan, M.T. and Chang, Y.Y. 2003. Over-expression of *Arabidopsis thaliana* heat shock factor gene (*AtHsfA1b*) enhances chilling tolerance in transgenic tomato. Botanical Bulletin of Academia Sinica, 44:129-140.

Lindbo, J.A. and Dougherty, W.G. 1992. Untranslatable transcripts of the tobacco etch virus coat protein gene sequence can interfere with tobacco etch virus replication in transgenic plants and protoplasts. Virology, 189(2):725–733.

Liu, H.H., Liu, J., Fan, S.L., Song, M.Z., Han, X.L., Liu, F. and Shen, F.F. 2008. Molecular cloning and characterization of a salinity stress-induced gene encoding DEAD-box helicase from the halophyte *Apocynum venetum*. Journal of Experimental Botany, 59(3):633-644.

Liu, Y., Schiff, M. and Dinesh-Kumar, S.P. 2002. Virus-induced gene silencing in tomato. Plant Journal, 31: 777-786.

Majoul, T., Bancel, E., Triboi, E., Hamida, J.B. and Branlard, G. 2003. Proteomic analysis of the effect of heat stress on hexaploid wheat grain: characterization of heat-responsive proteins from total endosperm. Proteomics, 3:175–183.

Miroshnichenko, S., Tripp, J., Nieden, U., Neumann, D., Conrad, U. and Manteuffel, R. 2005. Immunomodulation of function of small heat shock proteins prevents their assembly into heat stress granules and results in cell death at sublethal temperatures. Plant Journal, 41: 269–281.

Moriarty, T., West, R., Small, G., Rao, D., Ristic, Z. 2002. Heterologous expression of maize chloroplast protein synthesis elongation factor (EF-Tu) enhances *Escherichia coli* viability under heat stress. Plant Science, 163:1075–1082.

Neta-Sharir, I., Isaacson, T., Lurie, S. and Weiss, D. 2005. Dual Role for Tomato Heat Shock Protein 21: protecting photosystem II from oxidative stress and promoting color changes during fruit maturation. The Plant Cell, 17:1829-1838.

Nordborg, M., Borevitz, J.O., Bergelson, J., Berry, C.C., Chory, J., Hagenblad, J., Kreitman, M., Maloof, J.N., Noyes, T. , Oefner, P.J., Stahl, E.A. and Weigel, D. 2002. The extent of linkage disequilibrium in *Arabidopsis thaliana*. Nature Genetics, 30:190–193.

Panchuk, I.I., Volkov, R.A. and Schoffl, F. 2002. Heat stress- and heat shock transcription factor-dependent expression and activity of ascorbate peroxidase in Arabidopsis. Plant Physiology, 129:838–853.

Prasad, T.G. 2008. Influence of Moisture stress on Growth and Productivity. Training on *"Physiological and molecular basis of plant adaptation to drought"*. At Department of Crop Physiology, UAS, GKVK, Bangalore from 14th July to 03rd August 2008.

Prasanna, B.M. 2008. QTL Analysis and MAS for Improving Drought Tolerance in Crop Plants. IARI. New Delhi, Training on *"Physiological and molecular basis of plant adaptation to drought"*. At Department of Crop Physiology, UAS, GKVK, Bangalore from 14th July to 03rd August 2008.

Prasinos, C., Krampis, K., Samakovli, D. and Hatzopoulos, P. 2005. Tight regulation of expression of two *Arabidopsis* cytosolic *Hsp90* genes during embryo development. Journal of Experimental Botany, 56(412): 633-644.

Queitsch, C., Hong, S.W., Vierling, E. and Lindquist, S. 2000. Heat stress protein 101 plays a crucial role in thermotolerance in *Arabidopsis*. Plant Cell, 12: 479–492.

Raju, M. 2008. Temperature Induced Response. Training on *"Physiological and molecular basis of plant adaptation to drought"*. At Department of Crop Physiology, UAS, GKVK, Bangalore from 14th July to 03rd August 2008.

Reich, D.E. and Goldstein, D.B. 2001. Detecting associations in case-control studies while correcting for population stratification. Genetic Epidemiology, 20:4-16.

Ribaut, J.M. and M. Ragot. 2007. Marker-assisted selection to improve drought adaptation in maize: The backcross approach, perspectives, limitations, and alternatives. Journal of Experimental Botany, 58: 351–360.

Ribaut, J.M., Bänziger, M., Setter, T., Edmeades, G. and Hoisington, D. 2004. Genetic dissection of drought tolerance in maize: a case study. *In* Nguyen H and Blum A (Eds.). Physiology and biotechnology integration for plant breedingNew York Marcel Dekker Inc. pp. 571–611.

Ristic, Z. and Cass, D.D. 1992. Chloroplast structure after water and high-temperature stress in two lines of maize that differ in endogenous levels of abscisic acid. International Journal of Plant Science, 153:186–196.

Rohini, V.K. and Sankara Rao, K. 2002. In planta Strategy for Gene Transfer into Plants: Embryo Transformation. Physiology and Molecular Biology of Plants, 8(2):161-169.

Sairam, R.K. and Tyagi A. 2004. Physiology and molecular biology of salinity stress tolerance in plants. Current Science, 86:407–421.

Sanan-Mishra, N., Pham, X.H., Sopory, S.K. and Tuteja, N. 2005. Pea DNA helicase 45 overexpression in tobacco confers high salinity tolerance without affecting yield. Proceedings of the National Academy of Sciences USA, 102:509-514.

Schöffl, F., Prandl, R. and Reindl, A. 1999. Molecular responses to heat stress. In: K. Shinozaki K, Yamaguchi-Shinozaki K *Editors, Molecular Responses to Cold, Drought,Heat and Salt Stress in Higher Plants*, R.G. Landes Co., Austin, Texas pp. 81–98.

Senthil, K.M. and Udayakumar, M. 2006. High-throughput virus-induced gene-silencing approach to assess the functional relevance of a moisture stress-induced cDNA homologous to *lea4*. Journal of Experimental Botany, 57(10): 2291-2302.

Senthil, K.M., Rame Gowda, H.V., Hema, R. and Udayakumar, M. 2008.Virus-induced gene silencing and its application in characterizing genes involved in water-deficit-stress tolerance. Journal of Plant Physiology, 165(13):1404-21.

Senthil, K.M., Kumar, G., Srikanthbabu, V. and Udayakumar, M. 2007. Assessment of variability in acquired thermotolerance: Potential option to study genotypic response and the relevance of stress genes. Journal of Plant Physiology, 164 (2):111-125.

Sharp, R.E., Poroyko, V., Hejlek, L.G., Spollen, W.G., Springer, G.K., Bohnert, H.J. and Nguyen H.T. 2004. Root growth maintenance during water deficits: physiology to functional genomics. Journal of Experimental Botany, 55:2343-2351.

Sheshshayee, M.S., Bindumadhava, H., Ramesh, R., Prasad, T.G., Lakshminarayana, M.R. and Udayakumar, M. 2005. Oxygen isotope enrichment ({Delta}^{18}O) as a measure of time-averaged transpiration rate. Journal of Experimental Botany, 56: 3033-3039.

Southern, E.M. 1975. Detection of specific sequences among DNA fragments separated by gel electrophoresis. Journal of Molecular Biology, 98(3):503-517.

Tondelli, A., Francia, E., Barabaschi, D., Aprile, A., Skinner, J.S., Stockinger, E.J., Stanca, A.M. and Pecchioni, N. 2006. Mapping regulatory genes as candidates for cold and drought stress tolerance in barley. Theoretical and Applied Genetics, 112(3):445-454.

Tuberosa, R. and Salvi, S. 2006. Genomic based approaches to improve drought tolerance of crops. Trends in Plant Science, 11:405-412.

Tuteja, N. 2008. "Novel role of plant helicases in abiotic stress tolerance. International Centre for Genetic Engineering and Biotechnology, New Delhi, Training on *"Physiological and molecular basis of plant adaptation to drought"*. At Department of Crop Physiology, UAS, GKVK, Bangalore from 14th July to 03rd August 2008.

UAS. 2008. Reading material of Training on *"Physiological and molecular basis of plant adaptation to drought"*. At Department of Crop Physiology, UAS, GKVK, Bangalore from 14th July to 03rd August 2008.

Udaya Kumar, M. 2008. CO_2 enrichment mechanism as a drought adaptive strategy. Training on *"Physiological and molecular basis of plant adaptation to drought"*. At Department of Crop Physiology, UAS, GKVK, Bangalore from 14th July to 03rd August 2008.

Vierling, E. 1991. The roles of heat shock proteins in plants. Annual Review of Plant Physiology and Plant Molecular Biology, 42:579-620.

Vinocur, B. and Altman, A. 2005. Recent advances in engineering plant tolerance to abiotic stress: achievements and limitations. Current Opinion in Biotechnology, 16:123–132.

Visioli, G., Maestri, E. and Marmiroli, N. 1997. Differential display-mediated isolation of a genomic sequence for a putative mitochondrial LMW HSP specifically expressed in condition of induced thermotolerance in *Arabidopsis thaliana* (L) Heynh. Plant Molecular Biology, 34:517–527.

Wahid, A. and Close, TJ. 2007. Expression of dehydrins under heat stress and their relationship with water relations of sugarcane leaves. Biologica Plantarum, 51:104–109.

Wang, J.P.Z., Lindsay, B.G., Leebens-Mack, J., Cui, L., Wall, K., Miller, W.C. and dePamphilis, C.W. 2004. EST clustering error evaluation and correction. Bioinformatics, 20(17):2973-2984.

Yamaguchi-Shinozaki, K. and Shinozaki, K. 2005. Organisation of cis acting regulatory elements in osmotic and cold stress responsive promoters. Trends in Plant Science, 10:88-94.

Young, T.E., Ling, J., Geisler-Lee, C.J., Tanguay, R.L., Caldwell, C. and Gallie, D.R. 2001. Developmental and thermal regulation of the maize heat shock protein, HSP101. Plant Physiology 127:777-791.

Zhang, W., Ruan, J., Ho, T.H., You, Y., Yu, T. and Quatrano, R.S. 2005. Cis-regulatory element based targeted gene finding: genome-wide identification of abscisic acid- and abiotic stress-responsive genes in *Arabidopsis thaliana*. Bioinformatics, 21:3074-3081.

Molecular Plant Breeding: Principle, Method and Application
Eds : R.K. Singh, Rajesh Singh, Guoyou Ye, A. Selvi and G.P. Rao
Studium Press LLC, Texas, USA, 2009, pp. 409-419

Molecular Markers in Biosystematics

TARIQ HUSAIN, PRIYANKA AGNIHOTRI, PUSHPI SINGH, J.N. SINGH, SUSHIL KUMAR and AMISH TIWARI

ABSTRACT

Molecular markers have proven to be powerful tools for discerning biosystematic, biogeogrophic, and phylogenetic relationships. Biosystematic information can be important for guiding traditional breeding programmes, gene transfer, interspecific hybridization, and gene conservation for their further use in crop improvement. In this chapter, phylogenetic analysis and other experiments related with biosystematics with the help of molecular markers like RAPD, RFLP, SSR, AFLP, and especially with the DNA barcodes have been elaborated. Discussion about the use and scope of organelle DNA and rDNA in biosystematics and the various factors, which affect DNA technology, has also been provided. Singnificance of molecular markers which provide useful insights and interpretation on the disjunctive distribution of closely related taxa in different parts of the world is also explained.

Key Words: Molecular markers, biosystematics, matK, RFLP, RAPD, AFLP, DNA Barcode.

Angiosperm Taxonomy & Herbarium Division, National Botanical Research Institute, Rana Pratap Marg, Lucknow 226001, India e-mail: hustar_2000@yahoo.co.uk

1.0 INTRODUCTION

Biosystematics is the statistical analysis of data obtained from genetic, biochemical, and other observational studies to assess the taxonomic relationships of organisms or populations, especially within an evolutionary framework. Molecular markers are widely used in biosystematics, molecular biology and biotechnological studies to identify a particular sequence of DNA. As the DNA sequences are highly specific, they can be identified with the help of the known molecular markers which can find out a particular sequence of DNA from a group of unknown (Soltis *et al.*, 1999).

Molecular markers are any kind of molecules indicating the existence of a chemical or physical process. It can also be a molecule used as fiscal marker to distinguish lower-taxed items from higher-taxed ones when they otherwise can be substituted. Molecular markers include biochemical constituents and macromolecules. With the advent of molecular markers, a new generation of markers has been introduced over the last three decades, which has revolutionized the scenario of biological sciences. DNA based molecular markers have acted as versatile tools and have found their own position in various fields like taxonomy, physiology, embryology, genetic engineering etc. Ever since their development, they are constantly being modified to enhance their utility.

The method of molecular biology is increasingly being employed in systematic and evolutionary research to study genetic relationships and phylogeny. There are a variety of choices in initiating research in 'molecular biosystematics' first, a gene or genome that provides a level of genetic resolution appropriate for the materials under study must be selected. Common choices in plants include the chloroplast genome (cpDNA) or components of the chloroplast genome, the nuclear ribosomal RNA genes (rDNA), or nuclear-encoded, single-copy genes. A second consideration is that several methods can be employed to provide direct or indirect measures of DNA sequence divergence. One widely used method determines genetic divergence based on restriction site changes. A restriction site analysis typically requires some knowledge of the physical map of the DNA molecule under study. Empirical studies indicate that restriction site analysis of cpDNA provide good resolution for systematic investigations at or below the family level (Chase *et al.*, 2007).

Biochemical and molecular markers have proven to be powerful tools for discerning biosystematic, biogeographic, and phylogenetic relationships. Biosystematic information can be important for guiding traditional breeding programs, gene transfer, interspecific hybridization, and gene conservation. A phylogenetic framework is usually necessary, but frequently ignored, for making valid statistical tests in studies of adaptive evolution. Several studies have indicated a strong correlation between biochemical "races" and traits important to growth and adaptation, suggesting that evolutionary legacy may affect genetic architecture of fitness traits with consequences for seed transfer, breeding strategies, and tolerance of climate change. A number of methods for phylogenetic analysis exist, but differ in their assumptions. Use of an inappropriate method, such as, a method that assumes constant rates of evolution when rates vary, can lead to incorrect phylogenies. Because of their complexity, phylogenetic topologies are often difficult to determine unambiguously; estimates of statistical confidence should therefore accompany phylogenetic trees if they are to be regarded as providing new knowledge, or strong confirmation, of relationships. Molecular genetic markers are more expensive than biochemical markers such as allozymes and terpenes, but they provide increased accuracy and expanded scope of biosystematic inference, and facilitate statistical analysis of phylogenetic trees.

Molecular markers play on important role in Biosystematics as they not only quantify breeding systems and assess levels and patterns of gene flow but also help in studying hybridization and establishing phylogenetic relationships. Implications of these molecular markers can be seen in testing biological hypotheses, finding markers to characterize individuals and comparing morphological and molecular evoluation.

2.0 DNA TECHNOLOGIES

The data generated for the study of biosystematics may be qualitative or genetic in nature. The technology being used is depended upon many factors, which are as follows: (i) The extent of variation expected include the level of polymorphism, which depends on the type of marker being studied. Some may be highly polymorphic (*e.g.* microsatellites), while others may exhibit low polymorphism (*e.g.* expressed loci such as some RFLPs). (ii) Techniques that study highly mutable loci may not be ideal for phylogenetic studies. (iii) The level of skill required differ with the technique, RAPDs are least demanding whereas STMS and DNA sequencing require high technical skills. (iv) Some techniques such as STMS and RFLP analyses require higher financial input for establishment of basic infrastructure and analyses. (v) The PCR based techniques have high sample throughput whereas RFLP is more time consuming. (vi) The markers may be co-dominant as in case of RFLPs and STMS or dominant in nature as in case of RAPDs and AFLPs. (vii) The technique is also affected because of homology of characters due to common ancestry.

2.1 Organelle DNA

Organelle DNA differ from nuclear DNA in inheritance and tempo of evolution leading to different expectations of population genetic structure.

It is also observed that presence of adequate levels of selectively neutral genetic variation in organelle DNA make it suitable for phylogenetic studies. Molecular markers are designed and developed from chloroplast DNA (CpDNA) and mitochondrial DNA (mt DNA). More than 15 loci and 10 loci have been identified and primers are designed from CpDNA and mt DNA regions based on PCR-sequencing, respectively. Detail of three genome are given in the Table 1.

Table 1. Properties of three plant genomes

Attribute	Nuclear genome	Cp genome	Mt genome
Structure	Linear	Circular	Circular, internally recombining
Ploidy level	2n in higher plants	Haploid	Haploid
Inheritance	Biparental	Uniparental Maternal (Angiosperms) Paternal (Gymnosperms)	Uniparental Maternal
Recombination	Present	Absent *rbcL, atpB*	Absent

2.2 rDNA Regions

Genes coding for ribosomal genes (5.8S, 18S and 26S) are highly conserved and are interspersed with highly variable regions (*ITS*1 and *ITS*2). They code for ribosomal units. They are organized in the secondary constriction region of chromosomes. They are phylogenetically most informative. rDNA sequences exhibit a broad range of phylogenetic signal – differential levels of homogenisation. The primers developed from different regions of rDNA plays major role in biosystematics. The primer developed from 18S are relatively slow evolving and shows inter-familial relationship among the plant system. The *ITS*1 and 2 are rapidly evolving and namely used for studying the interspecific and intergeneric relationships, speciation and biogeography of plant systems. The marker developed from 26S showes moderate rate of evolution and bridge the gap between 18S and *ITS* while *ETS* shows faster rate of evolution and toxonomic levels. The markers devloped from 5S region shows higher homoplasy and is not useful for biosystematic.

2.3. Genomic DNA Based Markers

Various molecular markers are like AFLP, RAPD, RFLP, *trnL-F*, *ITS* along with DNA barcodes such as *matk* and Cp-DNA-RFLP etc being used

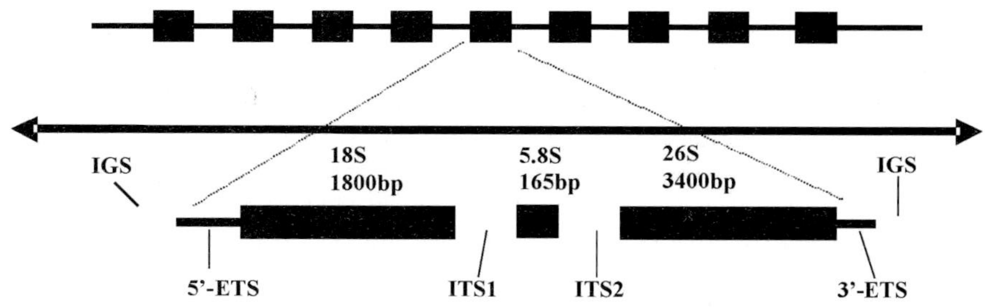

Fig. 1. rDNA in Biosystematics (Source : Soltis *et al.*, 1999)

in biosystematics. Many scientists of the world have proposed different plant barcodes like rbcL (Kress, *et al.*, 2005), *matk*, *rpoC1*& *rpoB* (Chase, *et al.*, 2007), and *matk*, *atpF/H* (Pennisi, 2007).

3.0 CASE STUDIED

Molecular markers are very useful in understanding the complex geographic distribution of plants as described in the study by Noyes (2000). He proposed that complex geographical distribution patterns and agamospermy have united to obscure systematic relationships in the genus *Erigeron* (400+ species). He analyzed the sequence data from the internal transcribed spacer region of nrDNA (*ITS*) and 5.8S cistron for 77 taxa including 63 *Erigeron* species. Results support the North American origin of *Erigeron* and document that *Aphanostephus, Conyza,* and three small genera restricted to South America (*Apopyros, Hysterionica, Neja*) are all derived from *Erigeron*. Phylogenetic data indicate that agamospermy has arisen at least three times in *Erigeron* and that autogamy has evolved independently in *Conyza* and in *E.* sect. *Trimorpha.*

In one study by Kardolus *et al.* (1998) on *S. microdontum* and Ser. Megistacroloba, the classifications based on the molecular markers was generally in agreement with current taxonomic opinions. Unexpectedly, *S. microdontum* was associated with Ser. Megistacroloba rather than with Ser. Tuberosa, and *S. demissum* (Ser. Demissa) and species of Ser. Acaulia appeared closely affiliated. AFLP is an efficient and reliable technique to generate biosystematic data and therefore a promising tool for evolutionary studies. In same way, molecular analysis has clarified many relationships within Poaceae, Solanaceae, Fabaceae, Brassicaceae etc.

Manoko *et al.* (2007) studied the relationships between the African material of *Solanum americanum* (also designated as *S. nodiflorum*), accessions of this taxon from other geographical areas, and American *S. americanum* using AFLP markers. 96 individuals representing 39 accessions of *S. americanum sensu lato* and related diploid species from the widest possible geographical range, and one accession of *S. dulcamara* (as outgroup) were used. The AFLP results suggested that American *S. americanum* differs from *S. nodiflorum* and that the material investigated in this study can be assigned to three different species: *S. americanum sensu stricto*, *S. nodiflorum* and a *Solanum* species from Brazil. These species can be differentiated based on a combination of floral and fruit characteristics.

The study of Lambertini *et al.* (2006) concluded that within the genus *Phragmites* (Poaceae), the species *P. australis* (the common weed) is virtually cosmopolitan, and shows considerable variation in ploidy level and morphology. Genetic variation in *Phragmites* was studied using AFLPs, and analysed with parsimony and distance methods. *P. australis*, supported in the analysis, was a group of South American clones and its was totally distinct from the groups of US Gulf coast and E. Asian and Australian octoploids. Among the other species, the paleotropical *P. vallatoria* is supported as monophyletic and most closely related to the paraphyletic *P. mauritianus* and to the Gulf Coast and S. American groups. The E. Asian species *P. japonicus* is closely related to a group of *P. australis* clones mostly from central North America. Tetraploidy predominates in the genus, and optimisation of chromosome numbers on to the phylogeny shows that higher ploidy levels have evolved many times.

Bakker *et al.* (1999) carried out a *trnL-F* based phylogeny for species of *Pelargonium* (Geraniaceae) with small chromosomes. This study included phylogenetic analysis, which was performed on 921 positions of *trnL* (UAA) 5' exon - *trnF* (GAA) exon chloroplast DNA regions from 68 representatives of *Pelargonium* sectt. *Campylia, Cortusina, Glaucophyllum, Hoarea, Isopetalum, Ligularia, Otidia, Pelargonium, Peristera, Polyactium,* and *Reniformia,* together with five putative outgroup species from sections *Ciconium, Chorisma* and *Jenkinsonia.* The total data set therefore comprised 67.2 kb of DNA sequence. Two main ingroup clades were identified: one clade contains sections *Peristera, Reniformia,* and *Isopetalum,* the other contains sections *Campylia, Cortusina, Glaucophyllum, Hoarea, Ligularia, Otidia, Pelargonium, Polyactium* and two species currently grouped in sect. *Peristera.* Branching order among five main clades within the latter clade was not resolved. The *trnL-F* sequence data support monophyly only for sections *Reniformia* and *Hoarea,* the remainder of the currently recognized sections of *Pelargonium* being either paraphyletic or polyphyletic. The data further suggest that sect. *Polyactium* is diphyletic and that sect. *Glaucophyllum* is nested within sect. *Pelargonium.* One relatively derived clade, which represents half of the genus, contains predominantly geophytic and succulent species, occurring in the geographically restricted winter rainfall region of the South African Cape. This pattern is interpreted as reflecting explosive radiation, possibly as an adaptive response to recent aridification in the western Cape.

Banfer *et al.* (2004) carried out AFLP analysis of phylogenetic relationships among myrmecophytic species of Macaranga (Euphorbiaceae) and their allies. Their study concluded that the palaeotropic pioneer tree of *Macaranga thouars* (Euphorbiaceae) is characterized by various types of mutualistic interactions with specific ant partners (mainly *Crematogaster* spp.). About 30 species are obligate ant-plants (myrmecophytes). They used amplified fragment length polymorphism (AFLP) markers to assess phylogenetic relationships among 108 *Macaranga* specimens from 43 species, including all available taxa from the three sections known to contain myrmecophytes. Eight primer combinations produced 426 bands that were scored as presence/absence characters. Banding patterns were analyzed phenetically, cladistically and by principal coordinate analysis. Monophyly of section Pruinosae is clearly supported. There is also good evidence for a monophyletic section Pachystemon that includes the puncticulata group. The monophyly of section Winklerianae and relationships between the three sections remain ambiguous. Section Pachystemon is subdivided into four well-supported monophyletic sub-clades that presumably correspond to taxonomic entities.

RAPD markers also, are very helpful in analyzing the phylogenetic relationships. In a study by Poczai *et al.* (2008), phylogenetic relationships and genetic variation were examined in the genus *Solanum* based on the random amplified polymorphic DNA (RAPD) technique. Genetic distances were estimated for 42 accessions from five subgenera (*Archaesolanum, Minon* (Syn. *Brevantherum*), *Leptostemonum*, *Potatoe*, and *Solanum*). This investigation provided new information and reinforced some suggestions from previous phylogenetic studies. Analysis with random markers from the total genome clearly separated *Solanum* sect. *Dulcamara*, from the other members of *Solanum* subg. *Potatoe*, and indicated that among the analysed *Solanum* subgenera subg. *Solanum* is most closely related to it. The results suggest that *Solanum* sect. *Dulcamara* should be excluded from *Solanum* subg. *Potatoe.* The subclusters formed by *S. rostratum* and *S. citrullifolium* appear to be distinct from the subcluster formed by the two accessions of *S. sisymbriifolium.* This topology indicates that *Solanum* sect. *Androceras* and *Solanum* sect. *Cryptocarpum* are fairly closely related, although the data suggest that the two sections should not be maintained.

Meimberg *et al.* (2008) tested relationship between carnivorous Droseraceae, Nepenthaceae and Dioncophyllaceae and ten other families of the Caryophyllidae using the comparative sequencing of Chl. *matk* gene and compared it with the previously published cladograms based on *rbcL, 18S* rDNA and *ORF2280* sequences and they found that carnivory within the order Caryophylladae has a monophylatic origin and this carnivorous character was lost in the taxa of the family Dionophyllaceae and Ancistrocladaceae with the exception of the genus *Triphypophyllum.* They excluded the genus *Drosophyllum* from the family Droseraceae on the basis of the presence of intron *rpl2* in chloroplast genome. The separation of the genus *Drosophyllum* from the family Droseraceae is also supported by DNA analysis. On the basis of *rbcL* sequences, Schlauer (1996) grouped Droseraceae, Drosophyllaceae, Dioncophyllaceae, Ancistrocladaceae, Nepanthaceae, Plumbaginaceae, Polygonaceae, Simmondsiaceae together in one clade and he kept Tamaricaceae/ Frankeniaceae in the closest sister clade. Studies were striking as this clade unites families from different orders and it brings several carnivorous families into close proximity.

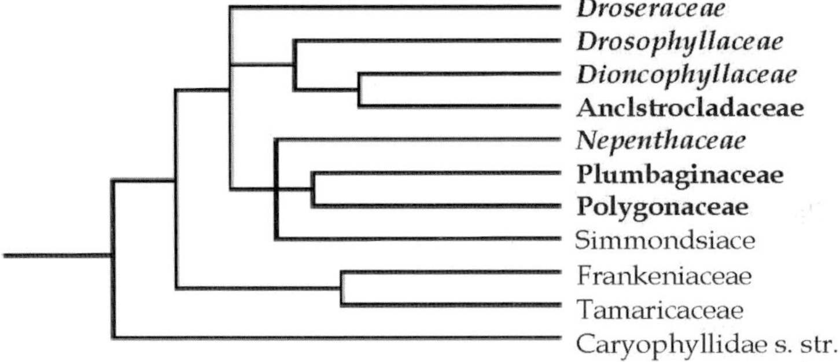

Fig. 2. Phylogenetic reconstruction by *rbcL* sequence comparison (Source : Memberg *et al.*, 2008)

DNA barcoding is a new term, which is also related with biosystematics. DNA barcoding is the use of a short DNA sequence or sequences from a standardized locus (or loci) as a species identification tool. The barcode, which can be used to identify plants using a small sample, could lead to new ways of easily cataloguing different types of plants in species-rich areas like rainforests. It could also lead to accurate methods for identifying plant ingredients in powdered substances, such as in traditional Chinese medicines, and could help to monitor and prevent the illegal transportation of endangered plant species. The team behind the discovery found that DNA sequence of the gene "*matk*" differs among plant species, but are nearly identical in plants of the same species. This means that the *matk* gene can provide scientists with an easy way of distinguishing between different plants, even closely related species that may look the same to the human eye. The researchers made this discovery by analyzing the DNA from different plant species. They found that when one plant species was closely related to another, differences were usually detected in the *matk* DNA. Lahaye *et al.* (2008) used this *matk* gene to identify 1600 species of orchids collected from Costa Rica. In the course of this work, they discovered that what was previously assumed to be one species of orchid was actually two distinct species that live on different slopes of the mountains and have differently shaped flowers adapted for different pollinating insects. In South Africa, the team was able to use the *matk* gene to identify the trees and shrubs of the Kruger National Park, also well known for its big game animals. They explained that in the long run the aim is to build on the genetic information his team gathered from Costa Rica and South Africa to create a genetic database of the *matk* DNA of as many plant species as possible, so that samples can be compared to this database and different species accurately identified.

The internal transcribed spacer (*ITS*) region has been proven a milestone in phylogenetic studies. According to Bottini *et al.* (2007), the phylogenetic relationships between 13 Patagonian species of the genus *Berberis* (Berberidaceae) can be determined using the sequence analysis of the *ITS*. He found that the divergence values between the pair wise sequences in the studied Patagonian species were in the range 2.9-22.9 %. The lengths of the *ITS* 1 and *ITS*2 sequences were in the range of 227-231 bp and 220-224 bp, respectively and the 5.8S sequence was 159 bp throughout all species. Thus, he concluded that *ITS* sequences along with data obtained from morphological, biochemical, AFLP and cytological characterizations, support the existence of diploid and polyploid hybrid speciation in the genus.

Many countries have undertaken DNA barcoding on plant species in a big way. In India, Department of Biotechnology, New Delhi has initiated several projects on DNA barcoding. They are as follows: (i) National Botanical Research Institute (NBRI), Lucknow- DNA barcoding in the Indian taxa of *Berberis* L. (Berberidaceae) (ii) UAS, Bangalore - DNA barcoding of Rattans and *Phyllanthus* developing fingerprints and securing IPRs, (iii) TERI, New Delhi - DNA barcoding in *Bambusa* spp. (Bamboos), (iv) DU, Delhi - DNA barcoding f selected species of *Dendrobium,* an orchid for checking the applicability of the concept to plants, (v) NCPGR, New Delhi - DNA barcoding of *Dalbergia* species, (vi) RGCB, Thiruvananthanpuram - DNA barcoding of the selected genera of the family Zingiberaceae: *Alpinia, Zingiber* and *Globba,* (vii) DU, New Delhi - Development of DNA barcode for amphibian fauna of the Western Ghats, (viii) WII, Dehradun - barcoding anurans of India, (ix) NCCS, Pune - DNA barcoding of butterflies from Western Ghat, (x) IISc, Bangalore - Identification of satyrine butterflies of peninsular India through DNA barcodes.

As mentioned above, NBRI, Lucknow, is involved in developing DNA barcode system for 15 species of Indian *Berberis* L. (Berberidaceae) through screening and evaluation of nuclear (*ITS* region of nr DNA) and plastid (Cp DNA) genomic regions. They are also involved in construction of a DNA-based tree for identification of species/species complexes for detailed taxonomic and phylogenetic studies. The strategy involves the investigation of the internal transcribed spacer region (*ITS*) of the nr DNA and the plastid intergenic spacers – trnH- psbA, followed by *rpoC1*, *rpoB*, accD, YCF5 and *matk* and rbcL in *Berberis* taxa. The outcome of the project will be in form of getting the validation of reported loci for barcoding in *Berberis*, DNA-based identification system for Indian taxa of *Berberis*, a digitalized geo-referenced database of DNA barcode and specimen record of Indian *Berberis* and novel variable loci with potential for phylogenetic analysis. Few case studies involving the Cp DNA-RFLP analysis shows that the *ITS* sequence analysis of 79 taxa of *Berberis sensu lato* (including 29 species of *Mahonia*) reveals low rate of sequence divergence at species level, but provides useful insights and interpretation on the disjunctive distribution of closely related taxa in Asia, Africa and South America (Kim and Jansen, 1996; 1998).

4.0 CONCLUSION

These studies including the techniques such as AFLP, RAPD and RFLP analysis of the natural population of species, along with data obtained from morphological, biochemical and cytological characterizations will serve as a major application in biosystematics. Using molecular markers, a genetic database of many plant species can be created and can be accurately identified by comparing with the existing databases. Molecular markers also help to assess and interpret the distribution and occurrence of various plant species of the world and their resemblance with other plants. The information obtained from biosystematics using molecular techniques can serve as a major tool in guiding traditional breeding programs, gene transfer, interspecific hybridization and gene conservation.

5.0 REFERENCES

Bakker, F.T., Culham, A., Daugherty, L.C. and Gibby, M. 1999. A *trnL-F* based phylogeny for species of *Pelargonium* (Geraniaceae) with small chromosomes. Plant Systematics and Evolution, 216:309-324.

Banfer, G., Fiala, B. and Weising, K. 2004. AFLP analysis of phylogenetic relationships among myrmecophytic species of *Macaranga* (Euphorbiaceae) and their allies. Plant Systematics and Evolution, 249:213-231.

Bottini, M.C.J., Bustos, A. De., Sanso, A.M., Jouve, N. and Poggio, L. 2007. Relationships in Patagonian species of *Berberis* (Berberidaceae) based on the characterization of rDNA internal transcribed spacer sequences. Botanical Journal of Linnaean Society, 153:321-328.

Chase, M.W., Cowan, R.S., Hollinsworth, P.M., van der Berg, C., Madrinan, S., Peterson, G., Seberg, O., Jorgsensen, T., Cameron, K.M. and Carine, M., 2007. A proposal for a standardized protocol to barcode all land plants. Taxon, 56: 295-299.

Kardolus, J.P., van Eck. H.J. and van den Berg, R.G. 1998. The potential of AFLPs in biosystematics: A first application in *Solanum* taxonomy *(Solanaceae).* Plant Systematics and Evolution, 210: 87-103.

Pennisi, E. 2007 Taxonomy- Wanted: A barcode for plants. Science, 318:190-191.

Kim, Y.-D. and Jansen, R.K., 1996. Phylogenetic implications of *rbcL* and ITS sequence variation in the Berberidaceae. Systematic Botany, 21:381–396.

Kim, Y.-D. and Jansen, R.K. 1998. Chloroplast DNA restriction site variation and phylogeny of the Berberidaceae. *Amer. J. Bot.* 85:1766–1778.

Kress, W.J., Wurdack, K.J., Zimmer, E.A., Weigt, L.A. and Janzen, D.H. 2005. Use of DNA barcodes to identify flowering plants. Proceedings of the National Academy of Sciences, USA, 102:8369-8374.

Lahaye, R., van der Bank, M., Bogarin, D., Warner, J., Pupulin, F., Gigot, G., Maurin, O., Duthoit, S., Barraclough, T.G. and Savolainen, V. 2008. DNA barcoding the floras of biodiversity hotspots. Proceedings of the National Academy of Sciences, USA, 105:2923-2928.

Lambertini, C., Gustafsson, M.H.G., Frydenberg, J., Lissner, J., Speranza, M. and Brix, H. 2006. A phylogeographic study of the cosmopolitan genus *Phragmites* (Poaceae) based on AFLPs. Plant Systematics and Evolution 258:161-182.

Manoko, M.L.K., van den Berg, R.G., Feron, R.M.C., van der Weerden, G.M. and Mariani, C. 2007. AFLP markers support separation of *Solanum nodiflorum* from *Solanum americanum* sensu stricto (Solanaceae). Plant Systematics and Evolution 267(1-4):1-11.

Meimberg, H., Dittrich, P., Bringmann, G., Schlauer, J. and Heubl, G. 2008. Molecular phylogeny of caryophyllidae s.l. based on *MatK* sequences with special emphasis on carnivorous Taxa. Plant Biology 2:218-228.

Noyes, R.D. 2000. Biogeographical and evolutionary insights on *Erigeron* and allies (Asteraceae) from *ITS* sequence data. Plant Systematics and Evolution, 220: 93-114.

Poczai, P., Taller, J. and Szabó, I. 2008. Analysis of phylogenetic relationships in the genus *Solanum* (Solanaceae) as revealed by RAPD markers. Plant Systematics and Evolution, 275:59-67.

Schlauer, 1996. A dicotomous key to genus *Drosera* L. (Droseraceae). Carnivorus Plant Newsletter, 25:67-68.

Soltis, D.E., Soltis, P.S. and Doyle, J.J. 1999. Molecular systematics of Plant II. DNA sequencing, vol. 2. Springer.

Molecular Plant Breeding: Principle, Method and Application
Eds : R.K. Singh, Rajesh Singh, Guoyou Ye, A. Selvi and G.P. Rao
Studium Press LLC, Texas, USA, 2009, pp. 421-444

CHAPTER 16

Formulating Marker-Assisted Selection Strategies Using Computer Simulation

GUOYOU YE[*1] *and MAARTEN VAN GINKEL*[2]

ABSTRACT

With the rapid development of molecular technology and statistical tools for QTL mapping and association mapping, countless marker-trait associations are established in many crops for many key agronomic traits. The development of highly efficient genotyping systems significantly reduces the cost of genotyping. With that marker-assisted selection will become more and more accessible to plant breeders. The challenge faced by breeders is how to best use the rapidly accumulating genotypic information to breed better varieties faster. Computer simulation, which models the process of plant breeding within the context of defined gene-environment systems, has been used to assist the development of more efficient breeding strategies. In this chapter, the requirements for designing a marker-assisted selection strategy using computer simulation were discussed. Examples including parental selection using known-gene information, foreground and background selection in advanced backcrossing and gene pyramiding were used to demonstrate how an efficient MAS strategy can be designed.

Key Words: QTL, Statistical tool, MAS, Computer simulation.

[1] *Bundoora Centre, Biosciences Division, Department of Primary Industries Victoria, and Molecular Plant Breeding Cooperative Research Centre, 1 Park Drive, Bundoora Vic 3086, Australia*
[2] *The International Center for Agricultural Research in the Dry Areas (ICARDA), P.O. Box 5466, Aleppo, Syrian Arab Republic*
**Corresponding authore-mail: guoyou.ye@dpi.vic.gov.au*

1.0 INTRODUCTION

The potential benefits of using molecular markers linked to genes of interest in breeding programs, thus moving from merely phenotype-based towards a combination of phenotype- and genotype-based selection, have been obvious for many decades (Ruane and Sonnino, 2007). The use of markers in the identification of Mendelian components underlying both simple and complex agronomic traits is currently a key component of breeding programs in most crops (Ribaut and Hoisington, 1998; Dekkers and Hospital, 2002). Recent developments in plant molecular genetics have lead to the development of large numbers of molecular markers in all major crop species, ready to be shown to be linked to genes of interest via a growing set of methodologies (*e.g.* bulk segregation analysis, association mapping, expression (transcriptome, proteome and metabolome) based information). Even for understudied minor crops a good number of markers can easily be developed thanks to the rapid development in gene sequencing, functional genomics and comparative genetics. Some previously considered orphan crops can now make large leaps in molecularizing their breeding programs. The development of highly automated genotyping systems not only significantly speeds up the genotyping process and quality, but also more importantly dramatically reduces the cost of genotyping. In the past few years in several crops genotyping has now become also cheaper than phenotyping, besides being faster and more accurate (Dekkers and Hospital, 2002).

The availability of abundant molecular markers greatly facilitates the construction of highly saturated linkage maps, which in turn speeds up the establishment of marker-trait associations and the quality of established associations. Great progress has also been made in developing efficient statistical methods for the identification of QTL using typical mapping populations (double haploids (DH), recombinant inbred line populations (RILs), F_2 populations, backcross generations, etc.) (Jansen, 1993; Zeng, 1994), and applying mixed model-based approaches for mapping using multiple populations (Jourjon *et al.*, 2005) and approaches accounting for epistatic or genotype-by-environment interactions (Jiang and Zeng, 1995; Vargas *et al.*, 2006). More recently, association mapping using populations consisting of cultivars, breeding lines and germplasm collections used in ongoing breeding program has started to be employed in the establishment of marker-trait association (Yu *et al*, 2007).

With all these break-through technical and methodological developments, underlying gene-based information about key agronomic traits are accumulating at an unprecedented speed. The challenge for plant breeders now is to determine how to best utilise this multitude of information in the improvement of crop performance (Ye *et al.*, 2007).

A breeding strategy consists of all the manipulations (breeding activities) a breeder applies to a breeding population to achieve the breeding objectives. This includes but is not limited to, mating options to create new recombination events, phenotyping schemes (trial test design), selection methods including the development of selection criteria based on phenotyping and genotyping data, handling seed from different plants [*i.e.* tracking family structure (pedigree selection) in inbred line development or mixing seeds from different individuals (bulk selection)]. The most effective strategy in reaching the target genotype depends on the characteristics of the breeding population, the complexity of the underpinning genetics of the target traits (gene-gene interaction, gene-by-environment interaction, heritability, pleotropic gene effects).

The number of factors affecting the efficiency of a breeding strategy is many and increasing, not in the least as market demands on traits is also growing. It is rapidly becoming impractical to identify the most efficient strategies by empirical approaches alone. In addition, epistasis, pleiotropy and linkage result in some of the best solutions no longer being immediately intuitive. The breeder needs help. Computer simulation can be used to model the outputs of many breeding strategies applicable to a particular situation faced by breeders, so that more efficient strategies can be identified, without expensive and time-consuming field studies. In this chapter, we demonstrate, by using examples, how to design efficient MAS strategies using computer simulation.

2.0 REQUIREMENTS FOR MODELLING BREEDING PROGRAM

To design a realistic MAS strategy through simulation we should be able to (1) define the genetics of traits at the gene, genotype and phenotype level for all relevant (testing and target) environments, (2) describe all feasible manipulation options breeders can use to achieve breeding objectives in genetic and environmental-effect terms.

The first task is required to ensure that the simulation results are relevant to crop at hand, the desired traits and the target environments. The essence of this task is defining the gene-environment (GE) system of the respective breeding program.

The second task is required to ensure that all the important factors affecting breeding effectiveness and efficiency can be investigated through simulation and the best breeding strategy identified.

2.1 Essential Information for the Definition of GE Systems

a. Genomics of the crop

(1) Number of chromosomes and their lengths (genetic or physical)

(2) Markers and genes, and their position (distance between genes) on the chromosomes

(3) Recombination characteristics between genes (different interference models)

b. Trait genetics

(1) Number of genes controlling the trait

(2) Number of alleles, allele effect and gene action model for each of the genes

(3) Heritability of individual trait in all environments.

c. Environment

The basic information is the number of environment types in terms of phenotypic expression of each genotypic allele(s) state, and their frequencies in the targeted population of environments.

Trait correlation is caused by linkage between genes for different traits and/or different effects of the same gene on different traits (*i.e.* pleiotrophy). Genotype-by-environment interaction is created by the same gene having different effects on the phenotype in different environment types. Therefore, by definition of the genotype-phenotype effects at the gene level all the genetic characteristics can be defined. This is most straightforward if we have the complete knowledge about the genetics of the traits. When we only have partial knowledge of the trait genetics, it is challenging to define the GE system model.

The known genotypic knowledge (*i.e.* number of QTL, their locations and effects, etc) can be incorporated into the definition relatively easily. However, the unknown part may have to be defined (or rather estimated) based on population and quantitative genetic theories using population and quantitative genetic parameters available, such as allele frequency, variance components, genetic correlations between traits, narrow- and broad-sense heritability, and the genotype-by-environment interaction. Since these parameters are estimated at the population level, the exact underlying details at gene and individual plant levels remain poorly defined.

2.2 Essential Information for the Definition of Breeding Strategies

A plant breeding strategy consists of recombination, followed by a reiterative testing and selection scheme applied to every generation of the breeding cycle. The recombination phase creates new recombinants by mating between all or a group of selected plant materials. Thus, mating method and mating units are required to define recombination. Testing, including both phenotyping in field or glasshouse trials, and genotyping for a number of markers provide the basic information for the development of selection criteria.

Although, as stated, genotyping is straightforward if the information is available, phenotyping is more complicated, even information is available. That is so because often information is collected with the empirical objective of either promoting or discarding plant materials, but no real observations are recorded neither in categories, n = nor or on a continuous scale.

As a start, the very basic parameters describe a testing schemes need to be recorded, such as the number of testing locations, number of replicates at each location and plot size. But especially the decision tree used, often intuitively, during selection needs to be noted. The selection phase aims at the identification of desirable recombinants for the next round allele/gene recombination, either through self- or cross-fertilization. The basic parameters that describe a selection scheme are (1) which selection criteria are used (2) how selection is conducted, (3) how many or what percentage of selection units are selected for promotion to the next generation. The selection criterion used can be the phenotypic performance of a single trait, a molecular score and a selection index combining information from multiple traits and even molecular scores. Selection can be for individuals with a higher or lower selection criterion value (*e.g.* shorter in height, or higher in grain yield, etc.). The selection unit can be individual plants, a population of related plants (*i.e* half-sib or full-sib families, etc). The number of selections can be a predefined number, or determined by selection proportion or threshold value of the selection criterion.

2.3 QuGene Simulation Platform

QuGene, developed by the University of Queensland, is a very versatile simulation software platform for the quantitative analysis of genetic models (Podlich and Cooper, 1998). To allow maximum flexibility QuGene has a two-stage architecture.

The functions of the first stage (called "engine") are to: (1) define the GE systems and (2) generate the starting population of individuals (initial breeding population) and estimate the experimental errors associated with traits.

The second stage is the development of application modules that allow modelling of different sets of breeding strategies. The application model is used to investigate, analyse, or manipulate the starting population of individuals within the GE systems defined by the user at the "engine' level. Special application modules can be developed to achieve a particular simulation objective.

The advantage of developing specific application modules with a defined objective is that modules can be developed very quickly. The drawback is that the breeding strategies that can be defined and compared are limited. More generic application modules can be developed to facilitate comparisons between a wide range of breeding strategies. The most comprehensive application module is QuLine (also known as QuCim) jointly developed by the staff at the University of Queensland and at the International Maize and Wheat Improvement Center (CIMMYT). It can model most of the breeding strategies for inbred line development, which covers breeding of self-pollinated crops and inbred line development in cross-pollinated crops intended to be used in hybrid development. Modules for population improvement and hybrid production are also under development. QuLine has been used to model both conventional and marker-based breeding (Wang *et al.*, 2003; Ye *et al.*, 2004; Kuchel *et al.*, 2005; Ye *et al.*, 2007). The definition of GE systems by the QuGene engine was detailed by Podlich and Cooper (1998) and the definition of breeding strategies by QuLine was detailed by Wang *et al.* (2003). All the simulations presented in the following section were conducted using QuGene.

3.0 EXAMPLES TO DEMONSTRATE THE DESIGN OF MAS STRATEGIES

3.1 Parental Selection for Inbred Line Development

In breeding programs aimed at developing superior inbred lines, it is well known that superior cultivars usually trace back to very few good crosses. Upon reflection it is often not immediately or intuitively clear why those crosses would be more successful in the end than other seemingly good combinations of parents. For example, in the bread wheat breeding program of CIMMYT, two major breeding strategies are commonly used and thousands of crosses are made every season.

Although breeders spend great efforts in choosing parents to make the targeted crosses including sophisticated information on the prospective parents, approximately 50-80% of the crosses are discarded in generations F_1 to F_8, following the selection for agronomic traits (*e.g.*, plant height, lodging tolerance, tillering, appropriate heading date, and balanced yield components), disease resistance (*e.g.*, stem rust, leaf rust, and stripe rust), and end-use quality (*e.g.*, dough strength and extensibility, protein quantity and quality). Then, after two cycles of yield trials (*i.e.*, preliminary yield trial in F8 and elite yield trial in F_9), only 10% of the initial crosses remain, from among which 1-3% of the crosses originally made are included as advanced lines for dissemination through CIMMYT's international nurseries (Wang *et al.*, 2003). Of these only a small percentage later is released by national breeding programs around the world. This is what has lead to describing breeding as "a numbers game". Therefore, using much more basic genetic information on the parents and/or highly distilled information from early generations to predict the eventual usefulness of a cross has long been recognized as an effective way to increase the efficiency of a breeding program (Fehr, 1991).

In inbred line development of self-pollinating crops, the purpose is to obtain the most desirable near-homozygous genotypes. Therefore, it would appear that cross selection is best based on the predicted performance of the progeny genotypes. This is possible only if the genotypic values of genes controlling the target traits are known, and most importantly how epistatic networks among genes affect the eventual genotype. In other words, what "gene-teams" make for the best phenotype? For most of the important agronomic traits, a complete knowledge of their genetic determination is not available and unlikely to be so in the near future. However, genes with large effects on some traits are known. For example, six glutenin genes have major effects on wheat dough quality (Eagles *et al.*, 2002a, b). Moreover, a great amount of studies on QTL mapping have been conducted for various traits in plants in recent years. The knowledge of trait genetics is thus accumulating at great speed.

Cross selection may be made using these known genes and/QTL. Cross selection for wheat quality improvement based on the six glutenin genes and two *Pin* genes have now been used in Australian wheat breeding programs for several years. A deterministic simulation tool, called CrossPredictor, was developed by (Guoyou ye) at the University of Queensland, the Department of Primary Industries, Victoria, and the Cooperative Research Centre for Molecular Plant Breeding to facilitate cross selection for wheat quality improvement based on six glutenin genes and two *Pin* genes (Ye *et al.*, 2005). The selection is based on the frequencies of satisfactory genotypes (defined based on selected quality parameters, such as Rmax, extensibility and water absorption performance) double haploid or recombinant inbred line populations generated from single- or three-way cross. The use of known genes for parental selection involves the following steps:

1. *Obtain the estimated genotypic values for all possible genotypes of the known genes*

Genotypic values are used as the basis for selection and must be predicted/estimated as precisely as possible. This requires extensive phenotyping in target environments of a comprehensive collection of genotypes that represent both the parental stocks available and the various allele combinations that may occur in individual genotypes. For example, for the glutenin genes in bread wheat Eagles *et al.* (2002a, b, 2004) applied the general linear mixed model with/without the pedigree information to predict the genotypic values of 1728 possible homozygous genotypes using data from multiple location tests conducted in different years in South Australia and Victoria, Australia.

2. Determine the desirable genotypes

A genotype is regarded as desirable if its performance meet breeders' requirements, which if properly defined also reflect farmers' and en-users' requirements. The desirable genotypes can be found by searching among all possible genotypes for ones with trait values within a specified range. It is, however, well possible that the very best genotype is not present among the parental stocks. Then the process requires the identification of pairs or trios of parents that would most effectively complement one another.

When two or more traits are used to evaluate the cross, the simplest method is the independent culling, by which the desirable (selected) genotypes meet the requirements for all the traits of interest. An even better method, especially when faced with breeding for multiple traits (reality), is to construct a selection index with user defined economic weights for each trait. Because only the genotypic values are used, the phenotypic variances and covariances are same as the genotypic variances and covariances. Therefore, the derivation of individual genotypic values from the overall average genotypic value can be used as the predictor of its genetic merit. The resulting selection index (SI) is simply defined as:

$$SI = \sum_{i=1}^{n} w_i(t_i - \vec{t}_i)$$

where w_i is the economic weight given to the i-th trait, t_i is the genotypic values of the i-th trait predicted using information on known-gene/QTL of an individual, and t_I is the average genotypic values of the i-th trait.

The number of desirable genotypes can be more than one.

3. Identify satisfactory crosses

A cross is regarded as satisfactory if it can produce progenies of the desirable genotype. Since the crossing parents must have alleles present to generate the desirable genotype, the desirable allele coverage of the crossing parents can be used to identify satisfactory crosses.

Desirable allele coverage is calculated as

$$\text{Coverage} = \frac{\sum_{j=1}^{n} S_j}{n}$$

where n is the number of genes, and S_j is an indicator variable to indicate whether the allele of the desirable genotype for the i-th gene is present in the parental lines. It is 1 if any of the parental lines has the desirable allele. Otherwise, it is 0.

When the number of desirable genotypes is 2 or more, desirable allele coverage needs to be computed for all the desirable genotypes. A cross is satisfactory if any of the coverage values is 1.

4. Identify the best crosses

The best cross is a cross that generates the highest frequency of desirable genotypes, and does that in the shortest span of breeding activities. The following two methods can be used to identify the best cross among all the satisfactory ones. If the number of satisfactory crosses is not too high, the best cross can be identified using the total frequency of desirable genotypes in their derived DH or RILs populations. When the number of satisfactory crosses is high, the average distance between the parental lines and the desirable genotypes is used.

The average distance between parental lines of a cross from the desirable genotype is calculated as

$$GD = 1 - \frac{\sum_{i=}^{p}\sum_{j=1}^{n} S_{ijk}}{np}$$

where p is the number of parents involved in a cross (e.g. 2 for single cross, 3 for three-way cross and 4 for double cross); S_{ij} is the number of shared alleles between the i-th parental line and the desirable genotype for the k-th gene.

For multiple desirable genotypes, the distances from each of the desirable genotypes are computed, and the minimum distance is used to compare crosses. Crosses with the smallest distances are selected as most promising.

The following example is used to illustrate the method outlined. Assume a breeder wants to develop lines with bread-making quality values Rmax between 300 and 350, extensibility values between 20 and 21, and dough development time between 3.5 and 4.6 minutes, and the ten parental genotypes available are those listed in Table 1. The genotype values for the three traits of all the 64 possible homozygous genotypes producible using these 10 parental lines were taken from Eagles *et al.* (2002a) and given in Table 2. Only six of the 64 genotypes satisfy these criteria and thus are the desirable ones (Table 2).

Among the ten parental genotypes 45 single crosses, excluding reciprocal crosses, are possible. However, only 20 crosses are satisfactory

Table 1. Allelic compositions of the 10 parental genotypes

Parental genotype	1A	1B	1D	3A	3B	3D
1	a	b	a	c	b	a
2	a	b	a	c	d	a
3	a	b	a	b	d	a
4	b	al	d	c	b	b
5	a	al	d	c	b	b
6	b	al	d	b	b	b
7	a	al	d	b	b	b
8	b	al	d	c	d	a
9	a	b	d	c	d	a
10	a	b	d	b	d	a

Table 2. Predicted genotype values for Rmax, ext and DDT of all 64 possible genotypes using six glutenin genes

Genotype	Rmax	Ext	DDT	Genotype	Rmax	Ext	DDT
a ala b b a	420.9	23.61	5.6	b ala b b a	429.6	23.6	5.65
a ala b b b	400.1	23.23	5.6	b ala b b b	408.8	23.22	5.66
a ala b d a	348.6	23.87	4.58	b ala b d a	357.4	23.87	4.64
a ala b d b	327.8	23.49	4.59	b ala b d b	336.6	23.48	4.65
a ala c b a	399.2	23.07	5.52	b ala c b a	407.9	23.07	5.58
a ala c b b	378.4	22.69	5.53	b ala c b b	387.1	22.68	5.59
a ala c d a	326.9	23.34	4.51	b ala c d a	335.7	23.33	4.57
a ala c d b	306.1	22.96	4.52	b ala c d b	314.9	22.95	4.58
a ald b b a	529.9	23.19	6.58	b ald b b a	538.6	23.18	6.64
a ald b b b	509	22.81	6.59	b ald b b b	517.8	22.8	6.64
a ald b d a	457.6	23.46	5.57	b ald b d a	466.3	23.45	5.63
a ald b d b	436.8	23.07	5.57	b ald b d b	445.5	23.07	5.63
a ald c b a	508.2	22.66	6.51	b ald c b a	516.9	22.65	6.56
a ald c b b	487.4	22.27	6.51	b ald c b b	496.1	22.26	6.57
a ald c d a	435.9	22.92	5.5	b ald c d a	444.7	22.91	5.55
a ald c d b	415.1	22.54	5.5	b ald c d b	423.8	22.53	5.56
a b a b b a	**319.6**	**20.8**	**4.25**	**b b a b b a**	**328.4**	**20.79**	**4.3**
a b a b b b	298.8	20.42	4.25	b b a b b b	307.6	20.41	4.31
a b a b d a	247.3	21.06	3.23	b b a b d a	256.1	21.06	3.29
a b a b d b	226.5	20.68	3.24	b b a b d b	235.3	20.67	3.3
a b a c b a	297.9	20.26	4.17	b b a c b a	306.7	20.26	4.23
a b a c b b	277.1	19.88	4.18	b b a c b b	285.9	19.87	4.24
a b a c d a	**225.6**	**20.53**	**3.16**	b b a c d a	234.4	20.52	3.22
a b a c d b	204.8	20.14	3.17	b b a c d b	213.6	20.14	3.23
a b d b b a	428.6	20.38	5.23	b b d b b a	437.4	20.37	5.29
a b d b b b	407.8	20	5.23	b b d b b b	416.5	19.99	5.29
a b d b d a	356.3	20.65	4.22	b b d b d a	365.1	20.64	4.27
a b d b d b	**335.5**	**20.26**	**4.22**	**b b d b d b**	**344.3**	**20.25**	**4.28**
a b d c b a	406.9	19.84	5.16	b b d c b a	415.7	19.84	5.21
a b d c b b	386.1	19.46	5.16	b b d c b b	394.9	19.45	5.22
a b d c d a	334.6	20.11	4.14	b b d c d a	343.4	20.1	4.2
a b d c d b	313.8	19.73	4.15	b b d c d b	322.6	19.72	4.21

Genotypes given in bold are the desirable ones

Table 3. Desirable allele coverage (DC) of all 45 single crosses between 10 parental lines calculated for five desirable genotypes, and Rogers distance between crossing parents (Rd)

CID	P_1	P_2	DC					Rd
			DG1	DG2	DG3	DG4	DG5	
1	1	2	0.83	0.67	0.83	0.50	0.33	0.17
2	1	3	1.00	0.83	0.83	0.67	0.50	0.33
3	1	4	0.83	0.83	0.83	0.67	0.67	0.17
4	1	5	0.83	0.67	0.83	0.67	0.50	0.33
5	1	6	1.00	1.00	0.83	0.83	0.83	0.50
6	1	7	1.00	0.83	0.83	0.83	0.67	0.33
7	1	8	0.83	0.83	1.00	0.67	0.67	0.50
8	1	9	0.83	0.67	1.00	0.67	0.50	0.33
9	1	10	1.00	0.83	1.00	0.83	0.67	0.33
10	2	3	0.83	0.67	0.83	0.67	0.50	0.50
11	2	4	0.83	0.83	1.00	0.83	0.83	0.67
12	2	5	0.83	0.67	1.00	0.83	0.67	0.83
13	2	6	1.00	1.00	1.00	1.00	1.00	0.50
14	2	7	1.00	0.83	1.00	1.00	0.83	0.67
15	2	8	0.67	0.67	1.00	0.67	0.67	0.83
16	2	9	0.67	0.50	1.00	0.67	0.50	0.67
17	2	10	0.83	0.67	1.00	0.83	0.67	0.67
18	3	4	1.00	1.00	1.00	1.00	1.00	0.50
19	3	5	1.00	0.83	1.00	1.00	0.83	0.83
20	3	6	1.00	1.00	0.83	1.00	1.00	0.67
21	3	7	1.00	0.83	0.83	1.00	0.83	0.67
22	3	8	0.83	0.83	1.00	0.83	0.83	1.00
23	3	9	0.83	0.67	1.00	0.83	0.67	0.83
24	3	10	0.83	0.67	0.83	0.83	0.67	1.00
25	4	5	0.50	0.50	0.67	0.67	0.67	0.83
26	4	6	0.50	0.67	0.50	0.67	0.83	0.83
27	4	7	0.67	0.67	0.67	0.83	0.83	0.67
28	4	8	0.50	0.67	0.83	0.67	0.83	0.83
29	4	9	0.67	0.67	1.00	0.83	0.83	0.67
30	4	10	0.83	0.83	1.00	1.00	1.00	0.67
31	5	6	0.50	0.50	0.50	0.67	0.67	0.17
32	5	7	0.50	0.33	0.50	0.67	0.50	0.67
33	5	8	0.50	0.50	0.83	0.67	0.67	0.50
34	5	9	0.67	0.50	1.00	0.83	0.67	0.17
35	5	10	0.83	0.67	1.00	1.00	0.83	0.17
36	6	7	0.50	0.50	0.33	0.67	0.67	0.17
37	6	8	0.50	0.67	0.67	0.67	0.83	0.17
38	6	9	0.83	0.83	1.00	1.00	1.00	0.33
39	6	10	0.83	0.83	0.83	1.00	1.00	0.33
40	7	8	0.67	0.67	0.83	0.83	0.83	0.33
41	7	9	0.83	0.67	1.00	1.00	0.83	0.17
42	7	10	0.83	0.67	0.83	1.00	0.83	0.50
43	8	9	0.50	0.50	1.00	0.67	0.67	0.17
44	8	10	0.67	0.67	1.00	0.83	0.83	0.50
45	9	10	0.67	0.50	1.00	0.83	0.67	0.67

Table 4. Total frequency of desirable genotypes (above diagonal) in the RILs populations derived from 45 single crosses between the 10 parental genotypes and the minimum RILs population sizes required to obtain at least one desirable line (below diagonal)

	P1	P2	P3	P4	P5	P6	P7	P8	P9	P10
P1		0.0	**25.0**	6.9	0.0	13.0	6.9	16.3	7.5	12.5
P2	nr		0.0	9.9	7.9	12.7	10.8	**25.0**	25.0	12.5
P3	**17**	nr		12.7	10.8	15.4	13.8	12.5	12.5	0.0
P4	65	44	34		0.0	0.0	0.0	0.0	0.0	0.0
P5	nr	57	41	nr		0.0	0.0	0.0	0.0	0.0
P6	34	35	28	nr	nr		0.0	0.0	0.0	0.0
P7	65	41	32	nr	nr	nr		0.0	0.0	0.0
P8	26	**17**	35	nr	nr	nr	nr		0.0	0.0
P9	60	17	35	nr	nr	nr	nr	nr		0.0
P10	35	35	nr	nr	nr	nr	nr	nr	nr	

nr: not relevant

Cross combinations highlighted in bold are the best crosses.

in terms of generating progeny that includes genotypes that give quality parameters within the above defined ranges (Table 3).

For this example, all the satisfactory crosses have a reasonable high frequency of the desirable genotypes.

Since genotyping for glutenin genes is still very expensive, the use of gene profiling in early generations of selection is not practical. Therefore, the frequencies of predicted desirable genotypes in RILs and DH populations (Table 4) are used for selecting the best crosses. Once large-scale genotyping becomes economically, selection in early generations can effectively enhance the frequencies of desirable genotypes in later generations and should be adopted (See the gene pyramiding example below.).

Parental selection using QuLine was demonstrated by Wang *et al.* (2005), also using the glutenin genes in wheat as an example. There are eight bread wheat sister lines derived from the variety Silverstar (numbered 1-8) with very similar morphological characteristics, but different values for two important bread-making quality traits, Rmax and extensibility. To breed for improved Rmax and extensibility, which of the four sister lines is the best parent for crossing with each of the four adapted cultivars, Westonia, Krichauff, Machete and Diamondbird? This is an example in which different "gene-teams" can give similar outstanding quality outcomes.

All possible 32 non-reciprocal single crosses were made by QuLine between the four selected parents and the eight Silverstar sister lines. For each cross, 1000 F_8 lines were developed from 1000 F_2 individual plants by single seed descent, using simulation. Forty F_8 lines were finally selected, based on line performance for Rmax and/or extensibility, resulting in a selected proportion of 0.04. Four selection schemes were considered: (1) the 40 lines were selected based only on line performance for Rmax (R0.04); (2) 200 lines were first selected based on line performance for Rmax and subsequently 40 lines were selected based on extensibility (R0.2E0.2); (3) 200 lines were first selected based on line performance for extensibility and then 40 lines were selected based on Rmax (E0.2R0.2); (4) 40 lines were selected based only on line performance for extensibility (E0.04).

Table 5. The best Silverstar sister lines for the four selected parents, under different breeding objectives

Cultivars	Objective	Selection schemes*			
		R0.04	R0.2E0.2	E0.2R0.2	E0.04
Westonia	High Rmax	3,7	3,7	3,7	1,3
	High extensibility (cm)	1	1,5	1,3,5	1, 3,5,7
Krichauff	High Rmax	3	5	7 1	3
	High extensibility (cm)	1,3,5,7	1,3,5,7	3,7	3,7
Machete	High Rmax	3,4,7,8	3,4,7,8	4,8	Nil
	High extensibility (cm)	1, 2,5,6	1,2,5,6	1,2,3	1,2,3,4
Diamondbird	High Rmax	1,2,3,4	1,3,4	3,4	3,4
	High extensibility (cm)	None	None	1,2,5,6	1,2,5,6

* R = Rmax; E = extensibility; trait followed by selected proportion.

When using crosses with Westonia, Silverstar lines 3 and 7 showed the largest improvement in Rmax, when Rmax was used in selection (*i.e.*, R0.04, R0.2E0.2, and E0.2R0.2) (Table 5). They can also improve extensibility in combination with Westonia, particularly when selecting for extensibility (*i.e.*, R0.2E0.2 and E0.2R0.2). When high Rmax and extensibility together are the required quality traits, but Rmax is more important, both lines 3 and 7 are parents of choice. However, Silverstar line 3 is the better of the two (Table 5).

For crosses with Krichauff, if selection is solely for Rmax, or if it is selected first when both traits are targeted for selection (*i.e.*, R0.04 and R0.2E0.2), Silverstar lines 1, 3, 5, and 7 can result in similar improvements in Rmax and extensibility. In crosses with Krichauff, if selection is solely for extensibility, or if extensibility is selected first, when both traits are targeted for selection (*i.e.*, E0.2R0.2 and E0.04), then Silverstar lines 3 and 7 are the best parents for improving both traits (Table 5).

For crosses with Machete, Silverstar lines 3, 4, 7, and 8 are the best parents to improve Rmax if it is the only trait selected, or if it is selected first when both traits are targeted (*i.e.*, R0.04 and R0.2E0.2). However, to improve extensibility simultaneously, Rmax should be selected first and then extensibility (*i.e.*, R0.2E0.2). If extensibility is selected before Rmax, then Silverstar lines 4 and 8 should be chosen to improve both traits in crosses with Machete (Table 5).

For crosses with Diamondbird, the use of Silverstar lines 1, 2, 3, and 4 can cause a slight increase in Rmax and extensibility, if Rmax is the trait targeted for selection (*i.e.*, R0.04 and R0.2E0.2). Only Silverstar lines 3 and 4 can improve both traits if extensibility is targeted (*i.e.*, E0.2R0.2 and E0.04).

3.2 Gene Introgression

Marker-assisted recurrent backcrossing (MARB) has been widely used to transfer a few genes/QTL from donor parents into recurrent parents. Theoretical and simulation results, and the application of MARB in cultivar improvement were reviewed by Ye *et al.* (2009) (Chapter 11). These results can be used to guide the design of an efficient gene introgression strategy relevant to the particular situations faced by the breeders. The genomics and the reproductive characteristics of the crop under study, the available markers that are polymorphic between the two parental lines and the linkage relationships among these markers, the financial resources available and the required recurrent genome percentage (RGP; generally aimed for to be high) are the key factors to be considered.

The genomics and markers available for foreground and background selections are fixed factors at any point of time and are not directly involved in the choice of a breeding strategy. Reproductive characteristics determine the number of seeds obtainable from a single plant by backcrossing, which may vary several orders of magnitude among crops, and thus the maximum family (population) size applicable. The RGP sets one of the key objectives of an introgression program. It usually is not a variable to be manipulated once the target level is defined. Therefore, progeny sizes at each generation, and when and how to conduct selection are the parameters breeders can manipulate to achieve the breeding objectives within the limit of available resources. Two examples are given below to demonstrate how computer simulation can be used in this context.

3.2.1 Foreground selection with markers surrounding a mapped QTL

When phenotyping is expensive or difficult to conduct markers linked to the target genes can be used to select for the presence of target genes. Figure 1 graphically represents the genotype for a QTL and five surrounding markers (M) of an inbred donor line with the distance between markers being 10 CM. The true location of the QTL is right in between markers M_2 and M3. Three strategies (STR) are compared. M2 and M3 will be used for foreground selection (strategy 1; STR1), if the QTL is mapped into the correct marker interval. If the QTL is mapped to the neighbouring position within the interval between markers M1 and M2, then M1 and M2 will be used if only flanking markers are considered (strategy 2; STR2). Since the precise location of a mapped QTL is unknown in practice, several markers surrounding the mapped QTL can be used for selection. For example, markers M1, M2, M3 and M4 can all be used in foreground selection (strategy 3; STR3).

To compare the three strategies we defined three virtual traits. Trait 1 is controlled by M1 and M2, trait 2 is controlled by M2 and M3, and trait 3 is controlled by M1, M2, M3 and M4. For all markers the genotypic values are 2, 1 and 0 for marker genotypes MM, Mm and mm, respectively. Four backcrosses to the recurrent parent were made, and 200 plants were generated at each backcrossing generation. Foreground selection was conducted in the first three backcross generations using makers and three plants were selected, while in the fourth backcross generation selection was for the QTL itself and all plants with the QTL were selected.

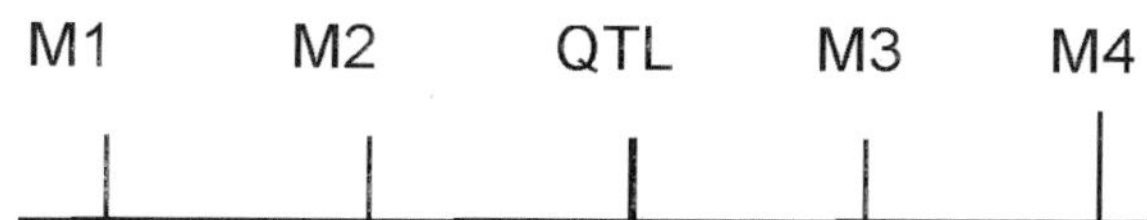

Fig. 1. The graphical genotype for the QTL to be transferred and surrounding markers

The best strategy was STR3 in terms of both the success rate defined as percentage of runs that lead to at least one individual of the desirable genotype in the third backcross generation, and the average number of desirable plants recovered at BC_4 across 1000 runs (Fig. 2). Therefore, it is critical that QTL must be located in the correct marker interval if only flanking markers are used, and including other markers surrounding the mapped QTL significantly increases the selection precision.

3.2.2 Background selection for recurrent genome recovery

Background selection to recover the recurrent genome is well established application of markers in applied crop breeding (Ye *et al*, 2008, Chapter 11). To demonstrate background selection, a genome with three chromosomes and 147 markers and one target gene was defined. The distance between a pair of neighbouring genes (markers) was 2 cM. Three traits were defined (Table 6). Target is a trait controlled only by the gene to be transferred. RGP is a virtual trait used for calculating percentage of recurrent genome and

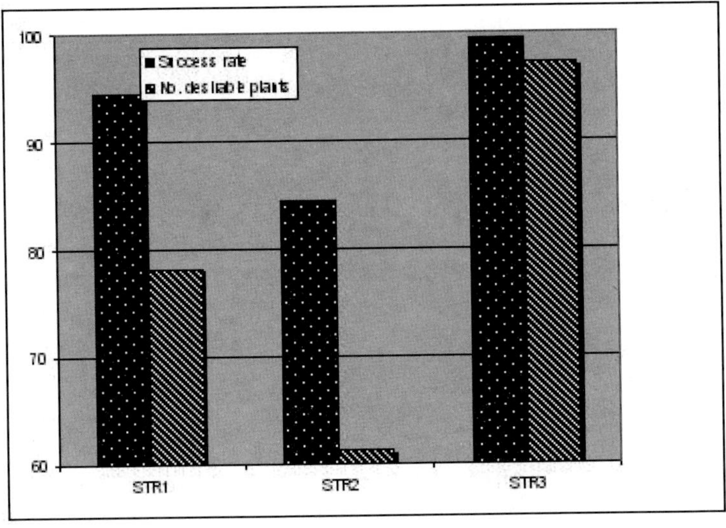

Fig. 2. The average number of plants with the target QTL at BC_4 and the probability of obtaining at least one BC_4 plants with the target QTL

BS is a virtual trait used for background selection. Heritabilities for all the three traits were assumed to be 1. Under this assumption, foreground selection can be conducted by directly selecting for the trait, which is the case when a functional or diagnostic marker for the target gene is available. The definition that RGP was controlled by many marker genes (147) ensured that the estimate of percentage of recurrent genome was precise. Markers genes for BS were situated sufficiently far apart from each other (roughly 20 cM) to closely reflect the situation in practical breeding programs. The effects of all the marker genes defined for RGP and BS were additive and thus individuals that were homozygous for the allele of the recurrent parent have higher RGP and BS values than those that were heterozygous.

Table 6. Defining a gene-environment system for modelling background selection in advanced backcrossing program

Trait	Number of genes	Which gene	Inheritance model (Defined as single-locus genotypic value)
Target	1	25	TT=2.0, Tt=1.0, tt=1
RGP	147	All 148 genes expect the 25th gene	MM=2.0, Mm=1.0, mm=1
BS	14	10,18,31,43,55 ,66,76, 87, 97,107,117 ,128,138,148	MM=2.0, Mm=1.0, mm=1

Although PRG at each generation of backcross without selection can be easily obtained using theoretical equation, RGP at each of the first six backcross generations were computed from the simulation results under our setting. Figure 2 presents the results from the theoretical prediction and simulation. It can be seen that the estimates of RGP from our simulation are always lower than the theoretical values. This may be caused by two factors, first, the theoretical prediction assumes an indefinite population, while the population size used in simulation was only 200 plants, and second, we made the distinction between the marker homozygote and heterozygote. A greater genotypic value was assigned to the homozygote.

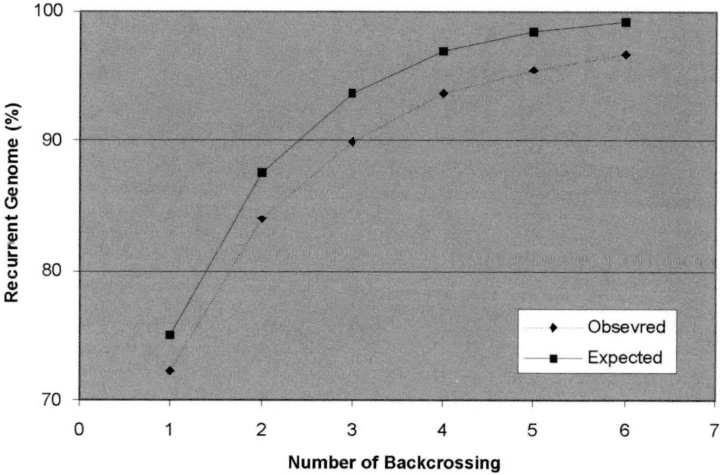

Fig. 3. Recurrent genome percentage predicted using theoretical equation and estimated under the simulation setting

According to the results from previous studies (Ye *et al* 2008, Chapter 11), with background selection three backcrossing are enough to have the recurrent genome percentage reach more than 95%. To simply the choice of a strategy the progeny size at each generation was fixed at 200 plants and three plants were selected at each backcross generation. Six selection schemes (Table 7) were defined and compared. Foreground selection for the presence of the target gene was conducted before background selection was carried out. When both foreground and background selection were conducted, all plants with the target gene were kept for background selection. When only foreground selection s conducted, only three plants were kept. The final number of selected plants was three for all generations.

As expected background selection increased the RGP effectively. Delaying background selection to later generations was beneficial. Selection should be conducted at the last backcrossing generation if one-step background selection is applied. Increasing the number of background selection steps increased RGP. The best strategy was BSBC123, for which

Table 7. The average recurrent genome percentages (RGP) resulted from different background selection strategies

Strategy	Background selection			RGP
	BC_1	BC_2	BC_3	
NBS	no	no	no	90.36
BSBC1	yes	no	no	94.95
BSBC2	no	yes	no	96.70
BSBC3	no	no	yes	97.73
BSBC23	no	yes	yes	98.24
BSBC123	yes	yes	yes	98.27

background selection was conducted at all three backcross generations. However, the differences between BSBC123, BSBC23 and BSBC3 were very small. Clearly, whether the increased RGP is worthwhile must be considered against the associated cost.

3.2.3 Gene pyramiding

The objective of a marker-based gene pyramiding approach in self-fertilizing crops is to obtain near-homozygous breeding lines that are fully homozygous for the desirable alleles of the target genes using the minimum number of generations of selection and the least genotyping and phenotyping (Ye and Smith 2009, Chapter 12). Therefore, the optimum strategy can be designed by finding the minimum number of individuals (*i.e.* population size) to be genotyped and/or phenotyped using the minimum number of generations, resulting in the maximum number of desired target individuals.

Barley breeders in the Department of Primary Industries, Victoria, Australia are using gene pyramiding method to develop two-row barely lines with six genes responsible for each of six important agronomic traits. The six genes responsible for each of six traits are photoperiod sensitivity (Ppd), Russian wheat aphid resistance (Rwa), leaf rust resistance (Rph), boron tolerance (Bot), earliness *per se* (EPS) and cereal cyst nematode resistance (Ha2). The order of these six genes on 2H and recombination frequencies between neighbouring genes is given in Figure 4. Three two-row barley lines with different desirable alleles at six genes of interest on chromosome 2H were chosen as parents. Sloop Vic (S) is an adapted commercial cultivar. HS078 (H) is a *H. spontaneum* accession with the desirable gene for leaf rust resistance, Rph16, while PI366444 (P) possesses the desirable Russian Wheat Aphid resistance gene RWA. Using the numbers '1' and '2' to represent the desirable and undesirable alleles, respectively, the allelic constitution of the three parental genotypes (S, H and P) can be represented as: 122111/122111, 221222/221222, and 212222/212222. Since S has more desirable alleles (four) than H and P (one each), and is also more

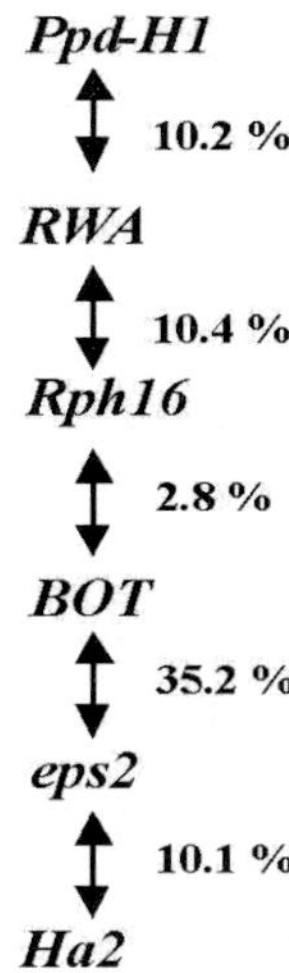

Fig. 4. The order of six genes on 2H to be combined, with recombination frequencies indicated between brackets

desirable in terms of other agronomic traits, The three-way cross H/P//S is the choice of crossing scheme. Ye *et al.* (2007) developed a selection scheme using simulation, which minimised the number of selfing generations after the three-way cross and the overall number of plants to be genotyped. Since H and P are not adapted cultivars, breeders would like to use backcross with S being the recurrent parent so that the resultant new breeding lines are more acceptable in other traits. The following steps were taken to develop an optimal selection scheme.

1). Determine the size of the progeny population of the three-way cross H/P//S

Table 8. The number of individuals with the desirable genotype, '211222/ 122111' (representing at least one desirable allele at each locus) recovered using different three-way cross (TC_1) population sizes

TC_1 Size	FRE (>2) %	Number of individuals		
		Mean	Min	Max
200	99.75	10.47	0	23
300	100	15.64	4	30
400	100	20.86	6	37
500	100	26.06	8	43
600	100	31.15	14	56

FRE (>2) % is the percentage of simulation runs that produces three or more individuals of the desirable genotypes (211222/122111).

The desirable genotype in the progeny population of H/P//S is 211222/122111. The minimum population size to ensure with 99.95% probability that 211222/122111 will be present at least once in the three-way cross population is 143, which is determined by using a binomial distribution. Simulations were run to determine the minimum population size to identify at least three individuals with the desired genotype (see Table 8). A population of 300 plants was shown to be sufficient in size. Please note the result is taken from Ye *et al.* (2007), since this is the common step between the current scheme and Ye *et al.* (2007)'s scheme.

2). Determine how many loci could be fixed for the desirable alleles at each subsequent backcrossing and selfing generation

As discussed by Ye *et al.* (2007), to reduce the number of generations required to achieve the selection objective, the selected individuals in early generations should have the highest number of loci fixed for the desirable alleles and at least one desirable allele (heterozygous status) on each of the remaining targeted loci. Assuming that three satisfactory TC_1 individuals were selected at the TC generation and used to backcross with S, a total of 500 derived BC_1 plants should be obtainable. At this population size, there was almost no chance to produce any individuals fixed at all the four loci for the desirable alleles and heterozygous on the remaining two loci. However, for all the simulation runs more than 3 individuals were fixed at three loci for the desirable alleles, while being heterozygous for the remaining three loci, and could be selected with a probability indistinguishable from 100% (Table 9). Indeed, selection for fixing three loci was practically achievable even with a population size as small as 300.

Table 9. Percentage of simulation runs obtaining different number of individuals with the desirable genotype using different BC_1 (211222/122111 x 122111/122111) population sizes

BC_1 Size	Desirable genotype	Percentage of simulation runs			
		NDG= 0	NDG= 1	NDG= 2	NDG> 2
300	211111/122111	0.05	0.25	1.1	98.6
400	211111/122111	0.05	0.00	0.35	99.6
500	211111/122111	0.00	0.00	0.00	100
300	111111/122111	81.55	16.70	1.60	0.15
400	111111/122111	73.60	23.05	2.85	0.50
500	111111/122111	69.05	24.35	5.60	1.00

NDG: number of individuals with desirable genotype

Knowing that three loci can be quite effectively fixed by selecting at the BC_1, the objective of BC_2 is to fix the fourth locus with the desirable allele being hosted by S. The objective can be achieved with high probability with a BC_2 size being 150 or more (Table 10).

Table 10. Percentage of simulation runs obtaining different number of individuals with the desirable genotype '111111/122111' using different BC_2 (21111/122111 x 122111/122111) population sizes

BC_2 Size	Percentage of simulation runs				NDG			
	NDG= 0	NDG= 1	NDG= 2	NDG> 2	AVE	MIN	MAX	STD
100	1.20	4.80	9.85	84.15	5	0	13	2.10
200	0.00	0.10	0.30	99.60	9	1	21	2.98
300	0.00	0.00	0.10	99.95	14	2	26	3.63
400	0.00	0.00	0.00	100.00	18	5	34	4.08

NDG: number of individuals with desirable genotype

After selection at BC_2, the lines obtained are all genotype 111111/122111. The RGP should be about 87.5%. Since all the six genes are closely linked, the selection conducted at the three-way cross and two backcross generations for the target genes would bring the RGP higher than the expected value. Background selection for other part of the genome was not considered in BC_1 and BC_2 due to the limited selection pressure applicable. If required background selection can be conducted by conducting one or two more rounds of backcrossing.

Self-pollination was then used to fix the two genes remaining segregating after backcross. These two genes are in the desirable linkage phase and the linkage distance is 10.4 cM, 11 plants should ensure that at least one individual with the desirable genotype to be obtained with a 99.9% probability. Therefore, several BC_2F_1 individuals of the desired genotype can easily be produced.

In summary, the best selection scheme was as follows.

1. In the TC_1 300 individuals are genotyped to obtain three or more individuals with a fully heterozygous genotype '211222/122111'.
2. In the BC_1 more than three individuals that are fixed for three desirable target loci and segregating for the remaining three loci ((211111/122111) are selected through genotyping of 400 BC_1 plants.
3. In the BC_2 150 or more individuals are genotyped for the fourth locus with the desirable allele be contained by S loci, and individuals fixed for four loci and segregating for the sixth locus are selected (111111/122111).
4. 11 or more individuals from self-pollinated seeds of the selected BC_2 plants are genotyped for the two segregating locus and individuals of the desirable genotype (111111/111111) are selected. The desirable lines are formed by collecting selfed seed from the selected plants.

Compared to the scheme using self-pollination after selection in the progeny of the three-way cross (H/P//S) (Ye *et al.*, 2007) the use of two rounds backcrossing to S reduced the number of plants to be genotyped. It also increases the recurrent genome content of the resultant lines (87.5% vs 50%).

4.0 CONCLUSION

As in designing a conventional breeding strategy, the description and quantification of the genomics of the crop under study and the genetics of traits using the most current knowledge is required to ensure the simulation results are relevant to the GE systems breeders are working with. The opportunities provided by identified marker-trait associations and markers distributed along the genome can be explored by using more targeted mating and selection schemes based on both phenotyping and genotyping information. Making mating and selection decision based on more information is more effective and also more complicated.

Several simple examples are used in this chapter to demonstrate the use of simulation in designing breeding strategies. Thought in essence simple, they are very relevant to cultivar development using well characterised genes/markers. Cross selection using known gene information makes it possible to concentrate limited resources on the most promising crosses and is the most obvious application of markers in crop breeding. Foreground and background selections using markers in advanced backcrossing program represents the majority of MAS implemented in practical crop breeding programs (Ye *et al.*, 2009, Chapter 11). Gene pyramiding has been and is still the method of choice for breeding for disease and insect resistance (Ye and Smith, 2009, Chapter 12).

Computer simulation can accommodate and manage infinitely more complex situations, such as multi-parent crossing and simultaneous selection strategies than those we demonstrated here.

5.0 REFERENCES

Dekkers, J.C.M. and Hospital, F. 2002. The use of molecular genetics in the improvement of agricultural populations. Nature Reviews, 3:22–32.

Eagles, H.A., Eastwood, R.F., Hollamby, G.J., Martin, E.M. and Cornish, G.B. 2004. Revision of the estimates of glutenin gene effects at the Glu-B1 locus from southern Australian wheat breeding programs. Australian Journal of Agricultural Research, 55:1093–1096.

Eagles, H.A., Hollamby, G.J., Gororo, N.N. and Eastwood, R.F. 2002a. Estimation and utilisation of glutenin gene effects from the analysis of unbalanced data from wheat breeding programs. Australian Journal of Agricultural Research, 53:367–377.

Eagles, H.A., Hollamby, G.J., Gororo, N.N. and Eastwood, R.F. 2002b. Genetic and environmental variation for grain quality traits routinely evaluated in southern Australian wheat breeding programs. Australian Journal of Agricultural Research, 53:1047–1057.

Jansen, R.C. 1993. Interval mapping of multiple quantitative trait loci by using molecular markers. Theoretical and Applied Genetics, 85:252–260.

Jiang, C. and Zeng, Z.B. 1995. Multiple trait analysis of genetic mapping for quantitative trait loci. Genetics, 140:1111–1127.

Jourjon, M.F., Jasson, S., Marcel, J., Ngom, B. and Mangin, B. 2005. MCQTL: multi-allelic QTL mapping in multi-cross design. Bioinformatics, 21:128–130.

Kuchel, H., Ye, G., Fox, R. and Jefferies, S. 2005. Genetic and genomic analysis of a targeted marker-assisted wheat breeding strategy. Molecular Breeding, 16: 67-78.

Podlich, D.W. and Cooper, M. 1998. QU-GENE: A platform for quantitative analysis of genetic models. Bioinformatics, 14:632-653.

Ribaut, J.M. and Hoisington, D. 1998. Marker-assisted selection: new tools and strategies. Trends in Plant Science, 3:236–239.

Ruane, J. and Sonnino, A., 2007. Marker-assisted selection as a tool for genetic improvement of crops, livestock, forestry and fish in developing countries: an overview of the issues. *In* E.P Guimaraes *et al* (*eds*) Marker-assisted selection: current status and future perspectives in crops, livestock, forestry and fish. Food and Agriculture Organisation of the United Nations. Rome, 2007.

Vargas, M., van Eeuwijk, F., Crossa, J., and Ribaut, J.M. 2006. Mapping QTLs and QTL x environment interaction for CIMMYT maize drought stress program using factorial regression and partial least squares methods. Theoretical and Applied Genetics, 112:1009–1023.

Wang, J., van Ginkel, M., Podlich, D., Ye, G., Trethowan, R., Pfeiffer, W., DeLacy, I.H., Cooper, M. and Rajaram, S. 2003. Comparison of two breeding strategies by computer simulation. Crop Science, 43:1764-1773.

Wang, J., Eagles, H.A., Trethowan, R. and van Ginkel, M. 2005. Using computer simulation of the selection process and known gene information to assist in parental selection in wheat quality breeding. Australian Journal of Agricultural Research, 56:465-473.

Ye, G., Eagles, H.A. and Dieters, M.J. 2004. Parental selection using known genes for inbred line development pp 245-248. *In*: Cereals 2004, Proceedings of 54th Australian Cereal Chemistry Conferences and 11th Wheat Breeders Assembly (*Eds*: C.K. Black,J.F. Panozzo, G.J. Rebetzke). Cereal Chemistry division, Royal Australian chemical Institute, Melbourne, Australia.

Ye, G. 2005. User manual of the CrossPredictor software. The Department of Primary Industries, Victoria, Australia.

Ye, G., Moody, D., Livinus, L. and van Ginkel, M. 2007. Designing an optimal marker-based pedigree selection strategy for parent building in barley in the presence of repulsion linkage, using computer simulation. Australian Journal of Agricultural Research, 58:243-251.

Ye, G., and Smith K.F. 2009. Marker-assisted gene pyramiding for cultivar development. Plant Breeding Review (In press)

Yu, J., Pressoir, G., Briggs, W.H., Bi, I.V., Yamasaki, M., Doebley, J.F., McMullen, M.D., Gaut, B.S., Nielsen, D.M., Holland, J.B., Kresovich, S. and Buckler, E.S. 2007. A unified mixed-model method for association mapping that accounts for multiple levels of relatedness. Nature Genetics, 38:203-208.

Zeng, Z.B. 1994. Precision mapping of quantitative trait loci. Genetics, 136:1457–1468.

Index

H

I

J

N

O

P

Q

R

S

T